# Phenomenology, Organizational Politics, and IT Design:

## The Social Study of Information Systems


Gianluigi Viscusi
*University of Milan Bicocca, Italy*

Gian Marco Campagnolo
*University of Edinburgh, UK*

Ylenia Curzi
*Marco Biagi Foundation & University of Modena and Reggio Emilia, Italy*



Information Science
REFERENCE

Managing Director: Lindsay Johnston
Senior Editorial Director: Heather A. Probst
Book Production Manager: Sean Woznicki
Development Manager: Joel Gamon
Development Editor: Myla Harty
Acquisitions Editor: Erika Gallagher
Typesetter: Russell A. Spangler
Cover Design: Nick Newcomer, Lisandro Gonzalez

Published in the United States of America by
Information Science Reference (an imprint of IGI Global)
701 E. Chocolate Avenue
Hershey PA 17033
Tel: 717-533-8845
Fax: 717-533-8661
E-mail: cust@igi-global.com
Web site: http://www.igi-global.com



Library of Congress Cataloging-in-Publication Data

Phenomenology, organizational politics, and IT design: the social study of information systems / Gianluigi Viscusi, Gian Marco Campagnolo, and Ylenia Curzi, editors.
p. cm.
Includes bibliographical references and index.
Summary: "This book offers a new look at the latest research and critical issues within the field of information systems by creating solid theoretical frameworks and the latest empirical findings of social developments"-- Provided by publisher.
ISBN 978-1-4666-0303-5 (hardcover) -- ISBN 978-1-4666-0304-2 (ebook) -- ISBN 978-1-4666-0305-9 (print & perpetual access) 1. Information storage and retrieval systems--Design. 2. Information technology--Social aspects. 3. Social media. 4. Communication and technology. 5. Organizational sociology. I. Viscusi, Gianluigi, 1972- II. Campagnolo, Gian Marco, 1977- III. Curzi, Ylenia, 1976-
QA76.9.S88P49 2012
302.23'1--dc23
2011044983

British Cataloguing in Publication Data
A Cataloguing in Publication record for this book is available from the British Library.

# Table of Contents

# Detailed Table of Contents

**Section 2**
**Phenomenology and IT Design**

*Section 2 investigates how the concepts proposed by the phenomenological approach make it possible to recast the idea and the use of formal representations as a challenge for the information systems design. Indeed, a phenomenological perspective on information systems design must let the complex infrastructure of intertwined phenomena behind the contraposition between formal representations and situated actions emerge. Thus, the final aim of Part 2 is to provide a contribution to a phenomenological foundation of design science rooted in the complex life world (Lebenswelt) of information systems.*



This chapter discusses how the current environment of information affluence and media convergence is constructing a comprehensive living and knowing habitat that induces the framing of life issues in terms of data availability and the concomitant data permutations this encourages. The analysis aims to let emerge how this environment differs from the work and professional settings in which information technology has been studied.



This chapter contributes to the debate about the concept of community of practice, considering the latter as a means for understanding the behavior of the people using in their practice artifacts and sharing spaces. The experience they share makes them a community and, when the space they share is a workspace, of a community of practice. What emerges from this analysis offers to designers of ICT-based applications, such as information systems, knowledge management systems, etc., some new hints on the nature of those systems.



This chapter addresses the question of how to study and design a technological support of work in organizational domains. The topic is addressed by investigating, on the one hand, the differences and complementarities between Information Systems (IS) and Computer Supported Cooperative Work (CSCW) perspectives/approaches to systems design; on the other hand, it addresses the problems of sense making in information quality management by providing evidence of these issues in the healthcare domain.



This chapter is about showing how artifacts impact engineering work processes by representing the important issues of individual and collaborative design work. After summarizing the state of the art in engineering as a design process, artifacts and their representational role in design and engineering, a selection of rich descriptions of artifacts' creation and use in engineering work including team-based coordination and decision activities is presented. The studies are based on ethnographic research carried out for several years in different design and engineering companies.



This chapter discusses some issues emerging from the phenomenological analyses carried out by Claudio Ciborra, in particular in the *Labyrinths of Information*. In particular, the chapter points out that concepts such as *Kairos*, *Drift*, and *Bricolage* unveil a specific path (*odos*) for the information systems as a discipline. This *odos* covers a *meth-odos* towards new opportunities offered to information systems design by answering the provocation coming from considering information systems as infrastructures (*Ge-stell*).

## Section 3
## Phenomenology and the Social Study of Information Systems

*Section 3 introduces a set of research projects, which the authors feel comfortable designating as being social studies of information systems. The common denominator of the chapters enclosed in this section is the search for the possibility of a sociology of technology, and of Information Infrastructures in particular. Each chapter contains an Information Infrastructure concrete case study but it also leaves space for a modification of sociology itself, and in particular to some sociological concepts inherited by Information Systems studies from the ethnomethodological and micro-sociological tradition of study.*



This chapter focuses on the Information Infrastructure's *shared* nature, its *openness*, its always being an extension of existing ones, the *installed base*, its *heterogeneity*, and its *evolving* nature.



By making reference to the 'fit-gap work' taking place during ERP implementations, this chapter identifies how participants found ways to *manage the reflexivity of their understandings*. In particular, the author makes reference to the "prioritizing/de-prioritizing" work. What this study suggests is that it is true that, according to ethnomethodology, people do not know what they mean when they produce accounts, but it is also true that they know that, and by knowing that, people's major occupation is to produce work-arounds in order to reduce the noise of the reflexivity of understandings.



A further modification of the received sociological notion of interaction via ethnomethodology concerns the principle of seeking a *perspicuous setting*, with its emphasis on bilateral relationships. This chapter recognizes the role of *other spaces* in stabilizing member's local streams of activity: how key tensions

# Preface

*"Our going back to the roots of phenomenology in order to restore the original notion of situatedness and compare it with the contemporary debate on situated action leaves us with three main research agendas. The first, leaning towards AI and cognitive science, states that situated action can be implemented through computer programs interacting with the environment and processing symbolic representations of what happens in the environment. The second, which claims to be an alternative grounded in the social sciences (phenomenology via ethnomethodology) is based on a social ecology of the mind: goals and plans are a vague guide to action. They must be complemented by the ad hoc improvisations of humans exploiting the circumstances and what the world offers at the moment of action. [...] Finally, we have Heidegger's research program where the notion of situation includes at all moments the inner life of the actor, his or her mind and heart, and where any form of understanding is situated, meaning 'affected.' It is the pathos that characterizes the whole person in his or her situatedness in the world (Heidegger, 2002, p. 192)."* (Ciborra, 2006, p. 139)

## INTRODUCTION

In February 2005, a composite group of scholars from different disciplines met in a nice Casa Rustica close to the Antica Vetreria on the Sarca River, at the entrance of Val di Genova, a small valley in the Italian Alps. It was the first of a series of events called Alpis: Alpine Ski Seminar on Information Systems. Since 2005, the purpose of that gathering of people has been to promote the social study of Information Systems in the Mediterranean region and the emergence of a European/Mediterranean identity for the scientific community on Information Systems. The Alpis ski seminar has been an "Institute des Hautes Ètudes" with a ludic/sport component, where high quality contributions from young researchers in the field have been presented in four subsequent days of discussion with peers, in an environment enriched by contributions of established senior researchers. Both the format, promoting interaction and sociality, and the scientific dimension, promoting interdisciplinarity while maintaining a specific focus, have been the distinct trait of the Seminar series. According to the invitation of Claudio Ciborra and Gianni Jacucci—the Scientific Coordinators of the Alpis Ski Seminars first edition—the research on Information Systems being presented and discussed in the Alpis community has been characterized by the reference to philosophical inquiry and to phenomenology in particular. The aim of this book is to be an 'ideal prolongation' of the endeavour brought forward by the Alpis community to analyze the concepts, hypotheses, and research strategies proposed by the work of Claudio Ciborra. It will do so according to the following three directions:

- **direction 1:** how phenomenological concepts make it possible to recast the idea and the use of formal representations in computer programming;
- **direction 2:** what are the changes in the practices of planning, designing, and deploying information systems in the case of global Information Infrastructures implemented across multiple sites within large branch-plan organizational structures, and how do these changes challenge core assumptions embedded in received notions of phenomenology via ethnomethodology;
- **direction 3:** what are the consequences of the phenomenological orientation for the form of applied research which combines theory and practice, research and action for change; and what are the possible alternative modes of approaching action research in organizations.

This aims to provide relevant theoretical frameworks and the latest empirical research findings in the mentioned three research directions of Social Study of Information Systems. Furthermore, the overall mission of this publication is to keep alive one of the most valuable teachings of Claudio Ciborra: "addressing the cultivation of alternative modes of approaching organizations as benchmarks for information system research." Values expressed in these words include pluralism and interdisciplinarity of scientific knowledge creation in the information systems research domain. This book proposal should be intended as a significant step towards that direction. The target audience of this book is composed of professionals and researchers working in the field of Organizations and Information Systems studies, especially those who are interested in a cross fertilization among different, yet complementary and strictly connected domains of scientific knowledge, consisting of information systems research, science studies, and organizational studies.

The book is structured in four sections: Phenomenology and Information Systems Research (Section 1); Phenomenology and IT Design (Section 2); Phenomenology and The Social Study of Information Systems (Section 3); and Phenomenology, Organizational Politics, and Organizational Design (Section 4). Each section is introduced by a dedicated section of the preface.

## ON THE RELATIONSHIP BETWEEN PHENOMENOLOGY AND INFORMATION SYSTEMS RESEARCH

The three chapters in Section 1 introduce the reader to the relationship between phenomenology and information systems research. The first chapter, authored by Giolo Fele, discusses the way information systems research adopts and encounters ethnography, focusing on the tensions between the two disciplinary areas. In particular, the analyses carried out by Fele show how these tensions and "dilemmas" uncover an inedited research area for social scientists and researchers in the human sciences; in particular, the relationship between information systems design and ethnography presents items of interests, "which may derive from long-standing inquiry but are still extraordinarily topical" (Fele, this volume). These items come out from a dialectic between two forms of knowledge, that Fele let emerge through the discussion of the tension between pairs of concepts: on the one hand, the tension between *metis* ('cunning knowledge' or 'knack') and *techne*; on the other hand between *bricolage* and *engineering*. Techne and engineering mainly rest on universal principles organized analytically into logical, decomposable, and verifiable steps. Whereas metis and bricolage rely on personal and tacit knowledge that resists simplification into transmissible and verifiable deductive principles. In a sense, this tension is reminiscent of the dialectic between the *science of the artificial* and the *science of the concrete*. Notwithstanding, the chapter claims for a critical approach to this tension grounded in the belief that "*techne* and *metis*, engineering and bricolage, are intimately bound up with each other." (Fele, this volume).

Liberman's chapter centers around the topic of situated knowledge, a growing concern in the field of information systems. Liberman adopts a phenomenological perspective, with a strong ethnomethodological orientation. A student of Peter Berger, Herbert Marcuse, and Hubert Dreyfus (1990), but especially of Harold Garfinkel (2002), Liberman discusses and presents his argument starting from an examination of Section 29 of the fundamental text of Heidegger's *Being and Time*. Essentially, this very complex text deals with the question of the adequacy of our modes of representation of social forms and with the ways in which these forms are experienced in our daily lives. It is well known that technological solutions create patterns and structures of social organizations, which impose life forms completely outside of our experience with which we must come to terms, often with difficulty. Much of the engineering culture, which is at the base of these technological solutions, is not oriented towards a social or sociological perspective. Recently, however, we have seen a growing interest in the social contexts of technological innovations (see Giolo Fele, this volume). This shift of attention by the specialists in business sciences, management, and information systems, has led to a deepening of social approaches into the constitutive mechanisms and the fundamental forms of social life (Dourish, 2001; De Michelis, 2008). Here we see the important role, on the theoretical level, of phenomenology and ethnomethodology, and on the methodological level, of ethnography, in identifying, recognizing, and describing the most profound and most subtle aspects of our social life (Fele, 2008). Heidegger's philosophy provides the ideas for a non-trivial reflection on the foundations of situated understanding (Dreyfus, 1995; Winograd, 1995; Ciborra, 2004). The issue of situated knowledge covered by Liberman's chapter goes far beyond the usual (although by no means obvious) importance attributed to context in the processes of communication. See the following passage from Winograd and Flores (1986): "The computer, like any other medium, must be understood in the context of communication and the larger network of equipment and practice in which it is situated. A person who sits down at a word processor is not just creating a document, but is writing a letter or a memo or a book. There is a complex social network in which these activities make sense. It includes institutions (such as post offices and publishing companies), equipment (including word processors and computer networks, but also all of the older technologies with which they may coexist), practices (such as buying books and reading the daily mail), and conventions (such as the legal status of written documents)" (pp. 5-6). Winograd and Flores's seminal perspective recognizes the role and value of the network of relations within which social action acquires meaning. Liberman's contribution invites us to look further and deeper. From an ethnomethodological perspective (Garfinkel, 2002), Liberman invites us to explore the depths of our ordinary social world, the primitive place of our experience. As an anthropologist who spent two years with some Australian Aboriginal tribes (Liberman, 1985) and three years in a Tibetan monastery (Liberman, 2004), he encourages us to reflect on that world taken for granted that we call reality. Similarly, as philosopher (Liberman, 2007), he sees the limits of reason and the difficulties we fall into when we overconceptualize our worldly relations, when we entrust entirely to what he calls "the formal analysis," when we don't recognize the very carnal, practical and experiential character of social life. Starting from this basis, the paper offers grounds for reflection on the field of information systems.

The chapter by Paolo Depaoli exemplifies the advantages that an in-depth examination of the basic outlooks and assumptions of theories and practices provide to promote novel methods and areas of investigation in IS research. The key argument is that researchers typically make reference to philosophy when they wish to address issues and topics considered crucial, but inappropriately treated or neglected by current theories and practice. Ciborra's journey in the information systems research world and his "escalation to philosophy" (and precisely to Heidegger's phenomenology) is reconstructed, interpreted, and proposed as an example of that. This journey starts from the empirical finding that the integration of information technologies in organizations never takes place as it was originally planned. Drifts normally occur. The theories and methodologies which dominate the scene of management, organization, and information systems are however unable to account for that. In fact, and in so much as they are based on the method of the natural sciences, and on the idea of prior absolute rationality, they aim at imposing uniformity, rather than addressing the complex reality and heterogeneity of the research settings under investigation. Depaoli discusses some directions that Ciborra drew from Heidegger's phenomenology to proceed in the study of organizations and information systems design and implementation. Firstly, the idea that information systems in organizations are particular singular worlds. Furthermore, the concept-word of *aletheia* as a methodological beacon to approach the way in which information and communication technologies become integrated in the organizational life. This approach lies in "un-veiling" the core characteristics of an organization, which typically remain unnoticed because of the "veiling" effect produced by the concepts drawn from dominant theories and methodologies. Finally, the assumption of the "human existence in organizational life" as a centre of gravity. This implies that the points of view of all stakeholders involved in the integration of information and communication technologies in organizations is the standpoint to assess the effectiveness of information systems design and implementation. The chapter highlights that the last point is especially crucial for a novel research area, i.e. individual information systems present in the workplace and addressed to the fulfilment of individual specific work related information necessities.

The contributions in the following Sections of the book move from some open issues concerning the relationship between phenomenology(ies) and the social study of information systems, namely the relationship with formal representations, with global phenomena, and with organizational (re)design and intervention for organizational change.

## THE USE OF FORMAL REPRESENTATION UNDER A PHENOMENOLOGICAL PERSPECTIVE

Section 2 investigates how the concepts proposed by the phenomenological approach make it possible to recast the idea and the use of formal representations as a challenge for the information systems design. In this section, a definition of representation in its relation with information systems together with a discussion of the perspective on phenomenology adopted in Section 2 are first presented. Then, the contribution of the different chapters are introduced and discussed. The starting point is the identification of the generic attributes of the current environment of information affluence and media convergence, where design takes place (Kallinikos, this volume). Subsequently, the focus moves from the general attributes of the environment to the analysis of what characterize our being in the world as different subjects having mutual relationships. Thus, the analysis considers from a phenomenological stance the general concept of community as the root of the concept of community of practice, as a relevant concept for understanding the social dimension of human life. This dimension often referred as situatedness open new paths and possibilities for design, as the practice through which something new is brought to existence (De Michelis, this volume). The following discussion considers two chapters aiming on the one hand to address the general problem of how to design and deploy effective computational tools that support actors of an organization domain in making sense of the information they manage (Cabitza and Simone); on the other hand, to show the roles artifacts and their representational role in design and engineering collaborative processes, through case studies based on ethnographic research (Tellioglu, this volume). Finally, the chapter proposing to integrate an *archaeology* and a *genealogy* of emergent representations in the early phases of information systems design is analyzed. The goal is to consider representations as a way to challenge the opposition of organisational and engineering/technological perspectives as not reducible facets of design (Viscusi, this volume).

As pointed out by Kallinikos (1995) *representation* is technology in terms of "an overall world orientation concerned with the objectification of the natural and social worlds in ways which render them amenable to calculation and mastery." Furthermore representation is technology in terms of "an ensemble of diverse and minute techniques which embody and realize such an orientation" (Kallinikos, 1995), namely an ensemble of *representations* (see Viscusi, this volume). As pointed out by Ciborra (2002):

*"When approaching an organizational phenomenon we encounter at least two distinct forms of evidence. First, we are ready to deploy a set of ideas and models taken for granted in the domain of organization theories or consulting models. These ideas and models come almost to be identified with the phenomenon we want to grasp, or at least they provide an unquestioned context in which to grasp it."*

A first challenge for information systems design is to deal with these formal representations, because "information systems is concerned with representation in general" (Boland, 1999, p. 239). Whereas in the literature formal representations are considered to be at odds with situated actions and analyses (Suchman, 1987; Winograd & Flores, 1986). As to these issues, the work of Claudio Ciborra (in particular, Ciborra 1999, 2002) shows how the adoption of phenomenological perspectives in information systems design allows to 1) challenge reified notions, such as the one of "situation," with living concepts as in the case of *situatedness* (Ciborra, 2006); 2) let emerge relevant concepts that are peculiar to the information systems domain, such as the critical concept of *infrastructure* (see also Section 3, this volume).

The above-mentioned contraposition between formal representations-situated actions reflects the "hybrid" nature (Avgerou, et al., 2004, p. 1) of information systems discipline and design (Ciborra, 2002; Currie & Galliers, 1999) as "an area of research positioned between management studies and applied computing, where it is influenced by numerous kindred of reference disciplines" (Avgerou, et al., 2004, p. 1). Despite this recognized hybrid nature of the field, it is still valid the claims that "a common and unified paradigm has been adopted across the board to deal with its human and natural dimensions: the paradigm of natural sciences and the collateral methodologies of measurement, formalization, and calculation" (Ciborra, 2002). The lack of a common and unified paradigm for information systems design is even more true if we consider the definition of design *as* science and the specific roles of representations as ways to simplify the view on the complexity of the context for problem-solving activities:

*"Design science is active with respect to technology, engaging in the creation of technological artifacts that impact people and organizations. Its focus is on problem solving but often takes a simplistic view of the people and the organizational contexts in which designed artifacts must function. As stated earlier, the design of an artifact, its formal specification, and an assessment of its utility, often by comparison with competing artifacts, are integral to design-science research. These must be combined with behavioral and organizational theories to develop an understanding of business problems, contexts, solutions, and evaluation approaches adequate to servicing the IS research and practitioner communities" (Hevner, et al., 2006).*

Thus, problem solving is the focus and the founding of design as science. Problem solving "simplistic view" of people and organizational concepts is seen simply has a contingent issue, due to a lack of integration with behavioral and organizational theories. Nevertheless, this claim for integration covers the reified and taken for granted issues of problem solving and related representations. For representations are the fundamental challenge for a unified perspective on information systems design and must be investigated as being reified and taken for granted concepts, having consequences on the context where design takes place and designed artifacts are deployed (Bowker & Star, 1999; Suchman, 1987; Winograd & Flores, 1986; Zuboff, 1988). Indeed, functional simplification and closure are not contingent issues of technology and information systems, but they are rooted in representations as the "objective" roots of problem solving as design focus and technology as self referential system (see Kallinikos, 2006). Thus, the need for a phenomenological investigation of representations as the root of design aims to challenge current design science perspectives. Furthermore, the relevance of the role and nature of representations as the root of design emerges from a preliminary investigation of what is named by design science literature as behavioural-science paradigm, namely a set of several and different disciplines, such as Anthropology, Organizational Behavior, Political Science, Psychology, Sociology (Bariff & Ginzberg, 1982). What is worth noting here, is the difference in classification of design science with respect to behavioural science disciplines classification: while design science class refers to a unique discipline, the behavioural-science class refers to at least five disciplines disregarding their internal (historical) complexity, their "qualitative" articulations (at least in the case of Anthropology and Sociology), and the difficult in framing them in a natural science paradigm (at its turn as taken for granted reification, despite the multifacets debate and perspectives of the epistemology debate, as branch of philosophy). What is expected these disciplines can provide to design science? Do they have to provide requirements for "artificial worlds (e.g. the 'enterprise models') made of deceivingly univocal and objective entities, data, processes, and activities" (Ciborra, 1999, p. 145)? How this requirements elicitation activity fits with each disciplinary framework and scopes?

The consequence of the design science perspective of the disciplines classified as behavioural science is once again a separation and a specialization of information systems research: the resulting information systems field consists in studies which analyse, on the one hand, the socio-economic and organizational aspects of information systems, thus reducing information systems to an organizational system; on the other hand, aspects more closely connected with the processing of data and information by means of technologies, thus reducing information systems to technological systems (Viscusi, this volume). As a consequence, we have sociological, organizational, economic, or technological analyses in the information systems area, but what an information systems research means remains veiled and forgotten. The consequence is the design of good organizations for ideal systems or the design of complex systems for unready organizations (Viscusi, this volume). Thus, a phenomenological perspective on information systems design must let emerge the complex infrastructure of intertwined phenomena behind the contraposition between formal representations-situated actions: "the unveiling of the platform organization required a different analytical approach from one common in industrial organization research" (Ciborra, 2002). Taking these issues into account, we have first to clear the position towards phenomenology adopted by the editor in Section 2 of this volume.

As pointed out by Lyotard (1954) the terms 'phenomenology' received "full and singular meaning" with the *Phenomenology of Spirit* of Hegel, where phenomenology is defined "the science of consciousness…in that consciousness is, in general, knowledge of an object, either exterior or interior." Here consciousness is the immediate being of spirit, possessing two moments, namely "that of knowledge and that of objectivity which is the negative with regard to this knowledge. When spirits develops itself in this element of consciousness…this opposition occurs at each particular moment, and they all appear as faces of consciousness. The science of this path is the science of the experience had by consciousness" (Phenomenology of Spirit, Preface, sec. 36)[1].

Thus as Lyotard claims "the very idea of phenomenology puts this question out of play: consciousness is always consciousness of, and there is no object which is not an object *for*. There is no immanence of the object to consciousness unless one correlatively assigns a rational meaning, without which the object would not be an object for. Concept or meaning is not exterior to Being; rather, Being is immediately concept in itself, and the concept is Being for itself" (Lyotard, 1954).

The reference to Lyotard recalls Hegel and his interpretation of phenomenology that allows the introduction of a perspective that does not adopt as starting point the phenomenological reduction which characterizes the Husserl phenomenology before the issue of the *Crisis of European Sciences and Transcendental Phenomenology* where the "radical foundation of truth reveals itself in the end as a return to *Lebenswelt*" (Lyotard, 1954). As noted by Giddens, the adoption of phenomenological reduction prevents Shutz and sociologists (and I would add some information systems scholars) to reconstitute social reality as an object-world (Giddens, 1976). This is a relevant point for information systems, where often scholars have integrated Husserl and Heidegger perspectives, without considering the fundamental philosophical differences and distances related to the issue of the phenomenological reduction[2].

The above discussion aims to introduce the perspective who guided the choice of contributions to Section 2 of this book, where "the term phenomenology refers only to the rules that will be found operating in any perspective—scientific, religious, aesthetic, practical, etc.—whereby certain areas of experience are regarded with a neutral attitude, i.e. are made non-relevant to the problem in hand, while others are accorded the accent of affirmation and just so, i.e. the accent of relevant matter of facts" (Garfinkel, 2008, p. 116).

Thus, the focus of the IS design from a phenomenological stance must not be on an individual whole person considered as an "actor," namely a problem solving situated perspective as the ones above introduced for design science; the focus must be on the "*rules* on which the irreducible character of data-experiences is based" (Garfinkel, 2008, p. 116) and to the conditions "under which the world of an experiencer appears as a closed rather than a open set of possibilities (Garfinkel, 2008, p. 130). As a consequence, Section 2 of the book deals with "Design as the practice through which something new is brought to existence" (De Michelis, this volume). The adoption of a 'phenomenological stance' allows to go beyond the traditional idea of design as science, to capture its complex social nature (De Michelis, this volume). Thus, the main goal of Section 2 is to provide a phenomenology of design *of* Information systems that lets emerge its main constitutive attributes and rules. Current design science considers information systems as characterized by *Information Systems* and *the organizations they support* as complex, artificial, and purposefully designed (Hevner, et al., 2006); thus, in some sense design science points out a neutral representation *of* information systems that must be related and instantiated (in and) by context representations coming from results of behavioral science. Contributions in Section 2 of the volume aim to challenge this neutral perspective by showing that Information systems *and* the organizations they support are the challenge of design, considering that "social seems to be diluted everywhere and yet nowhere in particular" (Latour, 2005, p. 2) and that "no information is possible except against the constancy provided by a background of invariant social expectations" (Garfinkel, 2008, p. 75). Thus, the final aim of Section 2 is to contribute to a further foundation of design science rooted in the complex lifeworld (*Lebenswelt*) of information systems.

As a starting point, Kallinikos's chapter considers how the current environment of information affluence and media convergence is constructing "a comprehensive living and knowing habitat that induces the framing of life issues in terms of data availability and the concomitant data permutations this encourages" (Kallinikos, this volume). The analysis aims to let emerge how this environment differs from 1) the work and professional settings in which information technology has been studied, and 2) from the "early internet," before the advent of social networks, the so called web 2.0 and mediated by the diffusion of smart phones, location services (GPS), and augmented reality technologies. Indeed, current information growth is characterized "by its remarkable ability to deeply penetrate the social fabric and increasingly induce the framing of life issues in terms of data availability, and sense making based on data, assembled into meaningful categories and structures by machines" (Kallinikos, this volume). Furthermore, the analysis aims to identify and exhibit the generic attributes of this environment "that cut across specific contexts of social and institutional life," namely the *prominence of cognition over perception* and *the preponderance of information and computational principles in defining reality*. These attributes define the inedited role of representations as "outcome of technological advances of reality documentation (satellite images, digital soil maps, weather forecasts), far beyond any human register capacity" (Kallinikos, this volume).

As a consequence, the challenge for design of information systems is to understand the attributes of the emerging prominence of "cognition qua computation" and the rendition of knowledge as outcome of agnostic statistical permutations performed upon mass of disseminated and distributed data tokens: "data comprehensively maps reality (the scans capture bone conditions) while computer-based data mining discovers its underlying principles (the mechanism of bone loss)" (Kallinikos, this volume). These attributes need to be studied and analysed in order to understand and expose "distinctive ways through which they are manifested, change and fuse into the fabric of social practices" (Kallinikos, this volume). Finally, the challenge for design is to understand the consequences and the way of acting on these representations that "create and establish particular versions of the world possible to act upon, perform, and manipulate. Kallinikos's contribution allows one to clarify the definition of information systems provided by design science as "complex, artificial, and purposefully designed" (Hevner, et al., 2006) through the emerging constitutive attributes of the current information environment. In order to understand the way these attributes manifest and intervene in social practices, a further investigation must consider social practices and question them as "the social milieu where organizations are formed, as well as the social dimension of learning." (De Michelis, this volume).

De Michelis's chapter investigates, from a phenomenological stance, the concept of community as a relevant concept for understanding the social dimension of human life. In particular, the study focuses on the concept of Community of practice as "a means for understanding the behavior of the people using in their practice artifacts and sharing spaces." Despite its relevance and diffusion, the concept of Community of a Practice (CoP) is characterized by an original ambiguity coming from its provenience from market rather than the academic field; indeed, "it becomes unclear if CoP are an analytical concept for understanding learning and practice or a powerful tool for developing effective education within organizations or for improving the professional skills of their employees" (De Michelis, this volume).

As a consequence, the chapter aims to root the concept of CoP in a more general concept of community as the first way through which we encounter something as a CoP. This concept emerge from the consideration by Jean Luc Nancy, "of 'being in the world' of Heidegger (*Dasein*) as a 'being with' (*Mitsein*), so that the community becomes 'la position réelle de l'existence'" (De Michelis, this volume). Furthermore, sense making and individuals emerge from the social experience of 'being with,' where "human beings appropriate the world, in the same moment they emerge as individuals with a specific identity" (De Michelis, this volume). Understanding social value as rooted in community as original appropriation of the world, provide a way to overcome the above mentioned "simplistic view" of people and organizational claimed by design science.

*"On the one hand, information systems should avoid the separation between organizational performances, like invoicing, ordering, producing, planning, monitoring, etc. and the practices within which they are embedded. On the other, even if any practice is situated within a community (of practice), we should remember that different human beings participate and contribute differently to any community of practice: this sheds a new light on situatedness of human practice emphasizing its distributed nature" (De Michelis, this volume).*

Furthermore, the artificial character of information systems design emerges as grounded in the "being with." Indeed, formalization "is the move through which human beings are able to interact even when they don't share a context" (De Michelis, this volume) and formal models are representations *of* the community: "modeling, de-contextualizing knowledge, allows to break the boundaries between communities: formal models are typically boundary objects" (De Michelis, this volume).

The two discussed contributions of Section 2 of this volume point out the challenge of Information systems design to question the *prominence of cognition over perception* and *the preponderance of information and computational principles in defining reality* as emergent attributes, emerging from community as a fundamental and original appropriation of the world. Taking these issues into account, we have now to re-consider work and professional settings in which information technology has been traditionally studied as an emergent human practice.

Cabitza and Simone chapter in this volume addresses the question of *how* to study and design a technological support of work in organizational domains. The topic is addressed by investigating on the one hand, the differences and complementarities between Information Systems (IS) and Computer Supported Cooperative Work (CSCW) perspectives/approaches to systems design; on the other hand, it addresses the problems of sense making in information quality management, by providing evidence of these issues in the healthcare domain.

The identification of a ground for 'reassembling' IS and CSCW perspectives is a relevant topic for information systems design. Indeed, while both the perspectives show the combination of technological concerns with the understanding of the setting in which the technology has to operate, fundamental differences seem to characterize their approaches to organization and work as social phenomenon: "CSCW focuses on users, as the sole source of knowledge about the cooperative settings and its needs; on the other hand, IS focuses mainly on management, and when it recognizes a role for end-users, it sees them mainly as one of the possible stakeholders at play, or social actors" (Cabitza & Simone, this volume). The starting point of what we can define a *second order* contribution by the authors is the conceptualization of work that is common to both IS and CSCW research.

*"work is an intrinsically (and probably irreducible) social phenomenon – and argument that nevertheless the phenomenon of collaboration is conceived differently in these two fields. This will bring us reconsider the role of "artifacts" in organizations (for informative and coordinative purposes), by distinguishing between the conceptualizations that see artifacts (and the representations they carry on)" (Cabitza & Simone, this volume).*

IT artifact is a widely investigated and debated topic in both IS and CSCW areas, and Cabitza and Simone propose a reconciliation that considers IT artifact as rooted "in a class of 'tools' that are widely used within organizations, let's say before any sort of digitalization: the paper artifacts that often take the shape of semi-structured forms" (Cabitza & Simone, this volume). Furthermore, the digitization of paper is an 'event' occurring along with the introduction of an information systems application, usually with aims "to preserve and enhance the role of artifacts as information *archives* and information *sources*, especially in their function of 'gateways' to the underlying information system." (Cabitza & Simone, this volume). Thus, the challenge posed by digitalization of paper as IT artifact is further rooted in the information-intensive domain where "data are produced *in the* practice, *during* the practice and *for* the practice itself" (Cabitza & Simone, this volume). Finally, through the discussion of the case study of a prototypical Electronic Patient Record (EPR), authors show the emergence of "sense-making networks of data that could reconcile their accumulation for the sake of coordination with the intrinsic (process and data) quality requirements" (Cabitza & Simone, this volume). As a consequence, the chapter points out how IS and CSCW perspectitves can be reconciled considering the emergent roles information plays, namely the archival and the coordinative one, that belong to two different layers of the information system. This connection is realized "by letting quality be interpreted differently in the two layers; and by supporting this interpretative act by augmenting the artifacts with meta-information" (Cabitza & Simone, this volume).

As to these issues, a relevant challenge is the role of artifacts in the work practices of people devoted to the design and development of information systems. Tellioglu study in this volume focuses on artifacts used in collaborative engineering processes. The aim is to understand work practices of engineers and designers in order to identify and provide them more suitable IT-tools. Grounded in the CSCW perspective, the study first summarizes the state of the art of engineering as a design process, artifacts, and their representational role in design and engineering. Then, the study analyses and discusses artifacts' creation and use in engineering work, focusing on team-based coordination and decision activities. The studies presented are based on ethnographic research carried out for several years in different design and engineering companies. Furthermore, the case studies consider artifacts from their representational role as *visualizations* of different kind. Visualizations emerge 'network-organizing devices' that "are used to illustrate the design ideas to different actors involved in the project, such as the clients, external professionals, or partners, convincing them of the design idea and mobilizing their cooperation" (Tellioglu, this volume).

The final chapter of Section 2 authored by Gianluigi Viscusi considers the issues for information systems design methods that emerge from the phenomenological analyses carried out by Claudio Ciborra (2002). In particular, concepts such as e.g. *Kairos*[3], *Drift*[4], *Bricolage*[5], unveil 1) a specific path (*odos*) for the information systems as a discipline and 2) a *meth-odos* towards new opportunities offered to design by answering the provocation of information systems considered as infrastructures (*Ge-stell*). This *meth-odos* aims to provide a preliminary support to uncover the foundation of representations in the readiness-to-hand of the being in the world. The aims is to define a way to move back from representations towards the unit of *Befindlichkeit* (i.e., affectivity), *Verstehen* (i.e., the understanding), and the *Rede* (i.e., speaking) (Viscusi, this volume). Thus, the chapter proposes to integrate this *meth-odos* and state-of-the-art methodologies in order to challenge the *bestand* of schemas and representations inscribed in the technological and engineering facets of the information systems. In particular, the chapter proposes to integrate an *archaeology* and a *genealogy* of emergent representations in the early phases of information systems design; these paths must be introduced either (1) in the observation phase in the social and participatory design oriented methods and (2) in requirements analysis in the engineering oriented methods (Viscusi, this volume). In particular, the integration of *meth-odos* in the early phases of systems design should support the uncovering of the opposition of organisational and engineering/ technological perspectives as not reducible facets, "which must be analysed and synthesized together in an evolutionary design of information systems" (Viscusi, this volume).

## ON THE FULLNESS OF SITUATION OR THE BIOGRAPHIC TURN IN THE SOCIAL STUDY OF INFORMATION INFRASTRUCTURES

Section 3 introduces a set of research projects, which the authors feel comfortable designating as being *social* studies of information systems. The common denominator of the chapters enclosed in this session is the search for the possibility of a sociology of technology, and of Information Infrastructures in particular. Information Infrastructures are long-lived complex technologies, that face the dilemma of aligning multiple stakeholders over extended periods (Ciborra, 2001), often implemented across multiple sites within large branch-plant organizational structures (Markus, et al., 2000). Uncomfortable with the many sociologies that have been proposed to study the Information Infrastructure phenomenon (micro-sociology and ethnomethodology in particular), the authors contributions in Section 3 seem to endorse an "empirical turn" in the social study of information systems, which deeply analyzes case studies and concrete issues concerning Information Infrastructure design and use. Assuming, according to many science studies programs, that artifacts have an agency, they look at how these particular technologies embed a view of interaction, society, and organization that may challenge core assumptions of received sociological notions. Each chapter contains an Information Infrastructure concrete case study but it also leaves space for a modification *of sociology itself*, and in particular to some sociological concepts inherited by Information Systems studies from the ethnomethodological and micro-sociological traditions of study. The authors of the chapters collected in Section 3 demonstrate to share the basic tenets of a phenomenological program in the social study of phenomena, in line with Ciborra (2006) and a vibrant seek for capturing "the total situation" (Heidegger, 1993, p. 185 and 262; in Ciborra, 2006) of the Information Infrastructure phenomenon. Therefore, social studies of information systems scholars abandon the concept of system and replace it with the concept of Information Infrastructure. The concept of Information Infrastructure drives our attention to its *shared* nature, its *openness*, its always being an extension of existing ones—the *installed base*, its *heterogeneity* and *evolving* nature (see Nielsen, this volume).

More specifically, the fact that each of the chapters concerns the same type of techno-scientific practice (the work surrounding Information Infrastructure development) contributes to elucidate how a specific research project of *doing* a sociology of Information Infrastructures might look like, in contrast with that of *applying* sociology to information systems studies or *importing* sociology to address information systems concerns.

One of the key aspect of the Information Infrastructure's phenomenological structure that is highlighted in all the five essays, is its unique relation with space, time and materiality. This specificity of Information Infrastructures puts some information system researchers at odds with sociologies that still give prominence to place by assuming space naturalness and its essential "thereness," and that underestimate the difference between materialities and their different capacity to order social relations. As stated by one of the contributors in this volume: "Zimbabwe water pumps, enterprise resource planning systems and many health ICTs have more flexible compositions from any other particular time to another than, for instance, pens, bicycles or electricity grids" (Hyysalo, this volume).

The empirical focus on technologies such as Information Infrastructures, makes the authors regard them as distinct and discontinuous with the technologies populating previous epochs of human history. One key assumption in social studies of interaction can be described, drawing on a concept from biology, as being the *isometric scaling* of social phenomena. Isometric scaling occurs when changes in size (during growth or over evolutionary time) do not lead to changes in proportions, i.e. types of interactions between parts of the organism. On the contrary, the social study of information systems introduces the notion of *allometry* in the understanding of social phenomena. Allometry is when changes in size can lead to discontinuities in the relation between properties (e.g. to stay within the biological metaphor: between mass and surface area-based properties or length-based properties). Therefore, the social study of Information Infrastructures addresses the need of a change in the concepts describing social interactions when the scale of their spatio-temporal distribution changes. Furthermore, given that each of our mundane affairs is increasingly information infrastructured (i.e. imbued if not directly at least potentially in larger-scale digitally-mediated social forms), the social study of Information Infrastructures may also suggest a rethinking of the concepts by which sociologists interested in interactions have addressed the way the orderliness of mundane affairs is achieved by members of a society.

The idea that Information Infrastructures embed modifications of sociology extends what Gian Marco Campagnolo noted in his doctoral dissertation (Campagnolo, 2007) in relation with new technologies of accountability with respect to the ethnomethodological distinction between "accountability" as what people provide as a description of their context and behavior and "accountability" as an intrinsic part of these activities. At least according to ethnomethodology, accountability belongs in the life-world of each of us as a radical property of social action from the start, 'since the immemorial,' as a fact in nature. However, some changes in the late modern conditions underscored by studies of new public management in the turn of the century (e.g. Marilyn Strathern book on the 'Audit Culture' or Micheal Power reflections on the 'Audit Society'), may have revealed the historicality of accountabilities. Take the example of an everyday conversation between a customer/client and an operator providing services through a call centre. The interaction begins with a warning that the conversation might be recorded for evaluation or training purposes. We know from then that the ways we will naturally act in order to make things visible-and-reportable are not at all circumscribed to the practical purposes impinging upon the present situation, but they will also be referred to in other context for other less practical purposes—namely, statistics purposes, training purposes. Furthermore, government actions are in place that expect to discipline subjects in being aware that their actions are not to be taken as having consequences only locally, but also on the way important decisions will be taken in spatio-temporally distant circumstances (e.g. with respect to socio-economic or environmental policies) with the help of technologies that record, save and export information from place to place. It is assumed here that what was said for the ethnomethodological concept of accountability goes also for other ethnomethodological notions. A modification of sociology that has been fostered by contemporary Information Infrastructures, as already suggested by Giddens (1994), is thus the addition of an institutional, strongly calculative, material dimension to the practice of making things visible-and-reportable for all practical purposes, as well as to the way these accounts contribute to constitute the observability of a setting (i.e. reflexivity). Recognizing this additional layer, makes it impossible an analysis of how members reflexively monitor their streams of action without strongly referring to the increasing capacity of institutions to code information to control action at a distance. One example of the heightened institutional appropriation of ethnomethod's reflexive nature emerges from Nicholas Rowland study of Enterprise Resource Planning software usage in American Universities (Rowland, this volume). His empirical case study demonstrates that the received ethnomethodological transcendental interest that *the social order is always a local accomplishment*—that is the result of a concerted activity of a community of co-operating fellows—requires some further discussion. Rowland's paper provides illustrations of how people involved in large-scale information systems implementation do not know what they mean when they produce accounts. He does so by making reference to the 'fit-gap work' taking place during ERP implementations, but he also identifies how participants found ways to deal with these uncertainties, *to manage the reflexivity of their understandings*. In particular, he makes reference to the "prioritizing/de-prioritizing" work. Something that cannot be decided, or evaluated on the basis of a sufficiently accountable manner, gets de-prioritized. The most useful feature of de-prioritization is that issues related to implementation that become de-prioritized are "removed without removing" and remain in the purgatory of prospective possibilities.

The process of natural objectification of practices for organizing the orderliness of events produces tools, instruments, artefacts, benchmarks that become available for future and distant accomplishments. What the illustration deriving from the "fit-gap work" reminds us is that these commodities are certainly re-enacted in each locale, but people do not re-invent the wheel all the same all the time. Prior than the transcendent re-enactment of social order, the chief interest of social researchers in information system is on the layers of customization encrusted in *organizing artifacts* and *ordering systems* that protect people from being disbanded every time they encounter even the most routine task. What Nicholas Rowland study suggests is that it is true that people do not know what they mean when they produce accounts, but it is also true that they know that. And by knowing that, people's major occupation is to produce work-arounds in order to reduce the noise of the reflexivity of understandings (e.g. the 'fit-gap work'). These accomplishments, that take the form of routines, tools, instruments, artefacts and institutions, are of central interest for a social researcher in Information Infrastructures.

A further specific aspect of Information Infrastructures phenomenological structure is that our experience of here and now in dealing with (and researching on) them has increasingly lost its immediate spatio-temporal referents and has become tied to and contingent on actors and actions at distance (Newmann, 2008). This allows for a further modification of the received sociological notion of interaction via ethnomethodology with respect to the principle of seeking a *perspicuous setting* (Garfinkel, 2002, pp. 181-182), with its emphasis on bilateral relationships. Perspicuous settings are settings where parties are actively engaged in making a situation intelligible (e.g. courtrooms, surgeries, classrooms, etc.). A perspicuous setting has to be able to tell us more about our interests than what we are able to imagine about it. However, to do ethnographic research, for example, on the social grounds that contribute to the production of an Information Infrastructure requires different practices and opportunities than does fieldwork among the situated communities (e.g. workplaces) such Information Infrastructures affect. While the default assumption of interactionist studies is that the relevant 'public' is the body of immediate co-present interactants, in the case of Information Infrastructures, the 'public' is a more diffuse set of (potentially significant) others, often dispersed temporally and physically. Thus, very different kind of research agendas are needed to address the multiple locations and the different timeframes in which technology operates. By emphasizing the central importance of local redesign for the success of information systems, interactionist studies of Information Systems implementation fail to recognize the role of *other spaces* in stabilizing member's local streams of activity: for instance, the production chain, the developers, the physical environment, the competent authorities, legislators, interest organizations, industrial analysts, and the scientific community. Gian Marco Campagnolo's chapter on the development of ERP systems addresses this problem by providing a map of how key tensions between distributed actors and social worlds contribute to the framing of a particular discourse on information technology development for the public administration, which to its turn affects workplace interactions between consultants, management, Staff personnel and line workers in an Italian public sector organization dealing with a large scale information system implementation. A concern with the space distribution of design-use relation in Information Infrastructure development is also masterfully raised by Sampsa Hyysalo contribution titled *How is situated design interaction situated*, when he says: " we hope to open up the question about the centrality of design process by examining what comprises the situated action taking place in design, in other words, examining how design interaction is situated within design-use relations that are temporally more long term and spatially more distributed" (Hyysalo, this volume).

The final chapter of Section 3 is authored by Neil Pollock and Robin Williams. Their argument relies on the empirical illustrations deriving from the development of Connecting for Health, an UK health infrastructure. There, Pollock and Williams maintain that local studies of immediate settings of action inevitably draw attention to the scope for discretion (user workarounds, appropriation strategies, resistance, etc.) but provide a poor vantage point for exploring longer-term processes of technology-organizational alignment (for example around common business process templates within enterprise systems). They find it odd that there is such a wide-ranging set of terms in sociological essays demonstrating interest in information systems (including Science and Technology Scholars) to describe the way standardised technologies are 'imported' ('domesticated,' 'appropriated' or 'worked-around') into user settings, while there is a comparative lack of emphasis on the reverse process through which an artefact is 'exported' from the setting(s) in which it was produced. This is striking since the bulk of organizational software in use today is produced in this way—the same systems are recycled from one context to another. Thus they attempt to develop the beginnings of a vocabulary to capture this exporting, what they describe as the practice of making software generic (*generification work*), including its various explicit and revealed generification strategies, as the process of "generification." Furthermore, sociological essays in information systems designed to focus on the 'importing' process are often accompanied by a particular temporal framing, which has been identified as snapshot type of study. Snapshot studies are often conducted a relatively short time after the introduction of a new technology, "arguably before the complete consequences of an innovation can be reasonably assessed" as Pollock and Williams say. According to our view, the research design and temporal framing of current sociological enquiries in information systems embed a tacit determinism with respect to the *nature and stability of the normality structure,* which derives from the interactionist perspective: it is a hidden naturalism, a tacit assumption that everyone has the same basic normality structure (Lynch & Bogen, 1994). The interactionist studies perspective derives from Garfinkel's critique of Durkheim's conception of the nature of social facts. Contrary to Durkheim's view that a fundamental principle of sociology is the objective reality of social facts, the ethnomethodological perspective assumes that the reality of social facts is a process of continuous accomplishment in the concert of daily life. Although theirs is no longer Durkheim's consensual rule-governed society, the picture emerging from interactionist studies is that of a society consisting of a vast but stable repertoire of fragmented micro-rules. As a consequence of the assumption of the stability of the normality structure, interactionst studies informed sociological enquiries in information systems may fail to fully realize that ambiguity of context in human societies is partially removed by a whole gamut of tools of which they analyze only a part.

Pollock and Williams (via the biography of artifacts framework—BoA), as well as Hyysalo (via the Biographies of Technologies and Practices study—BOTP), proposed modification of sociology is something that we can call a *biographic turn* in the social study of Information Infrastructures, which consists in suggesting that Information Infrastructures have biographies. It is, we think, a concern to reassemble macro and micro levels of analysis that explain the interest in biographical method by information systems researchers. The biographic method, as it has been used in different branches of social sciences, represents a way to clarify the connections between the personal and the socio/historical after the flat ontologies of postmodernism (see Wengraf, Chamberlayne, & Bornat, 2002, for a review), and so it does for the social study of information systems. Quoting Hyysalo, this turn in the social studies of information systems assumes that when the unit of analysis is design work, its biography can span different dispersed (in space) and successive (in time) organizations without loosing its unity:

*"any moment in human conduct is simultaneously a part of the unfolding of a task, the development of the individual doing it, the development of the work community, and the development of professional practice" (Hyysalo, 2004, p. 12).*

The biographic turn in information systems studies can also be seen as an attempt to write about the *fullness* of the *total* Information Infrastructure *situation*, while refusing to resort on the analytical movement from singular expressions to delocalized generalizations. The opportunity given by Information Infrastructure to social sciences scholars to *connect* (often concretely, by social bonds, the internet or global institutional affiliations) different social grounds into the same terrain of research (e.g. following the biography), feeds back on the legitimacy of grounding the critical force of the arguments social studies of information systems aim at making: if we can find ways to make connections between disparate events, we will have extended empirical analyses. While assuming that connections between historical players and ideas must be shown in the local setting, the biographical turn is also suggesting that there are no limits to the size of the analytical network we can create by concretely connecting multiple local analyses.

One final aspect of the phenomenological structure of Information Infrastructures as a research topic we want to flag is that of unifying different scholars from different countries into a common research program. Based on the critical comparison and juxtaposition of empirical aspects of distant but connected and converging phenomena, the social study of Information Infrastructures can solidify into its own specific methods and proposals of modifications of social science, of which the *biographic turn* is just an initial manifestation.

## PLURALISM IN ACTION RESEARCH FOR TECHNO-ORGANIZATIONAL CHANGE

Section 4 investigates the consequences of the phenomenological orientation for action research, and the possible alternative modes of approaching action research in organizations.

Action research can be generally defined as a form of applied research distinctively characterized by the intertwining between theory and practice, so that to shape an epistemology of knowing by changing and vice versa; a participative approach, according to the idea that all people involved in the social practices under investigation should actively cooperate in all research phases in order to produce knowledge actually able to change social practices; and participation extended to all decisional levels, according to the idea that change should concern the overall organizational design and regulation of social practices.

In the organizational field, the core characteristic of action research is the attempt of connecting the organizational analysis of empirical work settings, organizational change, and the evaluation of organizational re-design with each other.

In this framework, the chapters enclosed in Section 4 presents a number of methods and approaches for dealing with organizational re-design, and particularly with organizational re-design associated with the introduction of Information and Communication Technologies (ICTs) in work settings. And they provide examples of their application in different research settings, such as a medical Emergency Response Centre (Bassetti, this volume); an electronic small and medium sized enterprise (Jacucci & Campagnolo, this volume); an ambient assisted living project (Kai-Uwe, et al., this volume), a public administration (Albano, et al., this volume). In so doing, they leave space for reflections on the possible spaces for organizational re-design in such cases where: a) the action research starting point is the focus on concrete collective processes of work, each of which is seen as unique and unrepeatable (i.e. constantly different from itself, over time, and from other processes, at the same time); and b) the aim of action research is the organizational change effectiveness from the standpoint of the agents of the research setting under intervention.

The fist common denominator of the contributions enclosed in Section 4, thus, is that they are concerned with the issues of *whether* and *how* organizational re-design associated with the introduction of ICTs in work settings is still possible even though the starting point is the focus on unique and singular worlds.

In this section, I argue that answering the above questions requires a reflection on two further distinct, yet connected, controversial points. The first concerns the cognitive strategies employable to deal with ICT-related organizational change. The second one refers to what might be meant by organizational re-*design*.

With regard to the first point, the assumption that work processes are unique and unrepeatable should imply that they could not be organizationally re-designed accordingly to universal laws. As a consequence of that, answering whether and how ICTs-organizational re-design is still possible starting from that assumption requires first to reflect on the possible cognitive strategies that might be employed in place of the positivistic one for researching organizational change related to ICTs; and then to grasp which spaces they leave for organizational design.

Tony Hines's contribution offers preliminary insights to move some steps towards this direction of reflection. The chapter attracts attention on the phenomenological cognitive strategy rooted in the notion of "horizon of experience" drawn from Husserl's work. The premises are that social researchers consciously and intentionally set out to know the phenomenon of human existence that finds expression in human endeavours and human organizations. Knowing is understanding, i.e. making sense and meaning, through the lived experience, of the spatio-temporal relations between subjects and between subjects and objects. Knowing/understanding is an intentional and conscious act. One first experiences the world, and then conceptualizes it to understand personal experience. Every experience/knowledge has its own horizon, which, at least in part, limits it. According to Husserl, living together defines the horizon of the subjective experience. The researcher interested in the organizational change associated with the introduction of information technologies in work settings, should therefore adopt a natural standpoint. S/he should live in and experience the research setting, by setting aside any preconceptions and prior theories. His/her aim is to *understand* the *understanding* of change by people in the organization. Through understanding, change becomes an object of consciousness for the subjects, and the change in the research setting, as well as the subjects become objects of consciousness for the researcher. Both for the latter and for organizational members, understanding is the reconstruction of the points of view, and the subjective meanings of the agents in their relationships with other subjects and with objects. The researcher, and the research "objects" should therefore proceed in a way similar to Aristotle's "Posterior Analytics." A consistent application of this strategy for researching ICTs-related organizational change should therefore imply that the phenomenon at hand might be justified ex post only. Therefore, any possibility of guiding ICTs- related organizational change according to prior rules should be ruled out.

With regard to the latter point, a couple of contributions in Section 4 of this book points out that two cognitive strategies provide a chance for maintaining the possibility of guiding ICT-related organizational change in accordance with prior rules. One is that which typically underlies socio-technical systems approaches, and newer systems theories (cf. the chapter authored by Kai-Uwe Loser and colleagues in this book), that is the positivistic cognitive strategy. As I stated before, this strategy should face many difficulties in coping with the ICTs- related organizational re-design of unique and unrepeatable processes of work, and, indeed, Loser and colleagues' contribution in Section 4 of this book precisely attracts attention on the ways in which the method of socio-technical walkthrough, which is rooted in the above theoretical and methodological references, tries to overcome those difficulties. The other cognitive strategy which provides chances for guiding the ICTs- related organizational change accordingly to prior rules is the Weberian one. This epistemology underlies the approach to action research discussed in the chapter authored by Roberto Albano, Tommaso M. Fabbri, and Ylenia Curzi in this book. As this contribution highlights, the basic assumption of this approach is that work processes are ever changing processes.

As mentioned above, the second controversial point on which one should reflect to answer to *whether* and *how* ICT-related organizational re-design is still possible even though the starting point is the focus on unique and unrepeatable collective processes of work is "what the concept-word design might mean."

A number of contributions in Section 4 of this book suggests that the meaning of ICT-related organizational re-*design* varies accordingly to the cognitive strategy used for researching the ICT-related organizational change. They further suggest that this meaning, to its turn, strongly affects the way to proceed in real world work settings to organizationally re-design work processes affected by the introduction of ICTs.

The chapters enclosed in Section 4 of this book move from different epistemological viewpoints/ cognitive strategies. Therefore they provide different conceptualizations of ICT-related organizational re-design.

In the study by Chiara Bassetti in this book, the word "design" refers to the order(s) of the activities related to technologies, which emerge in the field through improvisation and become established gradually. In accordance with that, the "design" is recognizable/visible only after the fact. In particular, the author uses the word "design-in-use" to emphasize that techno-organizational design is a practical activity, concomitant to the use of the technology, and the word of "co-design-in-use" to stress that the improvisational process of techno-organizational change is collaborative, situated, interactional, and based on the workplace natives' collective know-ledge/how, which is both exploited and increased in the process.

Gianni Jacucci and Gian Marco Campagnolo's contribution in this book proposes to maintain the concept of design even thought the researcher assumes that social practices emerge by individuals' and groups' social construction, rather than being designed in an engineering way. The word should be used to emphasize the core characteristics of any process of organizational innovation, such as i.e. intentionality, proactive-ness, deliberation, and planning.

The study by Kai-Uwe Loser and colleagues in this book shows the limitations of upfront planning in coping with the complexity of technology-supported services processes for elderly people. This complexity is related to the fact that the same service process conducted by different stakeholders results in different processes, and to the fact that even with the same stakeholder, the service process changes, because the conditions and requirements of technology usage change rapidly. In such circumstances, the design of the technology, and of the coordination of cooperative work involved in the offer of domestic services to elderly people, should not be intended as the search for a complete and durable solution, for a strict control, and planning of every aspect of the process. Rather, it should include change, and provide the different stakeholders involved in changing environments and complex processes with a guidance, a point of orientation for local adaptations during processes conduction.

Finally, the last contribution in Section 4 of this book, authored by Roberto Albano, Tommaso M. Fabbri, and Ylenia Curzi, considers organizational phenomena as products of the intentional and bounded rationality of human beings. Accordingly, it conceives organizational design as the process of actions and decisions aimed at ordering, co-defining means and ends in accordance with bounded and intentional rationality, and with a logic of mutual consistency. "Organizational design" is a search under condition of uncertainty, embedded in any work processes and distributed among all those who are directly involved in it.

In connection with different conceptualizations of ICT-related organizational re-*design*, the chapters in Section 4 of this book propose different approaches to techno-organizational re-design.

Bassetti's study proposes a form of critical and dialogic action research, where the researcher lays the groundwork for the cooperative and emergent accomplishment of new order(s), and the self-reorganization of the setting. His/her task is to make explicit and intelligible to the other actors of the techno-organizational change process the tacit logics underlying the organization of the operators' practices of work. The author particularly stresses the role of ethnomethologically oriented ethnography as an useful methodology for analysing how different groups of people (i.e. designers, managers, future users) make sense of tools and artefacts, of the everyday technology-mediated work practices, and of the techno-organizational change process. The action researcher should therefore provide all subjects involved in the change process with additional understanding of themselves, of the relevant others, and of the specific ways in which orders are collaboratively and improvisationally achieved. This additional knowledge would increase their ability to identify, or to deal with techno-organizational problems, and to innovate work practices and instruments on their own.

The chapter authored by Jacucci and Campagnolo attracts attention on Social Practice Design (SPD). SPD is a form of action research based on phenomenology and counselling, which can be considered an extension of Participatory Design to the implementation phase of information systems. It creates a space for the design of social practices, and for the social design of organizational practices in the application of Participatory Design techniques in organizations with the final aim of "making room" for technology. Moreover, it is an approach to design based on improvisation, the latter intended as the reduction of the period of time between design and implementation. Within this approach, three external experts play a key role for the activation of staff and management people in the process of co-construction of innovation. The ethnographic researcher, who contributes to the identification and description of the social practices of the actors involved in the innovation process, and to the qualitative assessment of the change sustainability. The counsellor, who assists people in developing an attitude oriented to change, thereby fostering a process of co-construction of innovation in which all parties are involved and learn, and, finally, the action researcher, who supports the realization of innovation, by making people involved in the research group the authors of the organizational change, and by creating *paths* that the users of new technologies can follow to conceive and implement the visions of solutions for current problems in the social practice of work.

Loser and colleagues' study focuses on the method of socio-technical walkthrough. This method practically supports the idea of design put forward in the chapter, through the technical feature of vagueness, and by assuring that all relevant stakeholders (e.g. users, technology developers, staff) take part in the design efforts early on. The study also shows that within the ambient assisted living project, the method at hand was combined with several other ones in order to closely relate the process design with real world applications of the designed process, so that to foster a circle of mutual influence between them. First hand experience from real world applications provides users, process designers, and the other stakeholders involved in the process conduction, with deeper insight into the needs of elderly people, the technology use, and the coordination of the different parties involved in the services process. As a consequence, conditions and requirements of use change, and with them the service process, the supporting technology, and the design.

Albano and colleagues' contribution proposes an approach to action research based on the social theory of Elias, a theory of organizational learning, and a conception of action research in support of organizational learning, which share a non-dualistic view of the relation between individuals and organization. In this approach, the individuals involved in empirical organizational processes learn theoretical and methodological knowledge drawn from organization theory, and combine it with the knowledge drawn from their everyday practice of work, in order to analyse and change the structuration of their work processes on their own.

As mentioned at the beginning of this section, a further issue recurs in Section 4 of this book. It concerns *whether* and *how* ICTs- related organizational re-design is possible when the aim of action research is the organizational change effectiveness from the standpoint of the agents of the research setting under intervention.

These questions are related to the issue addressed, for example, in Bassetti's study, namely the issue of technology's usability and technology in use/in action in a complex, and changing situated configuration of people, artefacts, knowledge, information, activities and practices. Technology's usability and technology in use/in action particularly refer to the technology's appropriateness, usefulness, and coherence with respect to the endogenous organization of work practices and its underlying logic, which to its turn is related to the actual conditions of work, and the associated workers' needs. It is precisely in connection with this issue that the author claims for the necessity of techno-organizational co-design (-in use). The same questions just mentioned above are further related to the Ambient Assisted Living project discussed in Loser and colleagues' chapter. Ambient Assisted Living precisely concerns the technology support to the everyday life of elderly people; the final aim is to promote and prolong their well-being, and this to its turn implies to adapt the conditions of technology's use to their needs.

As these references show, the second recurrent issue in Section 4 of this book can be differently expressed in terms of *whether* and *how* it is possible to guide ICTs- related organizational change (i.e. it is possible to organizationally re-design) consistently with the organizational members' requirements.

These are crucial questions, in that they concern the evaluation of the organizational change/re-design, that is one of the key moment of any action research, and in this regard, the studies by Bassetti, Loser, and colleagues, and Albano and colleagues in this book precisely attract attention on the fact that in real world organizations and action research projects normally there is a variety of possible, and potentially conflicting viewpoints on what might be considered as a desirable organizational change.

*[Organizational] members, in fact, belong to different groups; each of them has a specific idea and representation of the (work in the) organization and its desirable future. These ideas, representations and desires are often conflicting, or at least diverging (Bassetti, this volume).*

Guiding ICT-related organizational change consistently with organizational members' needs thus requires to coordinate a number of potentially conflicting individual interests, purposes, and actions.

What the above-mentioned studies of this book suggest is that one might not answer to whether and how it is possible to guide ICT-related organizational change consistently with the organizational members' requirements, without thematizing the relationship between power, conflict, and organizational change.

In particular, the chapter authored by Bassetti and that authored by Albano and colleagues in this book suggest that power and conflict play an interpretative role with respect to organizational re-design.

Power accounts for the direction of organizational change. Bassetti (this volume) highlights that an evidence that we cannot ignore is that:

*participants, professional researchers, and local stakeholders represent a variety of value orientations and commitments that influence both the courses of action and the goal of a project. (Maurer & Githens, 2010; quoted in Bassetti, this volume) At the same time, different groups are equipped with different resources, power, and decision-making opportunities. (Bassetti, this volume).*

Albano and colleagues' study highlights that the extent to which the current regulation choices can be actually modified in order to meet individual goals and needs depends on the ability to sustain one's own point of view, interests, and choices in the regulation of interdependent processes.

This study also points out that conflicts result from some incongruence between the current structuration of mutual dependences in the work setting and human needs. Therefore, they can be overcome only through some structural/organizational change.

Bassetti's study similarly highlights that users group resistance to new information systems are usually due to their inconsistency with respect to the endogenous organization of activities and its underlying logic, which to its turn is associated with the actual conditions of work and the related individuals' needs.

Section 4 of the book also suggests that conflicts should be neither repressed, nor denied. The different groups involved in techno-organizational re-design should instead be provided with spaces for dialogical confrontation.

In this regard, Bassetti's chapter attracts attention on the role of the researcher in the form of critical and dialogic action research that she puts forward in her contribution. Here the action researcher acts as an agent of visibility and recognisability of the local organization of activities and its logic; as an agent of mutual awareness, by constructing the groundwork for the discussion among different groups of actors of the respective logics, and starting from his/her neutral and "expert" social position, s/he acts as an agent of legitimization of the standpoint of some actors—primary, even thought not exclusively, those more lacking in power—in front of the others.

Loser and colleagues' study in this book shows how the method of socio-technical walkthrough supports the representation of different perspectives for different stakeholders, enabling them to understand and discuss the influence of others on their behavior. It further provides a chance for agreement, if it is wanted or needed.

Finally, Albano and colleagues maintain that conflicts may be ruled, and precisely oriented towards an organizational innovation involving a power ratio surplus for all individuals involved in the work processes under analysis, thereby escaping from the "zero-sum" concept of power. The action research approach discussed in their contribution aims at this end. In this regard, the authors stress that if the individuals learn the theoretical and methodological knowledge provided by organization theory, and combine it with the knowledge drawn from their everyday practice of work, in order to analyse (on their own) the organizational structuration of their work processes, they will have greater chances of bringing the sources of conflicts (i.e. structural incongruencies) to light, and therefore of identifying more satisfycing organizational choices.

To sum up: Section 4 of the book proposes a number of different approaches and methods for dealing with the relations between power, conflict, and techno-organizational re-design/change of work settings, and some examples of their application drawn from action research fieldworks. The aim is twofold. On the one hand, to continue on the path suggested by Ciborra: i.e. cultivating pluralism in information systems and organizations research, and, on the other, to invite scholars and practitioners to reflect on the methods and approaches (that might be) employed to guide ICT-related organizational change consistently with organizational members' needs.

*Gianluigi Viscusi*
*University of Milan Bicocca, Italy*

*Gian Marco Campagnolo*
*University of Edinburgh, UK*

*Ylenia Curzi*
*Marco Biagi Foundation, Italy & University of Modena and Reggio Emilia, Italy*

## ACKNOWLEDGMENT

We would like to acknowledge all ALPIS participants from the 1st edition in 2005 to the 2010 edition. This book is based on their contributions and insights. Furthermore, we would like to acknowledge all the reviewers who contribute to the final version of the chapters in this volume. This book is in memory of Claudio Ciborra.

## REFERENCES

Avgerou, C., Ciborra, C., & Land, F. (2004). *The social study of information and communication technology: Innovation, actors, and contexts*. Oxford, UK: Oxford University Press.

Beck, U., Giddens, A., & Lash, S. (1994). *Reflexive modernization: Politics, tradition and aesthetics in the modern social order*. Cambridge, UK: Polity.

Boland, R. (1999). Accounting as a representational craft lessons for research on information systems . In Currie, W., & Galliers, B. (Eds.), *Rethinking Management Information Systems* (pp. 229–243). Oxford, UK: Oxford University Press.

Bowker, G. C., & Star, S. L. (1999). *Sorting things out: Classification and its consequences*. Cambridge, MA: The MIT Press.

Campagnolo, G. (2007). *Accountability in technology production: An ethnography of enterprise modelling*. Unpublished Ph.D. Dissertation. Trento, Italy: University of Trento.

Chamberlayne, P., Bornat, J., & Wengraf, T. (2000). *The turn to biographical methods in social science*. London, UK: Routledge. doi:10.4324/9780203466049

Ciborra, C. (1999). A theory of information systems based on improvisation . In Currie, W., & Galliers, B. (Eds.), *Rethinking Management Information Systems* (pp. 136–155). Oxford, UK: Oxford University Press.

Ciborra, C. (2002). *The labyrinths of information - Challenging the wisdom of systems*. Oxford, UK: Oxford University Press.

Ciborra, C. (2006). The mind or the heart? It depends on the (definition of) situation. *Journal of Information Technology*, *21*, 129–139. doi:10.1057/palgrave.jit.2000062

Ciborra, C. U. (2001). *From control to drift: The dynamics of corporate information infrastructures*. Oxford, UK: Oxford University Press.

Currie, & Galliers, B. (Eds.). (1999). *Rethinking management information systems.* Oxford, UK: Oxford University Press.

De Michelis, G. (2008). The phenomenological stance of the designer . In Binder, T., Loewgren, J., & Marlborg, L. (Eds.), *Re-)searching the Digital Bauhaus* (pp. 145–162). Berlin, Germany: Springer.

Dourish, P. (2001). *Where the action is: The foundations of embodied interaction*. Cambridge, MA: MIT/Triliteral.

Dreyfus, H. L. (1995). Heidegger on gaining a free relation to technology . In Feenberg, A., & Hannay, A. (Eds.), *Technology and the Politics of Knowledge* (pp. 97–107). Indianapolis, IN: Indiana University Press.

Fele, G. (2008). *In praise of thin descriptions: Ethnography, ethnomethodology and the study of practice*. Paper presented to Alpis Ski Seminar '08 - Alpine Information Systems Seminar - 4th Edition. Carisolo, Italy.

Garfinkel, H. (2002). *Ethnomethodology's program: Working out Durkheim's aphorism*. Lanham, MD: Rowman & Littlefield Publishers.

Garfinkel, H. (2008). *Toward a sociological theory of information*. Boulder, CO: Paradigm.

Giddens, A. (1976). *New rules of sociological method: A positive critique of interpretative sociologies*. London, UK: Hutchinson.

Heidegger, M. (1962). *Being and time*. Oxford, UK: Blackwell.

Hevner, A. R., March, S. T., Park, J., & Ram, S. (2006). Design science in information systems research. In King, J. L., & Lyytinen, K. (Eds.), *Information Systems: The State of the Field* (pp. 191–232). New York, NY: Wiley.

Kallinikos, J. (2006). *The consequences of information - Institutional implications of technological change*. Cheltenham, UK: Edward Elgar.

Latour, B. (2005). *Reassembling the social: An introduction to actor-network-theory*. Oxford, UK: Oxford University Press.

Liberman, K. (1985). *Understanding interaction in central Australia: An ethnomethodology of Australian aboriginal people*. London, UK: Routledge.

Liberman, K. (2004). *Dialectical practice in Tibetan philosophical culture: An ethnomethodological inquiry into formal reasoning*. Lanham, MD: Rowman & Littlefield.

Liberman, K. (2007). *Husserl's criticism of reason - With ethnomethodological specifications*. Lanham, MD: Lexington Books.

Lynch, M., & Bogen, D. (1994). Harvey Sacks primitive natural science. *Theory, Culture & Society, 11*, 65–104. doi:10.1177/026327694011004003

Lyotard, J. F. (1991). *Phenomenology*. New York, NY: SUNY Press.

Markus, M. L., Tanis, C., & van Fenema, P. C. (2000). Enterprise resource planning: Multisite ERP implementations. *Communications of the ACM, 43*(4), 42–46. doi:10.1145/332051.332068

Newman, S. E. (1998). Here, there and nowhere at all: Distribution, negotiation, and virtuality in post-modern ethnography and engineering. *Knowledge in Society, 11*, 235–267.

Power, M. (1994). *The audit explosion*. London, UK: Demos.

Strathern, M. (2002). *Audit cultures: Anthropological studies in accountability, ethics and academy*. London, UK: Routledge.

Suchman, L. A. (1987). *Plans and situated actions – The problem of human-machine communication*. Cambridge, UK: Cambridge University Press.

Winograd, T. (1995). Heidegger and the design of computer systems . In Feenberg, A., & Hannay, A. (Eds.), *Technology and the Politics of Knowledge* (pp. 108–127). Indianapolis, IN: Indiana University Press.

Winograd, T., & Flores, F. (1986). *Understanding computers and cognition – A new foundation for design*. New York, NY: Ablex Publishing Corporation.

Zuboff, S. (1988). *In the age of the smart machine: The future of work and power*. New York, NY: Heinemann Professional Publishing.

## ENDNOTES

1 All the citations are from the English translation of Lyotard (1954): Lyotard, J. F. (1991). *Phenomenology*. New York, NY: SUNY Press. Whereas we report in the text the reference to the original French volume.

2 This is not the case of Claudio Ciborra, who on the one hand considers Husserl mainly for the issues he raised in the *Crisis of European Sciences and Transcendental Phenomenology*; on the other hand, Ciborra develops a phenomenology of information systems mainly from the idea of phenomenology of Martin Heidegger, whose position refuses the idea of a subject (no matter how

pure or transcendental) as the original foundation of our relationships with reality (Viscusi, this volume).

3 'the moment of opportunity,' (see Ciborra, 2006, p. 138).

4 'a shift of the role and function in concrete situations of usage, compared to the planned, pre-defined, and assigned objectives and requirements that the technology is called upon to perform' (see Ciborra, 2002, p. 85).

5 '*Bricolage* (from the late Latin *bricola* catapult) means tinkering through the combination of resources at hand. These resources become the tools and they define in situ the heuristics to solve the problem' (see Ciborra, 2002, pp. 48-49).

# Section 1
# Phenomenology and Information Systems Research

# Chapter 1
# Why Is Information System Design Interested in Ethnography?
## Sketches of an Ongoing Story

**Giolo Fele**
*University of Trento, Italy*

## ABSTRACT

*The chapter reviews the main stages of the collaboration between ethnographers and information system designers, highlighting the reasons and motives for their mutual relationship. Is it possible to consider the ethnographic approach to information system design a "success story"? How is it that information system design— a field seemingly distant from the concerns, history, or tradition of ethnographic research—is today so interested in the approach, the methods, and the "philosophy" of ethnography? What has ethnography to offer information system designers?*

## INFORMATION SYSTEMS AND ETHNOGRAPHY

The study of information systems is a field of inquiry, which encompasses various disciplines, ranging from management to artificial intelligence, from informatics to software development. This article is devoted to the interest that information science—in particular, the design and evaluation of systems—has for some time shown in ethnography. Indeed, over the past twenty years the analysis of information systems has been one of the fields in which ethnographic approaches have received most attention, been subjected to serious consideration, and obtained the most significant results. But how has this happened? In what does the 'success' (if

DOI: 10.4018/978-1-4666-0303-5.ch001

it can so be called) of ethnography consist? How has a field so apparently distant from the history or tradition of ethnographic research been 'infected' by the philosophy, methods, and 'gaze' of ethnography? How does a 'hard' discipline like informatics consider a 'soft' one like ethnography? Does ethnography have something to offer to informatics? What in particular? Moreover, how can this interest among information technologists concern the disciplines that have traditionally drawn on ethnography? Is this a merely 'sectoral' or 'niche' interest, no more than a passing intellectual fashion, or does it reveal something profound concerning the value of ethnographic methods. Can it challenge what is often termed, with a certain emphasis, the 'interdisciplinarity' (in the weak sense) and sometimes 'hybridization' (in the strong one) of disciplines? Can discussion of such topics be made in a journal addressed to a broad audience and diverse academic communities, or must it be confined to specialist journals and conferences? The aim of this introduction is to clarify these questions[1].

The first novel feature of the interest shown by information engineers in ethnography is its direction, for it is information engineers who are interested in ethnography, not the other way round. There is long tradition of ethnographic studies on organizations[2], work in scientific laboratories[3], and technology[4] where the social or anthropological (and similar) disciplines have concerned themselves with the world of engineering and technology. Yet it is now the managerial and engineering disciplines, which are interested in the work of ethnographers. Of course, informatics deals with ethnography because it needs it for practical reasons: this concerns, not ethnography *by* engineers, but ethnography *for* or *with* engineers, which entails interdisciplinarity (a somewhat worn-out term but which seems to have acquired new currency). As Baba (2006, p. 37) writes:

*Designers are considered 'creatives' (i.e., there is an artistic aspect to their work), and their marriage with intellectuals and researchers (anthropologists) is an interdisciplinary challenge. The creation of 'design ethnography' thus represents the birth of a new interdisciplinary subfield that joins together anthropology and/or other qualitatively-oriented social sciences with the design profession.*

Engineers have always constituted an interesting case for study with ethnographic methods[5]. But anthropological and sociological accounts have frequently failed to consider the actual *work* of engineers (Bruni & Gherardi, 2007). In these cases, ethnographers (sociologists, anthropologists) are interested in describing phenomena such as the construction of identity or the self in the profession, or in reflecting on the boundaries (mobile, rigid, or mutable) that separate technology from 'humanity'[6]. In the case considered here, when information engineers design information systems jointly with anthropologists or sociologists, the ethnography examines the *organization* of knowledge and information. Yet the problems—and their technical, technological and expert solutions—cannot be ignored, for they constitute the core itself of the ethnography to be pursued. Whilst this approach may involve 'technicisms' difficult for a layman to handle, it should not bother an ethnographer, trained as he or she is to know all the 'technical details' of building a canoe or the 'technical details' of navigation by the natives of the Western Pacific (Malinowski, 1922)[7].

This article reconstructs the phases of collaboration between ethnographers and the designers of information systems[8], seeking to show the rationale for such collaboration. Of course, this introduction starts from the social sciences. It consequently represents the return of questions, which first arose within studies on information systems to the social sciences: it therefore conducts a reconstruction 'situated' in the social sciences, not in computer science.

## THE ENGINEERING MENTALITY

The classic engineering approach to the design of information systems presupposes a rationalistic model of problem solving, a 'nested' top-down approach in which decisions respond to the requirements of the technology and the purposes of the management[10]. Amid increasing complexity, the design of a new information system starts from a rigorous breakdown of the entire process into specific phases. One of the best-known models representing this philosophy is the 'waterfall model,' which was developed in the 1970s. An online document on the website of NASA, the American space agency, specifies the various steps in the standard 'waterfall model'[11]. It states that the development of information systems must move through the following stages: 1: the 'document system concept,' i.e., the design of the system expressed in a few words; 2: identification of the system requirements and their analysis; 3: breakdown of the system into pieces (the architecture design phase); 4: design of each piece (the detailed design phase); 5: coding of the system's components and their individual analysis (coding, debugging[12], and unit testing); 6: integration of the pieces of the system and testing the system's operation (system testing); 7: deployment of the system and its use. This is a model with a typically top-down logic of development (like a waterfall) where the move from one phase to the next is hierarchical and possible only when the previous phase has been completed[13]. What is denoted by the ugly word 'implementation' of an information system[14] is only the last aspect of this model: the application is the logical conclusion of a rational definitional process, which is already complete and defined theoretically and in abstract. It is possible to see this segmentation of phases as reflecting the principles of Taylorist scientific management[15]. It is a model which presumes the stability of the context being analysed and in which the system designer will operate. For social science researchers it resembles an idealized model of the phases of action-research. The practice of systems designers comprises a normative system according to which a team within the organization must guide the latter's operations. The 'reduction of complexity', which ensues, consists in containment of all the possible instances that arise in the context within the researcher/designer's analytical categories.

One distinctive feature of this model has interesting consequences for ethnography. In fact, the design proper is preceded by a phase called 'requirement acquisition and requirements analysis': that is, analysis of the properties and functionalities that the information management system must possess[16]. This is a very important stage in the design process, and potentially of great significance for ethnography. Substantially, an information system's designers must collect all the information relevant to constructing a new information management tool or to improving one already in use. This information is 'technical' in nature (connected with the logical-technological aspects of the information tool, in good part concerning how the software should be designed) but also largely contextual and socio-cultural, such as the organizational structure and culture, the work practices of the users, marketing strategies, economic and financial aspects, and so on[17]. The acquisition, definition, and specification (formal or informal) of requirements are the essential steps in describing the system's functionalities clearly, coherently and completely. These steps prescribe the most detailed and precise description possible of what the system will be able to do, and of how the organization will be involved in changing practices and procedures.

As the designers acquire and analyse the system's requirements, they consult the organization to determine the 'needs' of its users and their work practices. For this purpose, they should question the people inhabiting the field in which they have been called upon to intervene. But we know (Goguen, 1996) that one of the methods most widely used to collect information on requirements is based on introspection, i.e. on the common sense of a professional able to imagine problems and their solutions shorn of context and with an abstract and generic user in mind. Behind this user-model lies the idea that it is possible to identify problems according to their objective constitution—that they can be identified by applying a 'scientific' approach—and that these problems can be solved by applying a logical-deductive rational method. Science and method are properties of human beings, which seemingly spare them from having to dirty their hands with the practical, specific (often unique), and situated aspects of the organization of work. In the particular case of informatics and information systems, the differences of background, knowledge, and experience between experts and 'common mortals' may harbour nasty surprises: what an expert can imagine or remember of a problem and believes to be its optimal solution may be very different from what the non-expert (and actual!) user knows. Not infrequently, a non-expert user faced by rational solutions offering methodical answers to problems in the use of an information system is bewildered and simply "doesn't understand." This generates the recurrent experience of feeling that one must adapt to a way of 'reasoning' and 'doing' which is imposed and distant from one's own. In reaction to the unpredictable but recurrent impasses, which arise in practices with the computers, experts accuse non-expert users of being "naturally stupid"[18].

When experts do not use introspection, information is acquired from the statements of subjects interrogated by means of structured questionnaires or semi-structured interviews, or it is collectively elicited from focus groups. This information consists largely of accounts and explanations furnished orally; but this means that large part of practices, tacit knowledge, and covert strategies are left unexplored. The possible gap between 'saying' and 'doing' remains under-thematized.

The traditional 'engineering mentality' is therefore distant from ethnography for several reasons: it is based on abstract rational reconstructions and on "rationally idealized actors" (Ciborra, 2004, p. 27). Despite the complexity of contexts, circumstances, situations, and concrete practices of use, traditional information system designers produce technological solutions, which fail to recognize the importance of the context and of users' practices. Because their training has not taught these designers to go into the field, they seem to be in the condition described by Leo Tolstoy in *War and Peace*, when Prince Andrew finds himself in a real battle for the first time – an experience very distant from that taught in the schools of military strategy:

*The whole air reeked with smoke. The excited faces of the soldiers were blackened with it. Some were using their ramrods, others putting powder on the touchpans or taking charges from their pouches, while others were firing, though who they were firing at could not be seen for the smoke which there was no wind to carry away. A pleasant humming and whistling of bullets were often heard. 'What is this?' thought Prince Andrew approaching the crowd of soldiers. 'It can't be an attack, for they are not moving; it can't be a square for they are not drawn up for that'" (book I, part II, chapter XVIII).*

Prince Andrew's disorientation is figuratively the same as that of an information systems engineer with the traditional engineering mentality who, amid the complexities of work and organizational practices, is prevented from seeing 'what is' by 'what ought-to-be.'

There is increasingly widespread dissatisfaction in the community with the limitations of this mentality. Ciborra—a leading advocate of the separation of informatics from the mathematical and formal sciences, and its inclusion among the social and human sciences—argues that it is necessary to "drop the old methodologies, in order to be better able to see the new dimensions the technology is going to disclose to us. It is not time for calculation, but for a sort of deep contemplation of the everyday life surrounding the design and use of technology" (Ciborra, 2004, p. 42).[19] But informatics has only recently begun to move in this direction.

## FROM THE SYSTEM'S REQUIREMENTS TO THE 'USER' AND THE 'INTERFACE'

The engineering mentality of informatics just described places the computer at the centre of action: people serve computers, not the other way round. This engineering model used to work for particular needs (organizing systems for wage payments, flight bookings, or planning), but it was superseded by technological innovation. With the arrival of the third generation of machines—that is, with the advent of the personal computer (at the end of the 1970s)—an increasingly large number of people began to use computers for a variety of purposes. At theoretical level, there occurred a paradigm shift from a product-oriented perspective to a process-oriented one (Floyd, 1987). The personal computer generated demand for tools that were "easy to use, intuitive to understand, easy to learn cit. orig.?" (Crabtree, 2001, p. 146). The user became the centre of attention of programmers and the final referent for programming[20]. Programming no longer had to comply with the abstract, rationalistic, and general needs of the technological system's logic, but with those of 'ordinary' people, the average and non-expert users of the machine. Precisely because the model-user was not an expert, the technical aspects of the system's operation was of no interest to him/her, and they were no longer visible. The user could access the interface (another buzzword at the time, together with 'user'), but not the system itself, except in a mediated (graphical) way. All design efforts concentrated on a system's interface (its legibility, comprehensibility, and immediacy of use)[21].

The notion of a system's interface acquired great theoretical value. The interface was the point of encounter between two *models*: that of the system on the one hand, and that of the user, and especially his/her systems of mental representation, on the other. Understanding the user's mind and his/her mental processes (memory, attention, perception, association, imagination, and so on) became important for the design of systems and their interfaces. Computer science consequently turned to psychology for assistance. This accounts for the importance of cognitivism (Neisser, 1976) as an instrument with which to map the 'head' of the user, and of ergonomics as the practical application of the discoveries of cognitive science to human work with machines, the purpose being to improve the system's performance and to increase the user's satisfaction.

This was the field of studies on Human-Computer Interaction (HCI), Artificial Intelligence (AI), and Computer-Mediated Communication (CMC). Cognitivism was the perfect intellectual tradition for development of these studies because of the isomorphism that it predicated between the operations of computers and the mental activities of their users. The user was viewed as a constant information processor. According to a model of information coding and decoding, people are endowed with 'mental models' or 'cognitive schemes' which organize (and give meaningfulness to) experience. According to this approach:

*is the idea that human action is governed by rules, scripts and plans, and that through manipulation of symbols and the development of representations, individuals are able to execute intelligent action and interaction. The operation of the computer serves therefore both as a metaphor to characterize human reasoning and conduct, as well as a substantive domain, in which to discover cognitive processes." (Heath, et al., 2000, p. 302).*

According to cognitivist theory, if one examines how a user interacts with the technology, one can reconstruct the 'logic' of such use. This is equivalent to reconstructing the 'grammar of a person's mind' (often understood as the neural interconnections within the brain), and on the basis of this 'mental logic' computer programs can be constructed which mimic the cognitive processes of users.

Consequently, also the tools of the cognitivist approach could do without ethnography. Given their interest in the mind of the user, cognitive psychologists sought to map, model, and represent the knowledge, purposes, and activities comprised in the heads of people, not in the world. By means of essentially laboratory procedures, cognitive psychologists examined a single isolated user in a controlled environment. One of the most interesting aspects of this research method was its analysis of the verbal accounts furnished by subjects to the researcher as they performed tasks or solve problems (Ericsson & Simon, 1980). These verbal accounts were viewed as direct representations of the processes of the mind and thought. In a variant of the method, single and isolated subjects were interviewed while they perform recognized assignment and well-identified tasks, but the scientific bias of the classic engineering mentality was again apparent in this approach to modelling users' behaviour and reconstructing their mental models. Cognitivist models considered the user as having a *purpose*, pursued using *tools*, within a sequential structure of hierarchically organized *tasks*. Hence, first the cognitive scientist used this rigid analytical grid to translates the everyday activities of users, breaking the work down into a sequence of hierarchically organized tasks and sub-tasks (modelling activities). Then the information engineer reproduced this model in the specifications of the information system (often graphically as a flow diagram), 'reflecting' the information structure within people's minds in the technological system[22].

## WORK COOPERATION AND SOCIALITY

Before information engineers turned to ethnography, they had to perceive the crisis of normative or formal organizational models. They had to realize the problematic nature of the relationship among rules, plans, and models, on the one hand, and the reality of work on the other. They had to question the transparency and the obvious relationship between explicit, formalizable and representable aspects (propositional knowledge) and the tacit, contingent and contextual aspects (procedural knowledge) of work. They therefore had to consider the work of people as not reducible to the precise (and precisely reconstructable) tasks of an isolated user cut off from the system of organizational and working relations. This dissatisfaction with the traditional paradigms of human/machine interaction and computer-mediated communication gave rise a new, broadly interdisciplinary, field in which anthropologists and sociologists worked together with information engineers. The original purpose of studying Computer Supported Cooperative Work (CSCW)[23] was to create technological systems, which would facilitate collective decision-making and cooperative work through the construction of a media space shared in real time by users not necessarily in physical proximity. The field of CSCW analysis arose from awareness that the functional problems of technological systems do not reside in 'systems' themselves but rather in the scant understanding, among system engineers and designers, of social 'mechanisms,' interactions within groups, and the functioning of the organizations in which users work. Studies on CSCW initially consisted of analyses of small groups. They then gradually came to examine organizations "which are staffed by large, distributed, changing, differentiated, heterogeneous assembly's of people and are, as such, characterized by multiple viewpoints and competing if not conflicting interests" (Crabtree, 2001, p. 176). The field of CSCW arose as a systemic and holistic vision of work, which sought to move beyond the individualistic and compartmentalized view of classical cognitivist theory (the notion of interdependence). Moreover, it stressed that knowledge is not located in a single place but is distributed among diverse ones, and that responsibilities are connected and closely interwoven among actors[24]. CSCW conferences have become important forums for interdisciplinary discussion between technologists and social scientists[25]. *CSCW – The Journal of Collaborative Computing* was founded in 1992 and a thematic series of publications today consists of thirty-three titles[26].

CSCW draws on two intellectual sources: one the one hand, studies on 'situated cognition' (Resnick, et al., 1997) or 'distributed cognition' (Hutchins, 1996), which analyse socially shared representations and the coordination of people's actions in organizational settings; on the other, the sociological literature, in particular that on symbolic interactionism (Strübing, 1998; Star, 1998) as regards both analytical concepts and methodology.

Studies on 'situated' or 'distributed' cognition—which belong within the psychological tradition—maintain the models and metaphors used in the cognitive sciences and in studies on HCI: for example, the notion of 'representation' is shared by both the individual and the distributed approaches to cognition; and use is made of metaphors drawn from the world of computers to define technology-mediated cognition (Heath, et al., 2000, pp. 306-307). Studies of sociological orientation instead consider (and are inspired by) the analytical concepts first developed by Anselm Strauss (see Strauss, 1985, 1988; Corbin & Strauss, 1993; Star & Strauss, 1999) and Herbert Blumer (1969). Two notions in particular have been successful: those of 'articulation' and 'individuation.' For Strauss, the notion of articulation refers to a process similar to, but more flexible than, coordination consisting in the simple distribution of resources and roles. Work articulation concerns the way in which actors attribute meanings (they are bearers of beliefs, attitudes, perspectives, etc.), are absorbed in tasks which carry responsibilities (they have certain obligations, they must fulfil certain commitments), and possess cognitive structures in regard to a specific environment in spatial and temporal terms (Star, 1991, 1997). The notion of work articulation refers essentially to three kinds of social mechanisms: coordination mechanisms (Simone & Divitini, 1997; Schmidt & Simone, 1996), workflow management mechanisms (Schmidt & Bannon, 1992), and mechanisms for the construction and management of a shared information space (Schmidt & Bannon, 1992). For Elihu Gerson and Susan Leigh Star

*Articulation is a necessary part of due process because the definition of "adequate provision" shifts according to local circumstances. No matter how detailed requirements, they must be aligned with or tailored to a set of implementation conditions which cannot be fully specified ahead of time (Gerson & Star, 1986, p. 71).*

Moreover, Gerson, and Star write, "Every real world system thus requires articulation to deal with the unanticipated contingencies which arise. Articulation resolves these inconsistencies by packaging a compromise which 'gets the job done'; i.e., which closes the system locally and temporarily so that work can go on" (p. 76). The notion of articulation was introduced into CSCW by Schmidt and Bannon (1992) in order to supersede the strictures of psychological approaches with a sociological perspective. The concept of articulation has proved highly productive in some CSCW studies[27].

The notion of 'individuation' was first developed by Herbert Blumer to emphasise that actors are not the passive executors of cognitive programs. As well known, the interpretation by actors of the circumstances in which they operate is the basis of the interactionist approach. As Blumer writes:

*the human individual confronts a world that he must interpret in order to act instead of an environment to which he responds because of his organization. He has to cope with the situations in which he is called on to act, ascertaining the meaning of the actions of others and mapping out his own line of action in the light of such interpretation. He has to construct and guide his action instead of merely releasing it in response to factors playing on him or operating through him (Blumer, 1969, p. 15).*

The symbolic interactionism perspective expressed by Blumer enables CSCW studies not to lose sight of the individual while not engaging in the logical-formal reconstruction of his or her isolated behaviour.

Other important concepts of interactionist origin have circulated in CSCW: for instance, the notions of 'negotiation' (Strübing, 1992), of 'social worlds,' again developed by Anselm Strauss (1978) (see Clarke, 2006; Clarke & Star, 2007)[28], or 'boundary objects'[29], a term coined by Star and Griesemer (1989).

With CSCW, ethnography made its debut as an analytical tool used in the design of information systems. Often, however, the reference to ethnography merely disguised the use of non-standard methods for requirements acquisition, and it denoted a generic endeavour to get closer to organizational reality and work practices. In this case, too, symbolic interactionists, as the heirs to the Chicago School (Prus, 1995), provided the template. The insistence on working on the ground, on naturalistic analysis, Robert's Park appeals to "get the sits of your pants dirty" in the social worlds studied, were interpreted by information engineers as invitations to delve into social systems previously regarded as only 'noise,' or viewed as places of error perpetuation. Two members of the Chicago School warrant especial mention in regard to the discovery of ethnography by information engineers. The first is Everett Cherrington Hughes[30], whose researches in the sociology of occupations and 'professions' provided coordinates for the ethnographic study of work (Hughes, 1952, 1970). Between 1947 and 1957, Hughes supervised a remarkable series of studies—around sixty doctoral theses—on the most diverse of occupations. Subject to study were janitors, street sweepers, police officers, furriers, rabbis, taxi drivers, jazz musicians, funeral directors, boxers, school teachers, pharmacists, doormen, black trade unionists, restaurant managers, and yet others. Hughes was interested in analysing the relationship between work and identity; or the ways in which work constructs an image of the Self. Moreover, Hughes believed that the social nature of work entailed study of interaction patterns, the mechanisms of the control exerted by people over others and what they do, the sanctions which they can or want to apply, the negotiations that take place in groups of workers, etc. Hughes used a distinctive comparative method whereby two apparently very different situations, facts, or circumstances were connected in one particular respect.

The other outstanding member of the Chicago School was the anthropologist Lloyd Warner. The history of relationships between informatics and ethnography is also the history of applied anthropology. Warner was among the pioneers of the latter as the founder of *Social Research Incorporated* created in 1946 to study consumption patterns for the purpose of market research[31].

In short, informatics discovered ethnography, and by drawing on the sociological and anthropological tradition discovered the complexity of the settings in which technological innovation is embedded. The discovery was simple: even a 'perfect' information system may fail miserably if it does not take account of its setting. To parody Malinowski: "The ethnographic reality of an information system, and this is a truth too often overlooked, would not be brought much nearer to a scholar at home even by placing a perfect specimen right before him. An information system is made for a certain use, and with a definite purpose; it is a means to an end, and we, who study an organization, must not reverse this relation, and make a fetish of the object itself. In the study of the economic purposes for which an information system is constructed, of the various uses to which it is submitted, we find the first approach to a deeper ethnographic treatment. Further sociological data, referring to its ownership, accounts of who navigates in it, and how it is done: information regarding the ceremonies and customs of its construction, a sort of typical life history of an information system: all that brings us nearer still to the understanding of what an information system truly means to its user"[32].

## PRACTICES AND ORGANIZATIONS

Ethnomethodology has made the most important critique of the cognitivism underlying the dominant paradigm of artificial intelligence studies. The work of Harold Garfinkel, in particular[33], has highlighted that maps, plans and strategies—that is to say, everything which in cognitivism describes forms of 'structured expectations'[34] and constitutes the tools of systems designers—simply cannot describe (and therefore cannot replace or adequately represent) the concrete experience of people and the ways in which they conduct their everyday affairs. Maps, plans, and strategies may be forms with which we try to anticipate what we want to do in the future (prospective aspect), or explain or justify what we have done in the past (retrospective aspect); but they can never describe the real practices of the members of a social group, who must constantly deal with new, situated, and contingent aspects. On the one hand, say the ethnomethodologists, we can have a more or less clear account of the goals and purposes of action, of the strategies best suited to achieving certain goals, which may be prescribed before an actual course of action (in the form of imaginary scenarios that project future solutions). Or on the other hand, we may need a standard with which to judge an action after its performance (in the form of *ad hoc* reconstructions based on memory). But these devices do not represent the real (social and distributed) mechanisms that operate while we are doing something: action is much less reducible to execution of a pre-established plan than can be thus hypothesised. Rather than considering the performance of an action as the 'degraded' version (i.e. replete with uncertainties, false starts, blind alleys, lack of deliberate choices, reproduction of traditional behaviours, etc.) of a coherent, logically and scientifically representable, rational project—which is how the engineering mentality has traditionally represented the action of users—ethnomethodology maintains that action depends crucially on the material and social contingencies in which it occurs (Rawls, 2002).

Three consequences of this position should be stressed. Firstly, according to the ethnomethodological perspective, human action is rational, but it cannot be related to the requisites of formal logic. Rationality is a feature *exhibited* by the members of a social group to the extent that they are constantly required to show clarity and distinction (these being the fundamental requisites of rationality) of action to the community's other members by means of ingenious practices amid the unforeseen contingencies of the particular situation. Secondly, the 'mind' or the brain, which are the sanctuaries of cognitivism, are not the suitable places for the study of action. As Garfinkel used to say, there is nothing in the head but brains (Garfinkel, 1963, p. 190). From the analytical point of view, this requires a shift from an individualistic perspective, which envisages a single subject dealing with a more or less complex, but well-defined task to examination of the real conditions in which a task is performed as a social activity, and which envisages constant coordination and synchronization among the actors involved. Thirdly, this is not to exalt action and practices over plans and the strategies. Ethnomethodology does not deny that plans and strategies have a role in action (a role, as said, which may be anticipation, legitimation, or justification). Rather, ethnomethodology invites us to study the relationship between plans and action. As Lucy Suchman puts it:

*Rather than attempting to abstract action away from its circumstances and represent it as a rational plan, the approach is to study how people use their circumstances to achieve intelligent action. Rather than build a theory of action out of a theory of plans, the aim is to investigate how people produce and find evidence for plans in the course of situated action" (Suchman, 1987, p. 50).*

This last aspect is of special significance in the history of the relationships between informatics and ethnography.

Ethnomethodology has emphasised the nature as 'social constructs'—that is, 'practical accomplishments'—of all the objects, social facts and technologies which exhibit the features of order, stability, recognizability, logic, coherence, scientificity, and so on. Ethnomethodological research on scientific and laboratory work (Lynch, 1993) has highlighted the contrast between ordinary practice and *ex-ante* and *ex-post* idealizations. Numerous studies have evidenced that the work of scientists is chaotic and improvised, very distant from the idealized and rational procedures that science seeks to convey. Scientific accounts are written in compliance with the canons of scientific reporting. They do not describe at all what scientists actually do: indeed, scientists who obtain good results are often unable to explain how they have done so. A written account of observation procedures cannot be used as a self-sufficient set of instructions on how to replicate those observations. Something similar has happened in the field of information systems, where Garfinkel's ideas have found extraordinary application in the work of Lucy Suchman. Lucy Suchman is an anthropologist who obtained a doctorate at the University of California (Berkeley) in 1984 and in 1979 went to work at the Xerox Research Center of Palo Alto (California). She is currently a member of the Department of Sociology at the University of Lancaster in England. Her doctoral thesis, *Plans and Situated Actions: The Problem of Human-Machine Communication*, was published in 1987 and immediately became an essential reference for studies on design. The basic idea may seem commonplace today, but at that time, it was revolutionary. Suchman videorecorded two people trying to get a photocopier to work. Installed in the machine was an expert system, which the designers presumed would help the users of the photocopier. By simply comparing the actions of the users with the prompts from the photocopier's expert help system, Suchman was able to show the incompatibilities between two ways of 'reasoning' and the contradictions between the system's basic assumptions and the user's actual practice. Suchman's study still affords important critical insights into the models on which human/machine interaction is based.

In general, ethnomethodological studies have shown that both ordinary occupations and all the more so 'specialized' ones tied to particular work settings or particular professional communities (Garfinkel, 1986) comprise a set of complex activities that must be constantly monitored, with 'on-the-spot' evaluation of the actions of others, the anticipation and projection of events, and immediate responses which only in a very generic and metaphorical sense can be described as 'following instructions' or 'executing a project.' Large part of the results of ethnomethodological research consist in accounts of the concrete activities that develop *pari passu* with execution of a task, often in cooperative situations. Ethnomethodological theory raises two major methodological caveats, which should be stressed because they challenge the design of information systems. Ethnomethodology takes the anthropological principle of "respect for the native's point of view" (Geertz, 1983) to its extreme. Firstly, ethnomethodology warns against confusing the perspective of an external analyst, who must explain, describe or somehow represent the action taking place before his or her eyes, with the perspective of those who *in situ*, in first person, must perform certain tasks. Secondly, ethnomethodology stresses the importance of considering how people actually behave, regardless of how they are believed to behave by an analyst or a designer. For ethnomethodology this entails verifying, with respectful ethnographic procedures, how people behave as they perform tasks in natural situations, not ones reconstructed in a laboratory. Rather than relying on introspection, memory, or personal conviction—which instead obstruct understanding of the action—ethnomethodology invites the researcher to unravel the contingencies of situated action (Garfinkel, 2006).

With its emphasis on the radically 'indexical' (i.e. situated) nature of social action, ethnomethodology has developed forms of ethnography which grasp the local practices of the members of a social group: not because of a preference for small-scale phenomena, but because it is only in the particular situation, in the concrete circumstances in which people work together, that one can discern the particular sense acquired by general meaning structures like knowledge, information and technology. The Workplace Studies project pursued by Christian Heath and colleagues at King's College, London[35], is the most refined and elegant example of ethnography applied to high-tech work settings in which study is made of coordination and collaboration (social interaction), as well as forms of knowledge organization and management (the practical 'logic' of getting things to work), starting from their temporal configurations moment by moment in their constitutive details. In these cases, as we have seen in Lucy Suchman's research[36], the ethnography is assisted with video recordings[37]. Hubert Knoblauch (2005, 2006) has described the use of video as 'focused ethnography.' The use of video recordings in ethnography, or better ethnography conducted with video recordings, adds to ethnography's long-standing preoccupations: the problem of accessing the field, the question of participation in what is being observed, how to record and conserve significant information, the researcher's presumed subjectivity and freedom to interpret action, how to conduct analysis, how results should be written up and presented[38]. The use of video in ethnographic research, with the particular analytical mentality that derives from it[39], is one of the most interesting and fertile aspects of the ethnomethodological approach to the design of information systems.

One of the most radical outcomes of ethnomethodological reflection on information systems is what G. Button and P. Dourish have called, with an awkward neologism, 'technomethodology' (Button & Dourish, 1996; Dourish & Button, 1998; Crabtree, 2004). In Dourish and Button's words:

*Our goal, then, is not simply to look at how ethnomethodology can be used to critique technologies, crucial though that is, or even to apply ethnomethodological understandings in order to better understand the conditions in which technology comes to be developed (...). Rather, alongside those investigations, we have been engaged in a different one; to understand how ethnomethodological understandings of human social action and interaction can be used, directly, in designing interactive technologies (Dourish & Button, 1998, p. 14).*

As I interpret it, 'technomethodology' is an extreme form of interdisciplinarity that turns into hybridization. Here, a form of sociological knowledge represented by ethnomethodology is used 'directly' to design information systems 'with an ethnomethodological mentality.'

## ETHNOGRAPHY AND INFORMATION SYSTEMS

Ethnography is consequently an important tool and an indispensable analytical mentality for the design of information systems. Interdisciplinary projects are now in full swing. In England, the universities of Manchester (Psathas, 2008), Lancaster, Surrey, Oxford offer joint doctorate programmes in ethnography and informatics. There are many other places in Europe, especially in the Scandinavian countries, Germany, Austria, and also Italy. In the United States, Kling and colleagues at the University of Irvine, California, have developed an approach called Social Informatics (SI) (Davenport, 2008). The Xerox PARC (Palo Alto Research Centre, California) in the United States, and the Rank Xerox EUROPARC (Cambridge, England) in Europe, are the professional communities most concerned to involve anthropologists and ethnographers in the design of information systems (Suchman, et al., 1999; Orr, 1996). The theoretical references are much more nuanced and variegated than conveyed by this schematic overview. Here I can only cite some intellectual traditions, which I have not yet mentioned and which have contributed to this merger between the social sciences (with an 'ethnographic' approach) and informatics.

A particularly important tradition is that of Participatory Design (PD), a typically Scandinavian movement which began in the early 1970s. The ethnographer performs a crucial role in 'participatory design' by assuming the user's point of view in the design of information systems (Crabtree, 1998). The distinctive features of this approach is that workers are involved in the design from the outset, and that new technologies and IT solutions are used. The literature on technological innovation reports that computerization often impoverishes technical skills and reduces manning levels. The participatory design approach has developed action-research inspired by democratic ideals[40] in order to ensure that innovations are not conceived, developed and imposed by management, but are instead shared and appropriated by users[41]. Hence, the proponents of the participatory approach to the development of information systems share Becker's view that researchers should challenge constituted authority and give voice to positions expressed 'from below' (Becker, 1967). Ethnography thus becomes the means to activate a continuous feedback circuit among workers, designers and other institutional stakeholders in which information solutions are shared and appropriated (Simonsen & Kensing, 1994, 1997, 1998).

In Italy, Gianni Jacucci has promoted what he calls 'Social Practice Design' (SPD). This is inspired by the same democratic principles as participatory design and takes a phenomenological approach (Jacucci, 2007; Jacucci, et al., 2007; Jacucci, et al., 2008). When the SPD approach is used for the participatory design of socio-technical infrastructures for work, the ethnographer performs a fundamental role in parallel with that of the informatics/technology expert. S/he investigates the pre-existing work cultures and practices in the organization (Jacucci, et al., 2006), highlights discrepancies between the goals pursued by the designers and the real needs of users, and then validates the participatory approach employed (Jacucci, et al., 2008).

Besides the already-cited case of 'situated knowledge,' this rapid survey should also mention the 'Actor Network Theory' (ANT) developed by Bruno Latour (1992a, 2004) (Brighenti, et al., 2008) and his colleagues Michel Callon and John Law (Law & Hassard, 1999). Rather than studying the activities which take place in research laboratories (study to whose development Latour himself has contributed), the aim of actor network theory is to reconstruct the contexts lying beyond the laboratory's boundaries, seeking to identify inter-organizational links and networks, military, economic, and financial sponsorships, the rhetoric with which scientific texts and technological products are constructed, the historical frames in which scientific facts and technological solutions come into being and develop. This approach uses not only 'ethnographic' analytical tools but also, and especially, semiotic ones[42].

Finally, a certain influence has been exerted on the ethnographic study of technological artifacts by 'activity theory.' This derives from Soviet cultural psychology and the work of Vygotskij (Nardi, 1996; Kuutti, 1999; Redmiles, 2002), and it stresses that cultural practices and artifacts (among which technology is of great importance) are forms of mediation with the world. Activity theory places especial emphasis on the ways in which experience in real-life work settings structures mental processes and learning (see Alby & Zucchermaglio, 2005)[43].

Not only are there differences among the theoretical and philosophical assumptions behind the approaches counselling the use of an ethnographic mentality, but the word 'ethnography' covers a diversity of opinions on what should be done in the field. For some, ethnography is simply equivalent to fieldwork; for others, it entails espousing an entire philosophy; for yet others it is a form of activities description. Forsythe (1999) describes six misconceptions of ethnography relatively commonplace among design engineers: 1. ethnography does not require particular skills: anyone can do ethnography; it is only the application of common sense; 2. being a member of an organization qualifies you to do ethnography in your own work setting; 3. since ethnography does not involve a traditional research design, this means that ethnography does not have a systematic research method; 4. doing fieldwork means chatting with people and reporting what they say; 5. you only need to ask people if you want to find out what they do; 6. observation is simply a matter of collecting information in the field—the selection and interpretation of data is instead essential. These widespread misconceptions show that collaboration between ethnographers and engineers is by no means straightforward.

How can ethnography be adapted to satisfy the practical needs of designers? J.A. Hughes *et al.* (1994) describe four possible forms of research. The first is that of 'concurrent ethnography,' in which ethnography offers suggestions for the design by working simultaneously and in parallel with systems development. The second is 'quick and dirty ethnography,' similar to 'rapid ethnography' (Millen, 2000), in which the ethnography serves to provide an initial general description of a setting completely unknown to the designer. A third case is that of 'evaluative ethnography,' which has the specific purpose of verifying, validating, or evaluating already-decided design solutions. The fourth case is that of the 're-examination of past cases,' a form of 'clinical ethnography' in which previous research is reviewed in light of possible new interventions in other fields and settings.

Moreover, the kind of relation that should exist between ethnography and informatics is by no means unequivocal. Anderson (1997) describes three possible different relationships, ranging from integration between the two approaches, through complementarity between two distinct fields, to complete independence between ethnography and informatics. Ethnographic studies can be used to influence, pervade, guide, inspire, evaluate, and criticise the designing of information systems[44].

In substance, it seems that there are three main tensions between ethnography and information systems design which have not yet been resolved and are perhaps not resolvable.

The first tension is between research's practical and applied aspect and its dimension of reflection and analysis. Put briefly, the dilemma is that informatics must find practical solutions for problems concerning the organization and management of knowledge, whereas ethnography abides by the academic and disciplinary tradition by examining problems analytically. The two aspects should not be confused and they must be kept constantly separate. It is true that as early as the Chicago ethnographic tradition, and therefore at the beginnings of sociological fieldwork, there was no lack of analytical interest in practical aspects, or in finding solutions to social problems[45]. But the richness and depth of ethnographic research stem from its exemption—or at least momentary detachment—from having to deal with practical concerns and particular problems. This divergence of perspectives has been effectively described by Grudin and Grinter: "A distant designer may see the workgroup on one peak and dimly through the haze see a higher peak nearby, and imagine a straight path from the one up to the other, but an ethnographer sees the valleys in between, the difficulties in making the trip, the uncertainty of success, the possibility that the higher peak will not be as comfortable a place to work after all (or was an illusion)" (Grudin & Grinter, 1994, p. 56).

The second tension between ethnography and the design of information systems is connected with the first one. It concerns what can be termed the presumed 'gaze to the future' of informatics and the presumed 'gaze to the present' of ethnography. Informatics introduces change in organizations; ethnography has a knowledge-gathering purpose, and therefore 'photographs' what actually exists. The shift from description of a state of affairs and its change (in direction, form, timing, phases, etc.) – admitted that this is desirable – is far from being clear. I say "admitted that this is desirable" because the history of relations between ethnography and informatics has been littered with studies whose conclusion has been *to leave things as they are*. The most notable of these studies have been those by Suchman (1994) on a law firm, by Sommerville *et al.* (1993) on air traffic controllers[46], by Hutchins and Palen (1998) on aircraft cockpits, and by Harper and Carter (1994) on a firm of architects. How, therefore, can ethnography contribute to the development of information systems if one of the conclusions of ethnographic research is that it is better not to introduce any technological solution into settings which work very well as they are?

The last tension is substantive. It to some extent summarizes the two previous ones, and it concerns the underlying 'philosophies' which guide the work of ethnographers and that of information systems engineers. Ethnography does not produce formal models; it does not concern itself with functions, implementations, and interfaces. It often produces accounts, descriptions, and explanations, which require discursive elaboration by the ethnographer. Systems designers need formal methods, specifications, graphs, and models that 'close the circle.' Ethnography is not concerned with 'closing,' but rather with 'opening' the world, even if by only glimpsing alternative modes, subtleties, nuances of the social world which constantly 'reopen' our understanding of social relations[47]. Not only is this difficult to translate into encoded forms, it may be impossible. It is as if there is always a 'plus' in technological solutions, and it is this 'plus' which enables the system to work. The success of a system, therefore, does not lie internally to it, but rather in the web of social relations which makes it comprehensible, usable, and shareable. It is as if the trajectories of information systems and ethnography intersect but do not converge, as they would do if this were a necessary process.

## METIS AND TECHNE

Informatics has not responded with any great enthusiasm to ethnography's insistence on context, improvisation, situatedness, and sociality. For instance, Newell and Card (1985, 1986) argue that the only way forward is to stress the 'hard' nature of informatics, certainly not to shift to a 'soft' approach. With the 'hard' nature of informatics, Newell and Card refer to "the need for technical, engineering-style psychological theory" (Newell & Card, 1986, p. 257), declined in psychological terms, which must face the challenge of the complexity of settings. Their theory may take numerous forms: "rules, theorems, grammars, simulations, formulae, or other explicit means; and that it might operate on data that was symbolic, structural, or numerical" (Newell & Card, p. 258). In short, the response by informatics should consist in making instructions more accurate, more detailed, and more specific, the idea being that the problem is the boundedness of human thought, and that with greater elaboration of its plans informatics will be able to furnish responses without having to re-think its design methods. This brings to mind the Austrian lieutenant described by the Italian writer Italo Calvino, who stubbornly marched his men in close formation onto the sloping terrain of the battlefield as he had been taught to do at the military academy:

*A fair-haired Austrian lieutenant commanded a platoon of soldiers wearing perfect uniforms, with pigtails and bows, tricorns and gaiters, white cross-sashes, rifles and bayonets, and he made them march two-by-two, trying to keep them aligned on the steep paths. Unaware of what the wood held in store, but certain that he was following his orders to the letter, the officer proceeded in accordance with the lines drawn on the chart, so that his men constantly bumped into tree trunks, slipped with their spiked boots on smooth stones, or scratched their eyes in the undergrowth, but in constant awareness of the supremacy of imperial arms*[48].

Besides the solution of dilemmas and the emergence of problematic areas, social scientists and researchers in the human sciences can find numerous items of interest in the relationship between information systems designs and ethnography; items which may derive from long-standing inquiry but are still extraordinarily topical (Agre, 1995). This is the field[49] of comparison between two forms of knowledge and two ways of thinking. James C. Scott (1998) has described the dialectic between *metis* and *techne*[50]. He discusses cases of engineering design imposed, often with disastrous results, by the central authorities of the state[51]. These cases of huge engineering design, Scott says, are the fruit of *techne*, which is knowledge based on logical deduction from self-evident first principles. *Techne* rests on universal principles and is "organized analytically into small, explicit, logical steps and is both decomposable and verifiable [...] techne is characterized by impersonal, often quantitative precision and a concern with explanation and verification" (Agre, 1995, p. 320). *Metis*, which is translated into English as 'cunning intelligence' or 'knack,' and into French as 'ruse' (Detienne & Vernant, 1974), instead consists of practical abilities based on common sense, experience, knowing how to do things, procedural knowledge, 'personal knowledge' (Polanyi, 1958), tacit knowledge (Polanyi, 1967), an ability to get by, dexterity acquired with experience, and which are difficult to express with explicit propositional knowledge[52]. As Scott writes:

*Metis resists simplification into deductive principles which can successfully be transmitted through book learning, because the environments in which it is exercised are so complex and nonrepeatable that formal procedures of rational decision making are impossible to apply. In a sense, Metis lies in that large space between the realm of genius, to which no formula can apply, and the realm of codified knowledge, which can be learned by rote (Scott, 1998, p. 316).*

Claude Lévi-Strauss (1962) likewise contrasted the knowledge of engineers with what he called 'the primary science' of *bricolage*[53], the "science of the concrete." Lévi-Strauss (1962, p. 29) reminds us that: "In its old sense the verb *bricoler* applied to ball games and billiards, to hunting, shooting and riding. It was, however, always used with reference to some extraneous movement: a ball rebounding, a dog straying or a horse swerving from its direct course to avoid an obstacle." Yet whilst the engineer's mind, as characteristic of the modern world, appropriates the world by building abstractions on abstractions, is able to transcend its immediate present through numbers, codes and categories, through socio-technical processes made of standardized abstractions, the *bricoleur* proceeds by anomalous composition, "making do with whatever is at hand," where his action is "the contingent result of all the occasions there have been to renew or enrich the stock or to maintain it with the remains of previous constructions or destructions" (Lévi-Strauss, 1962, p. 30), where it is "impossible to separate [solutions] from the concrete situations in which they appeared" (Lévi-Strauss, 1962, p. 31).

I am not proposing a division of labour whereby information engineers are responsible for *techne* and ethnographers for *metis* and *bricolage*. Rather, I believe that *techne* and *metis*, engineering, and bricolage, are intimately bound up with each other. Information systems thus constitute a field of dialectical relations between two bodies of knowledge.

## REFERENCES


Agre, P. (1995). The soul gained and lost: Artificial intelligence as a philosophical project. *Stanford Humanities Review*, *4*(2), 1–19.

Alby, F., & Zucchermaglio, C. (2005). *Gruppi e tecnologie al lavoro*. Bari, Italy: Laterza.

Almklov, P. G. (2008). Standardized data and singular situations. *Social Studies of Science*, *38*(6), 873–897. doi:10.1177/0306312708098606

Anderson, R. J. (1994). Representations and requirements: The value of ethnography in system design. *Human-Computer Interaction*, *9*(1), 151–182. doi:10.1207/s15327051hci0902_1

Anderson, R. J. (1997). Work, ethnography and system design. In Kent, A., & Williams, J. G. (Eds.), *Encyclopedia of Microcomputers* (*Vol. 20*, pp. 159–183). New York, NY: Marcel Dekker.

Baba, M. (2006). Anthropology and business. In Birx, H. J. (Ed.), *Encyclopedia of Anthropology* (pp. 83–117). Thousand Oaks, CA: Sage.

Bannon, L. (1990). A pilgrim's progress: From cognitive science to cooperative design. *AI & Society*, *4*(4), 259–275. doi:10.1007/BF01894031

Bannon, L. (1992). Perspectives on CSCW: From HCI and CMC to CSCW. In *Proceedings of the International Conference on Human-Computer Interaction (EW-HCI 1992)*, (pp. 148-158). St. Petersburg, Russia: EW-HCI.

Bannon, L. (1993). CSCW: An initial exploration. *Scandinavian Journal of Information Systems*, *5*, 3–24.

Bannon, L., & Schmidt, K. (1991). CSCW: Four characters in search of a context. In Bowers, J., & Benford, S. (Eds.), *Studies in Computer Supported Cooperative Work: Theory, Practice and Design* (pp. 3–16). Amsterdam, The Netherlands: North-Holland.

Bauchspies, W. K. (2007). Science, ethnographic studies of. In G. Ritzer (Ed.), *Blackwell Encyclopedia of Sociology*. Malden, MA: Blackwell. Retrieved from http://www.blackwellreference.com/public/tocnode?id=g9781405124331_chunk_g978140512433125_ss1-40.

Becker, H. (1967). Whose side are we on? *Social Problems*, *14*, 239–247. doi:10.1525/sp.1967.14.3.03a00010

Bentley, R., Hughes, J. A., Randall, D., Rodden, T., Sawyer, P., Shapiro, D., & Sommerville, I. (1992). Ethnographically-informed systems design for air traffic control. In *Proceedings of the 1992 ACM Conference on Computer-Supported Cooperative Work*, (pp. 123-129). Toronto, Canada: ACM.

Berg, M. (1998). The politics of technology: On bringing social theory into technological design. *Science, Technology & Human Values*, *23*(4), 456–490. doi:10.1177/016224399802300406

Blomberg, J. (1995). Ethnography: Aligning field studies of work and system design. In Monk, A. F., & Gilbert, N. (Eds.), *Perspectives on HCL: Diverse Approaches* (pp. 175–197). London, UK: Academic Press Ltd.

Blomberg, J., Giacomi, J., Mosher, A., & Swenton-Wall, P. (1993). Ethnographic field methods and their relation to design. In Schuler, D., & Namioka, A. (Eds.), *Participatory Design: Principles and Practices* (pp. 123–157). Hillsdale, NJ: Erlbaum.

Blumer, H. (1969). *Symbolic interactionism: Perspective and method.* Englewood Cliffs, NJ: Prentice Hall.

Blythin, S., Rouncefield, M., & Hughes, J. A. (1997). Never mind the ethno stuff - What does all this mean and what do we do now? Ethnography in the commercial world. *Interaction, 4*, 38–47. doi:10.1145/255392.255400

Boden, D. (1994). *The business of talk: Organizations in actions.* Cambridge, UK: Polity Press.

Bowers, J., & Benford, S. (Eds.). (1991). *Studies in CSCW: Theory, practice and design.* Amsterdam, The Netherlands: North Holland.

Bowker, G., & Star, S. L. (1999). *Sorting things out: Classification and its consequences.* Cambridge, MA: MIT Press.

Brighenti, A., Giglioli, P. P., Ravaioli, P., Turina, I., & Venturini, T. (2008). Per un'etnografia dei moderni: Intervista a Bruno Latour. *Etnografia e Ricerca Qualitativa, 3*, 347–367.

Brun-Cottan, F., & Wall, P. (1995). Using video to re-present the user. *Communications of the ACM, 38*(5), 61–71. doi:10.1145/203356.203368

Bruni, A. (2003). *Lo studio etnografico delle organizzazioni.* Roma, Italy: Carocci.

Bruni, A., & Gherardi, S. (2007). *Studiare le pratiche lavorative.* Bologna, Italy: Il Mulino.

Bucciarelli, L. L. (1988). An ethnographic perspective on engineering design. *Design Studies, 9*(3), 159–168. doi:10.1016/0142-694X(88)90045-2

Burrell, G., & Morgan, G. (1979). *Sociological paradigms and organizational analysis.* London, UK: Heineman.

Büscher, M., Gill, S., Mogensen, P., & Shapiro, D. (2001). Landscapes of practice - Bricolage as a method for situated design. *Computer Supported Cooperative Work, 10*, 1–28. doi:10.1023/A:1011293210539

Button, G. (1993). The curious case of the vanishing technology. In Button, G. (Ed.), *Technology in Working Order: Studies in Work, Interaction, and Technology* (pp. 10–28). London, UK: Routledge.

Button, G. (2000). The ethnographic tradition and design. *Design Studies, 21*, 319–332. doi:10.1016/S0142-694X(00)00005-3

Button, G., & Dourish, P. (1996). Technomethodology: Paradoxes and possibilities. In *Proceedings of CHI 96*, (pp. 19-26). Vancouver, Canada: ACM.

Button, G., & Harper, R. (1995). The relevance of 'work-practice' for design. *Computer Supported Cooperative Work, 4*, 263–280. doi:10.1007/BF01846695

Button, G., & Sharrock, W. (1998). The organizational accountability of technological work. *Social Studies of Science, 28*(1), 75–102. doi:10.1177/030631298028001003

Chapoulie, J.-M. (1987). Everett C. Hughes and the development of fieldwork in sociology. *Urban Life, 15*(3-4), 259–298.

Chapoulie, J.-M. (1996). Everett Hughes and the Chicago tradition. *Sociological Theory, 14*(1), 3–29. doi:10.2307/202150

Ciborra, C. (2004). *The labyrinths of information: Challenging the wisdom of systems.* Oxford, UK: Oxford University Press.

Clarke, A. E. (1991). Social worlds /arenas theory as organizational theory. In Maines, D. (Ed.), *Social Organization and Social Process: Essays in Honor of Anselm Strauss* (pp. 119–158). New York, NY: Aldine.

Clarke, A. E. (2006). Social worlds. In Ritzer, G. (Ed.), *The Blackwell Encyclopedia of Sociology* (pp. 4547–4549). Malden, MA: Blackwell.

Clarke, A. E., & Star, S. L. (2007). The social worlds/arenas/discourse framework as a theory-methods package. In Hackett, E. (Eds.), *Handbook of Science and Technology Studies* (pp. 113–137). Cambridge, MA: MIT Press.

Corbin, J., & Strauss, A. L. (1993). The articulation of work through interaction. *Qualitative Sociology, 34*, 71–83.

Crabtree, A. (1998). Ethnography in participatory design. In *Proceedings of the 1998 Participatory Design Conference*, (pp. 93-105). Seattle, WA. Retrieved from http://www.mrl.nott.ac.uk/~axc/documents/papers/PDC98.pdf.

Crabtree, A. (2001). *Wild sociology: Ethnography and design*. Ph.D. Thesis. Lancaster, UK: Lancaster University.

Crabtree, A. (2003). *Designing collaborative systems: A practical guide to ethnography*. London, UK: Springer.

Crabtree, A. (2004). Taking technomethodology seriously: Hybrid change in the ethnomethodology-design relationship. *European Journal of Information Systems, 13*(3), 195–209. doi:10.1057/palgrave.ejis.3000500

Crabtree, A., Nichols, D. M., O'Brien, J., Rouncefield, M., & Twidale, M. B. (2000). Ethnomethodologically informed ethnography and information system design. *Journal of the American Society for Information Science American Society for Information Science, 51*(7), 666–682. doi:10.1002/(SICI)1097-4571(2000)51:7<666::AID-ASI8>3.0.CO;2-5

Davenport, E. (2008). Social informatics and sociotechnical research: A view from the UK. *Journal of Information Science, 34*(4), 519–530. doi:10.1177/0165551508091011

Deetz, S. (1996). Describing differences in approaches to organization science: Rethinking Burrell and Morgan and their legacy. *Organization Science, 7*(2), 191–207. doi:10.1287/orsc.7.2.191

Detienne, M., & Vernant, J.-P. (1974). *Les ruses de l'intelligence: La métis des Grecs*. Paris, France: Flammarion.

Diggins, T., & Tolmie, P. (2003). The 'adequate' design of ethnographic outputs for practice: Some explorations of the characteristics of design resources. *Personal and Ubiquitous Computing, 7*, 147–158. doi:10.1007/s00779-003-0226-y

Dillon, A. (1998). Cultural analysis and what designers need to know: A case of sometimes too much, sometimes too little, and always too late. *ACM Journal of Computer Documentation, 22*(1), 13–17.

Dourish, P., & Button, G. (1998). On "technomethodology": Foundational relationships between ethnomethodology and system design. *Human-Computer Interaction, 13*, 395–432. doi:10.1207/s15327051hci1304_2

Downey, G. L. (1998). *The machine in me: An anthropologist sits among computer engineers*. New York, NY: Routledge.

Erickson, F. (1988). Ethnographic description. In Ammon, U. (Ed.), *Sociolinguistics: An International Handbook of the Science of Language and Society* (pp. 1081–1095). Berlin, Germany: De Gruyter.

Ericsson, K. A., & Simon, H. A. (1980). Verbal reports as data. *Psychological Review, 87*(3), 215–251. doi:10.1037/0033-295X.87.3.215

Evan, W. F. (1964). On the margin: The engineering technician. In Berger, P. L. (Ed.), *The Human Shape of Work: Studies in the Sociology of Occupations* (pp. 83–112). New York, NY: MacMillan.

Finken, S. (2000). Bringing ethnography home: Reflections upon a style of enquiry. In *Proceedings of IRIS*, (vol 23), (pp. 267-277). IRIS.

Fjuk, A., Nurminen, M. I., & Smørdal, O. (1997). *Taking articulation work seriously: An activity theoretical approach. TUCS Technical Report, N. 120, August*. Turku, Turkey: Turku Centre for Computer Science.

Floyd, C. (1987). Outline of a paradigm change in software engineering. In G. Bjerknes, P. Ehn, & M. Kyng (eds.), *Computers and Democracy: A Scandinavian Challenge*, (pp. 191-208). Avebury, UK: Aldershot.

Forsythe, D. E. (1999). It's just a matter of common sense: Ethnography as invisible work. *Computer Supported Cooperative Work*, *8*(1), 127–145. doi:10.1023/A:1008692231284

Gamst, F. C. (1980). *The hoghead: An industrial ethnology of the locomotive engineer*. New York, NY: Holt, Rinehart & Winston.

Garfinkel, H. (1963). A conception of, and experiments with, 'trust' as a condition of stable concerted actions. In Harvey, O. J. (Ed.), *Motivation and Social Interaction* (pp. 187–238). New York, NY: The Ronald Press.

Garfinkel, H. (1967). *Studies in ethnomethodology*. Englewood Cliffs, NJ: Prentice-Hall.

Garfinkel, H. (1986). Introduction. In *Ethnomethodological Studies of Work* (pp. vii–viii). London, UK: Routledge.

Garfinkel, H. (2002). *Ethnomethodology's program: Working out Durkheim's aphorism. London, UK*. Lanham: Rowman & Littlefield.

Garfinkel, H. (2006). *Seeing sociologically: The routine grounds of social action*. Boulder, CO: Paradigm.

Geertz, C. (1983). From the native's point of view. In *Local knowledge* (pp. 55–72). New York, NY: Basic Books.

Gerson, E., & Star, S. L. (1986). Analyzing due process in the workplace. *ACM Transactions on Office Information Systems*, *4*, 70–78. doi:10.1145/214427.214431

Gherardi, S. (2006). *Organizational knowledge: The texture of workplace learning*. Oxford, UK: Blackwell.

Gherardi, S. (2008). Dalla comunità di pratica alle pratiche della comunità: Breve storia di un concetto in viaggio. *Studi Organizzativi*, *10*(1), 49–72.

Goguen, J. A. (1994). Requirements engineering as the reconciliation of social and technical issues. In Jirotka, M., & Goguen, J. A. (Eds.), *Requirements Engineering: Social and Technical Issues* (pp. 165–200). San Diego, CA: Academic Press.

Goguen, J. A. (1996). Formality and informality in requirements engineering. In *Proceedings of the International Conference on Requirements Engineering*, (pp. 102-108). Retrieved from http://sern.ucalgary.ca/courses/SENG/693/F00/readings/icre96.ps.

Goodwin, C. (1996). Transparent vision. In Thompson, S. (Ed.), *Interaction and Grammar* (pp. 370–404). Cambridge, UK: Cambridge University Press. doi:10.1017/CBO9780511620874.008

Goodwin, C., & Goodwin, M. H. (1996). Seeing as situated activity: Formulating planes. In Engeström, Y., & Middleton, D. (Eds.), *Cognition and Communication at Work* (pp. 61–95). Cambridge, UK: Cambridge University Press.

Goodwin, M. H. (1995). Assembling a response: Setting and collaboratively constructed work talk. In Ten Have, P., & Psathas, G. (Eds.), *Situated Order: Studies in the Social Organization of Talk and Embodied Activities* (pp. 173–186). Washington, DC: University Press of America.

Goodwin, M. H. (1996). Informings and announcements in their environment: Prosody within a multi-activity work setting. In Couper-Kuhlen, E., & Selting, M. (Eds.), *Prosody in Conversation: Interactional Studies* (pp. 436–461). Cambridge, UK: Cambridge University Press. doi:10.1017/CBO9780511597862.013

Grudin, J., & Grinter, R. (1994). Ethnography and design. *Computer Supported Cooperative Work*, *3*(1), 55–59. doi:10.1007/BF01305846

Haraway, D. (1991). *Simians, cyborgs, and women: The reinvention of nature*. New York, NY: Routledge.

Harper, R., & Carter, K. (1994). Keeping people apart: A research note. *Computer Supported Cooperative Work*, *2*(3), 199–207. doi:10.1007/BF00749017

Harper, R., & Hughes, J. A. (1993). What a f-ing system! Send 'em all to the same place and then expect us to stop 'em hitting: Managing technology work in air traffic control. In Button, G. (Ed.), *Technology in Working Order: Studies of Work, Interaction, and Technology* (pp. 127–144). London, UK: Routledge.

Harper, R. H. R. (1996). *The ethnographic turn: Why it has come about and how to do it*. Technical Report EPC-1996-109. Cambridge, UK: Rank Xerox.

Harper, R. H. R. (2000). The organization in ethnography: A discussion of ethnographic fieldwork programs in CSCW. *Computer Supported Cooperative Work*, *9*(2), 239–264. doi:10.1023/A:1008793124669

Harvey, D. (1990). *The condition of postmodernity*. Oxford, UK: Blackwell.

Harvey, L. (1997). A discourse on ethnography. In Lee, A. S., Liebenau, J., & DeGross, J. I. (Eds.), *Information Systems and Qualitative Research* (pp. 207–224). London, UK: Chapman and Hall.

Harvey, L., & Myers, M. D. (1995). Scholarship and practice: The contribution of ethnographic research methods to bridging the gap. *Information Technology & People*, *8*(3), 13–27. doi:10.1108/09593849510098244

Heath, C. (1984). Everett Cherrington Hughes (1897-1983): A note on his approach and influence. *Sociology of Health & Illness*, *6*(2), 218–237. doi:10.1111/1467-9566.ep10778393

Heath, C., Knoblauch, H., & Luff, P. (2000). Technology and social interaction: The emergence of workplace studies. *The British Journal of Sociology*, *51*(2), 299–320. doi:10.1080/00071310050030190

Heath, C., & Luff, P. (1996). Convergent activities: Line control and passenger information on the London underground. In Engeström, Y., & Middleton, D. (Eds.), *Cognition and Communication at Work* (pp. 96–129). Cambridge, UK: Cambridge University Press.

Heath, C., & Luff, P. (2000). *Technology in action*. Cambridge, UK: Cambridge University Press. doi:10.1017/CBO9780511489839

Heath, R. P. (1997). Seeing is believing: Ethnography gets consumers where they live and work, and play, and shop. *Marketing Tools*, *3*, 4–9.

Helmes-Hayes, R. C. (1998). Everett Hughes: Theorist of the second Chicago school. *International Journal of Politics Culture and Society*, *11*(4), 621–673. doi:10.1023/A:1025157424093

Hirschauer, S. (2007). Putting things into words - Ethnographic description and the silence of the social. *Human Studies*, *29*, 413–441. doi:10.1007/s10746-007-9041-1

Hodson, R. (2004). A meta-analysis of workplace ethnographies: Race, gender, and employee attitudes and behaviours. *Journal of Contemporary Ethnography*, *33*(1), 4–38. doi:10.1177/0891241603259808

Hughes, E. C. (1952). The sociological study of work. *American Journal of Sociology*, *57*(5), 423–426. doi:10.1086/221010

Hughes, E. C. (1970). *The sociological eye*. Chicago, IL: Aldine.

Hughes, J., O'Brien, J., Rodden, J., & Rouncefield, M. (1997b). *Ethnography, communication and support for design*. Retrieved from http://www.comp.lancs.ac.uk/computing/research/cseg/97rep.html.

Hughes, J., O'Brien, J., Rodden, J., Rouncefield, M., & Blythin, S. (1997a). Designing with ethnography: A presentation framework for design. In *Proceedings of DIS 1997: Designing Interactive Systems: Processes, Practices, Methods, & Techniques*, (pp. 147-158). DIS.

Hughes, J. A., King, V., Rodden, T., & Andersen, H. (1994). Moving out from the control room: Ethnography in system design. In *Proceedings of CSCW*, (pp. 429-439). New York, NY: ACM.

Hughes, J. A., King, V., Rodden, T., & Andersen, H. (1995). The role of ethnography in interactive system design. *Interaction*, *2*(2), 56–65. doi:10.1145/205350.205358

Hughes, J. A., Randall, D., & Shapiro, D. (1992). Faltering from ethnography to design. In *Proceedings of the 1992 ACM Conference on Computer Supported Cooperative Work (CSCW 1992)*, (pp. 115-122). Toronto, Canada: ACM.

Hughes, J. A., Randall, D., & Shapiro, D. (1993). From ethnographic record to system design: Some experiences from the field. *Computer Supported Cooperative Work*, *1*(3), 123–147. doi:10.1007/BF00752435

Hutchins, E. (1996). *Cognition in the wild.* Cambridge, MA: MIT Press.

Hutchins, E., & Palen, L. (1998). *Constructing meaning from space, gesture and speech.* Retrieved from http://goldin-meadow-lab.uchicago.edu/embodiedcogrg/Hutchins.pdf.

Iqbal, R., Jamesb, A., & Gatward, R. (2005). Designing with ethnography: An integrative approach to CSCW design. *Advanced Engineering Informatics*, *19*, 81–92. doi:10.1016/j.aei.2005.05.003

Jacucci, G. (2007). *Social practice design, pathos, improvisation, mood, and bricolage: The Mediterranean way to make place for IT?* Working Paper, AIS eLibrary MCIS2007. Retrieved from http://www.aisel.aisnet.org.

Jacucci, G., Cattani, C., Campagnolo, G. M., Wagner, I., & Tellioglu, H. (2008). *Facilitating co-construction by social practice design: A case of geodistributed employees facing the design of model-based enterprise systems.* Working Paper, AIS eLibrary MCIS2007. Retrieved from http://www.aisel.aisnet.org.

Jacucci, G., Tellioglu, H., & Wagner, I. (2006). *Practices and cultures of knowledge management in global virtual (software) and real component manufacturing.* Working Paper, AIS eLibrary MCIS2006. Retrieved from http://www.aisel.aisnet.org.

Jacucci, G., Tellioglu, H., & Wagner, I. (2007). *Design games as a part of social practice design: A case of employees elaborating organisational problems.* Working Paper, AIS eLibrary MCIS2007. Retrieved from http://www.aisel.aisnet.org.

Jordan, B. (1996). Ethnographic workplace studies and computer supported cooperative work. In Shapiro, D., Tauber, M., & Traunmüller, R. (Eds.), *The Design of Computer-Supported Cooperative Work and Groupware Systems* (pp. 17–42). Amsterdam, The Netherlands: Elsevier Science. doi:10.1016/S0923-8433(96)80005-0

Jordan, B. (1997). Transforming ethnography - Reinventing research. *Cultural Anthropology Methods Journal*, *9*(3), 12–17.

Jordan, B., & Dalal, B. (2006). Persuasive encounters: Ethnography in the corporation. *Field Methods*, *18*(4), 359–381. doi:10.1177/1525822X06292976

Jordan, B., & Henderson, A. (1995). Interaction analysis: Foundations and practice. *Journal of the Learning Sciences*, *4*(1), 39–103. doi:10.1207/s15327809jls0401_2

Jordan, B., & Lambert, M. (2009). Working in corporate jungles: Reflections on ethnographic praxis in industry. In M. Cefkin (Ed.), *Ethnography and the Corporate Encounter: Reflections on Research in and of Corporations*. New York, NY: Berghahn Books. Retrieved from http://www.lifescapes.org/Papers/7_Ch_4%20%20Jordan.doc.

Jordan, B., & Yamauchi, Y. (2008). Beyond the university: Teaching ethnographic methods in the corporation. *Anthropology News*, *49*(6), 35. doi:10.1111/an.2008.49.6.35

Karasti, H. (2001). Bridging work practice and system design: Integrating systemic analysis, appreciative intervention and practitioner participation. *Computer Supported Cooperative Work*, *10*, 211–246. doi:10.1023/A:1011239126617

Knoblauch, H. (2005). Focused ethnography. *Forum Qualitative Sozial Forschung*, *6*(3). Retrieved from http://www.qualitative-research.net/fqs-texte/3-05/05-3-44-e.htm

Knoblauch, H. (2006). Videography: Focused ethnography and video-analysis. In Knoblauch, H., Schnettler, B., & Raab, J. (Eds.), *Video-Analysis: Methodological Aspects of Interpretive Audiovisual Analysis in Social Research* (pp. 69–84). Frankfurt, Germany: Lang.

Knorr Cetina, K. D. (1983). The ethnographic study of scientific work: Toward a constructivist interpretation of science. In Knorr Cetina, K., & Mulkay, M. (Eds.), *Analytical Perspectives in Science Studies*. London, UK: Sage.

Koschmann, T., Stahl, G., & Zemel, A. (2006). The video analyst's manifesto (or the implications of Garfinkel's policies for studying practice within design-based research). In R. Goldman, B. Barron, S. Derry, & R. Pea (Eds.), *Video Research in the Learning Sciences*. Mahway, NJ: Lawrence Erlbaum Associates. Retrieved from http://edaff.siumed.edu/tk/manifesto.pdf.

Kunda, G. (1992). *Engineering culture: Control and commitment in a high-tech corporation*. Philadelphia, PA: Temple University Press.

Kuutti, K. (1999). Activity theory, transformation of work and information systems design. In Engestrog, Y., Miettinen, M. R., & Punamag Ki, R. (Eds.), *Perspectives on Activity Theory* (pp. 360–376). Cambridge, UK: Cambridge University Press.

Latour, B. (1992a). *Aramis ou l'amour des techniques*. Paris, France: La Découverte.

Latour, B. (1992b). Where are the missing masses? The sociology of few mundanc artefacts. In Bijker, W. E., & Law, J. (Eds.), *Shaping Technology / Building Society* (pp. 225–258). Cambridge, MA: MIT Press.

Latour, B. (2004). On using ANT for studying information systems: A (somewhat) Socratic dialogue. In Avgerou, C., Ciborra, C., & Land, F. (Eds.), *The Social Study of Information and Communication Technologies* (pp. 62–75). Oxford, UK: Oxford University Press.

Latour, B., & Woolgar, S. (1979). *Laboratory life: The social construction of scientific facts*. Los Angeles, CA: Sage.

Law, J., & Hassard, J. (Eds.). (1999). *Actor network theory and after*. Oxford, UK: Blackwell.

Law, J., & Singleton, V. (2000). Performing technology's stories: On social constructivism, performance, and performativity. *Technology and Culture*, *41*(4), 765–775.

Lévi-Strauss, C. (1962). *La pensée sauvage*. Paris, France: Plon.

Lévy-Bruhl, L. (1922). *La mentalité primitive*. Paris, France: Alcan.

Luff, P., Hindmarsh, J., & Heath, C. C. (Eds.). (2000). *Workplace studies: Recovering work practice and informing system design*. Cambridge, UK: Cambridge University Press. doi:10.1017/CBO9780511628122

Luff, P., Jirotka, M., Heath, C. C., & Greatbatch, D. (1993). Tasks and social interaction: The relevance of naturalistic analyses of conduct for requirements engineering. In Fickas, S., & Finkelstein, A. (Eds.), *Requirements Engineering* (pp. 187–190). IEEE Press. doi:10.1109/ISRE.1993.324818

Lynch, M. (1985). *Art and artifact in laboratory science: A study of shop work and shop talk in a research laboratory*. London, UK: Routledge & Kegan Paul.

Lynch, M. (1993). *Scientific practice and ordinary action: Ethnomethodology and the social studies of science*. Cambridge, UK: Cambridge University Press.

Lynch, M., & Bogen, D. (1994). Harvey Sacks' primitive natural science. *Theory, Culture & Society*, *11*, 65–104. doi:10.1177/026327694011004003

Macaulay, C., Benyon, D., & Crerar, A. (2000). Ethnography, theory and systems design: From intuition to insight. *International Journal of Human-Computer Studies*, *53*, 35–60. doi:10.1006/ijhc.2000.0376

Malinowski, B. (1922). *Argonauts of the western pacific*. London, UK: Waveland Press.

Martin, D., & Sommerville, I. (2006). Ethnography and the social structure of work. In Jones, C., Gacek, C., & Besnard, D. (Eds.), *Structure for Dependability: Computer-Based Systems from an Interdisciplinary Perspective* (pp. 169–188). London, UK: Springer. doi:10.1007/1-84628-111-3_9

McDermott, D. (1981). Artificial intelligence meets natural stupidity. In Haugeland, J. (Ed.), *Mind Design* (pp. 143–160). Cambridge, MA: MIT Press.

Millen, D. R. (2000). Rapid ethnography: Time deepening strategies for HCI field research. In *Proceedings of DIS, 2000*, 280–286.

Mintzberg, H. (1990). The design school: Reconsideration of the basic premises of strategic management. *Strategic Management Journal*, *11*, 171–195. doi:10.1002/smj.4250110302

Molotch, H. (2003). *Where stuff comes from: How toaster, toilets, cars, computers and many other things come to be as they are*. New York, NY: Routledge.

Morrill, C., & Fine, G. A. (1997). Ethnographic contributions to organizational sociology. *Sociological Methods & Research*, *25*, 424–451. doi:10.1177/0049124197025004003

Myers, M. D. (1999). Investigating information systems with ethnographic research. *Communications of AIS*, *2*(23), 1–20.

Myers, M. D. (2008). *Qualitative research in information systems*. Retrieved from http://www.qual.auckland.ac.nz/a>.

Nardi, B. A. (1996). Activity theory and human-computer interaction. In Nardi, B. A. (Ed.), *Context and Consciousness: Activity Theory and Human-Computer Interaction* (pp. 7–16). Cambridge, MA: The MIT Press.

Neisser, U. (1976). *Cognition and reality: Principles and implications of cognitive psychology*. San Francisco, CA: Freeman.

Nelson, D. (1980). *Frederick W. Taylor and the rise of scientific management*. Madison, WI: The University of Wisconsin.

Newell, A., & Card, S. (1985). The prospects for psychological science in human computer interaction. *Human-Computer Interaction*, *1*(3), 209–242. doi:10.1207/s15327051hci0103_1

Newell, A., & Card, S. (1986). Straightening out softening up: Response to Carroll and Campbell. *Human-Computer Interaction*, *2*(3), 251–267. doi:10.1207/s15327051hci0203_4

Noble, D. F. (1977). *America by design: Science, technology, and the rise of corporate capitalism*. New York, NY: A.A. Knopf.

Norman, D. (1988). *The psychology of everyday things*. New York, NY: Basic Books.

Norman, D. (1992). *Turn signals are the facial expressions of automobiles*. Cambridge, MA: Perseus Books.

Norman, D. (1993). *Things that make us smart: Defending human attributes in the age of the machine*. New York, NY: Addison-Wesley Publishing.

Norman, D. (1998). *The invisible computer*. Cambridge, MA: MIT Press.

Norman, D. (2004). *Emotional design: Why we love (or hate) everyday things*. New York, NY: Basic Books.

Norman, D. (2007). *The design of future things*. New York, NY: Basic Books.

Nussbaum, M. C. (1956). *The fragility of goodness: Luck and ethics in Greek tragedy and philosophy*. Cambridge, UK: Cambridge University Press.

Orlikowski, W. J., & Baroudi, J. (1991). Studying information technology in organizations: Research approaches and assumptions. *Information Systems Research, 2*(1), 1–28. doi:10.1287/isre.2.1.1

Orr, J. E. (1996). *Talking about machines: An ethnography of a modern job*. Ithaca, NY: ILR Press.

Polanyi, M. (1958). *Personal knowledge*. New York, NY: Routledge & Kegan Paul.

Polanyi, M. (1967). *The tacit dimension*. New York, NY: Anchor Books.

Prasad, P. (1997). Systems of meaning: Ethnography as a methodology for the study of information technologies. In Lee, A. S., Liebenau, J., & DeGross, J. I. (Eds.), *Information Systems and Qualitative Research* (pp. 101–118). London, UK: Chapman and Hall.

Prus, R. (1995). The ethnographic research tradition. In *Symbolic Interaction and Ethnographic Research: Intersubjectivity and the Study of Human Lived Experience* (pp. 103–140). New York, NY: SUNY Press.

Psathas, G. (2008). Reflections on the history of ethnomethodology: The Boston and Manchester 'schools'. *The American Sociologist, 39*(1), 38–67. doi:10.1007/s12108-008-9032-3

Randall, D., Harper, R., & Rouncefield, M. (2007). *Fieldwork for design: Theory and practice*. London, UK: Springer.

Rawls, A. W. (2002). Editor's introduction. In Garfinkel, H. (Ed.), *Ethnomethodology's Program: Working out Durkheim's Aphorism* (pp. 1–64). Lantham, UK: Rowman & Littlefield.

Redmiles, D. (2002). Introduction. *Computer Supported Cooperative Work, 11*(1-2), 1–11. doi:10.1023/A:1015215726353

Reese, W. (2002). Behavioral scientists enter design: Seven critical histories. In Squires, S., & Byrne, B. (Eds.), *Creating Breakthrough Ideas: The Collaboration of Anthropologists and Designers in the Product Development Industry*. Westport, CT: Greenwood.

Resnick, L., Säljö, R., Pontecorvo, C., & Burge, B. (Eds.). (1997). *Discourse, tools and reasoning: Essays on situated cognition*. Berlin, Germany: Springer Verlag.

Riesman, D. (1983). The legacy of Everett C. Hughes. *Contemporary Sociology, 12*(5), 477–481.

Rogers, Y., & Bellotti, V. M. E. (1997). Grounding blue-sky research: How can ethnography help? *Interaction, 4*(3), 58–63. doi:10.1145/255392.255404

Rosenberg, D. (2000). Three steps to ethnography: A discussion of inter-disciplinary contributions. *AI & Society, 15*(1), 295–315.

Ruhleder, K., & Jordan, B. (1997). capturing complex, distributed activities: video-based interaction analysis as a component of workplace ethnography. In A. S. Lee, J. Liebenau, & J. I. DeGross (Eds.), *Information Systems and Qualitative Research*, (pp. 246-275). London, UK: Chapman and Hall.

Ryle, G. (1966). *Dilemmas*. Cambridge, UK: Cambridge University Press.

Salvador, T., Bell, G., & Anderson, K. (1999). Design ethnography. *Design Management Journal, 19*(4), 35–41.

Sanday, P. R. (1979). The ethnographic paradigms. *Administrative Science Quarterly, 24*(4), 527–538. doi:10.2307/2392359

Schmidt, K. (1994). *Modes and mechanisms of interaction in cooperative work: Outline of a conceptual framework*. Roskilde, Denmark: Risø National Laboratory.

Schmidt, K. (2000). *The critical role of workplace studies in CSCW*. Retrieved from http://www.itu.dk/people/schmidt/papers/workplace_studies.pdf.

Schmidt, K., & Bannon, L. (1992). Taking CSCW seriously: Supporting articulation work. *Computer Supported Cooperative Work (CSCW). International Journal (Toronto, Ont.), 1*, 7–41.

Schmidt, K., & Simone, C. (1996). Coordination mechanisms: Towards a conceptual foundation of CSCW systems design. *Journal of Collaborative Computing, 5*, 155–200. doi:10.1007/BF00133655

Schuetz, A. (1943). The problem of rationality in the social world. *Economica, 10*(38), 130–149. doi:10.2307/2549460

Schuetz, A. (1951). Choosing among projects of action. *Philosophy and Phenomenological Research, 12*(2), 161–184. doi:10.2307/2103478

Schuetz, A. (1953). Common-sense and scientific interpretation of human action. *Philosophy and Phenomenological Research, 14*(1), 1–38. doi:10.2307/2104013

Schultze, U., & Leidner, D. E. (2002). Studying knowledge management in information systems research: Discourses and theoretical assumptions. *Management Information Systems Quarterly, 26*(3), 213–242. doi:10.2307/4132331

Schwartzman, H. B. (1993). *Ethnography in organizations*. Newbury Park, CA: Sage.

Scott, J. C. (1998). *Seeing like a state: How certain schemes to improve the human condition have failed*. New Haven, CT: Yale University Press.

Shapiro, D. (1994). The limits of ethnography: Combining social sciences for CSCW. In *Proceedings of CSCW, 1994*, 417–428.

Sharrock, W., & Anderson, R. (1994). The user as a scenic feature of the design space. *Design Studies, 5*, 5–28. doi:10.1016/0142-694X(94)90036-1

Sharrock, W., & Hughes, J. A. (2001). Ethnography in the workplace: Remarks on its theoretical bases. *Team Ethno, 1*. Retrieved from http://www.teamethno-online.org/Issue1/Wes.html.

Sharrock, W., & Randall, D. (2004). Ethnography, ethnomethodology and the problem of generalisation in design. *European Journal of Information Systems, 13*(3), 186–194. doi:10.1057/palgrave.ejis.3000502

Sherry, J. F. (2002). Foreword: Ethnography, design, and customer experience: An anthropologist's sense of it all. In Squires, S., & Byrne, B. (Eds.), *Creating Breakthrough Ideas: The Collaboration of Anthropologists and Designers in the Product Development Industry*. Westport, CT: Greenwood.

Simone, C., & Divitini, M. (1997). Ariadne: Supporting coordination through a flexible use of the knowledge on work process. *Journal of Universal Computer Science, 3*(8), 865–898.

Simone, C., Divitini, M., & Schmidt, K. (1995). A notation for malleable and interoperable coordination mechanisms for CSCW systems. In N. Comstock *et al.* (Eds.), *Conference on Organizational Computing Systems*, (pp. 44-54). Milpitas, CA: ACM.

Simonsen, J., & Kensing, F. (1994). Take users seriously, but take a deeper look: Organizational and technical effects from designing with an ethnographically inspired approach. In *Proceedings of PDC 1994*, (pp. 47-58). Palo Alto, CA: PDC.

Simonsen, J., & Kensing, F. (1997). Using ethnography in contextual design. *Communications of the ACM*, *40*(7), 82–88. doi:10.1145/256175.256190

Simonsen, J., & Kensing, F. (1998). Make room for ethnography in design! Overlooked collaborative and educational prospects. *The Journal of Computer Documentation*, *22*(1), 20–30.

Sommerville, I. (2004). Making ethnography accessible. In *Proceeding of the Workshop on Software Engineering/HCI*, (pp. 36-40). Edinburgh, UK: Institute of Electrical Engineers.

Sommerville, I., Rodden, T., Bentley, R., & Sawyer, P. (1992). Sociologists can be surprisingly useful in interactive systems design. In A. Monk, D. Diaper, & M. Harrison (Eds.), *Proceedings of the HCI 1992, Conference on People and Computers*, (pp. 341-353). Cambridge, UK: Cambridge University Press.

Sommerville, I., Rodden, T., Sawyer, P., Bentley, R., & Twidale, M. (1993). Integrating ethnography into the requirements engineering process. In *Proceeding of the IEEE International Symposium on Requirements Engineering (RE 1993)*, (pp. 165-173). San Diego, CA: IEEE Press.

Sommerville, I., & Sawyer, P. (1997). *Requirements engineering: A good practice guide*. Chichester, UK: Wiley.

Srinivasan, R. (2007). Ethnomethodological architectures: The convergence between an information system and the cultural landscape. *Journal of the American Society for Information Science and Technology*, *58*(5), 723–733. doi:10.1002/asi.20544

Star, S. L. (1991). The sociology of the invisible: The primacy of work in the writings of Anselm Strauss. In Maines, D. (Ed.), *Social Organization and Social Process: Essays in Honor of Anselm Strauss* (pp. 265–283). Hawthorne, NY: Aldine de Gruyter.

Star, S. L. (1997). Anselm Strauss: An appreciation. *Sociological Research Online*, *2*(1). Retrieved from http://www.socresonline.org.uk/socresonline/2/1/1.html doi:10.5153/sro.92

Star, S. L. (1998). Working together: Symbolic interactionism, activity theory and distributed artificial intelligence. In Engeström, Y., & Middleton, D. (Eds.), *Cognition and Communication at Work* (pp. 296–318). Cambridge, UK: Cambridge University Press.

Star, S. L., & Griesemer, J. R. (1989). Institutional ecology, 'translations' and boundary objects: Amateurs and professionals in Berkley's museum of vertebrate zoology, 1907–39. *Social Studies of Science*, *19*(3), 387–420. doi:10.1177/030631289019003001

Star, S. L., & Strauss, A. (1999). Layers of silence, arenas of voice: The ecology of visible and invisible work. *Computer Supported Cooperative Work*, *8*, 9–30. doi:10.1023/A:1008651105359

Strauss, A. (1978). A social world perspective. In Denzin, N. (Ed.), *Studies in Symbolic Interaction* (pp. 119–128). Greenwich, CT: JAI Press.

Strauss, A. (1985). Work and the division of labor. *The Sociological Quarterly*, *26*, 1–19. doi:10.1111/j.1533-8525.1985.tb00212.x

Strauss, A. (1988). The articulation of project work: An organizational process. *The Sociological Quarterly*, *29*, 163–178. doi:10.1111/j.1533-8525.1988.tb01249.x

Strauss, A. (1996). Everett Hughes: Sociology's mission. *Symbolic Interaction*, *19*(4), 271–283. doi:10.1525/si.1996.19.4.271

Strübing, J. (1992). Negotiation: A central aspect of collaborative work in software design. In *Proceedings of the 5ème Workshop sur la Psychologie de la Programmation*, (pp. 31-39). Paris, France: Institut National de Recherche en Informatique et en Automatique.

Strübing, J. (1998). Bridging the gap: On the collaboration between symbolic interactionism and distributed artificial intelligence in the field of multi-agent systems research. *Symbolic Interaction, 21*(4), 441–463. doi:10.1525/si.1998.21.4.441

Suchman, L. (1987). *Plans and situated actions: The problem of human-machine communication*. Cambridge, UK: Cambridge University Press.

Suchman, L. (1994). Working relations of technology production and use. *Computer Supported Cooperative Work, 2*(1-2), 21–39. doi:10.1007/BF00749282

Suchman, L. (1996). Supporting articulation work. In Kling, R. (Ed.), *Computerization and Controversy: Value Conflicts and Social Choices* (pp. 407–423). San Diego, CA: Academic Press.

Suchman, L., Blomberg, J., Orr, J. E., & Trigg, R. (1999). Reconstructing technologies as social practice. *The American Behavioral Scientist, 43*(3), 392–408. doi:10.1177/00027649921955335

Suchman, L. A., & Trigg, R. H. (1991). Understanding practice: Video as a medium for reflection and design. In Greenbaum, J., & Kyng, M. (Eds.), *Design at Work: Cooperative Design of Computer Systems* (pp. 65–90). Hillsdale, NJ: Erlbaum.

Sunderland, P. L., & Denny, R. M. T. (2007). *Doing anthropology in consumer research.* Walnut Creek, CA: Left Coast Press.

Taylor, F. W. (1911). *The principles of scientific management*. New York, NY: Harper & Brothers.

Theureau, J., & Filippi, G. (1994). Cours d'action et conception d'un système d'aide à la coordination: Le cas de la régulation du trafic du RER. *Sociologie du Travail, 4*, 547–562.

Turnbull, D. (2000). *Masons, tricksters and cartographers: Comparative studies in the sociology of scientific and indigenous knowledge*. London, UK: Routledge. doi:10.4324/9780203304587

Viller, S., & Sommerville, I. (2000). Ethnographically informed analysis for software engineers. *International Journal of Human-Computer Studies, 53*, 169–196. doi:10.1006/ijhc.2000.0370

Vinck, D. (Ed.). (2003). *Everyday engineering: An ethnography of design and innovation*. Cambridge, MA: MIT Press.

Wasson, C. (2000). Ethnography in the field of design. *Human Organization, 59*(4), 377–388.

Wasson, C. (2002). Collaborative work: Integrating the roles of ethnographers and designers. In Squires, S., & Byrne, B. (Eds.), *Creating Breakthrough Ideas: The Collaboration of Anthropologists and Designers in the Product Development Industry* (pp. 71–90). Westport, CT: Greenwood.

Wenger, E. (1999). *Communities of practice: Learning, meaning and identity*. Cambridge, UK: Cambridge University Press.

Wixon, D. (1995). Qualitative research methods in design and development. *Interaction, 2*(4), 19–24. doi:10.1145/225362.225365

Woodruff, A., Szymanski, M. A., Grinter, R. E., & Aoki, P. M. (2002). Practical strategies for integrating a conversation analyst in an iterative design process. In *Proceedings of the 4th Conference on Designing Interactive Systems: Processes, Practices, Methods, and Techniques*, (pp. 255-264). London, UK: IEEE.

Woolgar, S. (1985). Why not a sociology of machines? The case of sociology and artificial intelligence. *Sociology, 19*(4), 557–572. doi:10.1177/0038038585019004005

Woolgar, S. (1991). Configuring the user: The case of usability trials. In Law, J. (Ed.), *A Sociology of Monsters*. London, UK: Routledge.

## ENDNOTES

1 For useful introductions to ethnography and information system see Button (2000), Crabtree (2001, 2003), Randall *et al.* (2007), and Sharrock and Hughes (2001).

2 For a survey and discussion see Schwartzman (1993), Boden (1994), Morrill and Fine (1997), Bruni (2003), and Hodson (2004).

3 This strand of inquiry was inaugurated by at least three different independent and parallel research projects undertaken in California: Latour and Woolgar (1979) on a biochemistry laboratory at the Salk Institute; Knorr-Cetina (1983) on a biochemistry laboratory at Berkeley; and Lynch (1985) on an university biology laboratory. For a recent discussion see Bauchspies (2007).

4 For the point of view of Science and Technology Studies (STS) see Woolgar (1985), Vinck (2003), Berg (1998), and Davenport (2008).

5 For ethnographies on engineers see Evan (1964), Gamst (1980), Bucciarelli (1988), Kunda (1992), and Almklov (2008).

6 See Downey (1998) and Haraway (1991). The blending of 'human' and 'non-human' features it typical of 'actor network theory': see Latour (1992b).

7 Malinowski said that it was necessary to master the "technical details" of a canoe building in order to follow his analysis of the *Kula* (Malinowski, 1922).

8 Many things can be designed: work processes, organizations, towns, public policies, etc. An interesting perspective on product designers is that taken by Harvey Molotch (2003). The focus here is on the design of information systems.

9 See the recent special issue of *Organization Studies* (29:5, 2008) on the work of Erving Goffman, Harvey Sacks, and Harold Garfinkel.

10 Recognizable in my schematic (and idealized) description of the 'engineering mentality' is the positivist paradigm described by Burrell and Morgan (1979) and by Orlikowski and Baroudi (1991), or the 'normative discourse' described by Deetz (1996) and re-analysed by Schultze and Leidner (2002). See also Mintzberg (1990) or the notion of 'projectness' in Law and Singleton (2000). In general, this 'engineering mentality' expresses the modernist project of the "rational planning of ideal social orders under standardized conditions of knowledge and production" described by David Harvey (1990). See also Scott (1998, Chapter 3).

11 See http://asd-www.larc.nasa.gov/barkstrom/public/The_Standard_Waterfall_Model_For_Systems_Development.htm

12 Debugging is the process of finding and removing defects ('bugs') from software following use of the program: see http://it.wikipedia.org/wiki/ Debugging

13 Although no one would today endorse a rigid model, it is still an ideal of sorts that permeates engineering culture.

14 See http://it.wikipedia.org/wiki/Implementazione

15 See Taylor (1911), Noble (1977), and Nelson (1980).

16 For a critical discussion of requirements acquisition in standard planning see Goguen (1994, 1996), Luff *et al.* (1993), Anderson (1994), and Sommerville *et al.* (1993).

17 As Sommerville and Sawyer (1997) stress, the commissioners of an information system are often not those who will actually use it, while the 'requirements' often flatten a crucial aspect (the political and hierarchical dimension) of organizations into a single vague and idealized dimension.

18 See McDermott (1981). This is only a banal example of the complex relationship between common sense and the scientific interpretation of action. See Schuetz (1943, 1951, 1953) and Ryle (1966). The cases described by Turnbull (2000) are highly instructive

concerning the complexity of non-scientific knowledge.

19 Claudio Ciborra was Professor of Information Systems at the London School of Economics, and the advocate of a phenomenological approach to the design, implementation, management, and use of information systems.

20 See Woolgar (1991) and Sharrock and Anderson (1994).

21 Good illustrations of interest in the usability of design products are the books by Donald A. Norman (1988, 1992, 1993, 1998, 2004, 2007).

22 This view produces an odd shift of emphasis: whilst at first the computer was only a metaphor with which to describe the activity of the mind, now the computer mirrors the actual workings of the mind.

23 See the essays in Bowers and Bendford (1991) and Schmidt (2000).

24 See Bannon (1990, 1992, 1993) and Bannon and Schmidt (1991).

25 The first conference was organized in 1984, on which occasion the acronym CSCW was coined. Since then the conference has been held every two years: the last one in San Diego in November 2008, while the next is scheduled for February 2010 in Savannah, Georgia, USA.

26 The journal and the series are both published by Springer.

27 See Schmidt (1994), Simone *et al.* (1995), Fjuk *et al.* (1997), and Suchman (1996).

28 The term 'social arena' used by Clarke (1991) or that of 'community of practice' coined by Wenger (1999) have numerous affinities with the interactionist notion of 'social world.' The concept of community of practice has also been developed in sociology of organizations: see Gherardi (2008).

29 Boundary objects' are objects shared by different communities of practice for different purposes. Belonging in this strand of analysis are the studies by Susan Leigh Star (Star & Strauss, 1999; Bowker & Star, 1999) to which the essay by Rawls *et al.* in this issue of the journal also refers.

30 On Hughes see Riesman (1983), Heath (1984), Chapoulie (1987, 1996), Strauss (1996), and Helmes-Hayes (1998).

31 On anthropology applied to marketing see Blythin *et al.* (1997), R. P. Heath (1997), Reese (2002), Sherry (2002), Wasson (2002), Jordan and Dalal (2006), Jordan and Yamauchi (2008), Jordan and Lambert (2009), and Sunderland and Denny (2007).

32 The text of this quotation from Malinowski (1922) has been modified: I have replaced 'canoes' with 'information systems,' but it seems to me that the point remains the same.

33 See Garfinkel (1967, 1986, 2002).

34 'Frame,' 'script,' 'schema,' 'plan,' are some of the terms most frequently used to denote a set of structured expectations in cognitive science. These terms describe stereotyped representations of knowledge, which 'map' the information that we have acquired and collected during our lifetimes by means of concrete experience and formal learning through various socialization processes. Whenever we encounter a new situation, we do nothing other than retrieve from memory a sufficiently generic structure (entering a room, going to a restaurant, driving home, etc.) which enables us to recognize the current information by means of the knowledge that we already possess. How a particular case can belong in a general category, however, is not considered problematic.

35 See Luff *et al.* (2000), Heath and Luff (2000), Heath *et al.* (2000) and the special issue of *British Journal of Sociology*, *53*(2), 2002, edited by C. Heath and G. Button.

36 Lucy Suchman has promoted the Workplace Project, the purpose of which is to study so-called 'centres of coordination,' such as the air traffic control and ground operations tow-

ers at airports, emergency operations centres, the control rooms of public rail transport systems (trains and subways—see Heath & Luff, 1996). The work of Chuck and Candy Goodwin is of particular importance in this regard: C. Goodwin (1996), C. Goodwin and M. Goodwin (1996), and M. Goodwin (1995, 1996).

37 See Brun-Cottan and Wall (1995), Jordan and Henderson (1995), Suchman and Trigg (1991), Koschmann *et al.* (2006), and Ruhleder and Jordan (1997).

38 On the question of ethnographic description see Erickson (1988) and Hirshauer (2007).

39 In the study of forms of multimodal communication, partly derived from conversation analysis: Woodruff *et al.* (2002).

40 There are legal provisions in the Scandinavian countries, which stipulate that the trade unions must always be involved in the design and implementation of information systems.

41 Also in regard to the development of communities. For the situation in the USA see Srinivasan (2007).

42 This has the particular consequence that 'social' relations among scientists, interest groups, institutions and organizations, which sociologists would regard as the only 'actors' in the proper sense, are all placed on the same level (literary) with 'technical' relations among scientists, the tools used, 'natural' phenomena and 'non-human' agents (e.g., microbes, molluscs, cows, ocean currents, wind, etc.): they become 'actants.' A scientific venture thus becomes a story to be recounted, a text to be deconstructed.

43 Another approach is the French one consisting in analysis of 'courses of action' (Theureau & Filippi, 1994). This seeks to explain, from a social psychology perspective, how tools and technologies are used on the basis of the courses of action in which they are embedded.

44 See Diggins and Tolmie (2003), Rogers and Bellotti (1997), Button and Harper (1996), Anderson (1994), and Dillon (1998).

45 For that matter, also the anthropological tradition has a long history of 'collaboration' with management of the colonialism, as well as its 'criticism.'

46 This study on air traffic controllers was perhaps the first important instance of collaboration between information engineers and sociologists in a high-tech work setting. See Bentley *et al.* (1992), Harper and Hughes (1993), Hughes *et al.* (1993), and Hughes *et al.* (1992).

47 This aspect has often been confused with the problem of 'generalizability,' i.e. the fact that information systems need generalizable results that can be formulated at a high level of abstraction, whilst ethnography furnishes only local solutions relative to specific cases. I do not think, however, that this is a real problem.

48 Italo Calvino, *Il barone rampante* (*The baron in the trees*), 1957, cap. XXVII.

49 Here I use 'field' as a technical term denoting a dialectical domain which does not consider two isolated disciplines but their reciprocal relationship.

50 Scott reprises the argument of Martha Nussbaum (1956).

51 In the field of urban planning, as exemplified by the construction of Brazil's new capital Brasilia or the modern Paris re-designed by Le Corbusier; forced collectivization in Soviet Russia; the forced villagization of the Tanzanian population in the 1970s; the simplified, single-crop agriculture of contemporary scientific agronomy.

52 See Gherardi (2006).

53 On the notion of 'primary science' or 'primitive' science see also Lynch and Bogen (1994). On design as *bricolage* see Büscher *et al.* (2001) and Ciborra (2004).

# Chapter 2
# Experiencing Information Systems Research and Phenomenology:
## The Case of Claudio Ciborra and Martin Heidegger

**Paolo Depaoli**
*LUISS Guido Carli, Italy*


## ABSTRACT

*Given the ongoing debate on the explicit and implicit foundations of theories and methods within the IS discipline, this chapter examines the circumstances that appear to favor a close connection between Information Systems (IS) research and philosophy (phenomenology in the specific case examined). The question addressed is whether references to philosophy are typically made when a researcher is confronted with issues and topics that she considers crucial and which she believes inappropriately treated or neglected by current theories and practice. To answer this question the influence of Heidegger on Ciborra's work is explored. The method adopted is the review of Ciborra's publications and his references to Heidegger; meaningful issues were identified and significant passages of both authors compared. It appears that Heidegger's influence has been substantial in terms of support for both Ciborra's critical analysis and innovative solutions for the IS-related world. Given the multiplication and growing pervasiveness of information technology and IS and the consequent growing need for policies based on sound research, the "escalation" to philosophy (and especially the engagement with phenomenology), though demanding and difficult, will probably become increasingly useful.*


DOI: 10.4018/978-1-4666-0303-5.ch002

## INTRODUCTION: SETTING THE QUESTION

Philosophy, let alone phenomenology, is not explicitly mentioned in the Information Systems (IS) related disciplines mentioned by Baskerville and Myers (2002) in their article where they advocated a closer connection among different research branches interacting in a 'knowledge network.' In fact, they claimed that IS had become mature enough to be a reference discipline itself, able to contribute to the advancement of learning in several fields such as medical informatics or electronic commerce. Possibly for this very reason—the newly acquired awareness of the significance of IS—Lee and Baskerville (2003) the following year addressed (in depth) a crucial issue for any scientific specialty, the question of 'generalizability' of the results of a research endeavour. In their essay they point out that several articles published in leading journals relate a reduced sample size, or a limited number of cases, to a limited generalizability of findings. In order to show that the concept is misapplied and misguides both authors and reviewers, they conduct their analysis considering the relevance for IS research of the problem of induction raised by the eighteenth century Scottish philosopher Hume, and well known to epistemologists: no 'proof' can be reached through data unless one assumes the 'Uniformity of Nature' (UN). Uniformity cannot be proven by induction since what has to be explained (UN) cannot be the basis for its own explanation (UN itself). The authors used Hume's argument to show that in statistics it is not possible to generalize the estimates of a given 'sample' to the characteristics of the 'population' so that quantitative descriptions are not generalizable beyond the domain that the researcher analyzed; of course, larger sample sizes produce more 'reliable' results in that—within a given population—other samples based on the same criteria are more likely to show converging results. But 'reliability' of samples is quite a different concept from 'generalizability' of a sample to the population. For the same reason, results of qualitative research cannot be extended to other settings beyond the one observed. The authors conclude that a theory is generalizable to another context only if it survives an empirical test in that new context. This consideration, they comment, questions the possibility of generalizing to a comprehensive IS theory—for example the determinants of IS success, as in DeLone and McLean (1992) whom they cite—from results of different research studies conducted on different instantiations of information technology (ERP, Internet, packages, etc.). One possible solution is the adoption of 'action research' where theory can be both tested on a new setting and enriched by the interaction with the new context. Even though in their article Lee and Baskerville do not state explicitly to draw on phenomenology, their indication of action research is very close to a phenomenologist's position whereby "the positing of entities outside experience is ruled out as meaningless" (Moran, 2000, p. 13). In fact, Lee (in a 2004 article co-authored by Martenson) proposes a kind of action research named 'dialogical action research' that explicitly refers to phenomenology:

*Unlike other forms of action research, dialogical AR takes notice of and addresses heterogeneity in the forms of knowledge held by the scientific researcher and the real-world practitioner, where the heterogeneity is related to what phenomenology would call 'the scientific attitude' taken by the researcher and 'the natural attitude of everyday life' taken by the practitioner. (Martensson & Lee, 2004, p. 508).*

In the Hume-Lee and Baskerville case, the IS researchers used a philosophical conception both to criticize current practices and assumptions in the IS discipline (they uncovered misleading constraints in the evaluation of research, e. g. quantitative research is not more generalizable than qualitative research) and to propose solutions (they indicated an approach to be used by researchers to pursue both scientific significance and practitioners needs, namely action research). Certainly, there have been several other cases where the recourse to philosophy has had a significant role in IS research, the most famous probably being Dreyfus' (1992) criticisms of the computational model of the mind adopted in artificial intelligence research that he developed from the years sixties onwards drawing on phenomenology.

In 1991 Orlikowski and Baroudi through a research conducted on 155 articles published from 1983 to 1988 in four North American leading journals concluded that there was a dominant paradigm since 96.8% of the papers had a positivistic underlying epistemology. They defined a positivist information systems research approach in the following way:

*"Ontologically, positivist information systems researchers assume an objective physical and social world that exists independent of humans, and whose nature can be relatively unproblematically apprehended, characterized, and measured" (Orlikowski & Baroudi, 1991, p. 9).*

More than a decade later, Chen and Hirschheim (2004) found that the picture had not changed much since 81% (86% in US journals) of the 1893 articles they examined published from 1991 to 2001 on eight European and North American journals had a positivist nature.

Ciborra is one of the scholars that reacted to this state of affairs because he also believed that this lack of pluralism "has limited what aspects of information systems phenomena we have studied, and how we have studied them" (Orlikowski & Baroudi, 1991, p. 7). In fact, with respect to the dominant positivist approaches, he outlined a phenomenological information systems research vision to be considered as a different perspective on information systems [that] should be anchored to the unfolding "of the human process of encountering the everyday world" (Ciborra, 2002, p. 6). So, his "escalation to philosophy" (and precisely Heidegger's phenomenology) is based on his dissatisfaction with the dominant scientific paradigm (with its widely accepted assumptions) pointing to "an alternative centre of gravity: human existence in everyday life" (Ciborra, 2002, p. 1).

By reconstructing and interpreting Ciborra's journey in the IS research world supported by Heidegger's version of phenomenology, this chapter thus aims at exemplifying the advantages of an in depth examination of basic outlooks and assumptions capable of promoting novel methods and areas of investigation in IS research.

The chapter contains five sections inclusive of this introduction: the second section illustrates the method adopted; the third one examines Ciborra's peers perception of Heidegger's influence; the fourth one (divided into three sub-sections) identifies the most meaningful references made by the former to the German philosopher together with the key concepts and the quotations that the research work has identified as the most appropriate. The last section contains a discussion of the analysis that has been carried out and the conclusions that can be drawn.

## THE METHOD ADOPTED: THE INTERPRETATION OF THE RELEVANT WRITINGS

As it will be explained further on, Ciborra states to have gained a deeper and deeper understanding of Heidegger's work by successive readings and interpretations. Since 'interpretation' is one of the main differences of Heidegger's phenomenology with respect to Husserl's project of pure description (Moran, 2000, p. 20), it seems appropriate to utilize the 'hermeneutic circle' to try and reconstruct the encounters of the IS scholar with the philosopher. Thus, Heidegger's influence on Ciborra has been outlined by reading the relevant writings of the two authors (starting from Ciborra) and by trying to understand the possible interplay of approaches, ideas, and issues. Three main steps were followed. At first, articles, books, and parts of books authored or co-authored by Ciborra, published in English from 1994 to 2006 and containing references to Heidegger in the form of words or concepts that Ciborra used to either explicitly or implicitly refer to him were identified. Both the explicit references and the context of the discussion served as a basis to recognize implicit references. Next, Ciborra's texts were examined and passages containing Heidegger's concepts were singled out in order to understand how Ciborra had used them.

Finally, since Ciborra most only indicates the work of Heidegger rather than the section or page he refers to (with the exception of "The Mind or the Heart? It Depends on the (Definition of) Situation" published posthumously in 2006), it was necessary to trace back the references that he made to Heidegger's original writings. The part of Heidegger's text that more likely applied to the relevant issue discussed by Ciborra was then identified and consequently quoted. This was the most challenging part of the interpretation of the texts of both authors.

Furthermore, the essays in Ciborra's honour written by his peers after his death—contained in the 2005 issue of the European Journal of Information Systems (EJIS vol. 15, n. 5)—were reviewed, and references to Heidegger examined in order to gain a further insight on key concepts.

## CIBORRA'S PEERS INTERPRETING THE INFLUENCE OF HEIDEGGER ON HIS WRITINGS

Thirteen of the twenty-nine contributions published in the cited EJIS issue consider Heidegger has having affected Ciborra in a significant way. The influence concerns his approach as well as some specific topics.

On the one hand, an overall influence in terms of approach and vision is acknowledged by Kallinikos (2005) who, commenting on Ciborra's intellectual journey, points out that:

*He gradually became seriously engaged with Heidegger's thought, sought to understand Heideggerian philosophy in all its depth and let that understanding bear upon the ways by which we approach key issues associated with how technologies of information and communication become integrated in contemporary institutional life (Kallinikos, 2005, p. 469).*

Introna (2005) explicitly says that "[i]f we start to see information systems in the organisation as particular singular worlds (as Heidegger's ideas suggest) then Claudio's language opens up in very subtle and meaningful ways—as I believe he intended" (Introna, 2005, p. 514). In addition, De Marco (2005) believes that his fruitful approach to phenomena, organization, and IS stems from a substantial link with Heideggerian thinking:

*Particularly audacious was his insistence on drawing inspiration from Heidegger ... [which] at first appeared—to me, at any rate—as not just inappropriate but even incomprehensible.... [D] eeper reflection brought wiser counsel. Closer attention both to Claudio's own writing and a growing acquaintance with that of the German philosopher in front of whose house Claudio had even gone to the length of having himself photographed, revealed just how much the former drew meaning and significance from the latter (De Marco, 2005, p. 442).*

On the other hand, in Heidegger's work there are some crucial concept-words (such as *Aletheia* and *Gestell*) that have been specifically mentioned by his peers for their importance in Ciborra's work. Zuboff (2005, p. 471)—herself stating of having been "repelled by the geometrization of human experience" and thus having drawn to the "philosophers of the messy: Husserl, Heidegger, Merlau-Ponty, Gendlin"—notices that:

*when one is committed to giving a complete and accurate account of what one has experienced and observed, he speaks Aletheia, and that Aletheia was Claudio's remedy for the crisis in which he saw the disciplines of information technology and organizational theory and indeed of social science as a whole.*

In fact, the Greek word *Aletheia* is used by Heidegger in its etymological meaning of 'that which is disclosed' or 'the unveilness' and not only as 'truth' in the sense of 'being correct'; more precisely, in the philosopher's direct words: "The Greeks have the word *Aletheia* for revealing. The Romans translate this with *veritas*. We say 'truth' and usually understand it as correctness of representation." (Heidegger, 1993a, p. 318). As a matter of fact, Ciborra uses the terms 'unveiling' and 'to unveil' in many of his writings (and sometimes also the Greek word) especially when he wants to stress the existence of important phenomena that remain unnoticed because of the "veiling" effect produced by dominant theories. This is the case, for example, of the second chapter of *Labyrinths of Information*, where he considers the inadequacy of current methodologies derived from Cartesian generalizations, incapable of coping with the "messy everyday reality" that governs the evolution of organizations thus coming to the conclusion, which is at the same time an invitation, to "…drop the old methodologies in order to better see the new dimensions the technology is going to disclose to us.… Let truth be always our goal but understood as the Greek world *aletheia*: the unveiling of what lies hidden" (Ciborra, 2002, p. 27).

Another key word in the German philosopher's thought, *Gestell* (Heidegger, 1993a, p. 325), concerns the very essence of technology and has been acknowledged an important role in Ciborra's way of considering the "relationship between technology and organization. "…[It] constitutes a line of thought that continues to be explored by an entire generation of scholars." (De Marco, 2005, p. 443). The ordering effect of modern technology and its increasing pervasiveness which is a characteristic of the Heideggerian conception was realized by Ciborra (Zuboff, 2005) who based on it his understanding of information infrastructures (Whitley, 2005). Since Ciborra studied in depth this complex question in an article co-authored with Hanseth (1998) and then in the fourth chapter of his book (Ciborra, 2002), this topic will be examined in detail in the sub-section on "Gestell (the essence of technology): constraint – possibility."

There are other instances, evidenced by his peers, that point to the influence of phenomenology, and especially to the Heideggerian version of it, on Ciborra's thinking which enabled him to view information systems not as mere objects and tools or as the outcome of abstract theories and models but as specific complex worlds (Introna, 2005). This approach allowed him to take into account characteristics and traits of organizations usually neglected. It is the case of the 'red light zones' of organization examined by Ciborra (Whitley, 2005): the places that are barely tolerated and often overlooked by research, and that are, however, sources of new knowledge. Sources that can be leveraged just by going below the surface of the "systematic ways of organizing and executing work" and by taking into consideration "…the practice of *bricolage* and other kindred activities such as serendipity, hacking and improvisation" (Ciborra, 2002, p. 47). In fact in order to describe and explain the phenomenon of improvisation Ciborra drew, in the final chapter of his last book, on other important notions of Heidegger's which will be considered in the last section of the chapter before the discussion.

At the end of this however brief review, it is possible to conclude that the stimuli caught by Ciborra in Heidegger's thought are considered to be important by his colleagues and peers, related as they are to key concepts of both authors. It is therefore interesting to try and identify the 'places' in Ciborra's writings where Heidegger's influence is higher and the passages in Heidegger's works that explain such influence. This is the aim of the following section.

## HEIDEGGER AND CIBORRA'S *THE LABYRINTHS OF INFORMATION*

Within Ciborra's writings, *The Labyrinths of Information* (Ciborra, 2002) contains the most important references to Heidegger. In fact, eleven of his publications that have been examined were re-edited and contributed both to the *Methodological Appendix* and to seven of the eight chapters that articulate the content of the book. In his posthumous *The Mind or the Heart? It Depends on the (Definition of) Situation* (Ciborra, 2006), Ciborra makes further references to Heidegger's works which, however, do not change substantially the sense of the influence exerted by the philosopher on Ciborra's approach in defining the concept of 'situation.' As a matter of fact, this approach is described in the eighth chapter of Labyrinths: "*Kairos* (and affectio): seizing the opportunity (and moods and mental states)." In the last sub-section the relevant concepts (improvisation, moods, moment of vision, and Heidegger's *Befindlichkeit*) will be discussed.

### Heidegger's Phenomenology as a Way to Research for Ciborra

In the *Metodological Appendix*, Ciborra openly states that Heidegger not only influenced some of his ideas but he also gave him the key to understand his way to research: "...the introduction to *Being and Time*, specifically his definition of phenomenology ... (or rather my interpretation of it), at last helped me to make explicit the way that I understand how I proceed" (Ciborra, 2002, p. 173).

Ciborra uses the word 'interpretation' to refer to the hermeneutic circle, the process of going back and forth in a text, from details to an overall view, for an increasing understanding of it (Dreyfus, 1991, p. 36). Of course, he used it to gain an in-depth comprehension of Heidegger's writing ("I re-read, for the nth time, the introduction..."). However, as will be seen, the whole Appendix can be understood as a sign of agreement with Heidegger's statement that "...the meaning of phenomenological description as a method lies in *interpretation*" (Heidegger, 1962, p. 61). Interpretation is necessary because, on the one hand, Heidegger defines phenomenology as a science of phenomena that can "grasp its objects *in such a way* that everything about them which is up for discussion must be treated by exhibiting it directly and demonstrating it directly" (Heidegger, 1962, p. 59). But, on the other hand, since phenomena can be hidden because they are still undiscovered or because they are buried over, phenomenology should not stop at a descriptive stage and work its way through to the 'thing itself.' In fact "[e]ven in the concrete work of phenomenology itself there lurks the possibility that what has been primordially 'within our grasp' may become hardened so that we can no longer grasp it. And the difficulty of this kind of research lies in making it self-critical in a positive sense" (Heidegger, 1962, p. 61).

We can 'grasp' Ciborra's adherence to Heidegger's hermeneutical approach when he describes the two types of evidence that he encounters when approaching an "organizational phenomenon":

- "the set of ideas and models taken for granted in the domain of organization theories or consulting models...[which] following Heidegger we can refer to ... as illusory appearances" (Ciborra, 2002, p. 176), and
- "apparitions" which belong to a space that cannot be filled by any model and that surface in informal talks that "host the unexpected aspects of organizational life." (Ciborra, 2002, p. 177)

According to Ciborra, investigation often stops at the empty models instead of working on the apparitions that tell about the "underlying phenomenon to be unveiled" (Ciborra, 2002, p. 178). The two terms he uses, "apparitions" and "illusory appearances," remind of Heidegger's "appearance" and "semblance." In fact the philosopher gives an example to clarify the meaning of the two words: a person's red cheeks that announce a fever, and thereby a disease, are an 'appearance,' whereas the reflection of a red light on that face would be a 'semblance.' Thus the researcher has to try and pursue the apparitions that "announce" the core characteristics of an organization, that is the appearance that Heidegger defines as "that which does the announcing ... the bringing forth" (Heidegger, 1962, p. 51). He should therefore be able to distinguish "... 'phenomenon' from 'semblance,' which is the privative modification of 'phenomenon' ...." (Heidegger, 1962, p. 51). For Heidegger "the expression '*phenomenon*' signifies *that which shows itself in itself*, the manifest." (Heidegger, 1962)

It can be noticed, at this point, that the concepts just mentioned are basic features shared by a full range of approaches that can all be considered as belonging to phenomenology. The attempt to establish an 'orthodoxy' in phenomenology, to define a set of rules according to which it should be possible to determine its precise boundaries would probably amount to producing an oxymoron. In fact, in his account of phenomenology Moran (2000, p. 4) states that although

*there are a number of themes which characterize phenomenology, in general it never developed a set of dogmas or sedimented into a system. It claims, first and foremost, to be a radical way of doing philosophy, a practice rather than a system.*

As mentioned above, Ciborra found an affinity in his own practice with Heidegger's approach as described in the Introduction to *Being and Time* (BaT); yet, as will be seen in the following subsections, his references are not limited to such work but include, for example, *The Question Concerning Technology* (QCT) (Heidegger, 1993a) and the *Discourse on Thinking* (Heidegger, 1966), which cannot be considered a continuation of BaT. So, even though in Ciborra's writings there is no acknowledgment of a discontinuity of Heidegger's later works with his main *opus*, it seems plausible that Ciborra's recourse to Heidegger is not only due to a question of methodological resemblance (as he himself states) but to its being affected by the amplitude of vision and richness of themes that Heidegger offers to his readers. The two following sections confirm this idea: in the first one BaT is the main reference for Heidegger, whereas in the following one QCT and *Gelassenheit* (Ciborra prefers to include in his references the original work instead of the English translation *Discourse on Thinking*) are used by Ciborra to propose specific priorities in IS management.

## Tradition, Crisis, Aletheia

In 1997, Ciborra published "*De profundis*? Deconstructing the concept of strategic alignment." Deconstruction is still the central point of the version that was published five years later with the name "Krisis," the second chapter of Labyrinths. His starting point is that the "Internet has emerged as a flexible infrastructure outside any strategic master plan" (Ciborra, 2002, p. 13) whereas many "exotic, Artificial Intelligence (AI) based applications aimed at supporting decision making at the top of the organisation... [never] saw the light of the day" (Ciborra, 2002, p. 13). Furthermore, if we "consider system analysis and design methodology"—which he exemplifies with Computer Aided Software Engineering and Business Process Re-Engineering—we have to conclude that "we seem to be devoting much teaching (and consultancy) to something which to be generous, fails at least half of the time" (Ciborra, 2002).

These quotations—and others could be made from his work—show that dominant theories and consulting models have been considered by Ciborra not only as 'semblances' but as a consolidated 'tradition' that jeopardizes the accomplishment of a researcher's task; in Heidegger's words:

*When tradition ... becomes master, it does so in such a way that what it 'transmits' is made so inaccessible, proximally and for the most part, that it rather becomes concealed. Tradition takes what has come down to us and delivers it over to self evidence; it blocks our access to those primordial 'sources' from which the categories and concepts handed down to us have been in part genuinely drawn. (Heidegger, 1962, p. 43)*

Ciborra's criticism of prevailing methodologies—the following is another example: "No matter how practical or business oriented the current management and systems literatures are, they tend to rely … on methods that are simple, rational, and, especially, based on the natural science paradigm" (Ciborra, 2002, p. 5)—leads the way to his proposal of different approaches. Also in this respect a parallel can be drawn with Heidegger's thought; for example, one of Ciborra's suggestions for a change in attitude is to go to:

*the basics and encounter the world as it presents itself in our everyday experience. We rely on evidence, intuition, and empathy. We listen to managers, and we participate in their dealings with puzzles and riddles, and, on the other hand, we do not confer any particular relevance on words like 'strategy,' 'processes,' 'data,' 'systems,' or 'structured methodologies' (Ciborra, 2002, p. 25, italics added).*

The emphasis was added to show the connection with Heidegger's general construction which is founded on "being-in-the-world as the basic state of [man]." Heidegger warns the reader not to consider the preposition 'in' in its spatial meaning but as "derived from 'innan'—'to reside,' 'habitare,' 'to dwell'" (Heidegger, 1962, p. 80) connected also to "'I am accustomed,' 'I am familiar with,' 'I look after something'" (Heidegger, 1962). These expressions are certainly not far from Ciborra's invitation to rely on the world and on how we encounter it while going beyond consolidated concepts, so that as Heidegger puts it: "If the question of Being is to have its own history made transparent, then this hardened tradition must be loosened up, and the concealments which it has brought about must be dissolved" (Heidegger, 1962, p. 44).

Ciborra probably recognised in those words the contour of the very task of a researcher who is to find novel ways to approach phenomena, to take them into the open ("to unveil" them, as he often says), and to communicate them, thus contrasting consolidated and obstructing frameworks. It is worth considering the word 'to communicate' in BaT so that Ciborra's insistence on the concept of truth as 'un-concealment' (*aletheia*) can be better explained. For Heidegger there cannot be communication without the sharing of a common world where people, by 'dwelling' in it, are related to each other and through which a certain assertion can be understood. In his words, "As something communicated, that which has been put forward in the assertion is something that Others can 'share' with the person making the assertion" (Heidegger, 1962, p. 197).

However, "Assertion is not a free-floating kind of behaviour which, in its own right, might be capable of disclosing entities in general in a primary way; on the contrary it always maintains itself on the basis of Being-in-the-world" (Heidegger, 1962, p. 199)

Thus, as it was noted by Dreyfus (1991, p. 221), language cannot be considered a code that exists in itself connecting otherwise isolated minds, thereby explaining the difficulties in "programming computers to understand natural language"; the following is the quotation of Heidegger (1962, p. 205) used by Dreyfus to support his claim:

*Communication is never anything like a conveying of experiences...from the interior of one subject into the interior of another. Dasein-with is already essentially manifest in a co-affectedness and a co-understanding. In telling being-with becomes 'explicitly' shared; that is to say, it is already, but it is unshared as something that has not been taken hold of and appropriated.*

So people living in the same world (for example, the same organization) on the basis of their common understanding of it and of their being affected by it, and by one another, can share it 'explicitly' through speech: "Discourse 'lets something be seen'... In discourse...what is said ... makes manifest what it is talking about, and thus makes it accessible to the other party" (Heidegger, 1962, p. 56). But since discourse "is a letting-be-seen, it can *therefore* be true or false" (Heidegger, 1962). This point in the Introduction of BaT is probably at the origin of Ciborra's adoption of the concept of truth as un-veiling:

*The 'Being-true'... means that ...the entities of which one is talking must be taken out of their hiddenness; one must let them be seen as something unhidden (alethes); that is they must be discovered. Similarly 'Being-false' (pseudesthai) amounts to deceiving in the sense of covering up [verdecken]: putting something in front of something (in such a way as to let it be seen) and thereby passing it off as something which it is not (Heidegger, 1962).*

It seems thus clear that for Ciborra the 'illusory appearances' (the "ideas and models taken for granted" mentioned in the preceding section) constitute an obstruction to communication, a veil in approaching organizations as concrete worlds, an hindrance to the design of effective information systems. It is also possible to emphasize the view gained through the examination of the testimonies of Ciborra's peers: there are a number of key words and concepts that have a prominent role in Heidegger and that Ciborra has closely considered and utilized. The philosopher links the bringing forth of phenomena through appearance to the making manifest of beings through language. Ciborra uses this approach to look into the world and to overcome methodologies grounded in the natural sciences paradigm, which he believes dominate the scene of management, of organization, and of information systems. The crucial question Ciborra himself poses and the possible answers he gives referring to Heidegger should be addressed now: "What calls us to devise methodologies? Possibly it is technology itself, and its enframing effect" (Ciborra, 2002, p. 26).

## *Gestell* (The Essence of Technology): Constraint – Possibility

Technology, and its pervasiveness—including the power of "science-based, method driven approaches [that] can be misleading" when dealing with human and interdependent dimensions (Ciborra, 2002, p. 6)—was addressed by Ciborra for the first time in an article co-authored with Hanseth (1998) and subsequently in the fourth chapter of Labyrinths. Together with the *Methodological Appendix*, this chapter shows most explicitly Heidegger's influence. But while the Appendix seems to refer mainly to *Being and Time*, the chapter on "*Gestell*: The Power of Infrastructure" is based on later works of Heidegger's especially on the *Question Concerning Technology* (Heidegger, 1993a) which was published in 1954. Its content had been basically developed in the *Bremer Lectures* (Heidegger, 1994) which Ciborra included in his references together with "*Gelassenheit*" (Heidegger, 1992).

Ciborra's reasoning starts by relating the need to govern global business to the design and implementation of appropriate company-wide information systems which constitute a company's infrastructure. In these large and long lasting projects, the centre of attention is on the construction of a sophisticated management agenda where all possible variables and tools are sorted out and planned in order to assure the return of the investment made. A list of possible drawbacks is also worked out so that shortcomings can be taken care of. However, in spite of all the analyses and planning conducted in advance, such approaches fail to address the "central issue of *implementation;* the key transition between the conception of a vision and the realisation of that vision" (Ciborra, 2002, p. 59). And this is so because such management agendas are not only "highly abstract…[but] also too narrow" (Ciborra, 2002, p. 60). In fact, they are based on a centralized control and allocation of resources whereas the large number of stakeholders involved prevents any of them from being in control of the evolving situation. Furthermore, such large projects require time; with time, requirements change and adaptation has to take place. Different parts have to be made compatible so that interoperability can be assured. What happens is a significant drift form the original plan that had been neatly prepared and approved. In sum, a "whole infrastructure cannot be changed instantly." Also, "infrastructures… are never developed from scratch" (Ciborra, 2002, p. 61) as there always is an installed base.

Thus, Ciborra comes to a central point for this discussion: a "large information infrastructure is not just hard to change; it might also be a powerful actor influencing its own future life" (Ciborra, 2002, pp. 61-62). This concept is relevant for "tapping Heidegger's questioning about the essence of technology…able to evoke a new disposition and understanding of the technology itself" (Ciborra, 2002, p. 57). The crux of the matter is thus summarized by Dreyfus (2003, p. 40): "… scientific research is made possible by Descartes unthought, fully focused in Kant, that rationality consists in human beings imposing a total, systematic order on all that is. Heidegger calls this totalizing understanding of being *technological*."

The following are some of Heidegger's concepts put forth in *The Question Concerning Technology* (QTC) that show how the essence of technology (*Gestell*) was developed and how it possibly influenced Ciborra.

Heidegger starts by saying that the instrumental conception of technology is unable to uncover the essence of it; of course technology is a "means to an end" (Heidegger, 1993a, p. 312), and this definition is correct, but it is not enough. His argument begins by exploring the meaning of the Greek concept of cause: cause is that which is being responsible (Heidegger, 1993a, p. 315) for something. That which induces occasioning and manifestation, that which brings into appearance, which brings forth. *Poiesis*, the origin of the word poetry, means 'to make' or 'to do' and it is also interpreted by Heidegger as a 'bringing forth' (Heidegger, 1993a, p. 317). Nature brings forth: for example "the bursting of a blossom into bloom" (Heidegger, 1993a). Nature does that in itself. A silver chalice instead is brought forth by an artisan who uses instruments and tools to transform silver ore and to produce another tool, a container: the silver chalice. In fact 'to produce' derives from the Latin *producere* which is made up of *pro* (the adverb forth) and *ducere* (the verb to bring or to conduct) and in fact when something is produced, something that was not present before is brought into presence. The point is that if we think of a chalice as a 'sacrificial vessel,' we realize that it certainly can contain a liquid, for example wine, but that there is more to it that a mere container. In fact—and this example is given by Heidegger in his *Bremer* Conference—in a chain of thoughts we can consider that wine is brought forth by grapes grown in the land watered by rain and flooded with the sun; a wine that humans will offer as a gift to the divinities. There are four elements that are thus connected: the earth, the sky, the mortals, and the divinities. This is what has been revealed through the production of the artisan who has employed his technology.

Also modern technology is revelatory but in a different way. This way is a 'challenging' (Heidegger, 1993a, p. 320): somewhere "a tract of land is challenged into the putting out of coal and ore" and "the earth now reveals itself as a coal mining district, the soil as a mineral deposit." As the work of the peasant did not challenge the soil and the field he took care of, modern technology is something definitely different from the preceding one and even agriculture now is to be considered as "a mechanized food industry" (again in Heidegger words; Heidegger, 1993a). It compels us to view the world in a different way; this is so because:

*everywhere everything is ordered to stand by, to be immediately at hand, indeed to stand there just so that it may be on call for a further ordering. Whatever is ordered about in this way has its own standing. We call it the standing-reserve [Bestand] (Heidegger, 1993a, p. 322, italics added).*

This ordering and challenging characterizes modern technology and has been named by Heidegger *Ge-stell* (translated as enframing, the reason why Ciborra refers to the enframing effect of infrastructures). As Ciborra noticed, in German the verb *stellen* (to place or set) has a wide variety of uses. It can mean to put in place, to order, to arrange, to furnish or supply, and, in a military context, to challenge or engage. The prefix "ge" denotes gathering or collecting. We are thus confronted with "Enframing [*Ge-stell* that] ...challenges [man] forth, to reveal the real, in the mode of ordering, as standing-reserve." (Heidegger, 1993a, p. 325).

This is a general state of the world, of Being. Since for Heidegger man is not the owner of being (of reality, of the world) and we have to acknowledge the fact that we are not free and autonomous in the sense of having unbounded choice and limitless possibility in every situation, we cannot decide to simply avoid the dangers which belong to *Ge-stell* through an act of will. Heidegger mentions two dangers. The first one: "Man ... is continually approaching the brink of the possibility of pursuing and pushing forward *nothing but what is revealed in ordering*, and of deriving all his standards on this basis." (Heidegger, 1993a, p. 331, italics added). The second one: "...nature presents itself as a calculable complex of the effects of forces [*and*] can indeed permit correct determinations; but precisely through these successes the danger can remain that in the midst of *all that is correct the true will withdraw*." (Heidegger, 1993a, italics added).

So where can a solution be found? Even though men are not completely free, they are capable of thinking and of hearing the call of what is concealed. *Gestell* must be considered a destination for man, not a compelling fate. So that men have to be aware that "where the danger is, grows the saving power also" (this verse quoted by Heidegger at p. 333 belongs to Holderlin). They should practice thinking which entails to pay attention to (1) what is unconcealed, (2) what is concealed, (3) the coming to presence of technology (which can be interpreted as the evolution of technology).

Facing this complex construction built by Heidegger, Ciborra draws a parallel between information infrastructures and *Gestell* because of the danger he sees in them: "Information infrastructures can, as formative contexts, shape not only the work routines, but also the ways people look at practices, consider them as *being natural* and give them the overarching character of false necessity" (Ciborra, 2002, p. 76, italics added). Furthermore, specifically for his field of research and practice, he proposes some points of care in the management of organizations which draw upon the points of attention indicated by Heidegger and cited above, together with other concepts formulated by the philosopher in *Gelassenheit* (Heidegger, 1992): (1) to get out of mere calculative and instrumental thought; (2) to recognize the importance of moments of vision; (3) to adopt releasement which consists of saying 'yes and no' simultaneously to technology; (4) to promote a new sense of responsibility and not only of control; (5) to value marginal practices (Ciborra, 2002, pp. 77-78). Ciborra devised the first four points basically referring to *Gelassenheit*; the fourth one is a synthesis of the second (***Bricolage***) and of the eight (*Kairos)* chapters of Labyrinths. It is significant to consider *Kairos* because it is built around *Befindlichkeit,* one of Heidegger's three basic axes of human existence together with 'understanding' and 'speech.' The word has been translated into English using different terms because it was made up by Heidegger; therefore, in order to grasp the concept in a few lines, it is best to refer to the following interpretation (Gendlin, 1978, p. 45):

*In German a common way of asking 'How are you?' is 'Wie befinden Sie sich?' This literally says 'How do you find yourself?' One can also say to a sick person 'Wie ist Ihr Befinden?' ('How do you feel?') The same form can also be used to say that something or someone is situated somewhere, or in some way. For example, one can say, 'The White House finds itself in Washington, D.C.,' or 'I find myself in Chicago,' or 'I find myself in happy circumstances.'*

*'Sich befinden' (finding oneself) thus has three allusions: the reflexivity of finding oneself; feeling; and being situated. All three are caught in the ordinary phrase, 'How are you?' that refers to how you feel but also to how things are going for you and what sort of situation you find yourself in. To answer the question you must find yourself, find how you already are. And when you do, you find yourself amidst the circumstances of your living.*

This conception has been utilized by Ciborra to define the notion of 'situation' and to come to grips with 'improvisation,' one of the 'marginal practices' was mentioned above in his management agenda. Ciborra needs to look into "improvisation as a situated action carried out within the cognitive perspective in common currency…" (Ciborra, 2002, p. 170). The disappointment with this approach is due to the contraction of (clock) time so that improvisation is regarded as "quick problem solving" (Ciborra, 2002, p. 154) based on tacit knowledge. By ignoring moods and emotions, "they omit consideration of the situation of the actor. Implicitly they deal with the situation as a set of emerging circumstances but consider the actor as a passionless, problem solving robot" (Ciborra, 2002, p. 170). On this background, Ciborra's next step is to focus on moods because they "colour indelibly our being in the situation." Our rational capabilities are grounded in moods because the latter come first; in Heidegger's words:

*The fact that moods can deteriorate and change over means simply that in every case [a human being] always has some mood.... Why that should be, one does not know. And [a human being] cannot know anything of the sort because the possibilities of disclosure which belong to cognition reach far too short a way compared with the primordial disclosure belonging to moods... (Heidegger, 1962, p. 173).*

And this is exactly the point that Ciborra is trying to make: "the study of situated action in general, and of improvisation in particular, has focused so far … on the encounter between intentions and situations, but has systematically failed to reckon the (moody) situation of the actor" (Ciborra, 2002, p. 162). Ciborra's idea instead is to consider the usually neglected moods (he analyzes two opposite ones: fear and boredom) and the concept of 'our' time as opposed to 'clock' time. Improvisation happens when both lack of time or plenty of it lets people become aware of what really matters to them so that they become resolute and effective in pursuing what is important right there and then.

*In resoluteness, the Present is not only brought back from distraction with the objects of one's closest concern, but it gets held in the future and in having been. That Present which is held in authentic temporality and which thus is authentic itself, we call the 'moment of vision.' (Heidegger, 1962, p. 387)*

Ciborra then, through a different way of conceiving time and the possibility of a "moment of vision"—"suddenly the world, its resources, and people matter differently, so that they can be singled out and recombined anew" (Ciborra, 2002, p. 169)—can conclude that "the opposite of improvisation is not planned action, it is boredom (and sometimes panic)" because it is "extemporaneous i.e. it ruptures the way time entrances us in both situations…" (Ciborra, 2002). Also in this instance Heidegger had a decisive role in Ciborra's elaborated work: both anxiety and boredom, in its profound form, are connected to a possible moment of vision: the "Present of anxiety holds the moment of vision *at the ready*" (Heidegger, 1962, p. 394); in boredom "the lengthening is a *vanishing of the shortness of the while.*… In vanishing, *the moment of vision* still presses itself upon us…" (Heidegger, 1995, p. 153).

## DISCUSSION AND CONCLUSION

The results of the review conducted in the previous sections show that Ciborra has certainly found in Heidegger a substantive support for both (1) his questioning of consolidated theories and methods, which he considers largely inadequate, and (2) the generation of his 'multi-view' approach to research and to the exploration and understanding of his intuitions and findings. Key concepts and words employed by Ciborra have been found to be closely connected to 'concept-words' important in Heidegger's work, as is shown in specific quotations from relevant texts.

For what concerns the support to his critical evaluations of current theories, the Introduction to *Being and Time* has served as a methodological beacon to contrast the shortcomings of what he calls Cartesian or natural sciences based approaches that 'reduce' the complex reality of organizations (thus working as a 'hardened tradition' that has to be 'loosened up'), as opposed to the richness (and therefore the explanatory potential) of 'being-in-the-world.' Also, some specific concepts like 'illusory appearance' ('semblance') are employed which point out the *ex post* rationalizations of top management that 'cover-up' the "muddling through" which characterizes the everyday reality of organizations. The complex notion of *Gestell*, as an 'ordering' character and essence of contemporary technology, helps him explain the 'call to devise methodologies' as an everlasting (and apparently ever failing) attempt to control the evolving of information infrastructures through largely unrealistic management agendas.

The consequence of this unsatisfactory state of 'traditional' approaches is the underestimation or misinterpretation of 'apparitions' in organizations and IS research and practice. Here the supportive role of Heideggerian philosophy to some of Ciborra's findings emerges clearly. The notion of 'appearance' as the way to 'uncover' phenomena such as ***bricolage*** is relevant to understand and manage IS design and implementation. The basic function of moods (as opposed to mere rationality) he uses to explain improvisation (increasingly important in managing uncertainty) stems from Heidegger's concepts of *Befindlichkeit*, anxiety, boredom, and 'moments of vision.' *Gestell*, once again, and *Gelassenheit* are important cornerstones for the construction of alternative management agendas and policies. On the whole, some of Heidegger's key conceptions helped Ciborra in his research within the unusual, within the unexpected elements that can facilitate awareness, authenticity, and therefore change.

Ciborra's 'escalation to philosophy' appears therefore to be strongly grounded in an in-depth knowledge of the philosophical approach he drew upon. In his last published work he recommends "authentic commitment" when dealing with crucial and controversial issues so that references to philosophical elaborations are "not just a lip service to the founding fathers" (Ciborra, 2006, p. 130). The difficulty in understanding the specific and sometimes awkward language of philosophy—as the story of the interaction of Ciborra with Heidegger has probably shown—discourages unnecessary and possibly superficial "escalations" to philosophical grounds. At the same time though, the propagation and the ever growing pervasiveness of IT-IS and the consequent increasing need for appropriate policies and strategies—ample, effectively designed, and based on sound research—call for further explorations of philosophical territories.

In fact, for example, the appearing of new applications and the possibility of combining them to form different instantiations of information systems call for new possible fields of investigation. For example, Individual Information Systems (IIS) have recently been identified as a novel research area (Baskerville, 2011). IIS have been defined as activity systems "in which individual persons, according to idiosyncratic needs and preferences, perform processes and activities using information, technology, and other resources to produce informational products and/or services for use by themselves or others" (Baskerville, 2011, p. 2). Given the two billion Internet users (and their number has been growing at a very high rate), the IIS are likely to become an increasingly significant aspect of the overall information systems world. Two are the aspects of this emerging type of IS that are useful for the discussion developed in this chapter: (1) there is a myriad of hardware and of software packages (that are often combined on a trial and error basis by persons not educated in information technology or systems) plus the access to services via the Internet that are employed by the individual to respond to his/her individual requirements concerning leisure and business activities; (2) individual IS are also present at the work place and they differ from the IS designed to respond to organizational needs because they satisfy the individual specific work related information necessities. The first aspect responds to a phenomenon that was highlighted by Ciborra and mentioned above in this chapter: the practice of ***bricolage***. The second one points to drift, a concept developed by Ciborra (and mentioned earlier in this chapter) who is cited by Baskerville (2011, p. 7). The latter recalls that the individual practices and changes—which affect implementation and cause its drifting away from the originally planned functions—actually are able to shape change in organizations so that attention to IIS is one perspective to understand the evolution of larger IS.

Two further comments can thus be made. First, "encountering the everyday world" is the phenomenological anchor that Ciborra indicates for a more effective IS research activity and practice that was cited in the introduction to this paper and it is becoming increasingly helpful exactly because of the multiplication and of the spreading of information technology artefacts (as in the case of individual IS). Second, if new, emerging phenomena are to be described and interpreted, change of perspectives are necessary so that the members of the IS community and practitioners are confronted with unexpected proposals, analyses, and cases. It is a matter of awaking interest in the form of doubts on assumptions that are in current use and on the implications that experiencing new conjectures have on the present understanding of phenomena or for the discovery of new ones. This last remark is by Weick (1989) as he points out the critical role played by emotions in theorizing and as he mentions the work by Davis (1971) on interesting theories. The latter author notices that "the defining characteristic of a theory which some audience considers interesting is that it stands out in their attention in contrast to the web of routinely taken-for-granted propositions which make up the theoretical structure of their everyday life" (Davis, 1971, p. 311) and further on: the "'taken-for-granted world' includes not only this theoretical dimension but also a practical dimension as well. A theory will be considered truly interesting only if it has repercussions on both levels" (Davis, 1971).

Therefore, it seems that Ciborra has indeed worked his way through the IS research dominant paradigm in an 'interesting' manner, drawing rigorously on Heidegger's work with relevant implications for both IS theory and practice.

## REFERENCES

Baskerville, R. (2011). Design theorizing individual information systems. In *Proceedings of the Fifteenth Pacific-Asia Conference on Information Systems (PACIS 2011) on Quality Research in Pacific Asia.* PACIS.

Baskerville, R. L., & Myers, M. D. (2002). Information systems as a reference discipline. *Management Information Systems Quarterly, 26*(1), 1–14. doi:10.2307/4132338

Chen, W., & Hirschheim, R. (2004). A paradigmatic and methodological examination of information systems research from 1991 to 2001. *Information Systems Journal, 14*, 197–235. doi:10.1111/j.1365-2575.2004.00173.x

Ciborra, C. (2002). *The labyrinths of information.* Oxford, UK: Oxford University Press.

Ciborra, C. (2006). The mind or the heart? It depends on the (definition of) situation. *Journal of Information Technology, 21*(3), 129–139. doi:10.1057/palgrave.jit.2000062

Ciborra, C., & Hanseth, O. (1998). From tool to gestell - Agendas for managing the information infrastructure. *Information Technology & People, 11*(4), 305–327. doi:10.1108/09593849810246129

Davis, M. W. (1971). That's interesting. *Philosophy of the Social Sciences, 1*(4), 309–344. doi:10.1177/004839317100100211

De Marco, M. (2005). A portrait of a scientist. *European Journal of Information Systems, 15*(5), 442–445. doi:10.1057/palgrave.ejis.3000553

DeLone, W. H., & McLean, E. R. (1992). Information systems success: The quest for the dependent variable. *Information Systems Research, 3*(1), 60–95. doi:10.1287/isre.3.1.60

Dreyfus, H. L. (1991). *Being-in-the-world.* Cambridge, MA: The MIT Press.

Dreyfus, H. L. (1992). *What computers still can't do.* Cambridge, MA: The MIT Press.

Dreyfus, H. L. (2003). Being and power revisited. In Milchman, A., & Rosenberg, A. (Eds.), *Foucault and Heidegger: Critical Enconters.* Minneapolis, MN: University of Minnesota Press.

Gendlin, E. T. (1978). Benfindlichkeit: Heidegger and the philosophy of psychology. *Review of Existential Psychology & Psychiatry, 14*(1-3), 43–71.

Heidegger, M. (1962). *Being and time.* Oxford, UK: Blackwell.

Heidegger, M. (1966). *Discourse on thinking.* New York, NY: Harper and Row Publishers.

Heidegger, M. (1982). *The basic problems of phenomenology.* Bloomington, IN: Indiana University Press.

Heidegger, M. (1992). *Gelassenheit.* Pfullingen, Germany: Neske.

Heidegger, M. (1993a). The question concerning technology. In Krell, D. F. (Ed.), *Basic Writings.* London, UK: Routledge.

Heidegger, M. (1993b). The way to language. In Krell, D. F. (Ed.), *Basic Writings.* London, UK: Routledge.

Heidegger, M. (1993c). The end of philosophy and the task of thinking. In Krell, D. F. (Ed.), *Basic Writings.* London, UK: Routledge.

Heidegger, M. (1994). *Bremer und freiburger fortraege.* Frankfurt, Germany: Klostermann.

Heidegger, M. (1995). *The fundamental concepts of metaphysics.* Bloomington, IN: Indiana University Press.

Introna, L. D. (2005). Claudio Ciborra's way of being: Authenticity and the world of information systems. *European Journal of Information Systems, 15*(5), 513–517. doi:10.1057/palgrave.ejis.3000574

Kallinikos, J. (2005). The spirit never dies. *European Journal of Information Systems, 15*(5), 467–469. doi:10.1057/palgrave.ejis.3000582

Lee, A. S., & Baskerville, R. L. (2003). Generalizing generalizability in information systems research. *Information Systems Research, 14*(3), 1–14. doi:10.1287/isre.14.3.221.16560

Martensson, P., & Lee, A. S. (2004). Dialogical action research at omega corporation. *Management Information Systems Quarterly, 28*(3), 507–536.

Moran, D. (2000). *Introduction to phenomenology*. London, UK: Routledge.

Orlikowski, W. J., & Baroudi, J. J. (1991). Studying information technology in organizations: Research approaches and assumptions. *Information Systems Research, 2*(3), 1–28. doi:10.1287/isre.2.1.1

Weick, K. E. (1989). Theory construction as disciplined imagination. *Academy of Management Review, 14*(4), 516–531.

Whitley, E. A. (2005). Visiting the red light zones with Claudio. *European Journal of Information Systems, 15*(5), 477–479. doi:10.1057/palgrave.ejis.3000554

Zuboff, S. (2005). Ciborra disclosed: *Aletheia* in the life and scholarship of Claudio Ciborra. *European Journal of Information Systems, 15*(5), 470–473. doi:10.1057/palgrave.ejis.3000580

## KEY TERMS AND DEFINITIONS

**Bricolage:** is a set of informal activities that ”diverge from the formalized, pre-planned ways of operating, their outcome may well lead to serendipity, to the possibility of finding something valuable that was not sought for at the outset.” (Ciborra, 2002, p. 48).

**Dialogical Action Research:** ”takes notice of and addresses heterogeneity in the forms of knowledge held by the scientific researcher and the real-world practitioner, where the heterogeneity is related to what phenomenology would call ‘the scientific attitude’ taken by the researcher and ‘the natural attitude of everyday life’ taken by the practitioner” (Martensson & Lee, 2004, p. 508).

**Drift:** is a term indicating that “technology very often seems to shift away, to a greater or lesser extent, from the designed tracks and pre-defined goals.” (Ciborra, 2002, p. 88)

**Hermeneutic Circle:** is the process of going back and forth in a text, from details to an overall view, for an increasing understanding of it.

**Induction:** “refers to a reasoning process that begins with statements of particulars and ends in a general statement. Reasoning from data points in a sample to an estimate of a population characteristic is an instance of induction.” (Baskerville & Myerse, 2002, p. 224)

**Informations Systems:** are “complex social-technical phenomena .... These systems are fundamentally a type of human-computer system that positions [information and communication technologies] in a human context.” (Baskerville, 2011, p. 2)

**Individual Informations Systems:** are activity systems ” in which individual persons, according to idiosyncratic needs and preferences, perform processes and activities using information, technology, and other resources to produce informational products and/or services for use by themselves or others.” (Baskerville, 2011, p. 2)

**Phenomenology:** The authors belonging to the phenomenological movement have proposed different definitions; as Heidegger stated: ”There is no such thing as *the one* phenomenology...” (Heidegger, 1982, p. 328). Because of the case presented in this chapter, phenomenology is defined as .”..the science of the Being of entities−ontology.” (Heidegger, 1962, p. 61)

**Phenomenological Information Systems Research:** with respect to the dominant positivist approaches, can be viewed as “a different perspective on information systems [that] should be anchored to the unforlding of the human process of encountering the everyday world” (Ciborra, 2002, p. 6)

**Positivist Information Systems Research:** “Ontologically, positivist information systems researchers assume an objective physical and social world that exists independent of humans, and whose nature can be relatively unproblematically apprehended, characterized, and measured.” (Orlikowski & Baroudi, 1991, p. 9)

# Chapter 3
# Heidegger's Notion of Befindlichkeit and the Meaning of "Situated" in Social Inquiries

**Kenneth Liberman**
*University of Oregon, USA*

## ABSTRACT

*This chapter is based on a talk presented to the Alpis Information Systems 2009 Annual Conference Carisolo, Italy. It examines the role of phenomenology, most importantly Heidegger's notion of Befindlichkeit, in social situations. In discussing the meaning of "situated," the chapter covers the importance of in situ studies, the problems with over-conceptualizing and the limits of Conoscenza Teoretica, Befindlichkeit, itself, and the limits of formal analysis.*

Four months ago Gian Marco Campagnolo sent me an email inviting me to speak about phenomenology to the group here, and I emailed to Gian Marco, "Yes, I know a good deal about phenomenology, but I know nothing about Information Systems: how can I learn what would be relevant to say to them? I said I was interested because I had a long friendship with the chair of our Computer Science Department and also I had three colleagues that are ethnomethologists—Lucy Suchman, Jack Whalen, and Marilyn Whalen—and who have done interesting research for Xerox about how people actually use Xerox machines. I am aware of their work and I am also aware that I should know more.

I have, however, been spending most of the last twelve years studying the practices of reasoning of Tibetan monks, including spending three years in Buddist monasteries, so I am not very up to date on what information systems research has been accomplishing, so I asked Gian Marco

DOI: 10.4018/978-1-4666-0303-5.ch003

again, "How can my participation be useful?" and he replied that there is a group very interested in phenomenology. He sent me an article by Claudio Ciborra to introduce me to the interest of ALPIS in phenomenology, and so I said yes. It was then that he informed me that I was to speak the first night. I rejected that idea, saying that if someone does not know what the people are thinking, you want him to speak late in the program. I added that I was certain there would be something of relevance I could say after listening to all of the papers, but it was an unreasonable expectation for the first night.

Then I began reading Ciborra's article, "Getting to the Heart of the Situation: The Phenomenological Roots of Situatedness," and I ran across this passage from pages 5-6:

*References to phenomenology are often made, but never quite fully explored and exploited. Collateral aspects are mentioned, such transparency, ready-to-handedness, and so on. Yet nobody quotes Section 29 of Being and Time, where Heidegger (1962, pp. 172 – 182) introduces the notion of situatedness (Befinlichkeit), contrasting it with the privileged role attributed then (and now) to understanding, cognition, and the purely mental. ... Lack of proper references to phenomenology while using its ascendance may also induce the reader not versed in philosophy to believe that what these authors say about situatedness is indeed all that phenomenology has had to say on the subject.*

So I considered; if they want to know more about phenomenology and one of their founding thinkers has recommended reading Section 29 of Heidegger's *Being and Time*, then instead of offering more talk *about* phenomenology, our meeting could be an occasion for reading some phenomenology, and I could think of no better selection from Heidegger than the Section 29 cited by Ciborra.

Ciborra's paper "Getting to the Heart of the Situation," could be re-titled, "How did *Befindlichkeit* come to be 'Situated,' and what has it lost along the way?" Ciborra complains about what the Americans, like Suchman, do with Heidegger when they undertake "situated studies." I have some sympathy with Ciborra's lament here. As a social phenomenologist, I can say that I have spent much of the past four decades wincing whenever I heard or read sociologists, anthropologists, linguists, etc. use the term "phenomenological." During this time, the thin, shallow use of the term "phenomenology" is probably the principal reason I have rejected the manuscripts that I review for various journals and publishers.

So, let us examine Heidegger's section on "Being as Attunement." My discussion is divided into four parts:

1. What is the big deal about *in situ* studies?
2. The problem with over-conceptualizing, and the limits of *Conoscenza Teoretica.*
3. *Befindlichkeit*
4. The limits of formal analysis

## WHAT IS THE BIG DEAL ABOUT *IN SITU* STUDIES

How can something so subjective have had so much influence? For the significant reason that the way that most of our models represent the world is deeply flawed. This includes not only our models as analysts but also the lay models that ordinary people and professionals alike employ in their everyday life.

Our lives are lived subjectively, and our models do not account for the way we actually live our lives. They idealize matters, and the real work is made invisible; worse, the real work is obscured by the aggressive employment of our most cogent models. Worse still, the more cogent the models are made, the more obscuring they do. The actual social medium of our ordinary work of organizing local situations is rarely visible to the analyst, *or to the practitioners*, in just the way that water is unnoticed by the fish who swim in a stream. One of the reasons is that the influence of the Enlightenment survives today in the form of a compulsion to make things out to be more definite and certain than they really are, and here even the realists are lost within an idealism.

There is no fact, no "objective" that is not accompanied by an interpretation; therefore, there can only exist a subjective objective. There is objectivity, but it is always an interpreted objectivity. I ask my students sometimes, "What is the difference between a fact and an interpretation?" They have all these pop theories that they offer me, but by the end I manage to convince them that most facts are interpretations. We have a great need for objectivity, not only for pursuing reliable knowledge but also for making it possible to communicate with each other; but the objectivity we need is always and necessarily a subjective objectivity. You cannot find an objective objectivity. If you pretend that you have one and deny the subjective aspects of objectivity, then by denying the way the world really is you are in fact being less "objective" and more prejudiced than those who recognize the subjective role of understanding things. This is in fact "the Crisis" that Husserl writes about in his last great work, *The Crisis of the European Sciences, and Transcendental Phenomenology.* The "transcendental" here is a reference to the co-presence of subjectivity and objectivity.

More importantly, our lives are much more complicated than the purveyors of planful thought and rational choice theory would have it. The complexities of any ordinary life *in situ* renders most of the modeling of cognitive scientists foolish. There is time here just to mention one reason why, and that is the *reflexivity of understanding*: understanding is rarely the deductive process that Isaac Newton would want to talk about. Reflexivity refers to practices that *at the same time* describe *and* constitute a social framework because we have the ability to find whatever we describe—or rather, our practices of understanding constitute the framework *while* they describe it. That is, the describing is the constituting—the practice of employing an interpretation in our practical lives is what constitutes the framework, as it describes that same situation. Basically, our practices of understanding are finding themselves. The understanding is always describing itself, and I wish to direct you not only to researchers like ourselves, although it is certainly true for us, I am speaking of people in their ordinary lives.

In the case of the research of Lucy Suchman, I am talking about the people who use Xerox machines and try to understand the latest version of instructions on the LCD screen that the machine displays. People have no choice but to employ whatever understanding they have of xeroxing by fiddling around with the displays on the touch-screen and pushing buttons or whatever they think it is going on. No matter what, they are going to read into the situation what they think they know, and find a way to get the work done, and in most cases the work will get done, but there will be a great deal of serendipity to it. Most interestingly, the way they get the work done on the Xerox machine may never have been anticipated by the people who designed the machine or those who designed the instructions. The former Secretary of Defense of United States might have included what Xerox's designers didn't know among the "unknown unknowns." Reflexivity is that feature of comprehending some structure of social action that presupposes, *while it provides and provokes*, the conditions that make its own intelligibility sensible. Phenomenology is very much taken up with the practical tasks of finding and maintaining the intelligibility of local situations *in the course* of being in them, i.e. the *in situ*.

Most planful accounts miss the reflexivity of quotidian life, which is the moment-by-moment adjustment and feedback between situation and reflection, something so spontaneous that it cannot be predicted in advance. Accordingly, oftentimes planful accounts are unable to locate the real problems that people have to face, and so we get instruction manuals that no one is able to understand.

All this was something of an embarrassment to the organizational theory used by information systems researchers. So talk of "situated" studies began, in many instances led by ethnomethodological research projects that located the "troubles" that were to be found in local occasions, projects that exposed to analysts the work of the reflexivity of understanding.

Because of reflexivity and other phenomena like it (such as the indexicality of meanings) textbook versions rarely reflect reality. Let me give you three illustrations of what I am talking about:

1. The Oregon state highway department has a planning division that handles all the sign painting, traffic routing, painting of lanes, the setting up of signal systems for traffic flows of automobiles, et cetera, in the State of Oregon (USA). Around 1986, they came to the conclusion that all good ideas should be "put on the back-burner," which is American slang for "No matter how brilliant your theory sounds: don't implement it until you actually do a pilot study." The pilot studies are instituted for a short period of time (three – nine months is typical) in a test location before initiating more widespread application that could create messes all over the state that would then have to be cleaned up. The Oregon state highway department has learned, as part of their practical work, never to administer a new policy of lane-painting, sign-posting, signal systems, traffic-routing, etc. without first implementing it on a trial-only basis. This is for the very good reason that they have had long experience with the unanticipated consequences of their previous brilliant planning that created disasters that they were forced to undo. There are *so many* of these unanticipated consequences that they concluded that these consequences are *not anticipatable.*
2. The people who write computer programs have similarly learned not only that they cannot predict where the bugs in the program will occur, they also cannot predict what clever things they have devised, until the people who use the beta versions report back to them. The beta versions offer opportunities to tell the designers what they have designed. These reports on beta versions even teach them how they should market the programs and which people and groups to market it to. They find some bugs but they also learn what it was they really achieved. There is now a universal reliance on beta versions, and I interpret that as a frank admission that you cannot plan for everything.
3. The man who worked for the 3-M company who invented post-its was reading a technical book on an airplane and kept losing the place in the book where the footnotes were. So when he returned to the lab, he invented post-its. He didn't have the slightest idea he would change how every office in the world worked, and would also change the face of every refrigerator on earth; but he is happy to accept the credit for his wonder just the same.

These illustrations give you the idea that "situated" studies are oriented to discovering what the plans did not plan for, and they can only be discovered by going out into the world and looking. That is, you cannot recover it by restricting yourself to a review of your best theorizing. This is what the big deal about "situated" studies is about: situated studies are oriented to discovering what the plans didn't plan for.

## THE PROBLEM OF OVERCONCEPTUALIZING OR THE LIMITS OF *CONOSCENZA TEORETICA*

What Heidegger calls "Attunement" (*Il Trovarsi*) is one of the fundamental three existential facts of life of what he calls being-in-the-world—the other two fundamental existential facts of being-in-the-world are "Understanding" and "Discourse."

Attunement is the name for what is *being* more than *knowing*: attunement does not *know* why (Heidegger, 1996, p. 127) it is … *"Non si sa. Sono cose che l'esserci non può sapere"* (Heidegger, 2006, p. 134 & 389). The idea is that we already are in a situation doing something *before* we *know* about it, and the problem with cognitive scientists as well as with rational choice theorists in sociology and political science is that they have a model for how human beings act that examines phenomena that occur only from the chin up. For them people do have bodies, which is the point of Ciborra's "Heart" in the title of his work.

I have spent five years in India, and I have read a good deal of classical Indian scholarship, not only Tibetan scholarship in Tibetan but also a lot of Vedanta epistemology. India has a long tradition of scholarship going back before the time of Christ. They are very rigorous, almost too technical, but they are oriented to trying to find out about *being*, and they try to keep their *knowing* about being from preventing their winning reflective access to *being*. They have been dealing directly with the very problem I am addressing, and it is quite interesting to consider their work in this area. I think this is part of the reason that many philosophers of the West do not consider what Indians do to be "philosophy." For most European scholars, and for many of my colleagues, philosophy properly resides above the chin. When I once confronted a brilliant Sanskrit-speaking Vedanta scholar in India over a luncheon table, asking him whether he felt offended that Western scholars do not consider what he does to be philosophical, he told me, "They are right. It is not philosophy. We are not interested in philosophy: it is too mechanical and heartless a pursuit."

Derrida is following Heidegger's lead when he keeps seeking ways to infect his inquiries with what he calls "non-knowledge," and we social scientists need to do the same, since there is so much going on in social organization that is not conceptual. We need to put more effort into studying non-knowledge, which accounts for a majority of what people do.

According to Heidegger, in Section 29, "Discourse does not, as such, mean to be known" (Heidegger, 2006, pp. 127 & 134). And, "The possibilities of disclosure to *cognition* fall far short …" (Heidegger, 1996, p. 1217). He is not saying that there is not discourse or cognition, but there is so much else going on that we need to pay close attention to it as well.

At the same time these passages occur in Section 29, Heidegger sternly warns us against becoming 'touchy-feely.' Having taken a stand against common sense as well as against formal theoretical cognition, Heidegger still insists on *rigorous* attention to the just-what of the actual experience *in situ*. This is what is meant by phenomenology. Heidegger (2006, p. 130) writes, "We must not confuse demonstrating the existential-ontological constitution of cognitive determination in the attunement of being-in-the-world with the attempt to *surrender science to feeling.*"

*"Non si vorrà scambiare la dimostrazione esistenzial-ontologica che il determinare conoscitivo si costituisce nel trovarsi nell'essere nel mondo, col tentativo di consegnare onticamente la scienza nelle mani del sentimento".*

This rigor, our rigor, consists of paying attention only to "evidence" (German: *Evidenz*, Italian *Evidenza*), which is one of the basic notions of Husserl's program of rigorous inquiry. Formal analytic accounts *miss* this *Evidenz* due to the myopia created by their continuous preoccupation with their theorizing. Evidence grounds understanding: it is what comes *first*.

Heidegger also tells us, "Disclosedness does not mean 'to be known.'" We cannot reduce our experience to conceptual knowing. One must already have found oneself in a situation when one commences to know. We need to study the *how* of this *finding ourselves* as well as studying the knowing, and our problem is that we are studying only the knowing. Heidegger tells us that we are not to minimize the *Evidenz* of attunement "by measuring it against the apodictic certainty of the theoretical cognition" (Heidegger, 2006, p. 128). I do not have time, but I could give you the wonderful discussion that Emmanuel Levinas has provided us about the itinerary of ethics and morality in Western philosophy and how they have lost their connection to evidence and have been turned by philosophers into a strictly logical and formal analytic enterprise. So I will just mention it, rather than giving you the details on this occasion.

Merleau-Ponty, who is probably the phenomenologist who is most faithful to Husserl, expanded these inquiries in his project of non-dualistic reflection, especially in the amazing book he wrote at the end of his life, *Le Visible et L'Invisible*, where he explains his notion of "*sur-reflection.*"

*We are catching sight of the necessity of another operation besides conversion to reflection, more fundamental than it, of a sort of sur-réflection [that] would not lose sight of the brute thing and the brute perception and would not finally efface them, would not cut the organic bonds ... [of] our mute contact with the world when they are not yet things said.... It must plunge into the world instead of surveying it (Merleau-Ponty, 1969).*

Heidegger said something similar in Section 29: "Theoretical looking at the world has always flattened it down to the uniformity of what is purely objectively present." (Heidegger, 1996, p. 138). The Italian (Heidegger, 2006, p. 401) is "Lo sguardo teoretico ha già sempre schermato il mondo sull'uniformità del mero sottomano..." That is, there is a leveling off that theory does that is the "bottleneck" of standardization that Jannis Kallinikos has spoken about.

Heidegger is sophisticated enough to recognize that theoretical cognition brings benefits as well as constraints. Continuing the Italian, "... un'uniformità dentro quale, certo, è contenuta la nuova ricchezza di tutto quanto può essere svelato da un puro determinare" (Heidegger, 2006). Or in English, "... although, of course, a new abundance of what can be discovered in pure determination lies within that uniformity" (Heidegger, 1996). That is, it is also to be appreciated that there is an acknowledgement of the brilliance of formal analytic theorizing, a brilliance that we do not want to surrender. So the incredible situation that we are in as *Homo sapiens* (the humans who know) is how do we use formal knowing *without* letting it create a prison that we cannot escape. It seems there is more sapience that we *Homo*-s need to do. As human beings we are still at our task.

What is this abundance, this *richezza di coniscenza teoretica* that "pure determination" brings us? And why is it that formal reason brings such abundance at the same time that it closes us off to the complexity of real events? We are not dispensing with formal analytic reason here; we are only teaching ourselves how to use it more wisely.

Professor Giolo Fele and I are undertaking a study of coffee tasting. In particular we are examining how *assaggiatore di caffè* organize the intelligibility of the formal coffee descriptors they use. The formal terms they use, like "rich," "medium bodied," "acidic," "rotondo," fiorito," "vellutato," etc. all permit them to locate tastes and stabilize the intelligibility of their sensory experience. We are also studying the benefits and limits the use of quantitative scales for tastes play in the coffee industry. We are doing a real phenomenology of coffee tasting. We have already discovered that alongside making it possible for them to find, purchase, and verify shipments of first-rate coffee, these terms affect the tasting. They close off some tastes, which have a difficult time being noted, while other tastes receive abundant scrutiny, to the point that the tongue is taught a great deal about how to find and distinguish precise flavors.

One of Italy's principal *assaggiatore di caffè* speaks of the tasting card in *both* positive and pejorative terms: "The card certainly plays an important role in guiding the judging procedures that apply. However, we must remember that it is only a tool, and the taster is responsible for recognizing and evaluating organoleptic characteristics. ... The second [card] is based on thorough research aimed at modernizing the card by insisting that sensorial analysis is conducted with strict rules" (Odello, 2007). In Italian:

*Non disigna mai dimenticare che essa è solo uno strumento da capire e da utilizzare, la rilevazione delle caratteristiche organolettiche, la loro elaborazione in giudizi e la successiva espressione è di pertinenza del degustatore che mai potrà scaricarsi di questa responsibilità ... la seconda [scheda], realizzata sulla scorta di una forte ricerca volta all'innovazione e in considerazione delle severe regole dell'anilisi sensoriale (Odello, 2007).*

The seeming contradiction here is *not* a flaw in the procedure—formal analyses always and *necessarily* provide "*elaborazione in giudizi*" while at the same time cannot be made to substitute for the "*responsabilità*" to actually taste. As professional thinkers, we are continually trying to accomplish this very thing.

## *BEFINDLICHKEIT*

*Evidenz* demands that we first taste, and *Befindlichkeit*, or Attunement, pays respect to an understanding that is more doing than it is knowing. It involves rigorous discipline, and just because the discipline is in life and not in words does not mean that it is not vital. "Attunement" is rendered in Italian as *trovarsi*, which is a reflexive verb. Reflexive verbs are really splendid things, because they already undermine the human conceit that we are always in control of affairs. Take a verb like *annoiarsi*, "to become bored." The logic of reflexive verbs better represents the actual situation than the non-reflective English.

In the case of *Befindlichkeit*, it is a state one finds oneself in without any deliberate doing. As Heidegger (1996, p. 129) tells us, "[Mood] never comes from 'without' nor from 'within,' but rises from being-in-the-world itself." Here Heidegger is referring to times when we grow tired of ourselves, when our being has become manifest to us as a burden, such as when we are bored.

A closely related term that Heidegger uses is *Umsicht*, or "circumspective attention," "*la circumspectio*" which is not quite "*sapere,*" "*conoscenza,*" or "*conosciuto.*" This is a preliminary taking of bearings that people do to find a context before they settle into it. It implies a broad sweep, and Heidegger contrasts it with the more invasive, paternalist strategies of technological manipulation.

## THE LIMITS OF FORMAL ANALYSIS

Much of the foundational experience that Heidegger is describing is not readily made the subject of formal investigations, at least not without distorting its basic character beyond recognition. Ciborra (2004, p. 12) writes of Lucy Suchman's findings:

*Her empirical study confirms that the organization of situated action is an emergent property of the moment-by-moment interactions between actors and their relevant environments. Expert systems are built on a planning model of human action. 'The model treats a plan as something located in the actor's head.'*

The proponents of "planning models," which includes artificial intelligence designers and rational choice theorists but excludes the Oregon state highway planners, try to resolve difficulties by attempting to "*embed into expert systems more and more sophisticated plans*" (Ciborra, 2004).

An exclusive reliance on more expert systems loses sight of Heidegger's discovery that Dasein finds itself in a situation before beginning to reflect analytically, that a world *already* "matters to it" (Heidegger, 1996, p. 129) ("potere-essere-toccato," Heidegger, 2006, p. 399) before knowing. And this is where occurs most of the confusions experienced by the users of Xerox machines that are documented by Suchman (1987).

Dasein has always found itself "always already" (Heidegger, 1996, p. 128; "già sempre trovato," Heidegger, 2006, p. 391) in a finding which does not come from a direct seeking (trovato in un trovare che non scaturisce tanto da un diretto cercare). Heidegger comes down very severely on this '*diretto cercare,*' but this is just what most of our research consists of! Even my work in ethnomethodology is full of 'diretto cercare,' which will increase the likelihood that I miss the dimension that is most important, at least according to Heidegger.

The more that formal analytic investigators inspect a situation the less they will see. Heidegger (2006, p. 127) observes, "The that of facticity is never to be found by looking." The Italian is, "Il fatto-che della fatticità non è mai reperibile in un vedere intuitivo" (Heidegger, 2006, p. 391). That is because looking—un diretto cercare—is already looking for something it has in mind; that is, one already has the frame of the experience built—one's mind is so full of what one already knows that one cannot see what one does not know! Why do we want to do research, if we do not want to find out what we do not yet know? It may be called "positive science," but it is deluded just the same. The Tibetans I lived with have a great name for such delusion: they call it "ignorance" (*ma rig pa*). For Tibetans, it is not what you don't know that makes you ignorant, it is what you *do* know. That is, one's mind is so filled with the certainty of what one does know that there is no room to learn anything new. Especially, there is no room to learn what one does not already know.

There is no better model to be found for the imperialism of planful theorizing than in the social theories of Thomas Hobbes, John Locke, and John Stuart Mill. They offer an extreme rationalism whereby humans are born as separate individuals, come together out of their own free will, and then commence to abstractly negotiate their rights in a social contract. This is nothing more than a "just-so" story about the origin of society. People know nothing more than what these rationalist theorists put into their heads. It may be that people do not *know* more, but they *do* a great deal more than Hobbes et al. are able to account for, and this "more" is fairly well addressed by Heidegger in Section 29.

There is planful action as pre-determined rationalities, and then there is planful action as the actual course of the situated action, as *bricolage*, as *difference*, as reflexivity, and all the other ways we are learning to think—or not to think—about these quotidian activities that are the preoccupation of information system researchers. In his masterpiece *Negative Dialectics*, Theodor Adorno speaks of *mimesis*, a mode of social being that precedes formal reflection, and he considers it more critical to understanding than conceptual knowing (and we should not fail to note here that Adorno was a sworn opponent of Heidegger for all of his professional life).

Heidegger's fecund recommendation is instead of commandeering events, instead of "staring out at something" (Heidegger, 1996, p. 129) with "un guardar fisso"/"*Empfinded oder Anstarren*" (Heidegger, 2006, pp. 137 & 399), we should regard things in "a circumspectfully heedful way" (Il pro-curante ... circonspettivo). In this way, we can listen to the phenomena we are studying and not exclusively take charge of organizing the intelligibility of it in advance, like we are running the railways.

## CONCLUSION

This does not mean we have to leave off of formal analysis, but it does mean that we must first attune ourselves to what is there. Or if we are doing social research, we must first witness how others attune themselves to what is there, *and we must find their there* (not *our* "there"). That is, using Husserl's language, our professional task is to identify and describe what are the horizons of their experience, and especially what are the horizons of their understanding that are at work for the people we are studying. Only then will our professional advice be pertinent to the organizational tasks that people are facing in their quotidian life.

An important question for those who wish to work in a phenomenological way is how do we get access to the non-rational? This is a thorny question since the tools we have to work with are mostly rational tools. So how do we adapt them and make them sensitive to the actual "work" that people are doing in their everyday, practical lives? There is a vital clue for us in the concluding passage of § 29, in which Heidegger writes, "The phenomenological interpretation must give to Dasein itself the possibility of primordial disclosure and let it, so to speak, interpret itself" (Heidegger, 1996, p. 131). 'Letting things be' is the theme Heidegger writes a great deal about in his post-World War II writings, but it is only another way of reciting the principal phenomenological slogan from *Being and Time*, "To the things themselves!"

Our work of making social inquiries is not irrelevant, because we have the important descriptive task, Heidegger tells us, of raising "the phenomenal content of disclosure existentially to a conceptual level" (in Italian, "elevare al concetto l'importo fenomenale cosí dischiuso," Heidegger, 2006, p. 143). That is, our thinking must be kept appropriate to the events we are describing, and we must avoid submerging those events beneath our brilliant plans and theories, but we still need to reflect formally upon them.

So now have our most serious task—how do we teach each other speak objectively about these subjective things?

## ACKNOWLEDGMENT

This chapter is based on a talk presented to the Alpis Information Systems 2009 Annual Conference Carisolo, Italy.

## REFERENCES

Ciborra, C. (2004). *Getting to the heart of the situation: The phenomenological roots of situatedness*. Retrieved from http://projects-final.interactionivrea.org/2004-2005/SYMPOSIUM%202005/communication%20material/Getting%20to%20the%20Heart%20of%20the%20Situation_C_Ciborra.pdf.

Heidegger, M. (1996). *Being and time* (Stambaugh, J., Trans.). Albany, NY: SUNY Press.

Heidegger, M. (2006). *Essere e tempo and sein und zeit*. Milan, Italy: Mondadori.

Liberman, K. (2013). *MORE Studies in Ethnomethodology: organizing the local orderliness*. Albany, NY: State University of New York Press.

Merleau-Ponty, M. (1969). *Visible and the invisible*. Evanston, IL: Northwestern University Press.

Odello, G. (2007). *Espresso Italiano tasting*. Brescia, Italy: Centro Studi Assaggiatori.

Suchman, L. (1987). *Plans and situated actions: The problem of human machine communication*. Cambridge, UK: Cambridge University Press.

# Section 2
# Phenomenology and IT Design

# Chapter 4
# Communities of Practice from a Phenomenological Stance:
## Lessons Learned for IS Design

**Giorgio De Michelis**
*University of Milan, Italy*

## ABSTRACT

*This chapter contributes to the debate about the concept of community of practice considering the latter as a means for understanding the behavior of the people using in their practice artifacts and sharing spaces. The experience they share makes them a community and, when the space they share is a workspace, a community of practice. The unusual way through which one reaches a community of practice, and its being, first, a community, and second, a community of practice, makes it very important to understand what characterizes communities.*

*Community (Gemeinschaft in German) has emerged as a relevant concept for understanding the social dimension of human life, at the end of nineteenth century, when in a famous book by Ferdinand Toennies (1925), it was opposed to society (Gesellschaft in German). The debate that accompanied and followed Toennies' book at the beginning of the twentieth century opposed the irrationality of communities (where no utility value justifies membership) to the rational principle sustaining societies (that are ruled in order to balance costs and benefits of all members). More recently, the concept of community has been again at the center of philosophical debate after its deconstruction by Jean Luc Nancy: it is, therefore, interesting to situate the concept of community of practice within it. What emerges from this analysis offers to designers of ICT-based applications, such as information systems, knowledge management systems, etc., some new hints on the nature of those systems.*

DOI: 10.4018/978-1-4666-0303-5.ch004

## INTRODUCTION

Probably because of its special position among scientific disciplines, between natural and human sciences, informatics has paid a particular attention to philosophy and to its discourse; it has also raised the attention of philosophy, beyond its specialized branch, philosophy of science. This mutual interest has been alive and controversial, generating, on the one hand, sharp discussions with inflamed tones, and on the other, attention and curiosity[1].

The history of the relationship between informatics and philosophy, that began with the origin of computer science, in the forties, in the multidisciplinary environment of the Macy conferences on Cybernetics (Heims, 1980, 1991, 1993), is today rich with contributions and merits a specific attention, going beyond the scope of this chapter. I recall it because, when I began to pay attention to philosophy, as a computer scientist committed to design systems empowering human beings in facing the complexity of their (professional) lives, I discovered that other scholars were looking for inspiration and concepts in the European philosophy of the twentieth century. In particular, I was influenced by *Understanding Computer and Cognition* by Terry Winograd and Fernando Flores (1986), whose work in Computer Science was close to the one of my group and with whom I initiated a scientific friendship. Quickly my attention was focusing on the issue of design: what are the things we build when we design computer based applications, and how can we shape them in such a way that they are capable to give answers to human needs and desires? This work, bringing me to contribute to the birth of interaction design as a new discipline, led to several texts, including: *Aperto, Molteplice, Continuo* (1998), a book in Italian whose subtitle is (my translation) 'Artifacts at the End of the Twentieth Century'; 'The Phenomenological Stance of the Designer' (2009), my contribution to the book *Revisiting Digital Bauhaus* (Binder, Loewgren, Marlborg, 2008); and my coauthorship of *Design Things* (Telier, 2011), the book where I am one of the authors collected under the pseudonym of A. Telier.

In these texts, the focus is on design, from both the point of view of how we do design and what is its outcome. I adopted a 'phenomenological stance' showing how such a viewpoint allows one to go beyond the traditional idea of design as a practice, to capture its complex social nature. In all those texts, the social dimension of design is preeminent, but the social context of design is not discussed directly. This chapter wants to open, for me, this field of study, looking at the concept of community of practice (Lave & Wenger, 1991; Brown & Duguid, 1991; Wenger, 1998; Makoto Su, et al., 2011) as the means for understanding the behavior of the people using artifacts and sharing spaces. With reference to my interest on design, they are those who may be interested in what designers are doing. Design, as the practice through which something new is brought to existence, has several stakeholders, beyond its customers: the people whose space will be modified by the designed thing, the people whose potential for action will be affected by it, the people whose future may be changed by it, etc., and all these people have different interests and viewpoints with respect to the designed things. As a consequence, designers must be able to find a way through them, so that what they design is acceptable and sustainable by all of them. Even if it appears counter-intuitive, it is not a matter of negotiation, since the stakeholders of a design process share a space and the experience of living there together. The experience they share makes them a community and, when the space they share is a workspace, a community of practice. It is somehow unusual looking at communities of practice from the viewpoint of design, but, as a community, a community of practice shares a space that is impacted and transformed by the deployment of the outcome of a design process. The unusual way through which we reach a community of practice, and its being, first, a community, and second, a community of practice, makes it very important for us to understand what characterizes communities.

Community (*Gemeinschaft* in German) has emerged as a relevant concept for understanding the social dimension of human life at the end of nineteenth century, when in a famous book by Ferdinand Toennies (1935) it was opposed to society (*Gesellschaft* in German). *Gemeinschaft* refers to groupings based on feelings of togetherness and on mutual bonds, which are felt as a goal to be kept up, their members being means for this goal. *Gesellschaft*, on the other hand, refers to regulated groups that are sustained by it being instrumental for their members' individual aims and goals. The debate that accompanied and followed Toennies' book opposed the irrationality of communities (where no utility value justifies membership) to the rational principle sustaining societies (that are ruled in order to balance costs and benefits of all members).

In the next section, I review the CoP concept, as it emerges from the relevant literature on the subject, underlining its open questions; later I report some new contributions by contemporary philosophers, who have revived the debate, and I discuss what the new approach to community tells us about CoP. Finally, I discuss what all this tells to designers of ICT based applications, as information systems, knowledge management systems, etc. Some conclusive remarks end the chapter.

## THE CONCEPT OF COMMUNITY OF PRACTICE

The concept of Community of Practice (CoP) was introduced at the end of the eighties by Jean Lave and Etienne Wenger in order to define how apprenticeship took place (a rich account of the origins of communities of practice can be found in Makoto Su, et al., 2011, where I found most of the information I recall here). Its role was, probably, that of reorienting the research program of the Institute for Research on Learning (IRL), that, under the influence of AI people, looked for the creation of ICT based systems able to coach people. Lave and Wenger wanted to show how apprenticeship was inherently social and, therefore, not reducible to the interaction between the apprentice and a (intelligent) computer program. This explains the focus of the book introducing CoP (*Situated Learning: Legitimate Peripheral Participation*; Lave & Wenger, 1991) on the way newcomers learn how to become a member of a community of practice and, therefore, to practice a craft or a profession, but CoP appeared to be something more than the right context to understand apprenticeship, emerging as a new concept offering the right context for developing a social theory of learning.

As a matter of fact, the book of Etienne Wenger (1998), that can be considered the first systematic presentation of the concept, opens with four premises on the social nature of learning: "1) We are social beings. … 2) Knowledge is a matter of competence with respect to valued enterprises. Such as singing in tune, … 3) Knowing is matter of participating in the pursuit of such enterprises, that is, of active engagement in the world. 4) Meaning—our ability to experience the world and our engagement with it as meaningful—is ultimately what learning is to produce." (Wenger, 1998, p. 4). Learning is therefore strictly coupled with social participation, where participation refers to being an active participant in the practices of a community and to constructing identities in relation to these communities.

As a consequence, the social theory of learning Wenger proposes integrates the four components characterizing social participation as a process of learning and of knowing: "1) Meaning; 2) Practice; 3) Community; 4) Identity" (Wenger, 1998, p. 5): meaning allows us to give sense to our experience and to specify our ability to practice; practice allows us to talk about our mutual engagement in meaningful action; community allows to characterize the social context where practice is shared among different people with different competence; identity specifies who we are and become within our communities. For Wenger, communities are not social aggregates we build with an objective: rather they are the social aggregate within which we live: "We all ... belong to several communities of practice" (Wenger, 1998, p. 6), "Communities of practice are an integral part of our lives" (Wenger, 1998, p. 7).

It is natural, therefore, that learning becomes something that should be defined differently for individuals, communities, and organizations: for individuals it has to do with participating in the practices of a community; for communities, it characterizes innovation of practice and opening participation to new members; for organizations, it refers to the support they should give to the communities through which it performs and knows (Wenger, 1998, pp. 7-8).

IRL is a private entity operating in the market and not an academic institution: the professional environment where the concept was developed has, therefore, tried to build on it a new approach to professional education and human resource development. This has immediately led to confusion, since it becomes unclear if CoP is an analytical concept for understanding learning and practice or a powerful tool for developing effective education within organizations or for improving the professional skills of their employees. For many scholars, therefore, CoP is something in between a new concept that may help to renew our understanding of work practice and social life and a typical buzzword for selling some consultancy services (this has widely grown, since CoP has been associated with Knowledge Management and, later, with social networks and Web2.0). This ambiguity emerges, for example, in the principles worked out at IRL (Table 1) and, later at Xerox Business Services (Table 2), characterizing learning.

Differently from Wenger's principles characterizing CoP as the place of social learning, the latter outline some rules organizations should obey to, in order to enhance their performances, oversimplify the CoP concept to transform it into a tool for designing organizational structures and supporting knowledge sharing processes inside the organization and crossing its borders. In any case, I don't blame this fact. We must have a pragmatic approach to consultancy practice: if new concepts allow one to define ways to improve organizational performances and/or to design tools supporting their evolution, then there is no reason to complain. Today CoPs are frequently starving; their members don't perceive them as important contributors to their professional quality and to their identity: curing them may be a necessity, and helping organizations to do it is important. Therefore, we should not blame the development of professional practices based on CoPs. In the meantime, we should also continue to work, at the theoretical level, on the concepts of community and practice to deepen our understanding of the human experience.

In fact, disregarding the problem of its reification, I think that the concept of CoP introduces a new perspective in understanding work and its

*Table 1. IRL: principles for learning (Makoto Su, et al., 2011)*

| |
|---|
| 1. Learning is fundamentally social. |
| 2. Knowledge is integrated in the life of communities. |
| 3. Learning is an act of participation. |
| 4. Knowing depends on engagement with practice. |
| 5. Engagement is inseparable from empowerment. |
| 6. 'Failure' to learn is the result of exclusion from participation. |
| 7. People are natural lifelong learners. |

*Table 2. Xerox business services: principles for learning (Makoto Su, et al., 2011)*

| |
|---|
| 1. Learning is fundamentally social |
| 2. Cracking the whip stifles learning |
| 3. Leaning needs an environment that supports it |
| 4. Learning crosses hierarchical bounds |
| 5. Self-directed learning fuels the fire |
| 6. Learning by doing is more powerful than memorizing |
| 7. Failure to learn is often the fault of the system, not the people |
| 8. Sometimes the best learning is unlearning |

structures, becoming a way of analyzing, in a completely different way from traditional organizational studies, what happens within companies and institutions. In particular, I think that it is quite important that CoP allows one to recognize the social milieu where organizations are formed, as well as the social dimension of learning.

But if we want to develop CoP as a theoretical concept, understanding its implications with respect to the nature of organizations, then we need to read it within the more general concept of community that has influenced our ideas on social experiences.

## COMMUNITY IN RECENT PHILOSOPHICAL STUDIES

As I mentioned in the Introduction, the concept of community has been at the center of sociological and philosophical debate since the turn of the nineteenth century, as the irrational, identity making, social bound opposing the rational rule-based society constituted, for example, by European national states. Despite its parenthood with communism, community became, at the beginning of the twentieth century, a typical word of the conservative, right wing pole, in particular of Christian personalism and fascism. Later, the attention to it declined, probably for its assuming a negative connotation, since 'irrational' community bounds have been used to explain the dramatic events of the twentieth century, where national, racial, political, and/or religious distinctions motivated wars and genocides, but in recent years, some philosophers, further developing the phenomenological perspective, while trying to deal in new terms with the social nature of human beings, rediscovered the concept of community, and its ambiguity, and decided to go at its deepest roots, deconstructing it (Nancy, 1986, 1996; Agamben, 1990; Esposito, 1998). It has been a radical re-visitation of the concept, bringing forth a completely new imagine of communities, totally liberated from any ideology.

In 1986, *La Communauté Désoeuvrée* by Jean Luc Nancy appeared in France, raising immediately a great attention in the philosophical community. The book was quickly translated in German (1988), English (1991), Italian (1992), and many other languages, so that its diffusion has been all over the world. The main point of the book is the consideration of 'being in the world' of Heidegger (*dasein*; Heidegger, 1927) as a 'being with' (*mitsein*; Heidegger, 1927), so that the community becomes 'The real position of existence' (Nancy, 1986, p. 203). Being together with other people is not the situation when a person meets another person and exchanges with her knowledge, experiences, and values: human beings do not *share* the experience of being in the world, they live with other persons, and their individuality emerges from this experience. In fact, as Nancy writes: "There is no singular being without another singular being, and [...] there is, therefore, what might be called, in a rather inappropriate idiom, an originary or ontological 'sociality,' that in its principle extends far beyond the simple theme of man as a social being..." (Nancy, 1992, p. 28). Further on, "Communication is not a bond. The metaphor of the 'social bond' unhappily superimposes upon 'subjects' (that is to say, objects) a hypothetical reality (that of the 'bond') upon which some have attempted to confer a dubious 'intersubjective' nature that would have the virtue of attaching these objects to one another" (Nancy, 1992, p. 29).

Being with other people is constitutive of human existence, and human beings emerge as individuals from this 'being with.' As Nancy says, "Community is not a grouping of individuals, subsequent to the elaboration of individuality itself, because individuality as such cannot manifest but within such grouping" (Nancy, 1992, p. 256) and "Community, and communication, are constitutive of individuality, rather that its opposite (and individuality cannot be, in the final analysis, anything else than a limit of the community). But the community, further, is not an essence of all individuals, an essence that will be given in front of them. Community, in fact, is nothing else than the communication of distinct 'singular beings,' existing as such only through communication." (Nancy, 1992, p. 256). Even sense making, therefore, is not separable from this experience human beings live with other people: "Sense is common, communicating, communicated, in common, by definition. Assuming that my existence is meaningful is what lets it communicate and what communicates it to something other than me. Sense makes my report to me as reported to some other" (Nancy, 1992, p. 211).

From the quotations reported above, it should appear quite clear that Nancy's viewpoint modifies radically the idea of a community: a community is not the outcome of a goodwill of some people wanting to build something in common, neither is it what they build living together; rather, it is the condition of their existence. Whenever a person communicates with other people, with words or through any other medium, whenever she interacts with other people, a community takes form, and, in the same time, she, as a person, emerges together with it. Communication is not a way through which different people with different identities try to share their knowledge of the world; rather, it is the way through which a world emerges between them together with the knowledge they share about it.

Communities are, therefore, where human beings appropriate the world, in both the sense of taking control of it, and of getting knowledge about it. The social construction of communities has to do with recognizing, rather than with building them. Communities emerge in the life of human beings: no one can create any of them!

Since it is in the social experience of 'being with,' in the participation in a community, that human beings appropriate the world, in the same moment they emerge as individuals with a specific identity. The identity of a person has much to owe to the communities she is member of and, therefore, to their persistence and to their granting her the highest degree of appropriation of the world or, in other terms, allowing her an active participation in the process of creating and sharing knowledge.

Whatever we call 'social value' must, after all, refer to the world and its appropriation to the knowledge of the world, and, therefore, communities are where social value is created: when a community declines, when her social life (the interaction of its members) starves, then the creation of social value also starves and the quality of human life deteriorates.

Maintaining her identity, a person needs to also maintain the identity of the communities she is member of, generating the unnoticed paradox that many people believe their identity depends on the identities of the communities they are members of.

Communities live as their members interact with each other, as they create knowledge; otherwise they decline. This possibility is always present, menacing any community, due to both internal and external reasons: internal, because, if a community closes itself—and this happens frequently when communities perceive other people as enemies—then it lacks the access to sources of new knowledge; external, because the world is populated by a multitude of non-isolated communities that in some sense compete for the active participation of their members.

As a consequence, communities are unstable, ephemeral if their members do not keep them alive, if they do not interact with each other, giving rise to streams of knowledge creation and sharing. This explains why human beings have done whatever they can to protect communities, both institutionalizing their membership as well as their roles, rules, and hierarchies, and giving them clear boundaries separating their members from non-members. When sociologists oppose community to society, they oppose the institutions of a community defending its identity to those of a society ruling the coexistence of its members.

We have always to remember that the institutions of a community should not be confused with the community itself: the latter, in fact, is the condition of 'being with' of human beings and eludes human will, while the former are social constructs whose effectiveness depends on their capability to enhance social interaction.

## COMMUNITIES OF PRACTICE REVISITED

Within the above recalled philosophical perspective, the concept of community of practice emerges with new features bringing new light on it and on its relevance for human experience.

First of all, we can immediately underline that the concept of community of practice is coherent with the radical view of Jean Luc Nancy, since it is not a social construct, but where a practice is socially constituted and develops itself. Even if no institution has been created for supporting it (like professional associations, schools, etc.), even if practitioners themselves seem unaware of its existence, around a shared practice there is always a community (of practice), where practitioners share their abilities and with their competences concur to further develop the practice.

This means that when we claim that a community of practice is dead or starving, this is not due to a defect in its design, but to a lack of communication within the community and therefore to a lack of innovation in the practice. Designing and implementing institutions for and within a community of practice (like formal associations, professional schools, rewarding systems, etc.) is surely a relevant action for supporting it, but should never be confused with the creation of it. Their effectiveness cannot be taken for granted: it must be monitored, being ready to change the institutions when they are not able to sustain the community.

From this viewpoint, therefore, a community of practice cannot be created; as stated also by Etienne Wenger (1998), it can only be cured, so that it keeps itself alive. As a matter of fact, what we can do to support a community of practice is helping its members to interact each other.

What challenges the life of a community of practice? Why does it need to be cured?

Answering the above questions requires a deepening of our understanding of the social dimensions of human life.

First of all, let us recall that modernity has multiplied the number of communities in which a person is involved (e.g. a person can be part of one or more communities of practice, of a local community, of a religious community, of one or more communities of interest, etc.) and there is no way for organizing this multiple membership without avoiding conflicts among them (sometimes, the conflict is at the level of the time dedicated to each of them; sometimes, it is more radical, and there is a contradiction between two of more different memberships). In both cases, with respect to any community, the other communities are both generating value (because each member can bring in that community her experiences within them), and noise (because participating in them distracts members from their participation in that community).

Communities exhibit, therefore, a double complexity generated, first, by the tension of one community versus its many members, and second by the tension of one person versus the many communities that she is a member of.

Curing a community means helping its members, when needed, to situate themselves in it, without preventing their capability to transfer into it the experiences they live elsewhere. Since human beings are the main contact surface among communities, and they participate in several communities, then curing communities, means also creating 'continuity' (Brown & Duguid, 1994; De Michelis, 1998) among them, so that a fluid change of situation is possible. A person's identity, in fact, emerges through these movements from the communities she is member of.

Saying that people transfer knowledge between communities is not exact: we should say that they create the conditions for re-producing that knowledge in a different community.

Communities constitute, therefore, the contexts within which every human being lives her experiences. But we have to take into consideration that, in accordance with Nancy, human beings, through communication do not share anything: i.e. within their social experience human beings are not exchanging meanings related with an external reality; what they share is limited to their communication itself, to their linguistic experience. They share the language game (Wittgenstein, 1953); they play within the community, constituting the context of their interaction, creating the possibility of reproducing a linguistic behavior (when a person says 'yes' within a conversation, she is claiming that she could also use the phrase she has just heard, nothing more). Documents constituting the traces of past conversations are therefore of great importance in the life of a community, even if they cannot be considered repositories of the knowledge shared by their members: rather, they make accessible the necessary context for understanding to the people who are not aware of it. Written words are a secondary outcome of conversations, but conversations are almost impossible without making reference to them.

If written words play a fundamental role in the practice of a community, what happens when the conversation involves people not belonging to the same community? If the communicating persons are not participating in the same community, then there is no common language game, no shared experience: how can they communicate, how can they mutually reproduce linguistic behaviors?

Formalization is the move through which human beings are able to interact even when they do not share a context. Formal models—both if they are using heavily, a mathematical language, like many data and process models of computer science, and if they are simple intuitive devices like timetables in transportation systems, key-racks in hotels, etc.—are, in fact, de-contextualized knowledge, since every person can through them reproduce experiments and answer questions.

Modeling, de-contextualizing knowledge, allows one to break the boundaries between communities: Formal models are typically boundary objects (Star, 1989) or, better, they are at the core of any boundary object. This may seem surprising, since boundary objects are frequently underspecified objects allowing a certain degree of ambiguity, but, if we look inside them, they are always limiting their ambiguity with a core formalized knowledge, accessible to all their users (the key-rack example explains well this fact, since it is based on a formal and univocal correlation between guests and room numbers, but also a sketch of a house binding the architect and his customer is based on a univocal representation of some features of the draught building—the number of its floors, windows, rooms, etc. (on this point, the reader can also read Vinck, 2003).

Finally, it has to be noted that the life of a community is not dependent on its capability to conserve its knowledge, but on its capability to create continuously new knowledge. This is perfectly in line with the theory created by Nonaka and Takeuchi (1995) and has strong implications on the way we interpret knowledge management. From what we have said up to now, in fact, we can claim that there is an indissoluble link between communities of practice and knowledge for action, know-how: within organizations, curing the former is a necessary condition for preserving the latter. Not only does the hope of being able to extract knowledge from the heads of people to store it in digital devices not make sense, but storing knowledge is impossible: stored knowledge dies if it is not continuously renovated through human actions and interactions.

## DESIGNING FOR COMMUNITIES

Even if the survey of the concept of communities of practice I did in the previous pages is incomplete and partial, it tells us a lot with respect to the design of information systems and other applications to support organizations. Two major points should be underlined. On the one hand, information systems should avoid the separation between organizational performances, like invoicing, ordering, producing, planning, monitoring, etc. and the practices within which they are embedded. On the other, even if a practice is situated within a community (of practice), we should remember that different human beings participate and contribute differently to any community of practice: this sheds a new light on situatedness of human practice emphasizing its distributed nature.

The first point above proposes a new perspective with respect to Information Systems and, in particular, to Cooperative Information Systems. In 1998 (De Michelis, et al., 1998a, 1998b), I wrote together with several colleagues from Europe and Canada, a Manifesto for Cooperative Information Systems, where we proposed a three-faceted view on them: reading it today, it appears that it contains the fundamental features for conceiving a new type of Information System, capable to accompany the evolution of a company not only reflecting its strategic and organizational changes but also supporting at the same time the process through which changes are discussed, established, and implemented.

The second point above has inspired the design of a new operating system for workstations, *itsme* (De Michelis, et al., 2009), going beyond the desktop metaphor: while current workstations are designed as multitasking devices, *itsme*, reflects a new metaphor—stories and venues—configures a machine accompanying its user in any moment of her life, situating her in the context of the story (and of the related community) within which she is acting and interacting in that moment.

Finally, the two examples quoted above are only two instances of the possibilities opened to designers, if they adopt and develop the conceptual framework I have proposed insofar: it emerges in fact that situated computing is a perspective along which all the interactive applications of ICT can, and need to, be re-designed in a radical new way. Looking to user experiences, focusing on their situations, understanding the relationships between individuals and communities allows one to reinvent applications in such a way that they become capable of supporting users in the management of the complexity of their lives, without aiming to substitute them.

## CONCLUSION

This chapter is just a fragment of a serious development of the concept of communities of practice and of the analysis of the impact it can have on Information Systems and ICT applications. I do not have philosophical ambitions, and I work on these ideas from the viewpoint of an interaction designer willing to develop innovative ICT based systems. Therefore, the theoretical conceptualization will progress together with the systems it inspires.

## ACKNOWLEDGMENT

The author wants to thank the many colleagues, from different disciplines, with whom he has discussed these issues during the last ten years, because they helped him to enrich his understanding of communities and practice.

A particular thank you to the anonymous referees of this paper, whose comments were very useful for improving its readability.

## REFERENCES

Agamben, G. (1990a). *La comunità che viene.* Torino, Italy: Einaudi.

Agamben, G. (1990b). *The Coming Community.* Minneapolis, MN: The University of Minnesota Press.

Binder, T., Loewgren, J., & Marlborg, L. (Eds.). (2008). *(Re-)searching the digital bauhaus*. Berlin, Germany: Springer.

Brown, J. S., & Duguid, P. (1991). Organizational learning and communities–of-practice: Toward a unified view of working, learning, and innovation. *Organization Science*, *2*(1), 40–57. doi:10.1287/orsc.2.1.40

Brown, J. S., & Duguid, P. (1994). Borderline issues: Social and material aspects of design. *Human-Computer Interaction*, *9*(1), 3–36. doi:10.1207/s15327051hci0901_2

Ciborra, C. (2002). *The labyrinths of information*. Oxford, UK: Oxford University Press.

De Michelis, G. (1998). *Aperto, molteplice, continuo*. Milano, Italy: Dunod Italia.

De Michelis, G. (2008). The phenomenological stance of the designer. In Binder, (Eds.), *Re-) Searching the Digital Bauhaus* (pp. 145–162). Berlin, Germany: Springer.

De Michelis, G., Dubois, E., Jarke, M., Matthes, F., Mylopoulos, J., & Papazoglou, M. … Yu, E. (1998). Cooperative information systems: A manifesto. In M. P. Papazoglou & G. Schlageter (Eds.), *Cooperative Information Systems: Trends & Directions*, (pp. 315-363). New York, NY: Academic-Press.

De Michelis, G., Dubois, E., Jarke, M., Matthes, F., Mylopoulos, J., & Schmidt, J. (1998). A three-faceted view of information systems: The challenge of change. *Communications of the ACM*, *41*(12), 64–70. doi:10.1145/290133.290150

De Michelis, G., Loregian, M., & Moderini, C. (2009). Itsme: Interaction design innovating workstations. *Knowledge, Technology & Policy*, *22*, 71–78. doi:10.1007/s12130-009-9069-9

Dourish, P. (2001). *Where the action is: The foundation of embodied interaction*. Cambridge, MA: MIT Press.

Dreyfus, H. L. (1979). *What computers can't do: The limits of artificial intelligence*. San Francisco, CA: Harper & Row.

Dreyfus, H. L. (1992). *What computers still can't do: A critique of artificial reason*. Cambridge, MA: MIT Press.

Ehn, P. (1990). *Work-oriented design of computer artifacts*. Hillsdale, NJ: Lawrence Erlbaum Associates.

Esposito, R. (1998). *Communitas: Origine e destino della comunità*. Torino, Italy: Einaudi.

Ferraris, M. (2005). *Dove sei? Ontologia del telefonino*. Milano, Italy: Bompiani.

Ferraris, M. (2007). *La fidanzata automatica*. Milano, Italy: Bompiani.

Heidegger, M. (2006). *Sein und Zeit*. Tübingen, Germany: Niemeyer.

Heims, S. J. (1980). *John Von Neumann and Norbert Wiener: From mathematics to the technologies of life and death*. Cambridge, MA: MIT Press.

Heims, S. J. (1991). *The cybernetics group*. Cambridge, MA: MIT Press.

Heims, S. J. (1993). *Constructing a social science for postwar America: The cybernetics group, 1946-1953*. Cambridge, MA: MIT Press.

Lave, J., & Wenger, E. (1991). *Situated learning: Legitimate peripheral participation*. Cambridge, UK: Cambridge University Press.

Makoto Su, N., Hiroko, D., Wilensky, N., & Redmiles, D. F. (2011). Doing business with theory: Communities of practice in knowledge management. *Computer Supported Cooperative Work – The Journal of Collaborative Computing*. Retrieved from http://www.normsu.com/papers/Su-COP-JCSCW11.pdf.

McCorduck, P. (1987). *Machines who think*. New York, NY: WH Freeman & Co.

McCorduck, P. (2003). *Machines who think* (2nd ed). Natick, UK: A K Peters.

Nancy, J. L. (1986). *La communauté désoeuvrée*. Paris, France: Christian Bourgois.

Nancy, J. L. (1988). *Die undarstellbare gemeinschaft. Stuttgart, Germany: Legueil. Nancy, J. L. (1991). The inoperative community*. Minneapolis, MN: University of Minnesota Press.

Nancy, J. L. (1992). *La comunità inoperosa*. Napoli, Italy: Cronopio.

Nancy, J. L. (1996). *Être singulier pluriel*. Paris, France: Galilée.

Nonaka, I., & Takeuchi, H. (1995). *The knowledge creating company*. New York, NY: Oxford University Press.

Star, S. L. (1989). The structure of ill-structured solutions: Boundary objects and heterogeneous distributed problem solving. In Gasser, L., & Huhns, M. (Eds.), *Distributed Artificial Intelligence* (*Vol. 2*, pp. 37–54). San Francisco, CA: Morgan Kaufman.

Telier, A. (2011). *Design things*. Cambridge, MA: MIT Press.

Toennies, F. (1887). *Gemeinschaft und gesellschaft*. Leipzig, Germany: Fues's Verlag.

Toennies, F. (1912). *Gemeinschaft und gesellschaft* (2nd ed.). Leipzig, Germany: Fues's Verlag.

Toennies, F. (1935). *Gemeinschaft und gesellschaft* (8th ed.). Leipzig, Germany: Buske.

Toennies, F. (2005). *Community and society*. Darmstadt, Germany: Wissenschaftliche Buchgesellschaft.

Vinck, D. (2003). *Everyday engineering: An ethnography of design and innovation*. Cambridge, MA: MIT Press.

Wenger, E. (1998). *Communities of practice: Learning, meaning and identity*. New York, NY: Cambridge University Press.

Winograd, T., & Flores, F. (1986). *Understanding computers and cognition – A new foundation for design*. Norwood, NJ: Ablex.

Wittgenstein, L. (1953). *Philosophische untersuchungen*. Oxford, UK: Basil Blackwell.

## ENDNOTE

[1] For what regards the attention of computer scientists and practitioners to philosophy, let me remember: Winograd, Flores (1986), Ehn (1990), Dourish (2001), and Ciborra (2002). For what regards the attention of philosophy to Computer Science and Technology: Dreyfus (1979, 1992) and Ferraris (2005, 2007). Accounts on the debate that followed Dreyfus's work, in McCorduck (1987, 2003).

# Chapter 5
# Knowing and Living as Data Assembly

**Jannis Kallinikos**
*London School of Economics, UK*

## ABSTRACT

*The environment of information affluence and media convergence characteristic of our time is constructing a comprehensive living and knowing habitat that induces the framing of life issues in terms of data availability and the concomitant data permutations this encourages. It also promotes the understanding of knowledge as data assembled into meaningful categories and structures by algorithmic reasoning and statistical techniques. Placed in a wider historical purview, these trends attest to the prominence which cognition qua computation is acquiring and the concomitant retraction of perception as an axial principle of everyday conduct tied to palpable reality and situated encounters.*

## PERCEPTION, COGNITION, COMPUTATION

Human living and knowing are bound to waver between what is experienced through the senses (the sensible) and what is or can be thought (the intelligible), including counting and calculation, without immediate reference to palpable reality. Perception is a vital and inseparable component of living and, though shaped by culture, it is firmly anchored into the human sensorium. Yet, living and knowing always transcend the givens of perception and entail cognitive operations of an abstract nature that may lack ostensive reference (Cassirer, 1955; Flusser, 2000; Foucault, 1970).[1]

Integral as it is to all human knowing, however, the perception/cognition pendulum exhibits cultural and historical diversity while the trade-offs it entails are differently refracted across the variety of practices that make up human societies (Jay, 1994; Kallinikos, 2011). And yet, a case could be made for the fact that the diffusion of formal methods

DOI: 10.4018/978-1-4666-0303-5.ch005

of knowing in modern times has, in one way or another, systematically favoured abstract reasoning at the expense of the primordial dependence of living on information provided by perception and situated involvement (Borgmann, 1999; Jay, 1994; Cassirer, 1955). These observations acquire particular importance these days, as the result of the diffusion of computational models of cognition that inevitably accompany the growing involvement of information and information artefacts in domestic and institutional life.

In what follows I put forth the claim that the relatively recent developments that coincide with the pervasiveness of information and data tokens in most walks of living further perturb the balance between the sensible and the intelligible and accentuate the preponderance of cognition over perception. Essential strips of reality are increasingly mediated and acted upon by means of information and data tokens, produced and disseminated by technological means. In thus according data-driven correlations and pattern recognition resulting from bottom-up processes of data manipulation a central place in living and knowing, computation drives cognition away from the associative gestalts of perception, sensation and intuition, and the experiential knowledge built on observation, trial-and-error and world acquaintance (Ayres, 2007).

While recounting some of the earlier worries concerning the impact of computing on human practices (Dreyfus & Dreyfus, 1986; Winograd & Flores, 1986), the claim I put forward is not precisely concerned with the ill-fated prospects of experientially-based knowledge and skills that have commonly been subsumed under the label of tacit knowledge. The problem I seek to pinpoint is of a rather different nature, even though it may have far reaching implications for expert knowledge and the ways it is acquired and exercised (Ayres, 2007; Shirky, 2008). The environment of information affluence and media convergence that characterizes our time is constructing a comprehensive living and knowing habitat that stretches far beyond the work and professional settings in which information technology has commonly been studied (Kallinikos, 2011; Zuboff, 1988). This environment differs too from the early internet (Dreyfus, 2001; Turkle, 1995), most notably by its remarkable ability to deeply penetrate the social fabric and increasingly induce the framing of life issues in terms of data availability, and sense making based on data, assembled into meaningful categories and structures by machines.

In the rest of this paper, I try to provide reasons in support of this claim. Next section describes the importance, which data capture and data analysis assume in contemporary life, and the shifts in the practices of knowledge development and sense making these trends bring about. Following it, I endeavour to show how the proliferation of data tokens combines with the diffusion of computer-based devices across the social fabric to construct new forms of experience, in which a wide spectrum of life issues from trivial (e.g. shopping on the internet) to non-trivial ones (e.g. profiling and identity making) is increasingly framed in cognitive terms. In the last section, I revisit the double bind of the sensible and the intelligible and the intricate relation of cognition to perception, discussing some of the problems and paradoxes that are caused by the developments I refer to.

## THE ANALYTIC REDUCTIONISM OF COMPUTATION

Data availability is the distinctive mark of our age and its holy grail (Ayres, 2007; Kallinikos, 2006). If only available, data is assumed to be able to tell us of so many things: for instance, who we are, what exactly is the nature of our preferences beyond our delusions or how our body feels even if our awareness fails to register it; how our society and organizations work; what friends to choose and what communities to join; what travels, mortgage, or insurance to undertake this year; what flights may be cheaper today; what stocks to invest in over the coming few months; which films or plays to see this week; how to drive to a destination in the urban mazes of modern cities; and so forth (Ayres, 2007). The assumption is that if rightly perused, the data that is massively and meticulously captured and stored everyday in all walks of life would be able to illuminate human realities and innovatively address an impressive array of issues confronting contemporary people. A large range of technologies and services are already in place.

While many of these issues may appear trivial, their implications are not. As I claim in some detail in the next section, conducting an essential portion of the dealings of everyday living by relying on recommendations produced out of available data by machines signifies an important shift in core human habits, but the use of data has implications beyond the restructuring of everyday trivia. Reconstructing, for instance, the profile of people's preferences by means of computing data provided by Internet surfing habits and information on life styles, income or demographic characteristics compiled from other sources is an intricate practice with serious personal and legal repercussions (Hildebrandt & Gutwirth, 2008). The same holds true for contemporary science with all its sophistication and institutional complexity. There is some evidence to suggest that scientific practice increasingly becomes entangled with the production and computational manipulation of data in ways that far outstrip whatever prerogatives or reliance empiricism has commonly attributed to descriptive accounts of reality. Consider the following[2]:

*What can you learn from 80 million x-rays? The secret of aging, among other things, Sharmila Majumdar, a radiologist at UC San Francisco, is using an arsenal of computer tomography scans* ***to understand*** *how our bones* ***wear out from the inside****.*

*It works like this: A CT scanner takes superhigh-resolution x-rays and combines those individual images into a three-dimensional structure. The results are incredibly detailed; a scan of a single segment of bone can run 30 gigs.*

*Majumdar's method is to* ***churn through the data to identify patterns***[3] *in how the trabeculae-the material inside the bone-changes in people who have diseases like osteoporosis and arthritis. In one day of imaging, it's not uncommon for the lab to generate nearly a terabyte of data. Researchers also aggregate the data from many subjects, putting hundreds of terabytes to work. Majumdar hopes to learn why some patients suffer severe bone loss but others don't. 'We don't know the mechanism of bone loss,' she notes. 'Once we learn that, we can create therapies to address it.'*

The example provides a good illustration of wider shifts in core procedures of medical practice and research. It shows how the development of knowledge is increasingly made contingent on the extraction of patterns derived from permutations performed upon huge masses of data. The first step in this process is the comprehensive mapping of reality through the technological generation of huge amounts of data (superhigh-resolution x-rays) to be followed by computer-driven techniques of data reduction and interpretation that are assumed able to expose its underlying regularities (the mechanism of bone loss). The assumption is that more data leads to more adequate descriptions of reality and this is what renders millions of scans necessary.

Obviously, the conditions under which data is captured and aggregated far surpass the attention span, the register capacity and memory of humans (lay men or experts). The purpose of the bone scans, as this quote indicates, is not to provide ostensive evidence to the expert eye, at least not principally. Even though each patient's condition may be better illuminated by superimposed high resolution x-rays, the major use of the scans taken is to add data on bone conditions at a far more aggregate level. The value of particular scans is framed by the overall project of producing an adequate description of reality at the population level. This it would seem as a remarkable shift in the traditional use of scans and the medical practice of diagnosing through reading of x-rays. Accordingly, the scans taken amount to millions and a dizzying mass of data (counted in terabytes) is obtained through aggregation of individual x-rays in a common data pool.

Under these conditions, traditional, perception-attuned forms of making sense of the world may not seem well suited. Indeed, there is nowhere to go by just focusing on the perception and examination of individual scans even though the conditions of particular patients could be illuminated and some of its grave consequences perhaps alleviated, but again the purpose of scans is not to address particular patients. Lumped together, the resulting groups of scans make it extremely expensive and time consuming, if ever possible, to identify the mechanism of bone loss by means of the naked eye or other perception-based observations. The pattern of bone change (medical knowledge) that will eventually emerge from this data mass will be derived from the computational superiority of statistical correlations performed on terabytes of data generated by millions of scans, spanning over masses of people distributed over different ages.

In this regard, the development of knowledge by these means refigures not simply the role of perception in human life but also key conceptual habits and traditions of scientific research and practice. The statistical permutations performed upon the data mass are basically agnostic. In a nearly cunning degree, the process of discovery conforms to the canon of induction. No theory other than statistics is needed to detect patterns in the data. The pattern, if there is one, should emerge from bottom-up processes of data manipulation and clustering through statistical techniques. Data comprehensively maps reality (the scans capture bone conditions) while computer-based data mining discovers its underlying principles (the mechanism of bone loss).

In the same volume of *Wired* from which the excerpts are drawn, the information age guru and editor in chief of the journal Chris Anderson predicts the end of theory and science in the standard sense of conceptual development, based on empirical evidence of one or another type. Due to greater data availability and more accurate data, this pattern, he claims, will be intensified in the years to come and knowledge, equated with data configurations, will finally be derived inductively and exclusively through correlations performed upon huge masses of data.[4] The age in which data without theory meant noise is over, says Anderson, putting new oil in the fire maintained over centuries by the empiricism/rationalism divide. In this neo-positivist context, not only perception, even conceptual analysis (or at least a vital part of it) is rendered redundant. Cognition in the form of data analytics (or you may want to call it data alchemy) increasingly takes command. Matter and reality are regained, if ever, after a long analytic retreat, out of the cognitive dust of computational particles (data) (Kallinikos, 2009).

While science may seem as a context akin to the revival of the empiricism's dream, the trends I single out here are far from limited to the development of knowledge and science. Indeed, they extend over to an increasing range of vernacular practices some of which may carry important implications, pecuniary or otherwise. Here is another illustrative demonstration, this time taken from agriculture[5]:

*Last October, agricultural consultancy Lanworth not only correctly projected that the US Department of Agriculture had overestimated the nation's corn crop, it nailed down the margin: roughly 200 million bushels. That's just 1.5 percent fewer kernels but still a significant shortfall for tight markets, causing a 13 percent price hike and jitters in the emerging ethanol industry. When the USDA downgraded expecta-*

*tions a month after Lanworth's prediction, the little Illinois-based company was hailed as a new oracle among soft-commodity traders-who now pay the firm more than $100.000 a year for a timely heads-up on fluctuations in wheat, corn and soybean supplies.*

*The USDA bases its estimates on questionnaires and surveys-the agency calls a sample of farmers and asks what's what. Lanworth uses* ***satellite images, digital soil maps*** *and* ***weather forecasts*** *to project harvests at the* ***scale of individual fields****, it even looks at crop conditions and rotation patterns-combining all the numbers to determine future yields.*

*Founded in 2000, Lanworth started by mapping forests for land managers and timber interests. Tracking trends in sleepy woodlands required just a few outer-space snapshots a year. But food crops are a fast-moving target. Now the company sorts 100 gigs of intel everyday, adding to* ***a database of 50 terabytes*** *and counting. It is also moving to world production prediction-wheat fields in Russia, Kazakshtan and Ukraine are already in the data set, as are corn and soy plots in Brazil and Argentina The firm expects to reach* ***petabyte scale*** *in five years. 'There are questions about how big the total human food supply is and whether we as a country are exposed to risk,' says Lanworth's director of information services, Nick Kouchoukos. 'We are going after the global balance sheet.'*[6]

Similar to the medical example of bone loss, the case of Lanworth illustrates a dramatic shift in the whole practice of thinking about and predicting future yields. Such a shift is made again possible by 1) the remarkable expansion and aggregation of information sources about the conditions on which food crops are dependent and 2) the calculation of future yields on the basis of computational (statistical) techniques of data analysis and reduction. The remarkable expansion of information is the outcome of technological advances of reality documentation (satellite images, digital soil maps, weather forecasts) far beyond any human register capacity. They are also contingent on the (technological) ability to lump up that information into a single database or data repository of standardized data possible to crunch by statistical techniques. Equally, to the project of understanding the underlying reality of bone loss, data availability and algorithmic reasoning emerge in the case of predicting future crops as the milestones of a new practice.

There are undeniably blind spots, hidden assumptions, and simplifications in these stories of scientific journalism. One could certainly raise doubts with respect to the efficacy of the procedures of knowledge development described above and the hidden risks or even deceptions they may help disseminate (Nardi & Kow, 2010), but the trends of data affluence and algorithmic data reduction they depict are hard to dismiss. They are indeed supported by both wider evidence and common sense (Ayres, 2007; Gantz, et al., 2008; Lessig, 2008). At the very bottom, it would seem as a displaced gesture to question whether the procedures of data availability and computational rendition that underlie knowledge development in the contexts outlined above represent such an efficacious way to truth, as for instance Anderson claims. Most probably, they do not. If anything, these stories rather illustrate how truth is currently achieved; that is, "how truth is made true" or "adequate" by recourse to the instruments, techniques and solidified knowledge procedures of computation (Foucault, 1988; Hacking, 1985, 1999). Thus conceived, truth is not primarily a denotative or declarative game of verifiable or falsifiable propositions but a performative one. The issue is ultimately not whether technological representations truthfully map reality but rather how they create and establish particular versions of the world possible to act upon, perform and manipulate (Heidegger, 1977, 1998; Kallinikos, 1995).

## THE TECHNOLOGICAL MONITORING OF DAILY LIVING

While the implications of the developments I single out here for scientific and economic practices may appear evident, it is as yet unclear how they can carry over to everyday living and the making up of life styles or patterns in which cognition, in the form of computation, becomes prominent. Daily living is such that the documentation of reality produced by technological means (superhigh-resolution x-rays, digital soil maps, or satellite images) would seem not to be as well developed as it is in the delimited and more easily amenable to technological regulation institutional domains, considered in the preceding section.

However, upon a closer look, much of the technological and cultural change that has taken place over the last few decades casts everyday living in a similar context, saturated by information tokens and a variety of automated services that respond, *mutatis mutandis*, to the problem of information affluence characteristic of our age. The significant and continuously growing involvement of the Internet in everyday living is the most conspicuous manifestation of these trends. The standardized and for this reason commensurable information space, which the Internet and other information infrastructures help establish, makes possible the sharing and elaboration of data across the traditional (culturally and technologically based), divides of text, image, and sound, and the institutional conventions through which these have been produced and consumed such as films or video, books, photographs, games, journals, or magazines. Such a sharing and elaboration of data and the information-based culture and knowledge they mediate are not common offsprings of human activities. Rather, they are the outcome of a range of longstanding technological and institutional processes operating at different levels of aggregation whereby social and cultural life is made the object of mediation by an amazing ecology of artefacts and technologies controlled underneath by computational operations (software) (Hayles, 2005, 2006; Kallinikos, 2009; Manovich, 2001).

Often referred to as media convergence, these developments construct a common and trans-modal information space in which a variety of experiences, previously mediated by a wide array of incommensurable media/artefacts and deriving from different and mostly unconnected life orders, can be brought to bear upon one another (Jenkins, 2006; Kittler, 1997; Manovich, 2001). Private life can be made a YouTube video or a Facebook story and new experiences can be produced out of mixing an impressive range of texts and blogs with pictures and audios over the web (Lessig, 2008; Shirky, 2008). At the same time, companies and institutions meticulously record in databases, under still ambiguous legal conditions, the online habits of people spanning over widely different life compartments in their effort to produce commercially relevant behaviour profiles as the basis of the services they offer to individuals (Carr, 2008; Hildebrandt & Gutwirth, 2008).

In this respect, the dominant motif of the developments I pinpoint here pivots around the effects these may have in constructing or performing reality. The infiltration of human experience by the categories and schemes of technoscience (and power) is of course a pervasive modern theme (Borgmann, 1992, 1999; Heller, 1999; Mumford, 1934). However, the issues which I touch upon here may be traced back to a relatively recent and recurring Foucauldian theme in social sciences (Foucault, 1988), concerning the importance of statistics, actuarial techniques and accountancy, among others, on the constitution of particular forms of conduct. In a recent update of his acclaimed paper "Making up People," Ian Hacking lists counting, quantifying and correlating among the handful of techniques of fashioning human conduct that he now refers to as "engines of making up people."[7] In this respect, the methods, facilities and modes of capturing and acting upon reality that computation diffuses are made an intrinsic part of everyday living while the distinctions and classifications of computational techniques like profiling become embedded to social life through the interaction between people and profiling institutions (Hildebrandt & Gutwirth, 2008; Hildebrandt & Rouvroy, 2011). Profiling classifications are social classifications that turn out interactive (people assume the roles classification delineate) impinging on and shaping real life situations (Hacking, 1999).

An important consequence of all these developments, I feel, is the infiltration of an ever-greater range of details of everyday life by computation and the technological monitoring of the trivia of life. For perhaps the first time in history, technology in the form of computation is extensively involved in the framing of life issues, as predominantly issues of information and communication possible to address by computational means. Indeed, a new everyday seems to be taking shape (Lessig, 2008; Shirky, 2008). Routine, daily activities are increasingly cast in a technological context and conducted with the use of a range of computer-based artefacts via the World Wide Web, the Internet, or other communications networks. Some of these trends are well illustrated by IBM's programme of *autonomic computing* in which a considerable range of experiences spanning businesses and institutional compartments are thought possible to monitor through elaborate grids of intelligent technologies in which human agents are basically seen as the receiving end of a complex ecology of devices and computational processes (Gilat, 2005; Hildebrandt & Rouvroy, 2011).[8] More perhaps clearly than IBM's yet speculative programme, the infiltration of the personal space of experience by data technologies is manifested in the Microsoft-based research project *MylifeBits* in which human experience and the impressions/memories it mediates are made the object of deep-going technological mediation and monitoring[9]. Much of this data infiltration of everyday life is mediated by the spectacular diffusion of connectible devices (handsets and smart phones), so called location services (GPS), and augmented reality technologies that seek to provide navigational routes to the complex mazes of urban spaces. The invasion of urban, perception-based experience by these technologies represents a conspicuous manifestation of the trends I seek to depict in this chapter (Virilio, 2000).

But do these developments really differ from the effects industrial capitalism produced over the course of the last century by means of which domestic life and human experience in general became drastically redefined? Placed in a historical purview, the diffusion of industrial appliances and goods across the social fabric remade the infrastructural basis of everyday living and profoundly altered the contours of personal and cultural experience (Heller, 1999). However, the shifts I seek to describe in this paper are of a substantially different nature. The issue at stake is not so much whether industrial versus information artefacts have a smaller or bigger effect on living. More important seems to me to be the understanding of the distinct routes along which whatever effects they have are realized.

Information is a cognitive or communicative category and the framing of life issues (at the personal or institutional level) in terms of information implies a surreptitious but crucial shift. Life situations tend to be defined as cognitive problems of a computational and navigational nature (e.g. what to see or do, how to find a film/play, a friend, or even a partner). They are addressed by the complex and automated computations performed upon the affluence of data and information tokens modern technologies and the life styles they instigate make available. Highly diffused and generalized as they are, and will certainly be even more in the immediate future, these developments signify a deep-going redefinition of everyday living comparable perhaps to the life styles that diffused as a consequence of industrialism and mass production. However, they are realized via another route. The redefinition of everyday living coincides with the prominence of cognition leaning on technologically produced and disseminated data tokens, and the concomitant retraction of perception as the axial principle of everyday conduct tied to palpable reality and ongoing and situated encounters.

## ADDENDUM ON EMBEDDEDNESS

Even in the case the portrait I have painted in this chapter is taken as a valid depiction of contemporary developments, it would seem legitimate to raise objections as regards the kind of impact these developments may have on human practices and institutions. Can one really assume the unified refraction of these trends across the social and institutional fabric? Do not people and institutions appropriate these developments, interpret and remake them to fit their realities? After all, the changes I describe in this paper occur in a thick cultural and institutional context that variously conditions the introduction of new technologies and whatever habits these may propagate.

There is little doubt that the developments I have singled out in this paper are and will be variously refracted across fields and professional domains of contemporary life; medicine, for instance, is different from agriculture, architecture from law or the running of a business. Each of these institutional domains confronts specific problems. Each one addresses these problems by historically specific ensembles of knowledge, skills, traditions and practices (Kallinikos & Hasselbladh, 2009). Given, thus, the distinctive profile of problems, skills and practices underlying each one of these domains, it makes sense to assume that the implications of the developments I refer to here are bound to manifest differently across them (see Lessig, 2006; Terzides, 2005).

Reasonable as this assumption may be, it nevertheless does not exhaust the issues raised by the developments I seek to describe in this text. For, the comprehensive and deep-going character of these developments alters important premises on the basis of which reality is conceived and addressed. It does so in ways that are generic enough to transcend the specificities of particular domains of social life (Kallinikos, 2011). Contemporary computer-mediated architecture, to give another context of professional experience, makes its objects calculable, describable and manipulable in innovative ways that open up new spaces for expression and human intervention (Flachbart & Weibel, 2005; Terzides, 2005). Such calculability, describability, and manipulability are the outcomes of the distinctive ways by means of which computation assembles the composite pictures of reality we confront out of data tokens and the permutations they enter (Kallinikos, 2009). The same holds true for the other examples given earlier in this chapter.

A new reality is brought about by these techniques and ways of construing and assessing problems and situations in which data availability and data analytics prevail. In this regard, the reality outlook of computation and the techniques of capturing, storing, making available and analyzing data transcend the confines of particular professional practices (Kallinikos & Hasselbladh, 2009) and remake some of the basic premises on the basis of which facts are produced, made available, assessed and addressed. By the same token, the forms of expertise intrinsic to each one professional domain (e.g. medical, legal, economic, architectural) are changing in ways that seek to accommodate the computational rendition of reality I have been at pains to describe in this text.

Though differently, these developments carry implications, I have claimed, for the lifeworld and the patterns of everyday life. They cast human living in a complex environment in which data availability and the computations it occasions increasingly provide the coordinates of daily life patterns. Again, the individual or collective appropriation of these life styles and patterns inevitably entails the re-contextualization and situated interpretation of abstract and freely circulating data tokens. It is beyond any doubt that individuals and groups transform and complement the generic and de-contextualized solutions mediated by technologies (Esposito, 2004). Yet, by the same token, they become accomplices in the framing and sampling of the world these technologies and the services they enable mediate (Heidegger, 1977, 1998). Re-contextualization (and interpretation) is double act that changes not only what is re-contextualized but also the agents of re-contextualization. Seen in this light, the tasks and patterns of daily life that have predominantly been shaped by all those qualities and faculties we associate with human experience are subject to change, as computational techniques infiltrate the everyday living habitat. This seems to me as an important shift in which the front- and backstage (the tacit, unconscious, forgettable, beyond awareness) of human living change position (Kallinikos, 1995).

In other words, the technological and cultural developments I describe in this chapter exhibit generic attributes that cut across specific contexts of social and institutional life. These generic attributes—which I have identified with the prominence of cognition over perception, the preponderance of information and computational principles in defining reality—cannot be wished away by the widespread rhetoric of technological appropriation, technological malleability or contextual adaptation that has become common over the last few decades (Bijker, 2001; Bijker, et al., 1987; Orlikowski, 2000). They need to be conceptually decomposed and studied empirically with the view of exposing the distinctive ways through which they are manifested, change and fuse into the fabric of social practices.

## REFERENCES


Anderson, C. (2008, July). The end of theory. *Wired*.

Anderson, J. R. (1983). *The architecture of cognition*. Cambridge, MA: Harvard University Press.

Ayres, I. (2007). *Super crunchers: How everything can be predicted*. London, UK: Murray.

Bijker, B. (2001). Understanding technological culture through a constructivist view of science, technology and culture. In Cutcliffe, S., & Mitcham, C. (Eds.), *Visions of STS: Counterpoints in Science, Technology and Society Studies*. New York, NY: State University of New York.

Bijker, W. E., Hughes, T. P., & Pinch, T. (Eds.). (1987). *The social construction of technological systems*. Cambridge, MA: The MIT Press.

Borgmann, A. (1992). *Crossing the postmodern divide*. Chicago, IL: The University of Chicago Press.

Borgmann, A. (1999). *Holding on to reality: The nature of information at the end of the millennium*. Chicago, IL: The University of Chicago Press.

Carr, N. (2008). *The big switch: Rewiring the world, from Edison to Google*. New York, NY: Norton.

Cassirer, E. (1955). *The philosophy of symbolic forms*. New Haven, CT: Yale University Press.

Dreyfus, H. (2001). *On the internet*. London, UK: Routledge.

Dreyfus, H., & Dreyfus, S. (1986). *Mind over machine*. New York, NY: Free Press.

Esposito, E. (2004). The arts of contingency. *Critical Inquiry*, *31*. Retrieved from http://criticalinquiry.uchicago.edu/features/artsstatements/arts.esposito.htm doi:10.1086/427300

Flachbart, G., & Weibel, P. (Eds.). (2005). *Disappearing architecture*. Basel, Switzerland: Birkhauser. doi:10.1007/3-7643-7674-0

Flusser, V. (2000). *Towards a philosophy of photography*. London, UK: Reaktion Books.

Foucault, M. (1970). *The order of things*. London, UK: Tavistock.

Foucault, M. (1988). Technologies of the self. In Martin, L. H., Gutman, G., & Huttton, P. H. (Eds.), *Technologies of the Self.* London, UK: Tavistock.

Gantz, J. F., et al. (2008). *The diverse and exploding digital universe: A forecast of worldwide information growth through 2011*. IDC White Paper. Retrieved from www.emc.com/digital_universe.

Gilat, D. (2005). Autonomic computing-building self-managing computing systems. In Flachbart, G., & Weibel, P. (Eds.), *Disappearing Architecture*. Basel, Switzerland: Birkhauser.

Hacking, I. (1985). Making up people. In Heller, T. L., Sosna, M., & Wellbery, T. E. (Eds.), *Reconstructing Individualism*. Palo Alto, CA: Stanford University Press.

Hacking, I. (1999). *The social construction of what?* Cambridge, MA: The Harvard University Press.

Hayles, C. (2005). Computing the human. *Theory, Culture & Society*, *22*(1), 131–151. doi:10.1177/0263276405048438

Hayles, C. (2006). Unfinished work: From cyborg to cognisphere. *Theory, Culture & Society*, *23*(7-8), 159–166. doi:10.1177/0263276406069229

Heidegger, M. (1977). *The question concerning technology and other essays*. New York, NY: Harper.

Heidegger, M. (1998). Traditional language and technological language. *Journal of Philosophical Research*, *23*, 129–145.

Heller, A. (1999). *A theory of modernity*. Oxford, UK: Balckwell.

Hildebrandt, M., & Gutwirth, S. (Eds.). (2008). *Profiling the European citizen*. Berlin, Germany: Springer. doi:10.1007/978-1-4020-6914-7

Hildebrandt, M., & Rouvroy, A. (Eds.). (2011). *Law, human agency, and autonomic computing: The philosophy of law meets the philosophy of technology*. London, UK: Routledge.

Jay, M. (1994). *Downcast eyes: The denigration of vision in twentieth-century French thought*. Berkeley, CA: University of California Press.

Jenkins, H. (2006). *Convergence culture: How new and old media collide*. New York, NY: New York University Press.

Kallinikos, J. (1995). The architecture of the invisible: Technology is representation. *Organization*, *2*(1), 117–140. doi:10.1177/135050849521006

Kallinikos, J. (2006). *The consequences of information: Institutional implications of technological change*. Cheltenham, UK: Elgar.

Kallinikos, J. (2009). On the computational rendition of reality: Artefacts and human agency. *Organization*, *16*(2), 183–202. doi:10.1177/1350508408100474

Kallinikos, J. (2011). *Governing through technology: Information artefacts and social practice*. New York, NY: Palgrave/MacMillan.

Kallinikos, J., & Hasselbladh, H. (2009). Work, control and computation: Rethinking the legacy of neo-institutionalism. *Research in the Sociology of Organizations*, *27*, 257–282. doi:10.1108/S0733-558X(2009)0000027010

Kittler, F. (1997). *Literature, media, information systems*. Amsterdam, The Netherlands: OPA.

Lessig, L. (2006). *Code 2.0*. New York, NY: Basic Books. Retrieved from http://codev2.cc/download+remix/.

Lessig, L. (2008). *Remix: Making art and commerce thrive in the hybrid economy*. London, UK: Bloomsbury. Retrieved from http://codev2.cc/download+remix/.

Manovich, L. (2001). *The language of new media*. Cambridge, MA: The MIT Press.

Miller, & O'Leary, T. (1987). Accounting and the construction of the governable person. *Accounting, Organizations and Society*, *12*(3), 235-265.

Mumford, L. (1934). *Technics and civilization*. San Diego, CA: HJB.

Nardi, B., & Kow, Y. M. (2010). Digital imaginaries: What we (think we) know about Chinese gold farming. *First Monday*, *15*, 6–7.

Neisser, U. (1976). *Cognition and reality*. San Francisco, CA: Freeman.

Orlikowski, W. J. (2000). Using technology and constituting structures: A practice lens for studying technology in organizations. *Organization Science*, *11*(4), 404–428. doi:10.1287/orsc.11.4.404.14600

Paynter, B. (2008, July). Feeding the masses. *Wired*.

Rouvroy, A. (2011). Technology, virtuality and utopia. In Hildebrandt, M., & Rouvroy, A. (Eds.), *The Philosophy of Law meets the Philosophy of Technology: Autonomic Computing and Transformations of Human Agency*. London, UK: Routledge.

Shirky, C. (2008). *Here comes everybody*. London, UK: Penguin.

Terzides, K. (2005). *Algorithmic architecture*. London, UK: Elsevier.

Turkle, S. (1995). *Life on the screen: Identity in the age of the internet*. New York, NY: Simon and Schuster.

Virilio, P. (2000). *The information bomb*. London, UK: Verso.

Winograd, T., & Flores, F. (1986). *Understanding computes and cognition*. Norwood, NJ: Ablex Publishing Corporation.

Zuboff, S. (1988). *In the age of the smart machine: The future of work and power*. New York, NY: Basic Books.

## ENDNOTES

1 Some may find the distinction I make between perception and cognition overstated. Little wonder, perception is essential to cognition and the abstract operations that underlie cognition whereas the latter's cognitive templates furnish as important receptacles to the activities of attention and perception (Anderson, 1983; Neisser, 1976). By perception, then, I predominantly mean situated observations and experiences that differ from sheer sensations in the sense of involving cultural mediations that do not, however, evolve into deliberate mental exercises. The distinction cannot be drawn unproblematically but it does have the merit of placing perceptual and cognitive operations at a different distance from ostensive reality.

2 Goetz, T. (2008, July). Scanning our skeletons. *Wired*.

3 All bolds are mine.

4 Anderson, C. (2008, July). The end of theory. *Wired*.

5 Paynter, B. (2008, July). Feeding the masses. *Wired*.

6 All bolds are mine.

7 In a newer version of his "Making up People" with the same title in LRB, vol. 28, 17 of August 2006, Ian Hacking lists ten engines of making up people: Count, Quantify, Create Norms, Correlate, Medicalise, Biologise, Geneticise, Normalise, Bureaucratise, and Reclaim our Identity.

8 IBM. (2010). *Website*. Retrieved from http://www.research.ibm.com/autonomic/.

9 My Life Bits. (2010). *Website*. Retrieved from http://www.mylifebits.com.

# Chapter 6
# "Whatever Works": Making Sense of Information Quality on Information System Artifacts

**Federico Cabitza**
*Università degli Studi di Milano-Bicocca, Italy*

**Carla Simone**
*Università degli Studi di Milano-Bicocca, Italy*

## ABSTRACT

*This chapter addresses the general problem of how to design and deploy effective computational tools that support actors of an organization domain in making sense of the information these manage by means of those tools and technologies. To this aim, the chapter recognizes the complementary, but sometimes also diverging, approach of two related disciplines, Computer Supported Cooperative Work (CSCW), and Information Systems (IS). Contributions from both these academic communities are put in a common perspective to discuss a comprehensive solution to the apparently incompatible requirements of different communities of users that use the same information for different purposes. The authors take the theme of quality information standards, requirements, and users' expectations in information-intensive domains such as healthcare and hospital work as a paradigmatic case to discuss the characteristics of their proposal. This encompasses the conceptualization of a general-purpose architecture that they devised to support adequate exploitation by human actors of informative resources regarding how they perform their job and articulate their actions with others; and a specialized design-oriented construct, called Affording Mechanism (AM). An AM is a dyad composed by an artifact (i.e., the schema of a material information tool) and a dynamic relationship between the context of use and the artifact's affordances. AM relationships are expressed in terms of computable if-then statements that modulate the affordances conveyed through and by the artifact to evoke a "positive" and knowledgeable reaction in the actors' behavior. On the basis of observations performed in the hospital domain, the chapter discusses in a coherent constructivist light the role of artifacts and derives a set of general requirements for affording mechanisms that support situated behaviors.*

DOI: 10.4018/978-1-4666-0303-5.ch006

## INTRODUCTION

The introduction of an information technology within an organization involves two main aspects: the management of the information that supports the activities by which the organization fulfils its mission and the coordination of these activities in order to make their execution smooth and effective. Two academic fields have contributed to the definition of different and often complementary approaches that deal with the two aspects mentioned above: Information Systems (IS) and Computer Supported Cooperative Work (CSCW). Although the object of interest of these disciplines is often the same, i.e., supporting the organization as a socio-technical system through the comprehension of its "functioning," scholars that feel themselves closer to one of either fields look at information technology in organizations and at how it is conceived from substantially different and complementary perspectives.

Both IS and CSCW show the combination of technological concerns with the outcomes of disciplines oriented to the understanding of the setting in which the technology has to operate: Social Sciences for CSCW; Organizational Studies for IS. While in CSW there is a uniform agreement about the unavoidable strong relationship between technological design and the sociological field research that informs it, in IS the separation between technology design and organizational studies is much stronger since they are at most considered as complementary but still with abstract suggestions on how this should happen (Orlikowski & Barley, 2001) or on how organizational studies should take the technological issues more systematically (Orlikowski & Iacono, 2001). Another fundamental difference is the point of view from which the organization is looked at: CSCW focuses on users as the sole source of knowledge about the cooperative settings and its needs; on the other hand, IS focuses mainly on management, and when it recognizes a role for end-users, it sees them mainly as one of the possible stakeholders at play, or "social actors," as called by Lamb and Kling (2003); these authors tried to bridge the perspectives of CSCW and IS together, but still we believe they also underestimated the differences at play. Now, the question is whether CSCW and IS can coexist both conceptually and technologically. In regard to the conceptual level, the question is about the existence of sociological and organizational approaches that are compatible, while their specific domain of investigation is properly recognized. We leave the answer to the scholars of these disciplines: as designers, we can only advocate a positive answer and claim that the two disciplines can contribute to define a common vision of ICT where the management and end-users' perspectives can be reconciled with the same level of relevance. This chapter deals with the technological perspective, discussing architectural details only marginally but rather focusing on the concept of information quality: this means looking at the value of technological solutions through the requirements and needs that are contextually expressed by both the end-users and the management of an organization.

In order to build a bridge between IS and CSCW, which have so far had a lesser mutual influence than desirable, our main aim is to outline an approach that offers the positive contribution of both perspectives and tries to overcome some of the limits that can be observed when an information system is deployed in real organizations also to support work processes and collaboration. In this effort, we leave out the focus on information management that is patrimony of the IS field and concentrate on the issues related to collaboration and task coordination. To this aim, we start from the conceptualization of work that is common to both IS and CSCW research—work is an intrinsically (and probably irreducible) social phenomenon—and argument that nevertheless the phenomenon of collaboration is conceived differently in these two fields. This will bring us to reconsider the role of "artifacts" in organizations (for informative and coordinative purposes), by distinguishing between the conceptualizations that see artifacts (and the representations they carry on) as either "resources for action" (Suchman, 1987) or that, more common in ICT professional practice, of "gateways" to the underlying structures of an information system, respectively. This in turn will lead us to considering information in two complementary ways, as either a commodifiable asset that machines can process, or as a peculiarly human and culturally-embedded act of interpretation based on available and shared signifiers. This distinction is reflected in different ways to address the problem of "quality" in information, especially in those contexts where vital decisions are taken on the basis of available data. To depict this case, we take the healthcare case as paradigmatic although the same phenomena can be recognized in other domains, in any domain where work is documented in more or less structured forms and where the same data that compound the content of the information system are used for different purposes by different stakeholders. Finally, we propose a technological construct aimed at supporting the reconciliation between different expectations on information quality and its potential to "inform" action and decisional processes in organizational cooperative settings.

## RETHINKING ABSTRACTIONS OF WORK

As said in the previous section, both IS and CSCW consider work as a social phenomenon: as such, it is the place in which some interdependency between two or more actors occurs (Neff, 2006), or similarly, where a positive—though by no means necessarily harmonious—reliance of actor A "on the quality and timeliness of B's work and vice versa" (Schmidt, 1991) can be observed. Therefore, in complex work settings as today's organizations are, supporting work by means of computer systems entails supporting also coordination. Here however comes a basic difference between IS studies and the CSCW approach. IS interprets coordination in terms of organizational roles, patterns of communication, and task ordering sequences (i.e., procedures) (Malone & Crowston, 1994): accordingly, the conception of any support to coordination is based on the construction of various kinds of reliable and dependable models, in the broadest meaning referred also in Hevner et al. (2004) to be incorporated in a computer system as an infrastructure for that organization. This view is corroborated by several methodologies that have been conceived in the last decades to support this fundamental modelling step (Avison, et al., 2006) and, more specifically, by a plethora of languages, notations and models that have been proposed to define the (partial) order by which the activities are to be executed and the related information resources managed to reach a predefined goal: suffice it to mention the recent consolidation of common notations and languages to either (re)present organizational and procedural information to humans like the Business Process Modelling Notation (BPMN) or to have them processed by computational systems, like the Web Ontology Language (OWL). Here research typically follows a top-down approach: both work and how it unfolds in practice *can be* abstracted and represented in models; and these latter can bridge organizational stances and the practical design of information technologies that support work according to a reference "best one way" (Kanigel, 2005).

On the other hand, CSCW has been founded on the recognition that "the apparent stability of organizational roles and patterns of communication is a superficial hide beneath which a capricious beast is hidden," and that "work arrangements should be conceived as emerging formations that change dynamically in accordance with the requirements of the situation" (Schmidt, 1991). Accordingly, coordination is seen as a complex and dynamic phenomenon that can be understood only by focusing on how people competently behave in their everyday work. Here research follows a bottom-up approach: work, and how it unfolds, can still be modelled, but not in every case and in not every salient aspect of it; abstract models of work represent provisional *theories* (Barley, et al., 2004) of what has been observed and act as scaffoldings for the co-construction of technological supports that can diverge from their intended uses.

These radically different perspectives concretize in two different first-order concerns that well characterize the two fields. On the one hand, the focus on how to build "domain" models to be used in the design of technologies supporting coordination, given that coordination occurs only once entities have been "instantiated" in virtue of their attributes, mutual relationships, and viable operations. On the other hand, the focus on how coordination emerges as an acquired and continuously adjusting behavior that results from a situated learning process in which actors are necessarily involved while they interact with each other and fulfil their objectives.

In the IS literature, the term domain modelling is used to refer to the activity where business and software analysts elicit and describe the general knowledge that a technological system "needs to know" (as said in Olivé, 2007), as well as the body of knowledge that the designers and developers of that system have to keep in mind in the formal specification of the data and functional structures that compound the system (Hevner, et al., 2004). In today's practice and research, countless scholars and professionals deem these explicit and often formal descriptions of a portion of reality as nothing less than "necessary in order to develop an information system": every day and in every corner of the digitized planet, ICT professionals apply the tenets of conceptual modelling in the development of database management systems, network facilities, corporate application systems, and presentation systems (Thalheim, 2007) with very few and isolated defections from the mainstream (Nandhakumar, 2010). In this view, when procedures are instantiated in the real course of action the only admissible way to deal with problematic situations and model shortcomings is through some sort of guided exception handling or more or less partial circumvention of the system (e.g., Varpio, et al., 2006).

However, the IS picture is not uniform. In fact, different stances on domain modelling can be detected in different authors, which can be traced back to certain phenomenological discourses (cf. late Heidegger, Merleau-Ponty), or to some early contributions from the cybernetics debates that have been recently reconsidered (Pickering, 2004) and interpreted as anti-modellistic and *performative* in nature. For example, W. Ross Ashby (1956, p. 86) proposed the seminal concept of "black box" to denote those system whose functioning is not fully open to introspection but rather acquired "out of our performative experience of it" (Pickering, 2004, p. 20); and Stafford Beer proposed the concept of "exceedingly complex system" to apply to brains as well as to firms, to hint that they are "in principle unknowable" (Pickering, 2004). Ciborra declared himself to follow Heidegger's stances very closely when he referred to the ideal models and formal methods that are intended to capture "how things are and events flow" in terms of "illusory appearances" (Ciborra, 2002, p. 177); with this term, he meant the effect of an orientation to the world that Heidegger called "enframing" (cf. *Gestell*) and that can be traced back to the continuous human drive for a precise and scientific (or, rationalistic and "modern," in Foucault's words) knowledge of the world. In the correlated domain of user interface design, Merleau-Ponty is often mentioned as source for an alternative notion of time grounded on the claim that "time is not a line but a network of intentionalities" (Merleau-Ponty, 1962, p. 417): this dynamic network of intentionalities is applied to contrast the definition of process schemas that describe work as a series of well-defined activities that exist in time independently of the personal recollection of past events (Varela, 1999). These alternative stances can be read as a sort of distrust, or better yet, disenchantment, toward conceptual schemas, be they either classification schemas (like glossary, entity-relationship models) or process schemas (like workflows, organizational procedures)[1].

In the CSCW literature, the disillusion toward the formalization of information and organizational models is present from the very beginning (e.g., Suchman, 1987; Winograd & Flores, 1987). In some foundation papers authors provocatively asked if the abstract models in which work processes, information needs and functionalities are reified do not always end up by contributing in "automating a fiction" (Bannon & Schmidt, 1991).

*"the language of work is abstracted in a language of representation, useful to analysts. This is transformed again into an abstract formalism, chosen for its usefulness to the system implementers. The resulting system is then imposed on workers/users, taking a critical perspective, and changes the nature of the work that the representation was built on. This is a cycle that has clear potential for catastrophic change via a positive feedback loop" (Robinson & Bannon, p. 223).*

Also in the organizational literature, a point is made that "embedding fully-fledged and logically refined models in [... computer-based] tools [...] is not necessary—worse, they would more likely than not obstruct the work they are intended to support" (Button, 1992).

As anticipated above, the programmatic approach of CSCW research is to start from the observation of situated actions. From the ethnomethodologically informed analyses of these observations in the light of different analytical frameworks (e.g., Dieng, et al., 2000; Halverson, 2002), coordination is generally recognized as the situated outcome of a learning process where the meaning of actions and their resources is negotiated in relation to the context. In this ambit, analysts are called to identify those tools that support this continuous process, enable the sometimes idiosyncratic and often a-priori unpredictable alignment of activities, and finally are used by the cooperating actors to get mutual awareness of the current context in which they operate (Dourish & Belotti, 1992). In so doing, CSCW researchers come to consider and focus on those *distinct*, *persistent*, *physical*, i.e., *material*—but not necessarily tangible as rightly pointed out by Schmidt (2010)—*inscribed, situated,* and *symbolic* objects "which, in the context of a set of conventions and procedures, stipulate and mediate [...] work and thereby are instrumental in reducing its complexity" (Schmidt & Simone, 1996, p. 27), i.e., artifacts. The term "artifact" is one of the most often recurring in both CSCW and IS literature. Yet, while in the CSCW the meaning of this term has been thoroughly reconciled by a number of contributions (e.g., Schmidt & Simone, 1996; Sellen & Harper, 2001; Schmidt & Wagner, 2004; Schmidt, 2010) in the series of attributes that we have just mentioned above, in the IS field the term "artifact" has been associated with a pluralistic plethora of what also Gregor and Iivari (2007) define "somewhat convoluted definitions." Orlikowski and Iacono (2001) legitimized the expression "IT artifact" within the IS research community: they surveyed almost two hundred articles that had been published in the *IS Research Journal* in the 1990s, and from these contributions they could detect 13 different views for this concept, in their turn grouped together in five main categories: namely, the tool view, the proxy view, the ensemble view, the computational view, and the nominal view. Although the cooperation- and organization-oriented conceptualizations of the concept of artifact that, respectively, the CSCW and IS literatures have contributed to build in the last twenty years cannot be easily superimposed due to some of their subtleties, they certainly share some point of contact, being perhaps the coordinative perspective that is typically implied in CSCW contributions mapped by what Orlikowski and Iacono define the 'tool' and 'ensemble' views of technology. Perhaps due to the ethnographic and *technomethodological* (Button & Dourish, 1996) stances of much CSCW research, this latter has focused more than the IS field on the physical objects and the modalities in which coordination is achieved by the mutual arrangement of and interaction between human actors and their tools. As a matter of fact, when a technology has to be conceived to support coordination, the CSCW approaches are primarily aimed at uncovering these habitual and pre-existing tools first, rather than at conceiving novel ones from scratch as expression of an abstracting and rationalizing endeavour; these approaches are also aimed at observing *how* these habitual and often merely paper-based artifacts are put to work by the involved stakeholders to have their work done in the most effective way. The focus on "how" these tools are put to work, i.e., on technology-in-use (Button, 1992), is not incidental and certainly it is not the sole prerogative of CSCW; yet the focus on artifact that the CSCW tradition inherits from workplace studies, which are "concerned with the ways in which tools and technologies feature in work and interaction" (Heath & Luff, 2000, p. 4) and from anthropologically-informed analyses (e.g., Hutchins, 1996), makes a strong difference with those studies where the material artifact is just seen as the interface of a more complex "IT artifact" that is usually characterized in the terms that Orlikowski and Iacono (2001, p. 127) recognize in the "computational view" of information systems, that is as "gateway" to and "proxy of" the conceptual models of the underlying technology.

Conversely, in CSCW "artifacts are there for us when we are introduced into a certain activity, but they are also a product of our activity, and as such they are constantly changed through the activity" (Bannon & Bødker, 1991), and this is in line with the IS recognition that artifacts "are not static or unchanging, but dynamic bundles of material and cultural properties packaged in some socially recognizable form such as hardware and/or software" (Orlikowski & Iacono, 2001). In addition to that, we have also observed how these "changes" do not have to be necessarily physical modifications or tweakings, but rather they can regard how users use the artifacts, i.e., their conventional use (Cabitza, et al., 2009) as well as how these artifacts "react" to their interactions even after they have been deployed and adopted in an organizational setting, as we will discuss later.

The interpretations of coordination that we have outline above characterize the typical approaches of IS and CSCW scholars about the crucial objective of designing a technology that could help the organization achieve its mission, hopefully in an effective way. These perspectives differ in some points, obviously. The question is: is there any way to reconcile them? Are these both necessary for the success of the organization or, at least, for the success of a digitization initiative?

## RECONCILING SCYLLA AND CHARYBDIS ON ARTIFACTS

The answer to these questions has to do with the well-known tension between the global and the local, the top-down and bottom up, and the technologies that these different perspectives call for. We do not want to deal with this tension by negating one of the two sides. In fact, the global/top-down view is necessary to understand how local behaviors are to be harmonized to obtain results at the macro organization level. On the other hand, a local/bottom-up view is necessary to make these local behaviours efficiently and effectively contribute not only to the local mission but also to the global one. Then, the question is how to reconcile these opposites in an approach that recognizes their value, their diversity and finds a way to let them co-exist.

The reconciliation we propose in this chapter is rooted in a class of "tools" that are widely used within organizations, let's say before any sort of digitalization: the paper artifacts that often take the shape of semi-structured forms. IS and CSCW share an interest in these artifacts. On the one hand, it is straightforward for the IS research to consider organizational artifacts since these play the natural role of information sources, i.e., tools into which data are entered and from which information is fetched. On the other hand, there is a consolidated understanding, within CSCW, of the role of artifacts in cooperation with a specific focus on traditional and paper-based artifacts (e.g., records, forms, structured documents) in various application domains. However, IS and CSCW focus on complementary issues.

The digitization of paper-based forms usually occurs along with the introduction of an IS application where there is none or where existing applications have to be significantly enlarged or improved, e.g. to meet compelling administrative or economical needs. In this scenario, ICT introduction involves understanding not only what functionalities to endow users' artifacts with, but also, more specifically, what the role of these artifacts is in the actual practices of users. Usually digitization is intended to preserve and enhance the role of artifacts as information *archives* and information *sources*, especially in their function of 'gateways' to the underlying information system. In so doing, however, the role of the traditional artifacts as *coordinative tools* of the daily work can be irreversibly undermined (Luff, et al., 2000). For example, in the CSCW literature about the healthcare domain, there are several evidences (Heeks, et al., 1999) that reckless drives to make the EPR a better legal document and administrative tool, tend to jeopardise the patient record's fundamental role of tool for intra and inter ward communication, for horizontal and vertical cooperation and for the achievement of an accountable quality of care. Likewise, our studies in two wards of a large provincial hospital (Cabitza, et al., 2005; Cabitza, et al., 2009) confirmed other ethno-methodologically-informed accounts in the same literature (e.g., Heath & Luff, 1996; Berg, 1999; Fitzpatrick, 2000) that observed how paper-based patient records are used not only to store and retrieve information but also to support mutual learning, knowledge sharing and coordination of caring activities. The general conclusion these studies achieve is that paper-based patient records do not play a single, clear-cut role and that the artifacts that compound them cannot be simply 'replaced' by electronic counterparts, even if these latter were designed to mimic the traditional artifacts quite closely. Their introduction inevitably also leads to a complex and often unintended transformation of the whole clinical practice, since the logic governing how data are presented, in which order, how they are to be inscribed, etc., is shaped by procedures that are conceived by "looking at the global" picture and are hardwired in the application logic, so as to remain almost inaccessible to the users. This fact has serious consequences at the local level, e.g. over the quality of care, which cannot be underestimated. The case of Electronic Patient Record (EPR) is paradigmatic but other examples could be reported in other situations where human improvisation and creativity are important. All that said, the aim to reconcile the IS and CSCW perspectives toward a more comprehensive approach requires answering the following additional questions: can these multiple roles, especially the archival and coordinative ones, coexist within the same digitized form? and if they can, how can the digitized form be computationally augmented to better support them? The chapter tries to answer these questions by articulating the two roles in terms of the purposes they pursue and proposing a technological support based on our empirical investigations.

## TWO SIDES OF THE SAME COIN

Since both the archival and coordinative roles of records are equally important and mutually sustaining, the answer to the above questions implies that design must take them both into account and avoid the following possibilities: giving a single facet technological solution or thinking of a mere juxtaposition of unrelated solutions. This premise is not as obvious as shown by the EPR case. In fact, the view of the paper-based patient record[2] that upholders of its digitization generally advocate is that of a passive *repository of information* (Coiera, 1997; Dick & Steen, 1991). This point of view has so far heavily conditioned the design and development of EPRs from their first projects in the early 1970's. Even if EPRs have recently evolved in terms of flexible and good-looking web-based applications, their interfaces are still conceived as an access point to a data repository, and their business logic corresponds to some optimized procedure that maintains this repository up-to-date and consistent. However, it is known that EPR are not well accepted by practitioners: in fact interpreting the clinical record as a mere access to data repositories 'fail(s) to meet its essential purpose' (Korpman, 1990), i.e. to support practitioners in healing patients, since it would lack mechanisms for, e.g.., "focusing attention, patient-specific consultation, data analysis, and for implementing quality assurance and cost management policies" (Dick & Steen, 1991).

The CSCW field has so far contributed to this issue with several ethnomethodologically informed studies (e.g., Heath & Luff, 1996; Berg & Goorman, 1999; Hartswood, et al., 2003) of ward work that have convincingly and strongly countered the conceptualisation of the patient record as mere 'repository' of data. Fitzpatrick (2000), for instance, advocates the alternative conceptualisation of *record at work*, i.e., of an artifact supporting the practical delivery of care rather than just its documentation, and proposes to distinguish between the "archival patient record" and "the working patient record" (Fitzpatrick, 2004), which "is more than passive archival views over post hoc data." From a broader perspective, other studies (Heath & Luff, 1996; Bardram, 1997; Varpio, et al., 2006; Saleem, et al., 2009) report of hospital ward settings where clinicians have to invent ways to 'work around' archival (often electronic) systems that hinder their caring activities and where they give way to the double work of maintaining also informal and primarily coordinative artifacts.

That notwithstanding, depicting the archival function black and the work-supportive one white can result in an oversimplification of a much more blurred picture, especially in the most problematic contexts as hospitals are. In fact, any artifact conceived to play an archival role, once put to work, does not become merely the passive counterpart of a different supporting artifact: to the contrary, it can become 'working' in at least two ways. On the one hand, it coordinates tasks across different users of the same organization—e.g., between clinical and clerical practitioners. This role is usually characterized by the use of proper coding standards and classification schemas that aim at making communication across organizational borders possible and effective (Bowker & Star, 1999): this happens in general at the cost of missing some detail, whose importance is under lively debate (Winthereik, 2003). On the other hand, this artifact is definitely put *to work* also well far from the patient's bedside, by clerical practitioners and researchers who use (or better yet, reuse) clinical data to fulfil their specific goals. This is the way by which the global and the local coexist with the implications discussed in the previous section.

This evidence proves that solving the dichotomy of the two roles by means of a 'two artifacts' approach is less than appropriate. In fact, one should immediately dispel the idea that one of these artifact could be part of the other: neither they simply share some parts, nor a boundary between them can be easily traced. Unquestionably, work practices show that these complementary artifacts are the very *same artifact*. Berg and Bowker (1997) suggestively speak of a multiplicity of *layers* (or better yet, *bodies*) cohabiting the same patient record; layers where also time can flow at a different pace. For example, in the neonatology ward record we studied, the fluid balance exists in time measured in hours, the skull development curve in weeks; some pages capture a time span of two weeks, while other pages hold just a single day. The analysis of Berg (1999) goes even further, contributing to the understanding that these multiple layers do not share only the same structures and content, but also, to some extent, the same practices of reading and writing. Indeed, since the same artifact dynamically represents both the state of the patient's condition and the state of the work done or planned *around* and *on* the patient, "the coordination of activities is achieved through the accumulation of inscriptions, and vice versa" (Berg, 1999). This finding is also corroborated by informal surveys reporting that hospital practitioners can spend on 'documentation work' up to the same time they do in 'treatment work' (e.g., American Hospital Association, 2001); this is due to the fact that caregivers are documenting not only for themselves and their work, but also for someone else, typically the clerical (administrative) practitioners, for other purposes than "caring." Finally, also our field studies within two wards of different speciality and criticality (a Medicine Internal ward and a NICU—Neonatal Intensive Care Unit) leaded us to see how work accomplished in a hospital ward brings both the archival and the working and coordinative role of the clinical record artifact to life, at the same time. What is needed to find an appropriate reconciliation of these different layers is then a better understanding of the purposes involved in these two roles.

## ALL A MATTER OF DIFFERENT PURPOSES

Specialist literature distinguishes between *primary* and *secondary* purposes of the patient record (e.g., Mann & Williams, 2003) and this distinction can be adopted in the analysis of other domains.

The *primary purpose* of clinical records is to support direct patient care both by aiding medical decision-making and by "ensuring continuity of care by all providers" (cf. the JCAHO definitions). Decision-making is usually accomplished as simply as by aiding the physician's memory on the case in hand. Care continuity regards the need to support *articulation work* (see below) and communication among practitioners so that the caring trajectory can seamlessly unfold in time and space, even across responsibility gaps (e.g., within internal shifts and external transfers), during the patient's stay in hospital. The primary purpose is strictly related to the coordinative role of records.

In fact, articulation work (Strauss, et al., 1985) refers to the "meshing" of tasks, efforts, and actors involved in managing the numerous illness trajectories occurring simultaneously in the ward. Each of these trajectories includes not only the physiological unfolding of a patient's disease, but also the organization of the work done over that course and the impact for the ward staff and the whole hospital. Focusing further on the ward dimension, Strauss detected *three levels of articulation work* that refer to the primary purpose of caring patients as effectively and as efficiently as possible. At the first level, *physicians* collaboratively *plan* the so called 'illness trajectory' according to the big picture of the course of illness that they can draw from the peculiar signs manifested by the patient at each encounter; at the second level, the *head nurses organize* the physician's requested jobs and other activities according to the continuous and event-driven assessment of the interventions necessary to fulfil the patient's needs; at the third level, there are the *actual jobs*, accomplished to arrange resources so that tasks can be actually carried out by *all the involved people*. The last two levels regard vertical collaboration, i.e., collaboration between practitioners with different responsibilities and competencies. To this aim, the clinical record allows for orders to be explicitly expressed and transmitted and it encompasses adequate ways for executors to give officially some sort of feedback and confirmation of task accomplishment. Conversely, the first level of articulation regards a horizontal dimension of the group of physicians that are accountable for the medical dimension of care: from the first involvement of the physician who admits the patient in, up to the physician that will sign the discharge letter, through all the doctors that see the patient and judge her conditions over several shifts. To this aim, the clinical record is used both as a planning tool for the patient care and also as a record where to track the course of treatment and the changes in the patient's condition that could make the initial plan be reformulated. In our observations, for example, we have also seen as the physician's clinical diary, besides being a place for personal annotations, is above all a place where each medical decision is motivated both for legal reasons as well as to enable colleagues to understand the rationale behind a decision and possibly change the corresponding treatment regimen if necessary.

On the other hand, the archival role of records is calibrated on their *secondary purposes*. First of all, the legal one: the clinical record helps to secure the legal rights of both the patient and the clinician. In fact, on the one hand, it is supposed to be the faithful and neutral account of all the activities the patient has gone through, even the faulty ones; on the other hand it also permit to reconstruct the context in which the clinician had to take a life-saving decision, so as to release her from hindsight liabilities (Young, 1987). Accordingly, under the motto that "accurate charting is best defense," physicians and nurses have then been persuaded that spending time on accurate charting would make charts serve their legal interests, as well as ensure patient safety, reduce medical errors and contribute to a more effective risk management. In addition to the legal purpose, institutional manuals and policies state that clinical records must also serve to support healthcare service management and scientific research: the clinical record is in fact the main basis for billing and reimbursement and it is the main source of information both within a hospital (e.g., for care quality evaluation, resource planning and cost management) and outside any caring setting for a large number of scopes, like to enable clinical research surveys, epidemiology studies, clinician education, post-marketing surveillance of drugs, public policy making, facilities planning, and health care statistics.

The coordinative (primary purpose-oriented) and archival (secondary purpose-oriented) layers are about both information and the processes that accumulate, exploit and share this information. What differentiate them are the *purposes of the people using it*. In fact, practitioners perceive them from within situated practices, and make sense of data according to their interests and needs, as well as to the context where work is carried out. This sense-making activity reinvents each time the proper relevancy of data and establishes contingent quality attributes both for processes and data: i.e., respectively, "how things should be really done" and "which information really matters" to be the most suited to cope with a given situation. This high degree of context dependency makes attributes sometimes and to some extent paradoxical, in that they could require information be either redundant, provisional, or even ambiguous (e.g., Berg & Goorman, 1999; Hardstone, et al., 2004; Cabitza, et al., 2005). Examples of such paradoxes will be given in the next section.

## SOME CASES FROM HEALTHCARE WORK

From the early seventies, digitization enthusiasts have placed their rationalizing hopes in EPRs in the wishful assumption that all relevant clinical information could be held in a structured representation to be manipulated by a digital system (Rector, et al., 1991). Optimistic visions of EPR deployment began to see the clinical record as an artifact on which caring activities naturally "leave a trace," and hence as an artifact able to implicitly reconcile both primary and secondary purposes and make a high degree of integration possible.

The problematic nature of using and aggregating primary medical information for secondary utilization has been initially translated into a first drastic law: "data shall be used only for the purpose of which they were collected" (van der Lei, 1991). More recently, a more realistic law has been formulated to acknowledge that data are often used outside the context of their production, to be processed according to other purposes: "the further information has to circulate (i.e. the more diverse contexts it has to be usable in), the more work is required to *disentangle* the information *from* the context of its production" (Berg & Goorman, 1999). This same point is also taken by Goguen (1996), when he speaks of the need to *dry out* information of its contextual and social aspects, through abstraction and formalization, in all those cases it has to be moved among different contexts of use, i.e., to become what Latour (1987) called an "immutable mobile." In the clinical domain, this means that *other work than clinical one is required* to make clinical data suitable for their accumulation into an information system that could enable all the secondary purposes we have outlined above. From this plain consideration, a series of relevant questions arise: Who should be supposed to accomplish this additional work? How could ICT help? Is disentangling data from context a one-way and irreversible operation made upon original data, or rather it is something that is aimed at decoupling raw data (if such a thing can exist) from contextual data and presenting them both to the user? We will come back to this point later, as our proposal takes a clear stance on this latter point.

Here we want to show some cases in which it is clear that coping with this 'other work' is anything but a trivial or technological task, as it is even clearer taking the 'information quality' perspective. Practitioners use their artifacts to get 'information of good quality' to perform their daily activities, i.e., relying on the coordinative role of the artifacts themselves. For instance, clinical notes, vital parameters, even official orders reported on the clinical record are meant to be read and understood by small communities of colleagues that often work side-by-side—at arm's length (Schmidt & Bannon, 1992), that strengthen their relationships in highly stressful situations on a daily basis, that have frequent and extempore conversations to share opinions and knowledge and can count on official occasions to give each other full details about what they are working on (the handovers; cf. Cabitza, et al., 2005). These practitioners share a rich common information space (Bannon & Bødker, 1997), i.e., *not only the data* that could be easily stored within an IS database, *but also the meanings and interpretations* that are achieved over those data by single practitioners on specific occasions of use (Schmidt & Bannon, 1992).

Our studies in the hospital domain have confirmed the understanding (Berg & Goorman, 1999; Hardstone, et al., 2004) that care practitioners are able to coordinate with each other in a very effectively manner not *even though*, but rather *because* they do communicate in shorthand, omit irrelevant or self-evident information, and use discipline—if not team—specific jargon. In the coordinative layer of the patient record, care practitioners find it natural to manage conflicting statements, since disagreements among observers and changes of opinion reflect the reality of clinical practice (Rector, et al., 1991); they are also used to leverage the richness of descriptions that are recorded at an arbitrary level of detail and at the clinician's natural level of abstraction.

We give just two examples of this important point: let us consider the "RDS" and "infusion sheet" cases occurred within the NICU we studied. A common disorder affecting premature newborns is the Neonatal Respiratory Distress Syndrome (RDS), resulting in clogged respiratory passages and severe respiratory difficulties. In different situations, we observed clinicians perceive the presenting problem at different levels of abstraction, and indicate it in the specific "problem list" sheet which compounds the first part of their clinical records, as either "grunting breaths," "tachypnea," or "RDS." This practice makes the distinction between, respectively, "signs," "symptoms," and "diagnoses" blur into the background of the clinical observation but it perfectly makes sense according to the situation. In fact, in many cases, to "close a case" and give it a univocal label can prevent other physicians from expecting the "possible improbable," and hence avoid serious complications. As regards this point, Rector, Nolan & Kay (1991) warn that forcing clinicians to comply with a post hoc rationalization and standard coding of their observations could result in distorted records and failures in communicating decision rationales to other clinicians (i.e., the overall "context for interpretation" [Bannon & Bødker, 1997]). The infusion sheet is a pivotal artifact of neonatal care and the only one that is always at the bedside of premature infants. In this artifact, nurses are supposed to precompute the quantity of solution that should be administered to the newborn at certain hours; when it is time, they are also supposed to check the actual infused quantity and annotate the correct values aside the precomputed ones. Actual data serve two different purposes: to survey that the patient is given the right amount of drug; and, secondarily, to monitor how much drug the infusing machine is unpredictably retaining within its meanders. For the primary purpose, accurate data of infusion are almost irrelevant: in fact, sometimes nurses just dot a mark if the actual level is almost the same as the expected one; other times, if the two levels are more than slightly different, either they record the difference, or they report the infused value (as they would always be supposed to do), while not caring too much how precise their reading on the catheter scale is. Nurses valuate discrepancies and choose what to do according to several things: delays in the routine schedule, the newborn's conditions, infusion velocity and the "known" (or better yet, expected) behavior of the infusing machinery. The primary purpose comes first and the secondary purposes can be hindered or even completely thwarted. In the case of the infusion sheet, the intrinsic qualities of data accumulated during the care, e.g., accuracy and completeness would be almost completely unsuitable for any secondary processing, such as that of detecting either structural defects of the involved machineries, or their patterns of drug retention. These considerations bring us to the core point of this paper: information quality can be characterized differently in relation to the different situations where information is used.

## WHAT INFORMATION QUALITY TELLS OF INFORMATION

In the previous sections, we have considered complementary perspectives on how to study and design a technological support of work in organizational domains: we have purposely simplified a very complex matter by identifying recurrent antithetic concepts: top down vs. bottom up; global vs. local; conceptual modelling vs. ethnographic observation; process models taken as either maps for users or scripts to embed in computer-based information systems (Schmidt, 1997). All these complementary (if not conflicting in actual professional practice) approaches and perspectives are paradigmatically instantiated in the discourse on quality. It sounds perfectly reasonable to agree with Redman (1998) that the quality of processes, and in general the performance of a business, is affected by the quality of the data that are used by practitioners to take decisions and communicate; that notwithstanding, little research so far has been accomplished to address, or at least estimate, this relationship, with the exception of studies on a particular kind of work that is managerial decision making (e.g, Chengalur-Smith, et al., 1999). All in all, we also start from the point that almost any kind of work relies on the quality of information manipulated and shared within a cooperative setting, be this work performed in either secretarial and administrative domains (i.e., so called data crunching settings) or in those domains where practitioners rely on timely, accurate and complete information to make apt and appropriate decisions in critical domains, like in medical work, or to smoothly coordinate with other distributed teams on a tight schedule (e.g., civil engineering, enterprise resource planning, hospital work). Just to limit the scope to our reference domain, also in the healthcare sector information holds great importance, since the performance of almost any management systems—from prospective payment systems, legal guarantee systems, risk management systems, data management systems for clinical trials, and epidemiological research—is clearly affected by the quality of managed data (Gendron & D'Onofrio, 2001; Parker, et al. 2006).

In the white literature, as well as in the specialist one, Information Quality (IQ) is usually related to either the compliance to a reference model (i.e., a target level, a standard), or to the degree users feel satisfied in their cooperative environment and supported in achieving their goals, i.e., with the so called "user satisfaction" (DeLone & McLean, 1992). In the former case, the problem of quality is related to the *production* of the information; in the latter case, symmetrically, to the *use* of information. In what follows, we will discuss the main lines of these two approaches, which are usually tightly intertwined in practice, to recognize their common assumptions and then create the premises to propose a different perspective, which underlines and motivates our technological solution.

The theme of IQ and the related standardization efforts is one of the primary goals of consultancy initiatives focused on the rationalization of business and management and of several IS research projects as well. Historically, standardization has always been meant as one of the strongest factors that enables a tighter integration among businesses and a better interoperability between production systems; more recently, standardization is seen as a 'sine qua non' condition to rationalize and interconnect several value chains (Monteiro, 2003) into more efficient and valuable "networks of chains" (Poirier, 2003). Generally speaking, any sector where an uncontrolled increase of exchanges and transactions had occurred has been urged to find and implement new adequate means of control and standards by which to articulate, compare and assess heterogeneous provisions of either energy, material, or information (Beniger, 1986). More specifically, within the IS field, a prolific debate has recently got livelier and livelier about how to define, assess, guarantee and even improve information quality in diverse domains (Batini & Scannapieco, 2006) and the concept of *standard*[3] *for information quality* has been proposed for any domain where data really count.

Standard setting and the consequent job standardization is considered the "cornerstone of scientific management" (Nelson, 1992, p. 67); consequently, there is no wonder that one of the most influential contribution to the IQ literature has been the conception of *information as a product* (Wang, 1998); this is true for a number of reasons. The product-oriented view of IQ permits to see the whole information production process as a *manufacturing* process, which can leverage on decades (if not centuries) of best methods and optimized tools for its continuous enhancement toward higher and higher efficiency. At the same time, this perspective allows to see information itself as an end-product, and as such as something that is stored in a consolidated and reliable technology such as databases; that can be transmitted quickly and cheaply to almost everywhere through digital networks; and that is used—and often *re-used*—at need (Ballou, et al., 1998) thanks to very powerful retrieval techniques. As such, also the concept of "fitness for use," i.e., the extent to which an information successfully serves the purposes of its consumers (e.g. Strong, et al., 1997), can be operationalized as the extent a product meets either an agreed specification (Kahn, et al., 2002) or the favor of "data customers" (DeLone & McLean, 1992). It is important to stress that setting standards to comply with is not a means to improve quality indefinitely, but rather a means to specify—and hence detect, correct, prevent—a maximum tolerable number of "defects" that would hinder information fitness and imply "non-quality costs" (Batini, et al., 2008) that are deemed higher than the costs due to guarantee the desired level of (minimum necessary) quality (Beniger, 1986, p. 308).

Conceiving quality in terms of standards and information in terms of a product, be it either a *commodity*, i.e., as a substance that is present in the world and independent of its medium (Shanks, 1999; Nunberg, 1996), or a *service* (e.g., Pitt, Watson & Kavan, 1995), makes them measurable concepts but can also lead to overlook the active role that consumers can play in molding that "substance" and giving it representational concreteness on the artifacts they use to store and manage information as well as to support problem-solving and coordination. In other words, a product-oriented approach purposely disregards the situated goals information is either produced and consumed for; the quite obvious assumption lying here is that to consider all the possible kinds of production ambits, all the possible types of consumers of a piece of information (whenever producers and consumers do not coincide), and all possible uses a piece of information can be made of, would be an unfeasible (i.e., too costly), if not impossible, undertaking. Indeed, even if users are categorized in multiple profiles to filter data accordingly, e.g., in a hospital setting, the roles of doctor, nurse, clerical and laboratory employee, to set particular standards for each of these roles can be problematic, since we have observed that single users can use the same information in manifold ways for manifold purposes, even conflicting ones, and that their purposes can change over time. Thus, relying on bare categories to determine how to manage and supply good data still demonstrate how current engineering methods and digital supports lag behind in terms of necessary flexibility and adaptiveness.

In the light of the exemplificative cases described earlier, it is then clear that the notion of quality of a piece of information can be inflected in at least two ways: on the one hand, quality as an *intrinsic* concept, which is independent of the context in which data is produced and used; intrinsic aspects regard the syntactic (e.g., consistency) and semantic (e.g., accuracy and completeness) aspects of a datum (Stamper, 1992). On the other hand, quality as an *extrinsic* concept regards the pragmatic and social levels of information, i.e., its usage and its agreed understanding within or across communities of users (Shanks, 1999). In the specialist literature, extrinsic quality is also defined *contextual quality* (Strong, et al., 1997), to hint that it regards more "*relations* rather than *essences*" (Winthereik, 2003) and that it cannot do without taking into account the context where the information has been produced and used, i.e., the *context for interpretation* (Bannon & Bødker, 1997).

While intrinsic quality underpins most of the initiatives aimed at defining suitable metrics to quantify and evaluate quality levels in any work setting, contextual quality is usually left in the background of qualitative accounts addressing complex organizational problems (Strong, et al., 1997) without specifically considering the role of technology in these problems. Maybe this is due to the fact that contextual quality is much more difficult to assess and quantify than the intrinsic one. For example, the few surveys available for the clinical domain (e.g., Arts, et al., 2002) reach the conclusion that approximately the 5% of all clinical data extracted from clinical records and documents are *intrinsically* incomplete, inaccurate and inconsistent; yet, these figures regard only the kind of information that Garfinkle and Rawls (2008, p. 73) denoted as "material," i.e., pertaining to raw facts about the world, that is evaluated according to globally-good and top-down imposed reference values acting as a gold standard[4]. Conversely, in actual work and to make informed decision, we observed how another kind of information is involved, what Garfinkle denoted as "eidetic" information, which pertains to what the world actually *means* to someone involved in a communicative act. Thus, the contextual quality of the content to be found in clinical records and artifacts often eludes percentage-oriented measures (what a 100% would stand for?); rather, it regards the proteiform linkage between the data reported in the clinical record and the caring processes that the elementary tasks of reading and writing data are related to.

We can make a more specific example with reference to Section 6: even if the night shift nurse had not explicitly reported an RDS problem in the record, the network of links grouping together annotations such as "prematurity," "tachypnea," and "cyanosis" would make perfectly sense for the physician during the morning round and constitute a clear picture of complicating RDS. Likewise, the no-value marks reported in the infusion sheet could be deemed as incomplete only from the intrinsic quality perspective. Conversely, if they are related to other coeval parts of the same medical record where a stable and overall satisfactory health condition is reported, they would be anyway of great value for the following practitioners since they could rightly interpret those data as 'nothing to report,' 'everything as expected,' instead of an alarming 'missed to check.'

To achieve our goal to design a technology that does not disregard the pragmatic value of contextual quality but that rather conveys information enriched in some way to also guarantee better quality in both work and communication, we need to go beyond the mere contraposition of intrinsic/extrinsic IQ found in the specialist literature, as we detect it as still lying in the mold of the product-oriented conceptualization. We should then consider how the current debates on IQ are grounded on a concept of information that dates back to the historical period in which several scholars from different countries and disciplines met each other to discuss in more or less formal way of the nature of complex systems (like brains and business firms) at the so called Macy conferences in the early 50s. A short nutshell of history is in order here: It is well known that in the cultural milieu where the word cybernetics was also coined (not by chance), the conceptualization of information that assured to a sort of standard status was that proposed by Wiener and Shannon; these were certainly influenced by the coeval works in thermodynamics and quantum physics that contributed in having an entire generation of engineers, scientists and mathematicians accept uncertainty as a fundamental characteristic of human experience. Accordingly, Wiener and Shannon proposed to take any piece of information as the "choice of one message from among a range of possible messages" (Hayles, 1998, p. 52). In doing so, they were considering information as a probability function that is independent of the medium representing it and "different" from the noise with which it travels from a sender to a receiver. In the Hayles' (1998) words, this conceptualization, which encountered an extraordinary favor also because it allowed for the strictly quantitative treatment of information and for the formalization of information systems (Avison & Fitzgerald, 2003), contributed in having information "divorce from meaning" (p. 54). This posed also the basis for the very conception of the digitization of information and its "commodification" in a storable and transferable entity that keeps stable as it is moved from one context to another but that also exhibits the fundamental property of having intrinsic properties, such as a value, which can be transformed as output of automatic processing and which, most notably, can be improved till matching some reference value (cf. the concept of standard, mentioned above).

It is less known, yet, that at the same Macy conferences, and hence in the very same milieu where the foundations of computer science and information sciences were posed with such a cultural stance, a British researcher, Donald MacKay, argued for an alternative and complementary (if not opposite) conceptualization of information, which he called "structural." Structural information is eidetic information, which is inextricably linked with a context, and is related to specific changes in someone's mindset[5], according to her (both natural and nurtured) structures as a "receiver" (MacKay, 1969). MacKay was therefore proposing a concept of information whose peculiar nature regarded a relational dimension where meaning is *negotiated* and constructed every time "new" between communicants instead of being something that is selected from a range of admissible or predefined entities[6]. Representations, in their turn, are seen not only as information carriers about a reality of interest, but also as structures, which are interactive to the extent they affect changes in the observer's behavior[7] and are manipulated accordingly.

Thus, as noted by Marvin (in Slack & Fejes, 1987, pp. 49-62), the decontextualized conceptualization of information proposed by Shannon and Weiner, which is still very popular in professional practice and computer science, has ended up subsuming important ideological implications, like an unbalanced consideration that digital and disembodied information is more important than more context-bound, *analog* information (i.e., information as it is enjoyed by its users/consumers). On the contrary, MacKay's model puts in due consideration the interplay between the message and the receiver, form and content, meaning and representation. Even more importantly for our proposal, MacKay also proposed to recognize a strong and unavoidable correlation between the nature of a representation—that is, how an information looks like, how an artifact affords it—and its effects, that is, the capability of a piece of information to affect (inform) one's behavior in that it is able to evoke a knowledgeable response. To this regard, we like to mention also how Tzannes put this correlation down: whereas Shannon and Weiner defined information in terms of *what it is*, MacKay defined it in terms of *what it does* (Tzannes, 1968). In other terms, MacKay created the conceptual room for a pragmatic stance over information, well beyond the idea that information has to do solely with semantics: a perspective this latter that forty years later is still on the crest of the wave in terms of "semantic web," computable models of work and ontologies. Similarly, Hayles notes that where the representational definition of MacKay is adopted, information becomes more an action than a thing; rather than a product it becomes a process that someone has necessarily to enact in some context, be she either the producer or the consumer of concrete representations that affect someone else's behavior (Hayles, 1998, p. 56).

## DESIGNING SUPPORT TO MAKE SENSE OF INFORMATION

In the views of much of the literature about EPRs, the functionalities that correct 'bad' clinical data, store the good ones indefinitely, and almost instantaneously transport them in any place other than the point-of-care are among the most often mentioned (Institute of Medicine, 1997); these data-oriented functionalities are considered so important to justify the sacrifice of the whole paper-based record on the altar of the "paperless hospital" myth (Harper, et al., 1997). Berg and Goorman (1999), aware of the risks that such simplistic approach could imply, proposed to consider the commodification and transport of medical information as akin to *disentangling* information from its context of production. One of the disentangling methods to invest on, they suggested, is to code data, that is to make data suitable to be transported from the hospital wards, where they were created, to any other external facility[8], e.g., the clerical offices where they would be used for secondary purposes (Winthereik, 2003). Bannon and Bødker (1997) suggested a similar approach, when they spoke of "packaging information" to denote the *additional* task (with respect to the mere operation of documenting an event or fact) of enriching an information item with some explanation or rationale of its context of production; and, the other way round, when they spoke of "unpacking," the activity of readers who extract this meta-content and make sense of it *also* in virtue of local conventions, background knowledge and mutual acquaintance. *Disentangling* and *packaging* seem similar approaches, where the only difference is perhaps the extent the original and "local" data must persist once they have been normalized to be shared in a more "global" dimension: as less as possible in a disentangling operation; as much as possible in a packaging operation, as the terms themselves seem to suggest.

Both these operations can be accomplished on free text notes that are necessarily accumulated in EPRs, either automatically, (e.g., Baud, et al., 1992; Letrilliart, et al., 2000) or semi-automatically thanks to the explicit intervention of either doctors or specifically trained clerical employees.

To explore the feasibility of this latter case, we were involved in the validation of a prototypical EPR, ProDoc (Cabitza, et al., 2009b) where doctors could create annotations aside the content of their electronic records. In a later release of this prototype, and differently from other similar experiences (Bringay, et al., 2006), we also allowed doctors to annotate "semantically" the data they had produced: to this aim, doctors could select any portion of their notes or values of interest in the record and associate these data with one or more semantic "tags"; this could be an item chosen from a list of items previously populated from existing lexicons (e.g., the acronym list adopted within that hospital or imposed by the health regional authority), and standard coding systems (e.g., the WHO International Classification of Diseases, Functioning, and Disability), as well as from locally-defined or even user-specific tag lists that could grow through use. Obviously, any additional effort, especially if expected to come from doctors and nurses, must return something to the users involved: thus, this kind of textual annotation that is enriched with references to "semantic" resources can repay the annotators with functionalities that are specifically aimed at supporting alignment of meanings (e.g., Simone, et al., 1999); at facilitating the retrieval and the retention of intended meanings as well as some anticipated pragmatic implications (e.g., Cabitza, et al., 2006); as well as at raising the users' awareness of reciprocal expectations and potential misunderstandings, as we will present in the following.

The technological proposal that we are to outline is deeply influenced by the conceptualization of information seen in the previous section and that we traced back to MacKay, which conceives an information as a "potential for action," an opportunity to affect one's thought, on the basis of interpretations and interactions that are a priori unexpectable, and hence unmodellable. To this proposal we were brought by the trivial observation—also reported by Clarke et al. (2003)—of how doctors and nurses end up by really trusting their charts: they consider an information be "fit to use" and "fit to act upon," only when they feel *confident* about it. This usually happened when they could have a clear picture of what they can do with a piece of information (here adopting a pragmatic approach toward it, rather than a merely semantic one), and could get an even cursory glance of the many other contextual data that an information refers to: for instance, which one of their colleagues has recorded it (and therefore how "expert" and reliable she is); in which kind of situation she did it, e.g. routine, emergency, frantic arrangement (and hence with what chance of making a mistake or omission); which patient she was after; for which reason she decided to record that information (e.g., on either a mandatory or voluntary basis); and so on. Thus, we saw that each time quality pertained to the capability of the user to answer questions like the following ones: is the infusion going well? Does a nurse or a physician have to intervene? and if so, who has to? Is the new infusion machinery worth the investment? Is a 1.5% chloroprocaine infusion at 0.3 ml/kg/hr rate as effective as a 0.6 (provided that the less is administered to infants the better)? As the reader will understand, these questions are very difficult to address with the domain knowledge that can be reified into organizational structures, process models or even alerts embedded in the validation controls of an information system.

*Figure 1. A conceptual architecture of the proposal: a component called 'AM engine' monitors the content coveyed by the information system through its artifacts and executes modular mechanisms that change visual aspects related to the presentation of data (cf. the border depicted around datum 'D')*

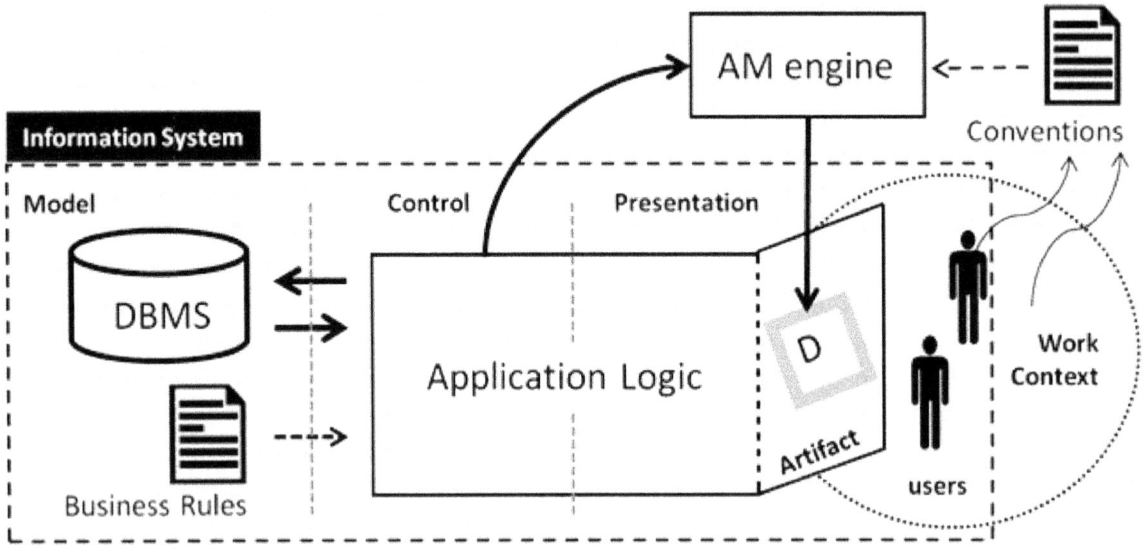


Thus, we thought to inscribe some of the contextual information that would allow competent users to become confident of their data on the artifact itself, just close or around those data, and convey at interface level in a conventional and necessarily purely graphical way[9] some clue for their interpretation, a sort of under-specified meta-content. Likewise, we thought that we could also convey in the same way suggestive indications about the objectives, interests, and quality requirements that had been expressed by the corporate management, so that different stakeholders could achieve a mutual awareness of possible differences in their local and idiosyncratic expectations, and hence achieve a better understanding of each other's viewpoints and purposes. In our view, the way to reconcile the coordinative and the archival roles of artifacts (or in other words, the primary and secondary purposes expressed in any work setting) is rooted in the design of a meta-content layer (see Figure 1), juxtaposed over the presentation layer of an information system, that could help users recognize and rediscover the essential hiatus between the context of information production (i.e., its intended pragmatic value) and the a priori unknown and potentially incompatible contexts of its later use (i.e., its support of action towards some goal).

## A Construct that Affords Clues for Interpretation

In this section, we discuss in better details our proposal for a technology that is grounded on the relational and performative stance toward information, quality, and work described in the previous sections. We decline this technology in the healthcare domain, also in the light of the cases discussed earlier of ontological categorization (cf. the RDS diagnosis reporting) and event recording (cf. the infusion sheet case). The conceptual construct of our proposal is that of *Affording Mechanism* (in what follows, AM), an evolution of a computational construct that has been widely discussed within the CSCW field, that of Coordination Mechanism (Schmidt & Simone, 1996). From the conceptual point of view, a single AM is a dyad logic-structure that, embedded into an electronic artifact within an information system, makes it "active" and capable of "affording" (some of) the knowledgeable uses that are associated with the artifact and its content; the AM logic is expressed in terms of if-then declarative statements, whose conditional patterns are closely bound to specific data structures of an organizational artifacts, so that we can also see an AM as the computational counterpart of a property that the artifact acquires during its use. Accordingly, at computational level, an AM consists of the *structural definition* of (a portion of) an artifact (i.e., its schema, with types, value ranges, constraints, etc.); and a conditional if-then statement that is computable by a rule-based engine; it is the then-part that specifies how the artifact's *affordances* (Dourish, 2001, p. 118) should change according to the evaluation of the conditions defined in the if-part, defined over the artifact's content and other contextual information, like what role is currently logged-in, what her expertise profile is (e.g., expert, novice, guest) or, in the healthcare domain, the patient under treatment and her main diagnosis. In Figure 2 we depict an abstract representation of the concept of AM, while for other architectural details we refer the reader to (Cabitza & Gesso, 2011).

More precisely, if we consider the phase of definition (of the structures) of the artifacts through which the user can interact with the computer-based information system (e.g., web forms, elec-

*Figure 2. A) An abstract representation of two affording mechanisms, with their dyadic nature: control (in black) and structure (depicted as a three-field stylized form). B) The same affording mechanisms at "run time" when their computational logic has been activated by the context: the artifact's structure changes affordance accordingly to trigger an interpretative act by the user.*

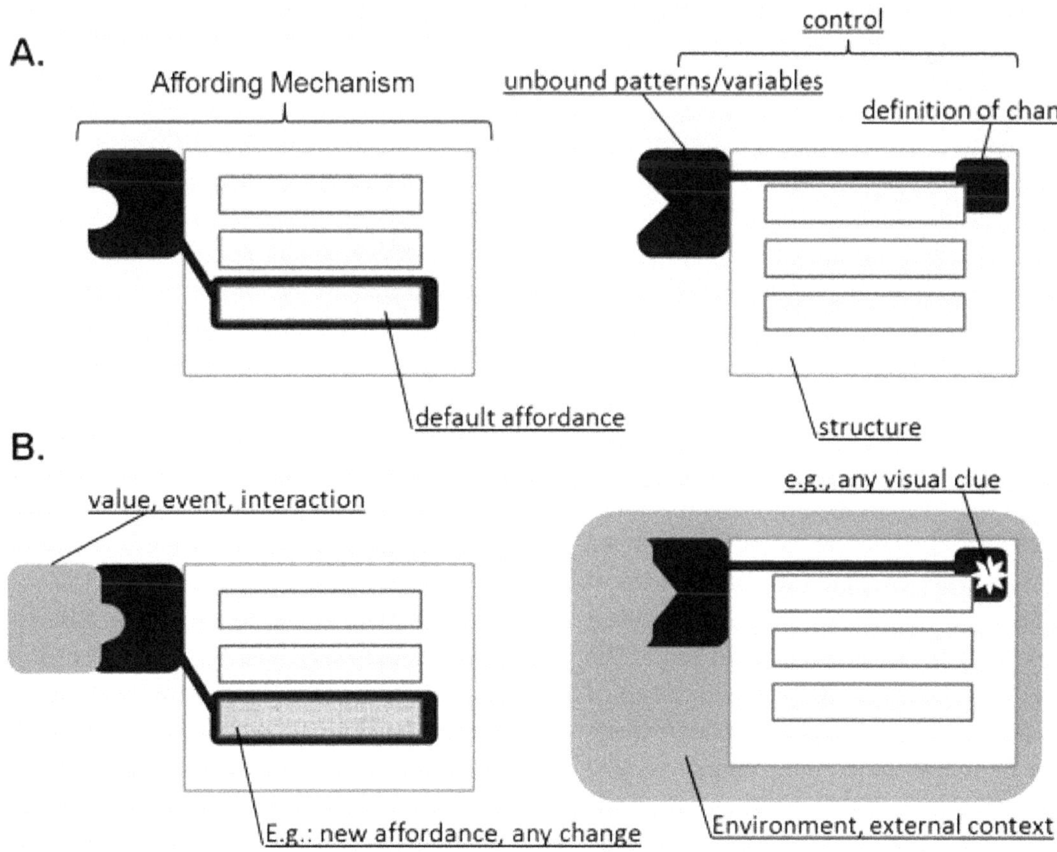

tronic documents), an AM *expresses how that structure changes its appearance according to the accumulated content and the context* (e.g., scheduled events, external messages from other systems). The reader should also note that more than one AM (as well as no one) can be defined over the same portion of an information system artifact. We denote this sort of modular rule-based construct as "affording" since we refer to the lesson learned in the characterization of coordination mechanisms, when it was observed that organizational artifacts, through their standardized format, always convey an intended and predefined protocol (Simone, et al., 1995, p. 44), which provides affordances and constraints to the coordinating and negotiating activities necessary to get the work done. A single AM is then conceived to tell the user what actions either can be or have been that far performed on a specific portion of the artifact, depending on the content this portion carries, and also to express "to some extent, how to do these actions" (Norman, 1988, p. 9). This can be done in virtue of a mapping that exists between the artifact and its function, i.e., the afforded action. It is important to note that the changes in content conveyance that AMs trigger would not *mean* anything (i.e., they would convey no extra-information) unless the target user is aware of this mapping, and that this mapping can not be imposed from above, but it is rather acquired through habit and use; accordingly, AMs define *dynamic relationships between potential contexts of use* of an artifact and *corresponding adequate uses* in virtue of the fact that they are not part of the intended information system at archival level, but rather they are external modules developed by the users with, or even without, the help of ICT professionals to enrich the information system at the coordinative and conventional level. The artifact potentialities are afforded, i.e., conveyed through the artifact in terms of its external appearance by means of explicit suggestions and more or less circumventable constraints "at interface level," what in Cabitza and Simone (2009) we called Awareness Promoting Information and Knowledge Evoking Information (respectively, API and KEI). This relation between recurring situations (the context) and potential uses of the artifact is conceived to produce context-driven dynamic affordances that can evoke in the users' minds conventional ways to cope with the artifacts' internal states, content, history of executed operations and current enabled functionalities. That said, it should lastly be noted that AMs do not "embed fixed notions of meaning within technologies" (Dourish, 2001). Rather, we conceive them as a pro-active and malleable support for users to recall conventional and local meanings "in their heads" *through* and *during* their embodied interaction with the artifact and to adequately exploit a possibly tacit knowledge that regards how to perform their job and positively articulate their actions with others.

## Affording Mechanisms in Situated Artifacts

In what follows, we report a simple case from our field work at the NICU where we got a preliminary validation of this approach. In that setting, we presented the AM construct as a sort of very short rule intended to change how data and content were presented on the "record at work." We defined a number of AMs with some of the practitioners we interviewed, and deployed them on ProDoc, the prototypical EPR mentioned above. Our aim was to augment two specific forms that the practitioners used to compile regularly during the treatment of patients by conveying additional meta-content on them and changing their affordances according to two different roles that would get access to the forms, i.e., either caregivers or administrative clerks. By enriching these two records in an unobtrusive manner, our declared aims were 1) to promote what is called collaboration awareness, that is the "elusive practices of taking heed of what is going on in the work setting" (Schmidt, 2002; Cabitza, et al., 2009); 2) to evoke proper and effective way to use the record on the basis of the best practices that had been described during our interviews; 3) to suggest the users conventional ways to articulate activities successfully within their community of practice; and, most notably, 4) to create a *two-way link* on-the-go, between the clinical (and hence primarily coordinative) and the administrative (primarily archival) functions of the clinical record, which could be seen as an artifact that is jointly shared between the hospital shop floor (i.e., the ward) and the hospital administrative floor.

On one way of this two-way information flow (see Figure 3, circle 1), the AMs are supposed to make explicit the quality expectations that secondary users bring about the very same clinical information that clinicians produce to document their job, keep track of the essential elements of the patient trajectory and coordinate with each other (Berg, 1999). These quality requirements are influenced by a quantitative use of these clinical data: for this reason well defined values of primarily intrinsic information quality can be reified within specific AMs that must activate at the clinical side of the artifact to convey a sort of meta-information that enhance data while they are filled in—or while they are still to be filled in—in the clinical record. These local AMs are intended to promote an awareness of quality targets in terms of *differences* between the actual data and the optimal data required for secondary purposes, or better yet, in terms of visual and simplified representations of *variance* from the minimal levels of accuracy/completeness/timeliness that are necessary to enable certain secondary activities. For instance, the form conveyed such indications in quite a similar way as usually electronic forms indicate what fields are mandatory or the robustness of a password *while* the user fills it in. Accordingly, a specific AM would automatically associate a piece of information, say A in Figure 3, with a side indication of 'poor' quality if it detected a too large variance from target values predefined for, e.g., statistical procedures in clinical trials; on the other hand, another AM would associate the same value with a 'quite good' quality for legal accounting and another with a 'discreet' quality for billing. Each AM would reify the requirement taken from a specific community of stakeholders and would inscribe a specific interest in the artifact (cf. Latour, 1991).

The fact that different AMs could be active at the same time on the same portion of the clinical record would just reflect the condition that the same clinical "fact" is used for different purposes, for each of which the information is more or less "fit." This apparent conflict is not directly addressed by the computational system, but it is rather conveyed as an additional information, like colored bends that overlap; in doing so, doctors are made aware of the underlying information policies regulating their textual output and, on the basis of this acquired awareness and of their professional autonomy, they can decide how to proceed: whether to temporarily disable the meta-layer; leave the layer on, but disregard its indications on the current level of IQ; or, conversely, amend A accordingly to reach some minimum level till the corresponding AM turns off. The important point made here is that it is the user that determines the relevancy of the meta-content provided by the AM and that acts accordingly: in this case, we can say that a constraint like, e.g., the mandatoriness of a specific field, is relaxed and disentangled from the artifact (instead of being embedded in it, as an intrinsic property of the data type as it is usually intended in top-down conceptual modelling approaches), and is explicitly proposed to the user as an indication of an organizational (i.e., external) specification.

In the same vein, we also designed simple AMs intended to convey *explicit alerts* and *reminders* on data either at the end of a working shift, or at the end of the patient stay, when the record must be consolidated and finalized at the patient's discharge. Therefore these AMs were supposed to activate only after that the record had been supportive of medical work to suggest the need to

*Figure 3. The two-way flow of the computational support for the disentangling work: "A" represents a datum that is produced on a form of the patient record artifact (on the right) and that relates to or is duplicated in other parts of the record. A is also used on another form of the same patient record (on the left) by the administrative clerks. Circle 1 represents an AM with some simple computational logic (e.g. if-then rules) that links administrative business rules (data constraints, quality requirements, etc.) to clinical record fields. Circle 2 represents another AM that link ward practice conventions to data that have been edited or imputed for secondary purposes.*

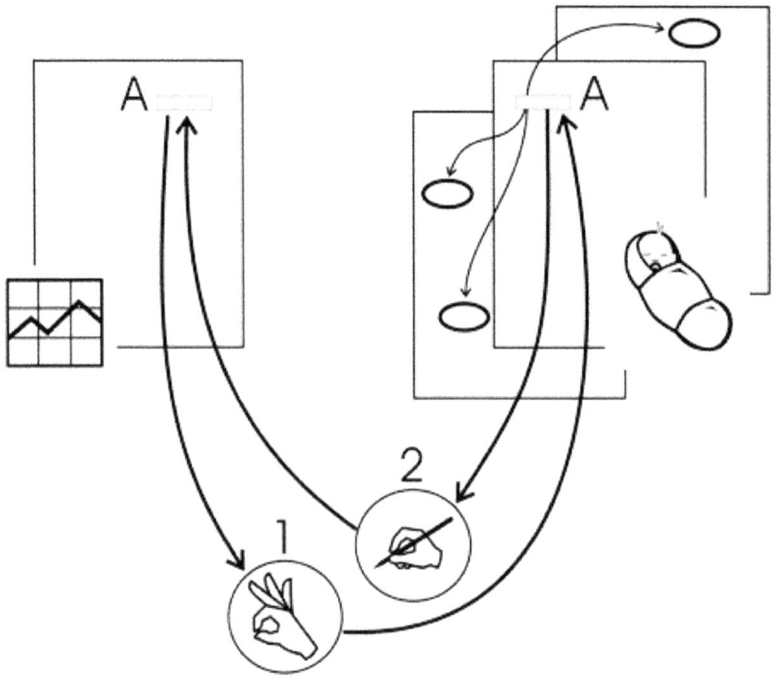


review and complete some forms for other reasons than the coordinative ones, e.g. for strictly legal compliancy. In this case, the field on which A, in Figure 3, is recorded could change its affordances accordingly, e.g. by being highlighted in red, so as to hint to doctors that they are supposed to review the record, before consolidating the report.

On the other way around (see Figure 3, circle 2), the artifact used by administrative clerks was endowed with AMs intended to provide these secondary users with meta-data enriching the clinical data, so as to promote what we call *bias awareness,* after Shanks and Corbitt (1999). In this case, the artifact provided data consumers, depending on their actual needs and the processes in which they are involved, with indications of how the clinical data have been produced (e.g., by who, at what time) and how their meaning is socially and contextually constructed in terms of norms, practices and values involved at the producers' side. Since this latter element is very difficult (or unrealistic) to convey to outsiders of a certain community of practice, we limited ourselves in designing AMs that showed some of the main relationships holding between the data scattered in the whole record and that doctors had accepted to make explicit in some use sessions (more details in Cabitza & Simone, 2008); moreover, other AMs gave access to some of the recurrent justifications we collected when we asked to doctors the reasons why some parts of the record were usually defective in IQ, by displaying short texts in pop-up boxes or help-like in-line inscriptions. In so doing, the raw data of the artifact would 'inform' secondary processes even if the context of production could not (and must not, to avoid unnecessary confusion) be reconstructed more faithfully than just giving the general idea of the situations in which data have been actually produced. In this step, the information system can be endowed with functionalities of semi-automatic *imputation* and data cleansing to improve, respectively, missing and inaccurate data (Winkler, 2004); these functionalities should not be allowed to change data, but they should rather provide a reference value that secondary users are called to interpret in the light of the additional context they are provided with to see whether these corrections are pertinent or not. For instance, the automatic correction of a value that is unusually larger or smaller in relation to the others (what is generally called an 'outlier') is generally considered a safe operation for statistical analysis, but it can also be highly misleading and bring to relevant mistakes if considered with respect to billing and accounting purposes.

Grounding on our experience with declarative end-user programming (Cabitza, et al., 2009c), we advocate an incremental and participatory design of a number of simple AMs, which express two macro-categories of relationships: on the one hand, rules expressing the relationship between a piece of information and all the others this directly refers to, according, for instance, to the relationship taxonomy we discussed in (Cabitza, et al., 2005). On the other hand, rules expressing how the above characterized conventional and local practices can affect and bias (the interpretation of) actual data with respect to an "ideal use." By declaratively representing how quality standards have been *localized* (i.e., both in the sense of 'made local' [Winthereik, 2003] and 'locally met') by ward practitioners, these mechanisms can be used in a sort of backward-chaining processing that can lead, to a first approximation, to better data or to practical recommendations of how to get purified data for secondary purposes. To some extent, these mechanisms would allow "dry information," à la Goguen, be efficiently "mobilized" across different settings all together with contextual indications that could facilitate its situated interpretation by secondary users.

## IN LIEU OF A CONCLUSION

In our case studies, we could often observe how doctors and nurses were more concerned with guaranteeing care in terms of participation, commitment, comfort and relief (what Strauss et al. [1985] called "sentimental work") than with strictly adherence to the protocols that were intended to make their care more effective and to use the clinical record more properly. This made us think of the importance to think of the artifacts of an information system as "artifacts at work," i.e., artifacts that must support goals and objectives that, often, evade most of the modellistic methods and rationalistic ambitions undertaking the design of a computerized information system.

The challenge posed by any hospital setting can be rephrased in almost any information-intensive domain where data are produced *in the* practice, *during* the practice, and *for* the practice itself. To this aim, this challenge can be rephrased in a twofold way: how to stimulate users in creating sense-making networks of data that could reconcile their accumulation for the sake of coordination with the intrinsic (process and data) quality requirements of an accumulation that is intended to enable scientific management and research, and, the other way round, how to convey the typical quality requirements of information systems into the same structures supporting—with a certain and controlled degree of provisionality, redundancy, and ambiguity—the articulation of partially unpredictable and incomparable work trajectories. These are mutually interrelated open questions that make sense for the research agendas of two ICT-related research fields: Information System

(IS) and Computer-Supported Cooperative Work (CSCW): in fact, from their foundation, the broadest aims of these complementary research fields have been, respectively, to improve and support practice by means of information technologies (Keen 1980) and to understand practice with respect to IT-based support (Schmidt & Bannon, 1992).

We tried to address these questions by discussing the notion of information quality and relating it to the healthcare domain as a paradigmatic domain where information is a key asset that is represented in a complex web of artifacts compounding a very heterogeneous information system. The key starting point of our discussion, corroborated by a number of field studies and by our own experience, is that information quality can be understood only in relation to the context and purposes where information is used. To this aim, the paper confronted how the social nature of work is traditionally interpreted by IS and CSCW scholars. We believe that a multidisciplinary approach to the design of information technologies aimed at achieving higher data and process quality in organizational settings is *indeed* necessary not to disregard any relevant dimension and practical purpose in which the "discourse on quality" really makes sense to users and actors (Shanks, 1999). To this aim, we observed how CSCW and IS disciplines have so far pursued essentially different and complementary approaches to ICT design for quality and that these approaches are manifested in their focus on different classes of information purposes: the primary and coordinative purpose for the CSCW scholars; and the secondary, archival purpose for the IS ones, respectively.

This chapter claims that the two approaches are both necessary and useful, unless one does not hinder the accomplishment of the purposes of the other in the design of computer-based technologies that support and flank organizational practice. In addition to that, we also believe that both approaches have to be integrated in a more comprehensive approach serving the two purposes, quite differently from what happens in today's research and professional ICT practice. Specifically, in regard to information quality, two worldviews have to be integrated: one that takes information as a commodity and evaluates its quality on the basis of extrinsic properties that are oriented to its usage far from where the information is produced; and one that considers information as a resource for action and conceives its quality in terms of its fitness for understanding and action, and in particular for collaboration.

The way this integration can be achieved is based on the fact that the two main roles information plays, the archival and the coordinative one, belong to two different layers of the information system, where requirements coming from IS and CSCW can be met, or at least addressed, separately. The two layers are however connected through the usage of the artifacts that support users often in seamless, situated, and unpredictable ways. Our point is that this connection is realized by letting quality be interpreted differently in the two layers; and by supporting this interpretative act by augmenting the artifacts with meta-information that, on the one hand, would allow the users to *disentangle* or *unpack* the context of production from the information that has been created and used for coordinative purposes; and on the other way round, with information that would make practitioners aware of the quality constraints and properties that are deemed necessary for archival purposes.

To avoid obvious problems of information overload, we intend this meta-information be dynamically generated either at need according to the context, or on demand by user request; moreover, we intend this meta-information be generated according to local and conventional policies that either users of the grassroots level feel their own and want to make explicit; or that only specific ethnomethodologically informed analyses can let emerge and make temporarily reified in computational modules. These modules, that we call Affording Mechanisms, are to be taken as complementing, not substituting, the functionalities an information system is endowed with at design time, and as something that is progressively added and modified only after an information system has been deployed and incorporated in habitual practices of work. It is the conventional and appropriated nature of this technological additional support that allow to glue the two layers mentioned above together and make the related activities a unique, although articulated, process where information is managed according to local needs, and according to peculiar visions of what information quality really is.

We conclude this chapter by mentioning that the integration we hope for requires first a change of perspective from both sides of the discipline boundary: IS and CSCW should ignore no more or contest the other point of view as "something that is not pertinent" to their interest and concerns. Rather, they should look for a new integration that, among other things, would facilitate both in achieving their "ontological" goals. Only once this to some extent groundbreaking change will have been happened, little difference will be made on what technical solution practitioners of any information-intensive domain will be provided with. Because we are confident that it will be a technical solution that will work for the success of its users as much as these latter ones will work for the success of the technical solution. We are working towards this challenging objective in our own domain.

## REFERENCES

American Hospital Association. (2001). *Patients or paperwork? The regulatory burden facing America's hospitals. PricewaterhouseCoopers Report*. New York, NY: PricewaterhouseCoopers.

Arts, D. G., De Keizer, N. F., & Scheffer, G. J. (2002). Defining and improving data quality in medical registries. *Journal of the American Medical Informatics Association*, *9*(6), 600–611. doi:10.1197/jamia.M1087

Avison, D., Elliot, S., Krogstie, J., & Pries-Heje, J. (Eds.). (2006). The past and future of information systems: 1976-2006 and beyond. In *Proceedings of the IFIP 19th World Computer Congress*, (vol. 214). Berlin, Germany: Springer.

Avison, D., & Fitzgerald, G. (2003). *Information systems development: Methodologies, techniques and tools* (3rd ed.). Maidenhead, UK: McGraw Hill.

Ballou, D. P., Wang, R. Y., Pazer, H., & Tayi, G. K. (1998). Modelling information manufacturing systems to determine information product quality. *Management Science*, *44*(4), 462–484. doi:10.1287/mnsc.44.4.462

Bannon, L., & Schmidt, K. (1991). CSCW: Four characters in search of a context. In Bowers, J., & Benford, S. (Eds.), *Studies in Computer Supported Cooperative Work: Theory, Practice and Design* (pp. 3–16). Amsterdam, The Netherlands: North-Holland.

Bannon, L. J., & Bødker, S. (1991). Designing interaction: Psychology at the human-computer nterface. In *Beyond the Interface: Encountering Artifacts in Use*. Cambridge, UK: Cambridge University Press.

Bannon, L. J., & Bødker, S. (1997). Constructing common information spaces. In *Proceedings of ECSCW 2007, The 5th European CSCW Conference*. Lancaster, UK: CSCW.

Bardram, J. E. (1997). I love the system - I just don't use it. In *Proceedings of the ACM Conference on Group Work (GROUP 1997)*, (pp. 251-260). ACM Press.

Barley, S. R., Dutton, W. H., Kiesler, S., Resnick, P., Kraut, R. E., & Yates, J. (2004). Does CSCW need organization theory? In *Proceedings of the 2004 ACM Conference on Computer Supported Cooperative Work (CSCW 2004)*, (pp. 122-124). New York, NY: ACM.

Batini, C., Cabitza, F., Cappiello, C., & Francalanci, C. (2008). A comprehensive data quality methodology for web and structured data. *International Journal of Innovative Computing and Applications*, *1*(3), 205–218. doi:10.1504/IJICA.2008.019688

Batini, C., & Scannapieco, M. (2006). *Data quality: Concepts, methodologies and techniques*. Berlin, Germany: Springer.

Batini, D. P., Lenzerini, & Santucci. (2002). *La progettazione concettuale dei dati*. Rome, Italy: Franco Angeli.

Baud, R. H., Rassinoux, A. M., & Scherrer, J. R. (1992). Natural language processing and semantical representation of medical texts. *Methods of Information in Medicine*, *31*, 117–125.

Beniger, J. R. (1986). *The control revolution: Technological and economic origins of the information society*. Cambridge, MA: Harvard University Press.

Berg, M. (1998). Medical work and the computer-based patient record: A sociological perspective. *Methods of Information in Medicine*, *37*, 294–301.

Berg, M. (1999). Accumulating and coordinating: Occasions for information technologies in medical work. *Computer Supported Cooperative Work*, *8*(4), 373–401. doi:10.1023/A:1008757115404

Berg, M., & Bowker, G. (1997). The multiple bodies of the medical record: Toward a sociology of an artefact. *The Sociological Quarterly*, *38*, 511–535. doi:10.1111/j.1533-8525.1997.tb00490.x

Berg, M., & Goorman, E. (1999). The contextual nature of medical information. *International Journal of Medical Informatics*, *56*, 51–60. doi:10.1016/S1386-5056(99)00041-6

Berg, M., & Winthereik, B. (2003). Waiting for Godot: Episodes from the history of patient records. In Berg, M. (Ed.), *Health Information Management: Integrating Information and Communication Technology in Healthcare Work*. London, UK: Routledge.

Bowker, G., & Star, S. L. (1999). *Sorting things out: Classification and its consequences*. Cambridge, MA: MIT Press.

Bringay, S., Barry, C., & Charlet, J. (2006). Annotations: A functionality to support cooperation, coordination and awareness in the electronic medical record. In *Proceedings of the 7th International Conference on the Design of Cooperative Systems*. COOP.

Button, G. (Ed.). (1992). *Technology in working order: Studies of work, interaction and technology*. London, UK: Routledge.

Button, G., & Dourish, P. (1996). Technomethodology: Paradoxes and possibilities. In R. Bilger, S. Guest, & T. Michael (Eds.), *Proceedings of CHI 96*. Vancouver, Canada: ACM SIGCHI Press

Cabitza, F., & Gesso, I. (2011). Web of active documents: An architecture for flexible electronic patient records. In Fred, A., Filipe, J., & Gamboa, H. (Eds.), *Communications in Computer and Information Science (Vol. 147)*. Berlin, Germany: Springer. doi:10.1007/978-3-642-18472-7_4

Cabitza, F., Sarini, M., Simone, C., & Telaro, M. (2005). When once is not enough: The role of redundancy in a hospital ward setting. In *Proceedings of GROUP 2005*. ACM.

Cabitza, F., Sarini, M., Simone, C., & Telaro, M. (2006). Torres: A conceptual framework for articulation work across boundaries. In *Frontiers in Artificial Intelligence and Applications* (pp. 102–117). Dordrecht, The Netherlands: IOS Press.

Cabitza, F., & Simone, C. (2008). Supporting practices of positive redundancy for seamless care. In *Proceedings of the 21th IEEE International Symposium on Computer-Based Medical Systems*. Jyväskylä, Finland: IEEE Press.

Cabitza, F., & Simone, C. (2009). Active artifacts as bridges between context and community knowledge sources. In *Proceedings of the 4th International Conference on Communities and Technologies*. University Park, PA: ACM Press.

Cabitza, F., Simone, C., & Sarini, M. (2009). Leveraging coordinative conventions to promote collaboration awareness. *Computer Supported Cooperative Work, 18*(4), 301–330. doi:10.1007/s10606-009-9093-z

Cabitza, F., Simone, C., & Zorzato, G. (2009). PRODOC: An electronic patient record to foster process-oriented practices. In *Proceedings of the 11th European Conference on Computer Supported Cooperative Work*. Vienna, Austria: Springer.

Cabitza, F., & Simone, F. (2009). WOAD: A framework to enable the end-user development of coordination oriented functionalities. *Journal of Organizational and End User Computing, 22*(1), 1546–5012.

Chengalur-Smith, I., Ballou, D. P., & Pazer, H. L. (1999). The impact of data quality information on decision making: An exploratory analysis. *IEEE Transactions on Knowledge and Data Engineering, 11*(6), 853–864. doi:10.1109/69.824597

Ciborra, C. (2002). *The labyrinths of information - Challenging the wisdom of systems*. Oxford, UK: Oxford University Press.

Clarke, K., Hartswood, M., Procter, R., Rouncefield, M., & Slack, R. (2003). Trusting the record. *Methods of Information in Medicine, 42*(4), 345–352.

Coiera, E. (1997). *Guide to medical informatics: The internet and telemedicine*. London, UK: Chapman and Hall Medical.

DeLone, W. H., & McLean, E. R. (1992). Information systems success: The quest for the dependent variable. *Information Systems Research, 3*(1), 60–95. doi:10.1287/isre.3.1.60

Dick, R. S., & Steen, E. B. (Eds.). (1991). *The computer-based patient record: An essential technology for health care*. Washington, DC: National Academy Press.

Dieng, R., Giboin, A., & Karsenty, L. (Eds.). (2000). Designing cooperative systems: The use of theories and models. In *Proceedings of the 5th International Conference on the Design of Cooperative Systems (COOP 2000)*. Amsterdam, The Netherlands: IOS Press.

Dourish, P. (2001). *Where the action is: The foundations of embodied interaction*. Cambridge, MA: MIT Press.

Fitzpatrick, G. (2000). Understanding the paper health record in practice: Implications for EHRs. In *Proceedings of Health Informatics Conference (HIC 2000)*. Adelaide.

Fitzpatrick, G. (2004). Integrated care and the working record. *Health Informatics Journal, 10*(4), 291–302. doi:10.1177/1460458204048507

Garfinkel, H., & Rawls, A. W. (2008). *Toward a sociological theory of information*. Boulder, CO: Paradigm Publishing.

Gendron, M., & D'Onofrio, M. (2001). Data quality in the healthcare industry. *Data Quality, 7*(1).

Goguen, J. (1996). Formality and informality in requirements engineering. In *Proceedings of the 2nd International Conference on Requirements Engineering*. Washington, DC: IEEE Computer Society.

Gregor, S., & Iivari, J. (2007). Designing for mutability in information systems artifacts. In Hart, D. N., & Gregor, S. D. (Eds.), *Information Systems Foundations: Theory, Representation and Reality*. Sydney, Australia: Australian National University Press.

Halverson, C. A. (2002). Activity theory and distributed cognition: Or what does CSCW need to do with theories? *Computer Supported Cooperative Work, 11*(1-2), 243–267. doi:10.1023/A:1015298005381

Hardstone, G., Hartswood, M., Procter, R., Slack, R., Voss, A., & Rees, G. (2004). Supporting informality: Team working and integrated care records. In *Proceedings of the CSCW, 2004*, 142–151. doi:10.1145/1031607.1031632

Harper, R. H., O'Hara, K., Sellen, A., & Duthie, D. (1997). Towards the paperless hospital? *British Journal of Anaesthesia*, *78*, 762–767.

Harper, R. H. R., O'Hara, K. P. A., Sellen, A. J., & Duthie, D. J. R. (1997). Toward the paperless hospital? A case study of document use by anaesthetists. *British Journal of Anaesthesia*, *78*, 762–767.

Hartswood, M., Procter, R., Rouncefield, M., & Slack, R. (2003). Making a case in medical work: Implications for the electronic medical record. *Computer Supported Cooperative Work*, *12*, 241–266. doi:10.1023/A:1025055829026

Heath, C., & Luff, P. (1996). Documents and professional practice: 'Bad' organisational reasons for 'good' clinical records. In *Proceedings of CSCW, 1996*, 354–362.

Heath, C., & Luff, P. (2000). *Technology in action*. Cambridge, UK: Cambridge University Press. doi:10.1017/CBO9780511489839

Heeks, R., Mundy, D., & Salazar, A. (1999). *Why health care information systems succeed or fail: Information systems for public sector management*. Working Paper. Manchester, UK: University of Manchester.

Hevner, A. R., March, S. T., & Park, J. (2004). Design science in information systems research. *Management Information Systems Quarterly*, *28*(1), 75–105.

Hutchins, E. (1996). *Cognition in the wild*. Cambridge, MA: The MIT Press.

Institute of Medicine. (1997). *The computer-based patient record: An essential technology for health care*. Washington, DC: National Academy Press.

Kahn, B. K., Strong, D. M., & Wang, R. Y. (2002). Information quality benchmarks: Product and service performance. *Communications of the ACM*, *n.d.*, 184–192. doi:10.1145/505248.506007

Kanigel, R. (2005). *The one best way: Frederick Winslow Taylor and the enigma of efficiency*. Cambridge, MA: MIT Press.

Keen, P. G. W. (1980). MIS research: Reference traditions and a cumulative tradition. In *Proceedings of 1st International Conference on Information Systems*, (pp. 9–18). ACM.

Korpman, R. A. (1990). Patient care automation: The future is now. *Nursing Economics*, *8*(5), 345–349.

Lamb, R., & Kling, R. (2003). Reconceptualizing users as social actors in information systems research. *Management Information Systems Quarterly*, *27*(2), 197–235.

Latour, B. (1987). *Science in action*. New York, NY: Open University.

Latour, B. (1991). Technology is society made durable. In Law, J. (Ed.), *A Sociology of Monsters: Essays on Power, Technology and Domination* (pp. 103–131). London, UK: Routledge.

Letrilliart, L., Viboud, C., Boelle, P., & Flahault, A. (2000). Automatic coding of reasons for hospital referral from general medicine free-text reports. In *Proceedings of the AMIA Symposium*, (pp. 487-491). AMIA.

Luff, P., Hindmarsh, J., & Heath, C. (Eds.). (2000). *Workplace studies: Recovering work practices and informing system design*. Cambridge, UK: Cambridge University Press. doi:10.1017/CBO9780511628122

MacKay, D. M. (1969). *Information, mechanism and meaning*. Cambridge, MA: MIT Press.

Malone, T. W., & Crowston, K. (1994). The interdisciplinary study of coordination. *ACM Computing Surveys*, *26*(1). doi:10.1145/174666.174668

Mann, R., & Williams, J. (2003). Standards in medical record keeping. *Clinical Medicine*, *3*(4), 329–332.

Maturana, H., & Varela, F. J. (1992). *The tree of knowledge: The biological roots of human understanding*. Boston, MA: Shambhala Publications.

Merleau-Ponty, M. (1962). *Phenomenology of perception*. London, UK: Routledge.

Monteiro, E. (2003). Integrating health information systems: A critical appraisal. *Methods of Information in Medicine*, *42*, 428–432.

Nandhakumar, J. (2010). Contrarian information systems studies. *European Journal of Information Systems*, *19*, 687–688. doi:10.1057/ejis.2010.49

Neff, W. S. (2006). *Work & human behavior*. Piscataway, NJ: Transaction Publishers.

Nelson, D. (1992). *A mental revolution: Scientific management since Taylor*. Columbus, OH: Ohio State University Press.

Nonaka, I., & Takeuchi, H. (1995). *The knowledge-creating company*. Oxford, UK: Oxford University Press.

Norman, D. A. (1988). *The design of everyday things*. New York, NY: Doubleday.

Nunberg, G. (1996). Farewell to the information age. In Nunberg (Ed.), *The Future of the Book*, (pp. 103-138). Berkeley, CA: University of California Press.

Olivé, A. (2007). *Conceptual modeling of information systems*. Berlin, Germany: Springer.

Orlikowski, W. J., & Barley, S. R. (2001). Technology and institutions: What can research on information technology and research on organizations learn from each other? *Management Information Systems Quarterly*, *25*(2), 145–165. doi:10.2307/3250927

Orlikowski, W. J., & Iacono, C. S. (2001). Research commentary: Desperately seeking the IT in IT research - A call to theorising the IT artefact. *Information Systems Research*, *12*(2), 121–134. doi:10.1287/isre.12.2.121.9700

Parker, M., Stofberg, C., De la Harpe, R., Venter, I., & Wills, G. (2006). *Data quality: How the flow of data influences data quality in a small to medium medical practice*. Paper presented at the Community Informatics for Developing Countries: Understanding and Organising for a Participatory Future Information Society Conference. Cape Town, South Africa.

Pickering, A. (2004). The science of the unknowable: Stafford Beer's cybernetic informatics. *Kybernetes*, *33*(4), 499–521. doi:10.1108/03684920410523535

Pitt, L. F., Watson, R. T., & Kavan, C. B. (1995). Service quality: A measure of information system effectiveness. *Management Information Systems Quarterly*, *19*(2), 173–188. doi:10.2307/249687

Poirier, C. (2003). *The supply chain manager's problem-solver*. Boca Raton, FL: CRC Press.

Rector, A. L., Nolan, W. A., & Kay, S. (1991). Foundations for an electronic medical record. *Methods of Information in Medicine*, *30*, 179–186.

Redman, T. C. (1998). The impact of poor data quality on the typical enterprise. *Communications of the ACM*, *41*(2), 79–82. doi:10.1145/269012.269025

Robinson, M., & Bannon, L. (1991). *Questioning representations*. Paper presented at ECSCW 1991. Amsterdam, The Netherlands.

Salber, D., Dey, A. K., & Abowd, G. D. (1999). The context toolkit: Aiding the development of context-enabled applications. In *Proceedings of the SIGCHI Conference on Human Factors in Computing Systems: The CHI Is the Limit*, (pp. 434-441). Pittsburgh, PA: ACM Press.

Saleem, J. J., Russ, A. L., Justice, C. F., Hagg, H., Ebright, P. R., Woodbridge, P. A., & Doebbeling, B. N. (2009). Exploring the persistence of paper with the electronic health record. *International Journal of Medical Informatics*, *78*(9), 618–628. doi:10.1016/j.ijmedinf.2009.04.001

Schmidt, K. (1991). Riding a tiger, or computer supported cooperative work. In *Proceedings of the Second Conference on European Conference on Computer-Supported Cooperative Work (EC-SCW 1991)*, (pp. 1-16). Norwell, MA: Kluwer Academic Publishers.

Schmidt, K. (1997). Of maps and scripts - The status of formal constructs in cooperative work. In *Proceedings of the International ACM SIG-GROUP Conference on Supporting Group Work: The Integration Challenge (GROUP 1997)*, (pp. 138-147). New York, NY: ACM.

Schmidt, K. (2002). The problem with "awareness": Introductory remarks on awareness in CSCW. *Computer Supported Cooperative Work*, *11*(3), 285–298. doi:10.1023/A:1021272909573

Schmidt, K. (2010). Keep up the good work! - The concept of work in CSCW. In *Proceedings of the 9th International Conference on Designing Cooperative Systems*, (pp. 265-286). Aix-en-Provence, France: Springer.

Schmidt, K., & Bannon, L. (1992). Taking CSCW seriously: Supporting articulation work. *Computer Supported Cooperative Work*, *1*(1), 7–40. doi:10.1007/BF00752449

Schmidt, K., & Simone, C. (1996). Coordination mechanisms: Towards a conceptual foundation of CSCW systems design. *Computer Supported Cooperative Work*, *5*(2-3), 155–200. doi:10.1007/BF00133655

Schmidt, K., & Wagner, I. (2004). Ordering systems: Coordinative practices and artifacts in architectural design and planning. *Computer Supported Cooperative Work*, *13*(5-6), 349–408. doi:10.1007/s10606-004-5059-3

Sellen, A., & Harper, R. (2001). *The myth of the paperless office*. Cambridge, MA: MIT Press.

Shanks, G. (1999). Semiotic approach to understanding representation in information systems. In *Proceedings of the IS Foundations Workshop*. Sydney, Australia: IS Founation.

Shanks, G., & Corbitt, B. (1999). Understanding data quality: Social and cultural aspects. In *Proceedings of the 10th Australasian Conference on Information Systems*. Wellington, New Zealand: IEEE.

Simone, C., Divitini, M., & Schmidt, K. (1995). A notation for malleable and interoperable coordination mechanisms for CSCW systems. In *Proceedings of Conference on Organizational Computing Systems*, (pp. 44-54). New York, NY: ACM.

Simone, C., Mark, G., & Giubbilei, D. (1999). Interoperability as a means of articulation work. In *Proceedings of WACC 1999*, (pp. 39-48). San Francisco, CA: ACM Press.

Slack, J. D., & Fejes, F. (1987). *The ideology of the information age*. New York, NY: Ablex Publishing.

Stamper, R. (1992). Signs, organisations, norms and information systems. In *Proceedings of the 3rd Australian Conference on Information Systems*. Wollongong, Australia: IEEE.

Strauss, A., Fagerhaugh, S., Suczek, B., & Wiener, C. (1985). *Social organization of medical work*. Chicago, IL: University of Chicago Press.

Strong, M. D., Lee, W. Y., & Wang, Y. R. (1997). Data quality in context. *Communications of the ACM*, *40*(5), 103–110. doi:10.1145/253769.253804

Suchman, L. (1987). *Plans and situated actions: The problem of human-machine communication*. Cambridge, UK: Cambridge University Press.

Thalheim, B. (2007). Challenges to conceptual modelling. In Krogstie, J., Opdahl, A. L., & Brinkkemper, S. (Eds.), *Conceptual Modelling in Information Systems Engineering* (pp. 59–74). Berlin, Germany: Springer. doi:10.1007/978-3-540-72677-7_4

Tzannes, N. S. (1968). The concept of 'meaning' in information theory. In *Warren McCulloch Papers*. Philadelphia, PA: American Philosophical Society Library.

van der Lei, J. (1991). Use and abuse of computer-stored medical records. *Methods of Information in Medicine*, *30*, 79–80.

Varela, F. J. (1999). The specious present: A neurophenomenology of time consciousness. In Petitot, J., Varela, F. J., Pachoud, B., & Roy, J.-M. (Eds.), *Naturalizing Phenomenology: Issues in Contemporary Phenomenology and Cognitive Science* (pp. 266–314). Palo Alto, CA: Stanford University Press.

Varpio, L., Schryer, C. F., Lehoux, P., & Lingard, L. (2006). Working off the record: Physicians and nurses transformations of electronic patient record-based patient information. *Academic Medicine: Journal of the Association of American Medical Colleges*, *81*(10), 35–39. doi:10.1097/01.ACM.0000237699.49711.35

Wang, R. Y. (1998). A product perspective on total data quality management. *Communications of the ACM*, *41*(2), 58–65. doi:10.1145/269012.269022

Winkler, W. E. (2004). Methods for evaluating and creating data quality. *Information Systems*, *29*(7), 531–550. doi:10.1016/j.is.2003.12.003

Winograd, T., & Flores, F. (1987). *Understanding computers and cognition: A new foundation for design*. New York, NY: Addison-Wesley.

Winthereik, B. R. (2003). We fill in our working understanding: On codes, classifications and the production of accurate data. *Methods of Information in Medicine*, *42*(4), 489–496.

Young, D. W. (1987). Discussion paper. In Anderson, J. G., & Jay, S. J. (Eds.), *Use and Impact of Computers in Clinical Medicine* (pp. 8–14). New York, NY: Springer-Verlag.

## ENDNOTES

1 The reason why this disenchantment towards formalized knowledge is often traced back to the phenomenological tradition could be justified by recalling the argument that Husserl made about the "mathematically substructed world of idealities [that have surreptitiously substituted] the only real world," his attention to the "pretheoretical, prerational world of everyday experience" (Dourish, 2001) and his claimed and radical departure from previous Cartesian and Galileian positions on knowledge and understanding that still seem to inform mainstream information system design. See also Liberman (2007) Husserl's criticism of reason—with ethnomethodological specifications. Lanham, Lexington Books.

2 In the following, the terms "clinical record" and "patient record" will be used without distinction. In literature, also the expression "medical record" is frequently used, especially when the medical dimension is considered predominant.

3 Here the term *standard* is widely meant in terms of both sets of reference properties, i.e., specifications, against which something can be compared and benchmarked, as well of policies that are supposed to guarantee those properties and have data meet the intended specifications.

4 Yet, in medicine the problem of identifying a good-for-all gold standard to assess the information quality of, say, a record is not trivial, as it necessarily relates to a living person and to an interpretative act that is often irreducible to post hoc analysis according to some "objective" criterion.

5 The term mindset is not to be confused with the concept of knowledge and hence the MacKay's conceptualization of information with other common definitions in the IS literarature (e.g., Batini, et al., 2002);

while knowledge usually indicates a set of notions that are expressed in some explicit or declarative form, a mindset here indicates an attitude for action and can be tacit as well as ineffable.

6 This seminal idea was then linked to the savvy and metaphorical definition of information that was later popularized by Bateson when speaking of initiation rituals, as "a difference that makes a difference."

7 This conceptualization is close to what later was denoted as "structural coupling," i.e., a coupling between mutually communicating agents that occurs whenever each agent involved is, with respect to the other(s), a source (and target) of perturbations that triggers structurally-determined and reciprocally-induced change (Maturana & Varela, 1992, p. 75).

8 Here it is clear the idea that standards are a powerful means to enable interoperability, also among organizations.

9 Remember the definition of information popularized by Bateson ("information is a difference that makes a difference") and the following passage by Garfinkel "no information is possible except against the constancy provided by a background of invariant social expectations" (Garfinkel & Rawls, 2008, p. 75).

# Chapter 7
# About Representational Artifacts and Their Role in Engineering

**Hilda Tellioğlu**
*Vienna University of Technology, Austria*

## ABSTRACT

*This chapter is about showing how artifacts impact engineering work processes by representing important issues of individual and collaborative design work. After summarizing the state of the art of engineering as a design process, artifacts, and their representational role in design and engineering, a selection of rich descriptions of artifacts' creation and use in engineering work including team-based coordination and decision activities will be presented. The studies are based on ethnographic research carried out for several years in different design and engineering companies. Artifacts used in these studies will be analyzed from their representational point of view to discuss their important role in design and engineering, by considering users' motivation to use and sometimes adapt them as well as internal and external constraints given in work settings which call for improvisations, before concluding this chapter.*

## INTRODUCTION

Engineering is about bringing together technologies to meet human needs or to solve problems (Khandani, 2005). Considering engineers as problem solvers, engineering design is about problem solving, which defers from other types of problem solving, in the nature of the problem and the solution. Design problems are open; they might have more than one correct solution. The solution to a design problem is normally a system that possesses specified properties, and the process of solving a design problem is usually iterative and cooperative, depending on its complexity.

DOI: 10.4018/978-1-4666-0303-5.ch007

A basic process of an engineering design can be described in five steps (Khandani, 2005, p. 5): defining the problem, gathering pertinent information, generating multiple solutions, analyzing and selecting a solution, testing and implementing the solution. In all these steps, several tasks are evolved; several actors are involved; several sub-problems occur. Interactions for actions, possibilities, clarification, and orientation (Tellioğlu, 2003), needed for communication, coordination, decision-making, and improvisation, make use of representations of problems and solutions on the one hand, and of tasks, responsibilities, restrictions, and dependencies on the other.

Artifacts in general have qualities to help representing problems, possible solutions, and the state of a process. Within the CSCW research, the role of artifacts in the everyday work of professionals was studied from several perspectives (Haas, 1996; Henderson, 1999; Nardi, 1993; Sellen & Harper, 2001; Schmidt & Wagner, 2004). Frameworks like organizational memory (Conklin & Begeman, 1988; Conklin, 1989; Ackerman & Malone, 1990; Ackerman, 1997; Conklin, 1993), common information spaces (Bannon, 1997; Schmidt & Bannon, 1992), workflow systems (Grinter, 1996; Bardram, 1997), coordination mechanisms (Carstensen, 1996; Divitini, et al., 1996; Schmidt & Simone, 1996; Simone & Divitini, 1997) or boundary objects (Star, 1989; Star & Griesemer, 1989; Bowker & Star, 1999; Lutters & Ackerman, 2002) have been developed to address specific aspects of coordinative practices. For instance, Bardram and Bossen studied non-electronic coordination artifacts at a hospital ward (Bardram & Bossen, 2005). They focused on "1) the material characteristics of these artifacts, 2) on how they order the world by providing templates, 3) on how they provide overview and signal ad hoc status, 4) on the importance of acknowledging the importance of meaning in collaboration, and finally 5) on the importance of supporting second order articulation work" (p. 168). The network of these highly interwoven artifacts operates as a resource for action. It supports information flow, status overviews, synchronous and fine-grained coordination, and articulation of status.

Artifacts used in collaborative engineering processes can be identified and their use can be supported by well-designed computer systems. Our ethnographic studies, we have carried out for almost last 10 years, show that these tools are not really addressing the issues, which are important for designers. Designers do not want to give up their conventions and standards they have established so far when new tools are introduced into their work environments. What they usually do is to adapt their work habits to make them keepable with the systems they have to use, or to find a work around in order to meet their goals quantitatively and qualitatively. In fact, they want to continue working with their old known artifacts they invented, modified, or composed. How artifacts are shaped is a result of experiences, cooperative work, and conventions established in work groups. It takes time to create and adapt especially common artifacts to meet differing needs and procedures in work groups and to avoid misunderstandings and frustration of users.

To design more suitable IT-tools for designers and engineers, we have to understand their work practices. Several ethnographic studies (Tellioğlu, 2010a, 2010b, 2009, 2007) in manufacturing, electronic engineering, system design, and especially multimedia production show that knowledge workers overcome the complexity of their engineering work by using different types of artifacts in different formats and views depending on the setting they are currently in. For instance, multimedia designers have developed a particular visual culture, which allows them create and read design representations in different media. Most of these visual artifacts are produced as part of discussions, as an integral part of explanations, developments, and arguments, and they are re-used in follow-up meetings and sometimes annotated with supplements, modifications, and comments (Tellioğlu & Wagner, 2005).

Another example is the use of models by engineers. Models help engineers in overcoming complexities and to create a common understanding about processes and products. Organizational, commercial, technical, and process-based circumstances have impact on models and modeling practices. In a previous work (Tellioğlu, 2009), several modeling practices are identified and described: 1) modeling to visualize several important issues in a cooperative project, 2) modeling to support collaboration and coordination among members, 3) modeling to support system engineering for individual and group work, and 4) models triggering automated actions in workgroups. Different types of models are created and used in these different modeling activities. Each type consists of different information, has a different format, and normally a different role in the course of work processes. Analyzing how models are visualized and how they are used in which context, informs us about the work habits of engineers, about the problems they try to solve, about the circumstances under which they work and cooperate.

Analyzing artifacts and their representational roles in design-based engineering can help to create an approach for studying engineers' work processes and, in this way, design systems to meet their requirements. The main purpose of this chapter is to study, on the one hand, how artifacts are constructed and shaped and how they evolve during a design process, and on the other hand, in which ways artifacts' presence and use have impact on design decisions and practices. Besides the theoretical background, studies of work practices in two engineering teams are presented and analyzed to show empirical evidence to the results derived.

The chapter is structured as follows: First of all, the state of the art about engineering as a design process, different types of artifacts, and their representational role in design and engineering will be summarized. Then, a selection of rich descriptions of artifacts' creation and use in engineering work including team-based coordination and decision activities will be presented. Artifacts used in these studies will be analyzed from their representational point of view to discuss users' motivation to use and sometimes adapt them as well as internal and external constraints given in work settings, before concluding this chapter.

## ENGINEERING AS A DESIGN PROCESS

In engineering, designer is an engineer who designs and develops systems, and design is an engineering activity affecting human life, using laws and insights of science, building upon special experience, and providing the prerequisites for the physical realization of solution ideas (Pahl, et al., 2007, p. 1). Engineering requires systematic procedures combined with so-called heuristic principles or creativity techniques. Certain conditions must be given for an engineer to provide the possibility to apply such systematic approaches, like definition of an overall goal and individual sub goals, clarification of conditions by defined initial and boundary constraints, dispelling of prejudice to ensure the wide-ranging search for solutions possible, search for variants to find solutions or combinations of solutions, evaluation based on goals and conditions, decisions made built upon objective evaluations to enable progress (Pahl, et al., 2007, p. 53).

A typical engineering design process consists of several steps, which can occur iteratively[1]: problem identification, research, requirements specification, concept generation, design, prototyping, system integration, and maintenance. So far several approaches for design are established in engineering: top down methodologies like functional decomposition, top-down iterative refinements of roughly finished designs to make it finer and more exact until the complete design is done, case-based methodologies focusing on similar design cases to model the own process, incremental re-design practices by unraveling an existing design from the bottom up, by modifying as required. Sometimes there is a need to combine both the top-down and bottom-up approaches to so called hybrid design processes. In case of creating design ideas, explorative methodologies may be very useful especially in the initial design and specification phases.

## REPRESENTATIONAL ARTIFACTS

No matter individual or cooperative, actors usually need to use *artifacts* to carry out their work. They produce, read, annotate, modify, check, evaluate, communicate, and delete artifacts in the course of their individual or cooperative work. Besides mediating contacts among people by helping to search for people with specific competencies (Hertzum, 1999), artifacts mediate articulation work by acting as an intermediary with a specific material format between actors. For Kuutti "instruments, signs, procedures, machines, methods, laws, forms of work organization" are examples of artifacts (Kuutti, 1996, p. 26). Furthermore "an object can be a material thing, but it can also be less tangible (such as a plan) or totally intangible (such as a common idea) as long as it can be shared for manipulation and transformation by the participants of the activity" (Kuutti, 1996, p. 27). In a usual work environment actors interact by using several artifacts. Where an artifact is placed, when it is accessed, by whom it is modified, what the exact modifications are, how these modifications are represented, etc. are important questions related to cooperation and especially coordination issues among the parties involved.

As being permanent symbolic constructs artifacts act as mediators of the coordination. At the same time, they are used to clarify ambiguities and to settle disputes. Mediating requires accurate representation of a certain level of details, which need to be exchanged. Representations are not real (Suchman, 1987), are local and temporary (Gerson & Star, 1986) and are "conventionalized practices based on rules of mapping and translation between representation and the object that is represented" (Schmidt & Wagner, 2004, p. 15). Representational artifacts are immediate objects of the work (Schmidt & Wagner, 2004). They are objectifications of things-to-come and of things-in-the-process-of-becoming.

To understand the impact of artifacts being as representational objects we have to study artifacts used in work settings. Professionals interact usually with three types of artifacts in their current work environment (Cole, 1982; Hertzum, 1993). Some artifacts are readily at hand and often piled up on the desk. They mainly consist of action information. Others are within reach but put on shelves or binders. These artifacts can be considered as personal work files. Some artifacts are kept away from the office. They are used to archive relevant information for eventual future use. Hertzum focused on the relations "between the use of documents in the individual professionals' execution of their current activities and the use of documents for information sharing among professionals who are separated in time or space" (Hertzum, 1999, p. 57). In this sense, artifacts as personal work files are mainly accessed by an individual person and through her/him by other actors within the work environment. These artifacts are arranged and maintained personally and not organized by categories defined in the work setting.

Another distinction is made by artifacts' properties of mediating. Wartofsky defined three types of mediating artifacts [255], which has been further developed by Engeström (1990) into a three-level hierarchy using Leont'ev's hierarchy of activity (1981). "Primary artifacts are tools used directly in production to mediate the relationship between the subject and object of activity; secondary artifacts are representations of modes of action—models—used to preserve and transmit skills in the production and use of primary artifacts; tertiary artifacts are imaginative or visionary and give 'identity and overarching perspective to collective activity systems'" (Guy, 2003, p. 3). This scheme has been further developed within activity theory (Engeström, 1999; Collins, et al., 2002) and includes what-artifacts, which contribute a means of achieving the object, how-artifacts that contribute to understanding how to achieve the object, why-artifacts motivating achievement of the object, and where-to-artifacts motivating evolution of all elements in the activity system (Guy, 2003, p. 3).

Lundberg and Sandahl have investigated artifacts' properties in work by analyzing and identifying how artifacts' peripheral properties become common resources within a community of practice and how artifacts are active within a work environment (Lundberg & Sandahl, 1998). Actors need to see any time what others are doing, which is normally provided by a common artifact. Besides facilitating implicit communication within a team (Berlage & Sohlenkamp, 1999), a common artifact serves as a template representing a limited model of the work to be done by leaving enough space to be filled by its users. If this template is filled in then it becomes a record of the work. Common artifacts may coordinate activities by their visibility or by their arrangements. "Documents' material and visible presence, in a shelf or on a table according to their structured trajectory in the radiology department and news agency afford the linking of actions and events over different sites and times without personal interaction between staff. ... the coordination of work is indicated by 'who is holding the document.' The paper acts as a token and the shelf in which the documents are placed represents the state of work" (Lundberg & Sandahl, 1998, p. 7). These artifacts are reminders of things to do and help manage one's work activities (Hertzum, 1999). They are arranged spatially in particular ways. Some artifacts are formatted in ways that enable coordinating activities by tracking the work between cooperating actors within a work group. They are used then as tokens related to material objects, which are accessible to all involved in a shared work process (Schmidt & Wagner, 2002). The location of material artifacts includes some relevant information. Some of the actors can probably make sense of it and some others not. One can also see the history on a material artifact, the history of past work as well as the contributions of different actors. The materiality of artifacts is also referred to affordances (Gibson, 1979; Sellen & Harper, 2001) or to immutability of inscriptions on paper and their mobility (Latour, 1986).

On the other hand, there are coordinative artifacts, which are material artifacts that have a coordinative role in carrying out work practices. In this sense they are communication objects and persuasive (Wagner, 2000). They help actors in several ways in their daily work: They create a common understanding of a task. They enable talking about a task in a rich way. They can be used consequently to share information with some, yet withhold it from others, e.g., for the reason to avoid affecting the power structures and privileges in a work environment (Hertzum, 1999). Sometimes it is important to reduce the multiple voices of an artifact. This makes them to convey meaning among the people sharing them. For using them, no continuous interpretation of their meaning is necessary.

Coordinative artifacts also remind principles, approaches and methods applied, questions that are still open. They also help keeping track of activities and materials (Schmidt & Wagner, 2002). Some artifacts usually used for coordination purposes contain work plans. They include work to do, project phases, how to proceed in a specific project phase, material to collect or create which is necessary to represent the work, methods defining rules and conventions within the work group, illustrations like sketches, images or photos to explain things or to generate new meaning, references to material to look for, names of actors responsible for certain tasks. As boundary objects (Star, 1989; Star & Griesemer, 1989) coordinative artifacts are accessed and modified by all responsible actors. They enable crossing organizational and professional boundaries many times. All decisions made in a project can be recorded and available in a common design artifact. Actors annotate their artifacts by circling certain areas, adding notes, or marking certain parts. This type of annotations makes artifacts multi layered. Multi layered artifacts "facilitate coordination between activities (and the people who are responsible for them). They, for instance, provide a collective or individual space for experimentation and change" (Schmidt & Wagner, 2002, p. 10). That implies that artifacts are interrelated to work activities.

There are several types of coordinative artifacts (Schmidt, 2001, p. 7):

- Traces are artifacts that indicate aspects of past activities of coordinating actors.
- Templates are artifacts that specify the properties of the result of individual contributions, like product standards, drawings, style sheets, etc.
- Maps are artifacts that specify interdependencies of tasks or objects in a cooperative work setting like organizational charts, classification schemes, taxonomies, etc.
- Scripts are artifacts that specify a protocol of interaction in view of task interdependencies in a cooperative work setting, like checklists, production schedules, office procedures, bug report forms, etc.

The changing of static state of all types of coordinative artifacts offers cues and an array of signals to other actors as to the intentions, challenges, and problems of the actor carrying out the changes (Schmidt & Wagner, 2002).

In the next section, we will present two cases of design-based engineering projects to illustrate different ways of working with representational artifacts in design processes. This depends on the type of the product, on conventions and standards established in the project team, on project circumstances being planned, foreseen or situated ad-hoc. We selected one multimedia company and one engineering company to show differences in approaches to design and engineering in the course of the project evolvement.

## CASE *ARCHCOM*: A MULTIMEDIA COMPANY

*Archcom*[2] is a multimedia company providing web applications with database integration for contemporary architecture, by combining web engineering with architectural design. It was founded in 1996. The main product of this company is an Internet forum for contemporary architecture. Its manager is an architect with a strong interest and advanced skills in multimedia production and use. He, in parallel, works on small architectural design and building projects. Four professionals with different backgrounds, like web production, designing multimedia architecture and web information, developing web content strategies, web programming, including two free lancers with skills in HTML, Java Script, Microsoft Access, Perl programming and image processing, and external programmers for the development of database integration, and graphic designers with experiences in design and implementation of Flash animations build the team of the project that we[3] were studying in the scope of an international project[4].

The Internet forum builds on a central database managing architectural data, containing images, journal articles, as well as technical data about architects and buildings. Its technical implementation is complex because it gets the data from distributed sources, each of them using slightly different technology. Interfaces to these sources need to be flexible enough to enable regular data transfer without failure. The multi-media character of the data adds to the complexity of the system.

We characterized this design process as best practice. This applies to planning activities, to the use of methods, and the documentation. The project manager uses a project plan in form of

*Figure 1. Part of a to-do-list created and updated regularly after weekly meetings showing the dynamic character of a coordinative artifact (Archcom)*

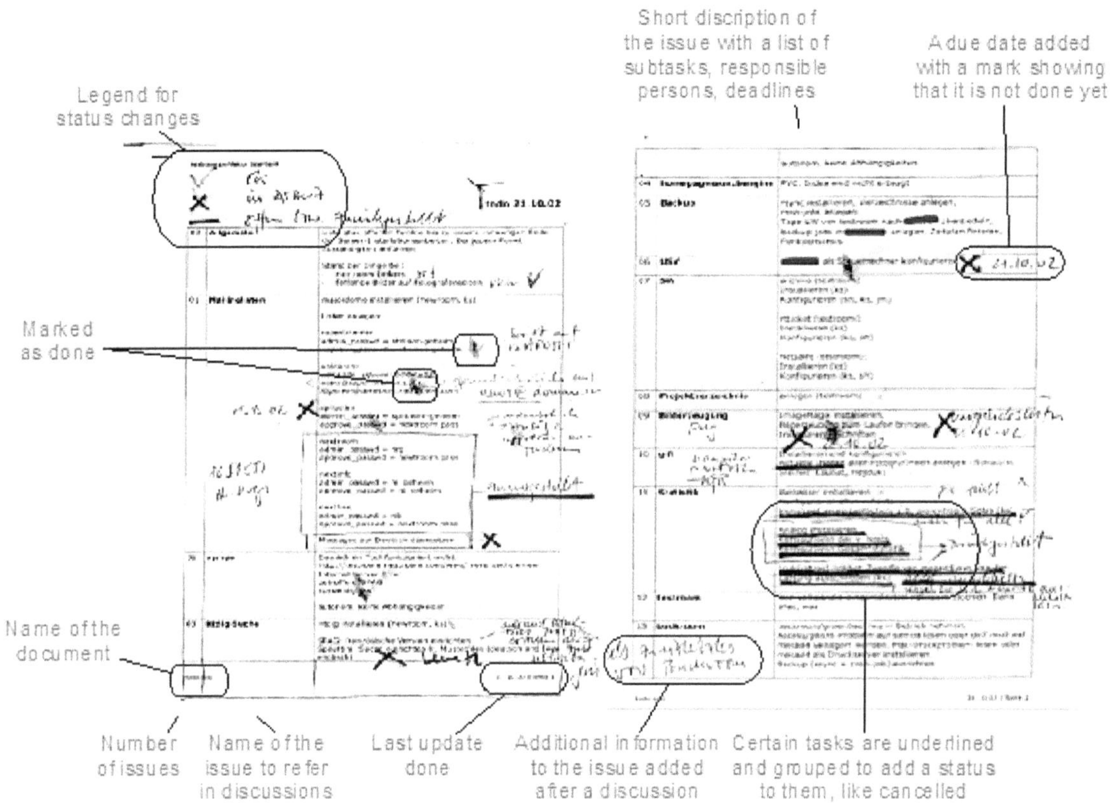


a spreadsheet for the management of deadlines, responsibilities, and workflows, which he updates regularly. During the weekly project meetings detailed to-do-lists are generated, which afterwards are used to organize the cooperative work within the team (Figure 1). These to-do-lists are essential for continuous project work. They contain a lot of data representing the status of the project, the distribution of work between project members, decisions made so far or added newly to the open issues, the importance of certain issues by explicitly showing their due date, status of some issues (either done, cancelled, or scheduled to the next milestone). The owner of these to-do lists is the project manager. He updates the actual list, which contains only open issues, during the regular project meetings by discussing the single issues with all team members. These to-do lists are also used as meeting minutes. They provide crucial coordinative data for team members to follow all decisions made about the functionality and user interface of the system, about priorities made to issues including certain tasks, about changes and enhancements of the underlying database model, about modifications of the web pages, about the business logic, about open questions and the responsibilities of team members for different tasks.

With regard to important issues, for example concerning the design of the search function of the forum, thorough research is conducted in order to prepare a solid base for decision-making. This ended in detailed documents evaluating different approaches and technologies for searching. To integrate a search facility into the website, detailed sketches of navigation and graphic layout of the pages are produced, partly before and

*Figure 2. Linking and navigation designed for the web site by Archcom*

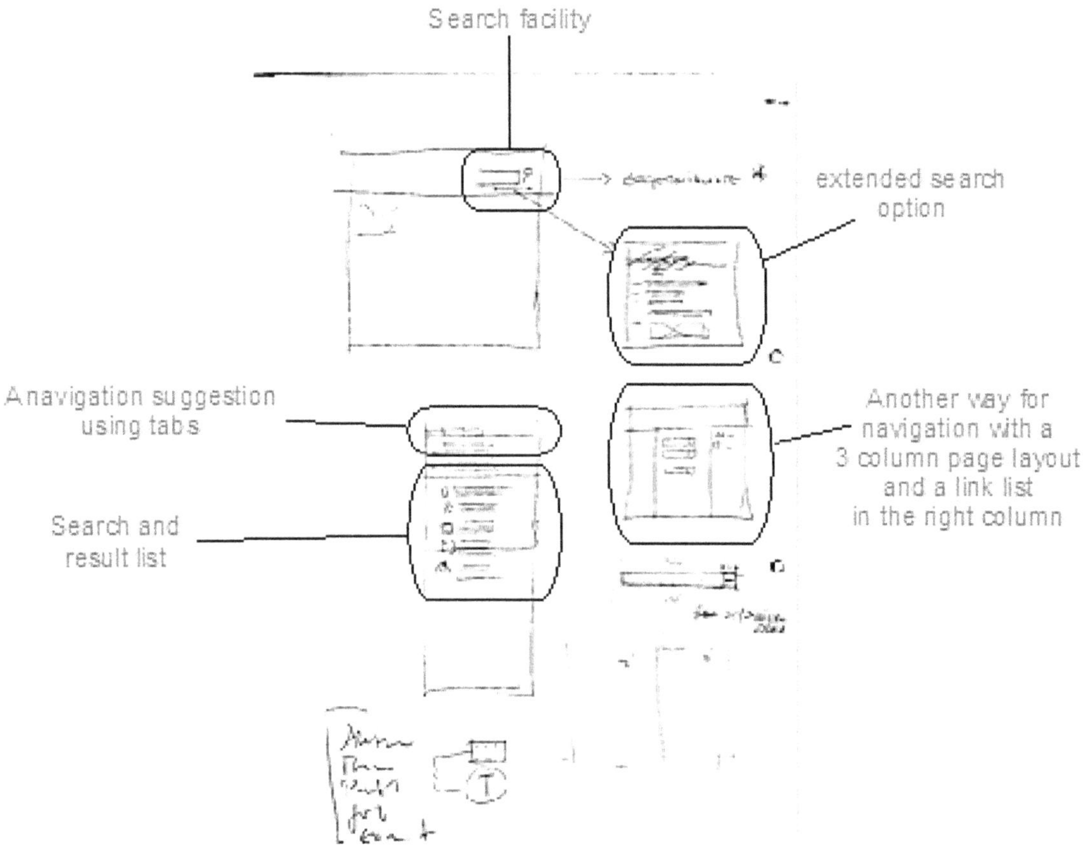


partly during the meetings cooperatively (Figure 2).

Equally detailed and thorough are the descriptions of the data for the persistence layer of the system, which are discussed before integrating the database components (Figure 3).

Besides the multi layered sketches of the database, the final data structure is represented by a UML class diagram. UML is also used for modeling the system and to describe the use cases. The team considers representations of the system-in-development with UML particularly useful in convincing the customer of the technical design and of its quality. The team member responsible for the technical concept tries to follow a 'best practice example' in software engineering. He applies coding conventions and tries to keep the code simple, readable, understandable, and easy to maintain. At any possibility, he shows the insights of his code and models to convince others of his professional approach to design and development.

Design representations including the layout and functionality issues are prepared in different media: sketches on paper, computer-based drawings and lists, storyboards, videos, and HTML prototypes. In meetings these are used to communicate the design-in-progress, considering constraints, such as deadlines, available resources, and user requirements (Figure 4).

Printouts of the website (representing a part of the HTML prototype) are created as part of the design process and annotated with a short description and open questions. These composite visu-

*Figure 3. The database model with additional data attached after a discussion in a meeting (Archcom)*

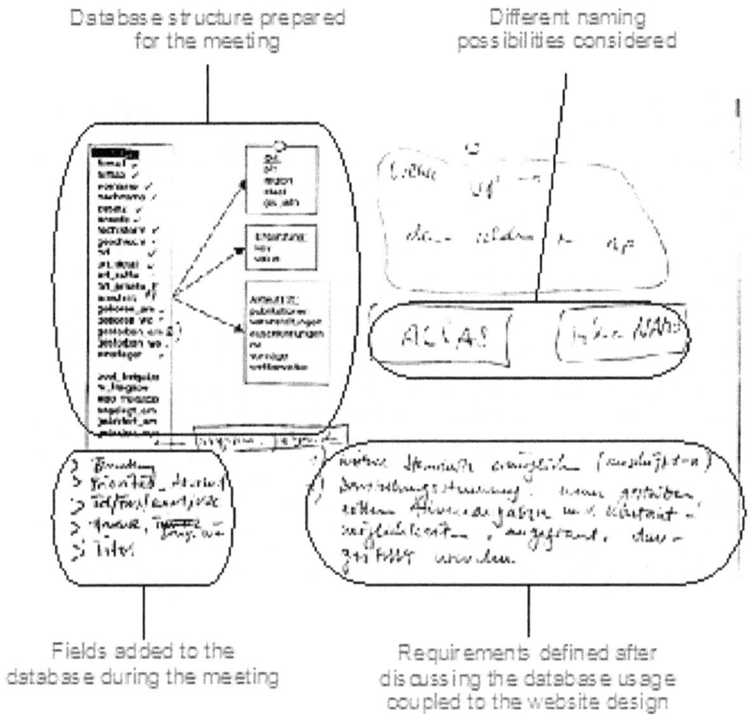


*Figure 4. A page printout of the HTML prototype with a description of design steps and open questions added on bottom of the page (Archcom)*

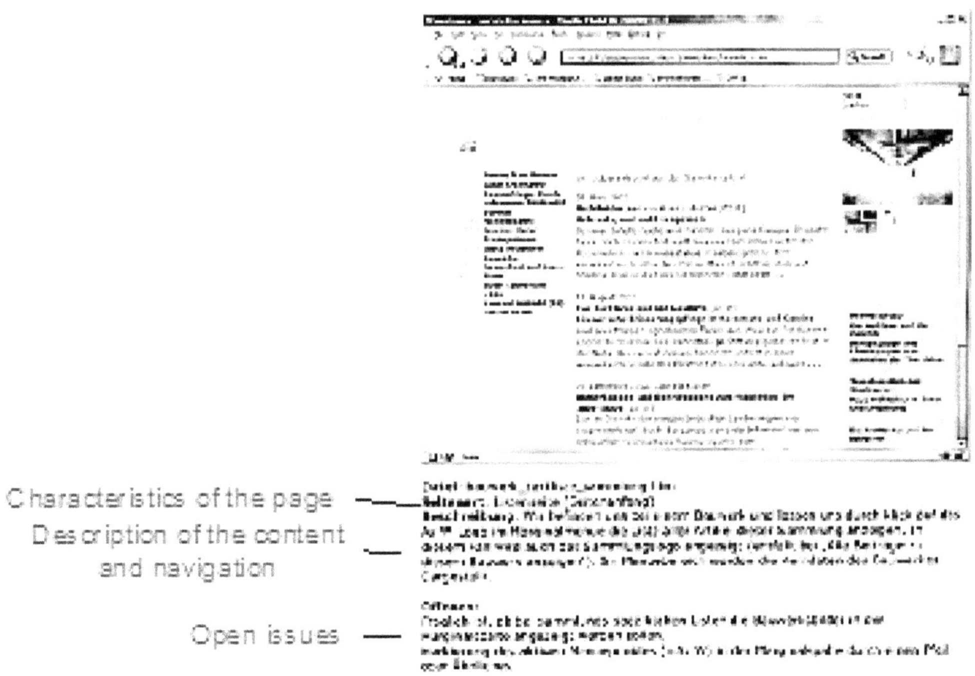

alizations represent a design step together with all the issues that need to be resolved in a way that is visible and understandable to all. The collection of printouts serves as an important base for planning of further activities in the project.

As has also been observed by Newman and Landay (2000), team members use all these artifacts simultaneously, switching between formats and media with ease. For the presentation of the whole system to the customer, a HTML prototype is built together with a video movie illustrating the use of the system.

## ANALYSIS AND DISCUSSION OF REPRESENTATIONAL ARTIFACTS AT *ARCHCOM*

Multimedia production performed by *Archcom* combines design with web engineering. Behind the goal of creating attractive multimedia interfaces for WWW, highly technological engineering decisions are made in an interdisciplinary team. Decisions are embedded in processes that are carried out very professionally to be examples of 'best practice.'

The to-do list used at *Archcom* (Figure 1) is, on the one hand, the personal work file of the project manager having it readily at hand in project meetings, and on the other, the coordinative artifact used in the project. It is the primary artifact of the manager, which makes it to the what-artifact showing coordination issues in the whole project. It partly contains the evolution of the issues listed, and is then a where-to artifact. It is material, because it represents the contributions of actors and history of past work, e.g., by status marks, due dates, names of responsible persons attached to single issues and single sub-tasks listed by the issues. As a common artifact the to-do list makes coordination arrangements in the team visible. Acting as a token, it is a resource for action by facilitating implicit communication between project members. At the same time, if it is added marks like done, cancelled, or postponed with a specific due date, it becomes a record of work and articulates the state of the work in the group, and so synchronizes the coordination within the team.

Coordination of the design and production of the multimedia platform is mainly carried out by the manager. He uses his to-do list that he shares with the other team members to define (sometimes synchronous) tasks and dependencies in work activities before they collaboratively assign team members and deadlines to them. This shared artifact is then referred to by all in case of misunderstandings, troubles to keep the deadline, contingencies occur unexpectedly and call for improvisations, etc. The to-do list shows at the same time the status of the work progress in the project. It is not only triggering actions, it keeps also track of work done, cancelled, or postponed. It grows through the project and helps remember decisions made, alternatives considered, and actions taken. Without a to-do list *Archcom* would not be able to coordinate and document its projects.

Detailed sketches of layout and navigation (Figure 2) of the web pages are what-artifacts crated cooperatively during project meetings at *Archcom*. Based on skills and experiences of team members, different suggestions are discussed to fulfill the requirements defined for the product. Ideas are visualized easily on a piece of paper, making their communication ad-hoc and simple for others. Alternatives can be represented one by one put next to each other to make variants available to all involved. Design ideas become more tangible for the team of designers, and this way a common understanding can be created for the decision of choice. Such a sketch can structure and guide the individual design work of team members by being a template and specifying the properties of the results of each.

Sketches support the creativity of the *Archcom* team as a whole. Each team member can present his design idea easily in form of a sketch usually in regular meetings. This helps not only understanding each other's solution to a current problem but also creating alternatives and makes them visible for decision-making. Ad-hoc exchange between team members is also made possible by sketches between the meetings, just to find a solution to a design problem cooperatively. As templates, sketches structure the work all by guiding the design of each member and providing a common approach to design in the team.

The database model (Figure 3) is a primary artifact of the database engineer. The objects in the database representing the primary what-artifacts used in the Internet forum are at the same time common objects enabling crossing boundaries between professionals, like the graphic designer, who designs the presentation of these objects in the web site, or the database engineer, who takes care of the storage, retrieval, and consistency of these objects at the backend, and the web programmer, who accesses database services to retrieve and display database objects at the frontend.

The database model is a typical boundary object enabling communication between different disciplines in the design team. The common problem how to make web data persistence becomes technical. This interface needs to be presented and negotiated between the database engineer and the graphic designer, because the implementation in the database can have impact on the graphic design and vice versa. The model is an abstract representation of a possible solution to agree on in the team. At the same time it protects the competence areas of different professionals because it does not show the complexity and the details of the implementation.

The page printout of the HTML prototype (Figure 4) has again a coordinative character like the to-do list, by grouping open issues by the structure of the web site pages. At the same time, it provides an overview of all data composed at one page, but managed and maintained separately in different system components. Such a page represents the integration of the data processed with different mechanisms in the same system. It is the only interface to the user and visible for evaluation and further negotiation. So, it supports data exchange between the design team and the customer by enabling implicit articulation of the work progress.

Multimedia producers use a variety of techniques, from sketches on paper to mock-ups, prototypes, and flow charts. Newman and Landay (2000) have described some of these artifacts in detail. High fidelity mock-ups for instance "contain(ed) images, icons, rich typography, and sophisticated color schemes, and these details of the visual presentation were meant to be taken literally" (p. 266). This type of visualizations is typically the ones used to present the user interface of a product as they can be easily understood by the client. In *Archcom*, linking and navigation are usually presented as sketches on paper or in the form of site maps, as shown in Figure 4. These are "high level visualizations of site structure in which web pages or entire subsections of the site are represented by textual labels" (Newman & Landay, 2000, p. 268) and are created and readable only by professionals.

## *CHIPCOM*: AN ENGINEERING COMPANY

In a European STREP project called MAPPER[5] (Model-Based Adaptive Product and Process Engineering) (IST-016527) we[6] investigated technological and methodological possibilities to support designers in collaborative engineering, which involved participative engineering methodologies to enable joint product and process design, interdisciplinary and inter-organizational collaboration throughout multiple product life cycles. We carried out an ethnographic study (Jordan, 1993) in one of the industrial partners we call *Chipcom*[7] in this paper. We visited *Chipcom* three times to observe participatory the workplaces, meetings, and design sessions, what we recorded as video and audio files. We interviewed in-depth some of the key actors. Additionally, we gathered and analyzed *Chimcom*'s artifacts like documents, source codes, test reports, simulation results, project plans, charts, organization diagrams, and models of different kinds. This qualitative study resulted among others in field visit and validation reports (Jacucci, 2005).

*Chipcom* provides IP (Intellectual Property) cores since 1997, VHDL and Verilog model development services as well as hardware and software development for microcontroller-based systems. In this paper we are referring to the production of the USB High Speed OTG Transceiver (PHY, physical layer) IP core with UTMI+ interface, shown in Figure 5. Analog block is implemented by the group A, digital by B, where as the prototype of the whole product is prepared by the group C. The arrows between analog and digital blocks represent the data flow between the components and by this the interfaces, which must be defined and agreed on by both groups A and B.

At *Chipcom* we could find other types of presentations as well. Several models are created to capture the as-is work processes and then to define the to-be processes. All three groups (A, B, C) generated their internal design workflows as Active Knowledge Models (AKMs) with IDEF (ICAM Definition Language) notation. Besides showing the interfaces to other workflows, this AKM-based workflow spanned over specification, development, verification, and product preparatory phases in all groups. It was a result of numerous consultations between managers and engineers of all groups. Next, actors responsible for particular design phases in the design flow were identified, technologies for components manufacturing and tools to be used were decided.

The design flow represented as a visual model (Figure 6) comprises sets of design tasks performed at these three geographical locations. The design object is a mixed-mode component. Design pro-

*Figure 5. Chipcom presents the company's product and its parts, which at the same time represents implicitly for the insiders of the project the organizational structure around the production*

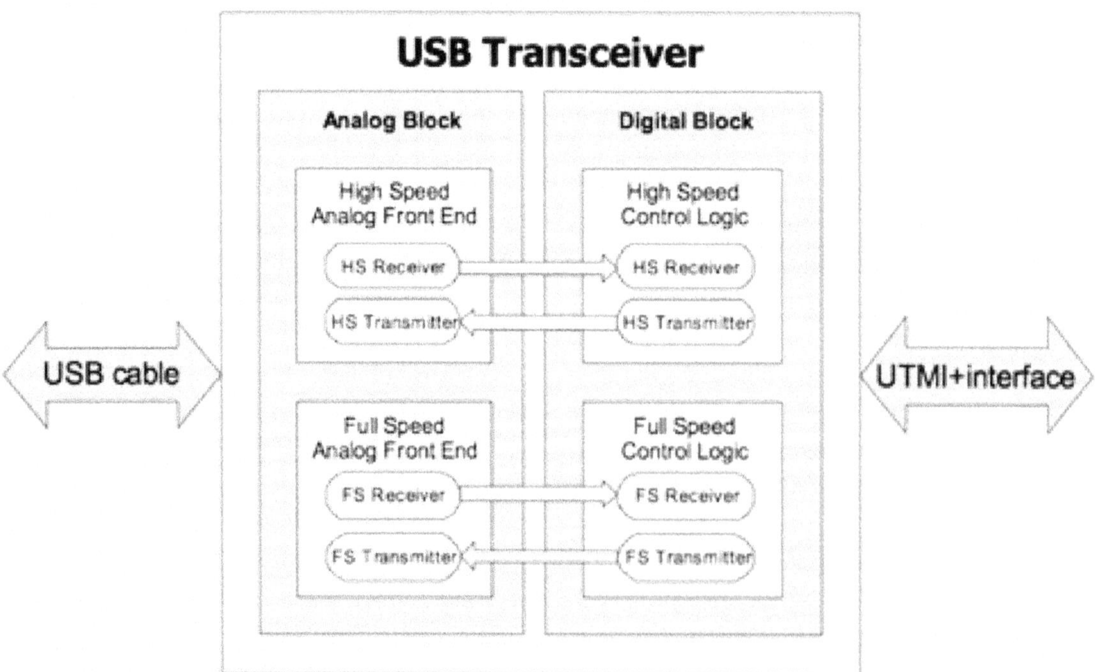

*Figure 6. A view of a part of the design processes distributed into three locations (A, B, C): AKM in IDEF notation of "design and verification" process*

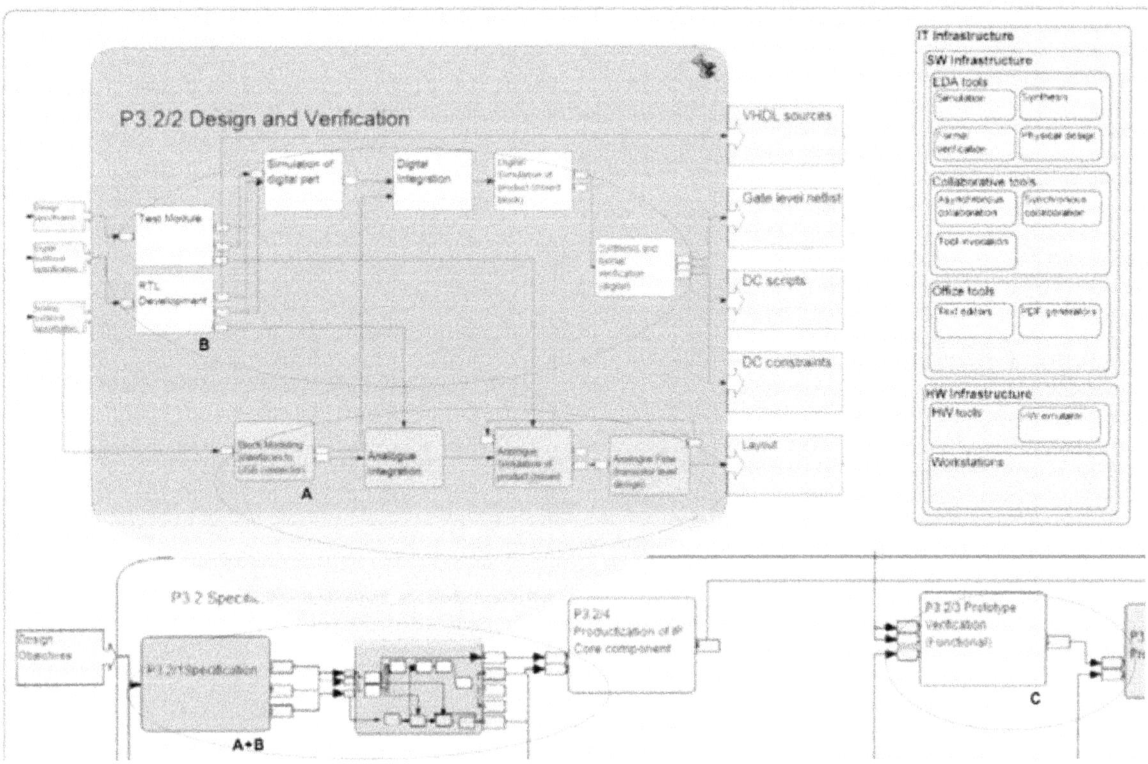


cesses are split among the team members, taking the engineering competences into account, in such way that an analog design takes place at A, a digital design at B, and a board-level testing at C. This model represents details of the process needed to create the product USB Transceiver shown previously (Figure 5). "Design and Verification" is a part of a larger engineering process, starting with design objectives, and resulting in delivering the final product (Figure 6). After the process "Specification," which is the co-responsibility of the groups A and B, the process "Design and Verification" can be started. It contains ten tasks, which are connected to each other. Some tasks can be carried out parallel; some are sequential and depending on the outcome of a previous task. The model shows the logical and temporal dependencies between all tasks. Additionally it represents the assignment of tasks to the groups. Six of the tasks are assigned to B, four to A, marked by different colors. How the IT infrastructure is associated to the process "Design and Verification" is not presented in this model. It shows also a wide spectrum of other information related to the current shared product, namely information on: the internal organization of involved companies (e.g., company structure, locations, human resources, staff competence skills), the available IT infrastructure (e.g., design automation, administration, and office tools), the current project organization (e.g., project responsibilities), the detailed structure of the common product, and the project plan (e.g., management and design workflows). These details are not illustrated in the figure.

As one of the research results in MAPPER, *Chipcom* has installed a system called TRMS (Tool Registration and Management Services) to support distributed design. The workflow used

*Figure 7. The TRMS workflow system used at Chipcom*

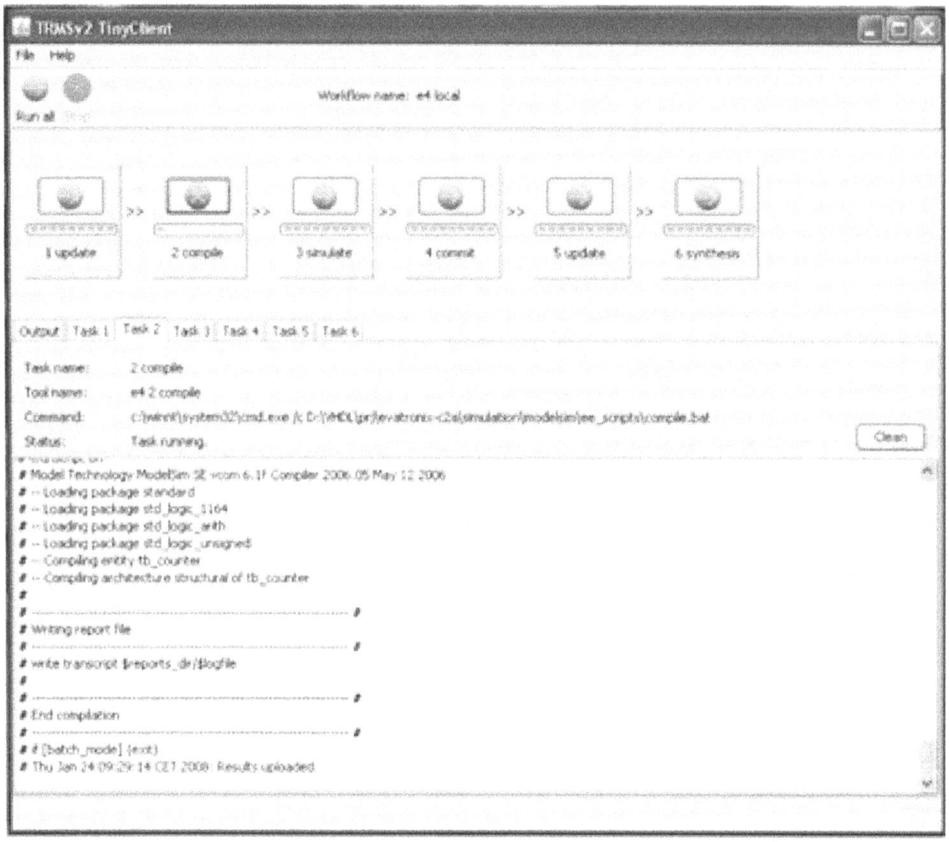

represents the involvement of different persons working on distributed teams implicitly. Multiple users can access TRMS at the same time, and long jobs can be invoked and reported. A sequential workflow is integrated by automating some steps in simulation and evaluation. The aim was to support distributed design between the location A, B, and C.

TRMS can be started by anyone remotely, which displays then the different steps in the workflow in their sequential order (Figure 7). If a task has been successfully finished, its status is visualized green and a textual report is displayed in the appropriate task tab. If there was an error, this is marked red to get the attention of the engineers. Also a detailed list of execution steps is provided to follow the process invoked by that erroneous task.

A scenario illustrates how TRMS is used between different teams. One of the engineers at B working at the USB Transceiver performs a synthesis with support of TRMS. After the synthesis has been completed, TRMS sends the results to the CVS repository, which is also accessible from C, where the engineer at C can download the file and start FPGA prototyping. One of the advantages is that the synthesis can be done on one machine (with one single license); the other benefit is that if the engineer at C finds a small error, he can fix it on his computer and invoke the synthesis again remotely.

The engineer at B has found a bug, which he needs to correct in the source file. He updates the actual source file locally, uploads the modification to the repository, updates the file in TRMS, and starts the synthesis again.

Different artifacts are created in this process: log files from the synthesis, where engineers can look for warnings; the result, which is a binary file in MCS format, used for FPGA prototyping, with the software only being available at C; and the scripts, which are evoked by TRMS, which are shared with A in the form of a Netlist.

When the synthesis has been completed, which is hard to predict, the engineer at B switches to the internal chat forum GaduGadu to inform his colleague at C that the update has been completed. He does not chat in the system attached to TRMS "because it does not work'.'

The engineer at B, who has used TRMS for a few weeks for hardware verification together with his colleague at C, comments: "This is a good thing that everything is automated. There is no possibility to run the wrong script, you are sure that the actual sources are used, you don't have to think about which file to update, it is simplified now, TRMS does the CVS update."

TRMS does not provide awareness information like who else is currently accessing it or any notification when a task or workflow is completed. The feedback delivered about the state of the synthesis is in most cases not sufficient. The processes invoked by different actors are not synchronized. Engineers must synchronize their invocations manually.

## ANALYSIS AND DISCUSSION OF REPRESENTATIONAL ARTIFACTS AT *CHIPCOM*

The visualization in the Figure 5 shows a model of the product USB Transceiver, what different teams create cooperatively. This is an image presented inside and outside the company. It is used not only to communicate the product's components within the company between cooperating branches, but also to represent the product to customers, external partners, and other potential consumers or supporters. At the same time it is a representation of the actual distribution of work and interfaces between cooperating teams. One can see easily the areas of responsibilities and who to blame if there are errors or feature lacks. It shows also, no matter who is responsible for the single components of the product, the product must provide two interfaces, one via a USB cable and one via UTMI. This is a concern of all groups participating in the design of this device. In case of interface definition or enhancements to the product, this representation helps talking about possibilities, alternatives for solutions, and impacts among components.

Besides the practical and work-related issues, this representation of the USB Transceiver (Figure 5) serves as a tertiary artifact illustrating the vision of a transceiver implemented by *Chipcom*. It is a goal, a product, which has not been implemented yet, a challenge, embedding an evolution of a new product. At the same time, it is common, which makes it also a where-to artifact used by all three groups, to arrange their work by keeping interfaces to others transparent and visible all the time. The visionary character of this product representation supports focusing on the product by providing a common artifact to refer to.

The AKM model of work processes, including single tasks in their relation to each other and showing human and non-human resources, is the major secondary artifact shaping work arrangements around the product (Figure 6). Not only in planning processes to produce a product but also in reorganizing existing design and production for collaboration can be modeled and communicated in such a secondary artifact. It is created cooperatively and shared among groups involved. It can also be used to identify problems in the process, like deadlocks, lack of integration or interfaces, not considered interdependencies between tasks and subtasks and furthermore to interfere in processes by, e.g., assigning other persons or groups to certain tasks or moving non human resources to support the performance of certain activities. Besides facilitating orientation, the process and resource model in AKM format engages all project members, who have access to it, just by making them aware of work processes planned, optimized, or rearranged. It makes process management transparent for all, which invites people to think about it and perhaps to make suggestions for improvement.

Having a common workflow system implemented in TRMS (Figure 7) makes it for all engineers at all locations possible to run and debug others' code. The code becomes a primary artifact of the person using it. This makes collaborative engineering possible, especially by communicating explicitly via other channels like instant messaging. So, TRMS is a common artifact. It is a well-accepted tool, which standardize and automate scripts for synthesis or compiling, linking and versioning. Unfortunately, it was not coordinative at the time of our study. All coordinative features were replaced by the engineers with additional tools.

TRMS impacts the work processes at *Chipcom*. As a secondary artifact, it shows a model of engineering work in temporal and logical order. Furthermore, it forces engineers to run scripts in a very certain predefined way. It not only shows a map how to run, compile, and link code simultaneously, it also acts as a script. By means of TRMS, it is possible to access the code of others and debug and upload them to the common server. However, due to ownership reasons no one tries to fix a bug in the code of a colleague. The convention established at *Chipcom* is to report or discuss bugs found in others' code. Fixing it is the responsibility of the owner of the code. Even the technologies make the handling otherwise possible, the norms and work protocols agreed upon established in the company determine how to act and setup the constraints for work processes. This shows that artifacts and the technology cannot determine how to act at work places. At the end, it is the human who decides what to do and what to avoid.

So, synchronization, awareness, and configurability are features needed by engineers at *Chipcom*, which is barely provided by the system used. To overcome this gap, engineers use texting to communicate with each other, for notification but also for exchange of pieces of code or makefiles to decide how to change them. Not only in the process of simulation, also in design and development phases, engineers at *Chipcom* are used to contact their colleagues of other branches to ask them questions about coding (if there is a special case, which must be considered by all), compiling (the exact compiler options), or debugging (when they debug each others' code) issues in developing electronic devices.

At *Chipcom*, a kind of representation to show the state of a task across the organization is needed. Erickson et al. (2004) called this "task proxy," who not only considered visualization on technical level, but also focused on task information in a way intended to support social inference. Making task proxy visible to participants causes emerging of social dynamics, like peer pressure or proactively helping those who need support. This type of visualization of the task proxy and through this the social proxy is one of the examples showing the interest on social visualization in the literature. Shared representations and body lists, profiles, and changes of status of people as we are familiar with in social software are further developments of social visualizations playing a very important role in social interaction and collaboration in teams. Halverson et al. (2006) explored this in change management.

## CONCLUSION

In this chapter, we presented first a trajectory of artifacts by focusing on their representational role in cooperative design-based engineering. After illustrating two cases from real work environments we described some of important artifacts we could identify in our ethnographic studies. Using the concept of artifacts we analyzed the artifacts used in these cases by stressing their types, roles in individual and collaborative design work, and impacts on work practices. We mainly focused on their representational role as visualizations of different kinds. Henderson (1999) looks at visualizations as 'network-organizing devices.' They support individual and cooperative thinking and organizing. Visualizations help to keep the design concept present in teams and to coordinate the work around it. They are used to illustrate the design ideas to different actors involved in the project, such as the clients, external professionals, or partners, convincing them of the design idea and mobilizing their cooperation (Schmidt & Wagner, 2002).

In their study of designers in a software company, Muller and Carey describe all these visualization strategies and tools as designers' artifacts and typical of their techniques of creating representations of the various aspects of their work and communicating them. They identified prototypes (which may take a number of forms) and graphics as one of the strongest media, arguing: "Some highly influential designers use very low-technology prototyping tools, such as basic painting programs. Others dig deeply into what appears to be simple business software, reinventing presentation systems into elaborate prototyping environments. Still others develop competences that are very similar to those of programmers, albeit in environments that support rapid idea expression rather than production software performance" (Muller & Carey, 2002, p. 387). This is an example of style that is representative of a particular occupational community, which members preserve and culture even if they work in other contexts, e.g., as a minority discipline in a software company.

In this chapter, we showed how representational artifacts could be identified and analyzed in cooperative engineering environments. Studying real work environments by means of ethnographic methods provides rich empirical material, which is not easy to analyze, compare, and use to develop design implications. Applying the corpus of different types of artifacts to ethnographic study material results in better understanding of work processes, tasks, and artifacts used, and furthermore cooperation practices between actors and groups. This is a good start to design or redesign systems, to introduce new artifacts of any kind to support work processes.

## ACKNOWLEDGMENT

We are grateful to project partners for their cooperation in MAPPER, especially to our users at *Chipcom*. Many thanks to the company *Archcom*, where I spent a lot of time observing their meetings and design sessions.

## REFERENCES

Ackerman, M. S. (1997). Augmenting the organizational memory: A field study of answer garden. In T. Malone (Ed.), *Proceedings of the Conference on Computer-Supported Cooperative Work*, (pp. 243-252). Chapel Hill, NC: ACM Press.

Ackerman, M. S., & Malone, T. W. (1990). Plans and situated actions: The problem of human-machine communication. In *Proceedings of the Conference on Office Information Systems*, (pp. 31-49). Cambridge, MA: ACM Press.

Bannon, L. (1997). Constructing common information spaces. In *Proceedings of the Fifth European Conference on Computer Supported Cooperative Work*, (pp. 81-96). IEEE.

Bardram, J. E. (1997). Plans as situated action: An activity theory approach to workflow systems. In *Proceedings of the Fifth European Conference on Computer Supported Cooperative Work*, (pp. 17–32). IEEE.

Bardram, J. E., & Bossen, C. (2005). A web of coordinative artifacts: Collaborative work at a hospital ward. In *Proceedings of GROUP, 2005*, 168–176.

Berlage, T., & Sohlenkamp, M. (1999). Visualizing common artefacts to support awareness in computer-mediated cooperation. *Computer Supported Cooperative Work, 8*, 207–238. doi:10.1023/A:1008608425504

Bowker, G. C., & Star, S. L. (1999). *Sorting things out: Classification and its consequences*. Cambridge, MA: MIT Press.

Carstensen, P. (1996). *Computer supported coordination*. Roskilde, Denmark: Risø National Laboratory.

Cole, I. (1982). Human aspects of office filing: Implications for the electronic office. In *Proceedings of the Human Factors Society 26th Annual Meeting*, (pp. 59-63). Santa Monica, CA: ACM.

Collins, P., Shukla, S., & Redmiles, D. (2002). Activity theory and system design: A view from the trenches. *Computer Supported Cooperative Work: The Journal of Collaborative Computing, 11*(1-2), 55–80. doi:10.1023/A:1015219918601

Conklin, J. (1989). Design rationale and maintainability. In B. Shriver (Ed.), *Proceedings of 22nd Annual Hawaii International Conference on System Sciences*, (Vol. 2), (pp. 533-539). IEEE Computer Society.

Conklin, J. (1993). Capturing organizational knowledge. In *Readings in Groupware and Computer-Supported Cooperative Work: Assisting Human-Human Collaboration* (pp. 561–565). San Mateo, CA: Morgan Kaufmann Publishers.

Conklin, J., & Begeman, M. L. (1988). gIBIS: A hypertext tool for exploratory policy discussion. In *Proceedings of the Conference on Computer-Supported Cooperative Work*, (pp. 140-152). Portland, OR: ACM Press.

Divitini, M., Simone, C., & Schmidt, K. (1996). ABACO: Coordination mechanisms in a multi-agent perspective. In *Proceedings of the Second International Conference on the Design of Cooperative Systems*, (pp. 103-122). Antibes-Juan- les-Pins, France: INRIA Sophia Antipolis.

Engeström, Y. (1990). When is a tool? Multiple meanings of artifacts in human activity. In *Learning, Working and Imagining: Twelve Studies in Activity Theory* (pp. 171–195). Helsinki, Finland: Orienta-Konsultit Oy.

Engeström, Y. (1999). Innovative learning in work teams: Analyzing cycles of knowledge creation in practice. In *Perspectives on Activity Theory* (pp. 377–404). Cambridge, UK: Cambridge University Press.

Erickson, T., Huang, W., Danis, C., & Kellogg, W. A. (2004). A social proxy for distributed tasks: Design and evaluation of a prototype. In *Proceedings of the CHI 2004*, (pp. 559-566). ACM Press.

Gerson, E. M., & Star, S. L. (1986). Analyzing due process in the work place. *ACM Transactions on Office Information Systems, 4*(3), 257–270. doi:10.1145/214427.214431

Gibson, J. J. (1979). *The ecological approach to visual perception*. Boston, MA: Houghton-Mifflin.

Grinter, R. E. (1996). Supporting articulation work using software configuration management systems. *Computer Supported Cooperative Work: The Journal of Collaborative Computing, 5*(4), 447–465. doi:10.1007/BF00136714

Guy, E. S. (2003). Patterns as artifacts in user-developer collaborative design. In *Proceedings of ECSCW 2003*. ECSCW.

Haas, C. (1996). *Writing technology: Studies on the materiality of literacy*. Mahwah, NJ: Lawrence Erlbaum.

Halverson, C. A., Ellis, J. B., Danis, C., & Kellogg, W. A. (2006). Designing task visualizations to support the coordination of work in software development. In *Proceedings of the 20th Anniversary Conference on Computer Supported Cooperative Work*, (pp. 39-48). New York, NY: ACM Press.

Henderson, K. (1999). *On line and on paper: Visual representations, visual culture, and computer graphics in design engineering*. Cambridge, MA: MIT Press.

Hertzum, M. (1993). Information retrieval in a work setting: A case study of the documentation part of chemists' work. In J. P. Bansler et al. (Eds.), *Proceedings of the 16th IRIS Information Systems Research Seminar in Scandinavia,* (pp. 786-798). Copenhagen, Denmark: IRIS.

Hertzum, M. (1999). Six roles of documents in professional's work. In S. Bœdker, M. Kyng, & K. Schmidt (Eds.), *Proceedings of the Sixth European Conference on Computer-Supported Cooperative Work,* (pp. 41-60). Dordrecht, The Netherlands: Kluwer Academic Publishers.

Khandani, S. (2005). *Engineering design process: Education transfer plan*. Retrieved from http://iisme.org/etp/HS%20Engineering-%20 Engineering.pdf.

Kuutti, K. (1996). Activity theory as potential framework for human-computer interaction research. In *Activity Theory and Human-Computer Interaction* (pp. 17–44). Cambridge, MA: The MIT Press.

Latour, B. (1986). *Visualization and cognition: Thinking with eyes and hands. Knowledge and Society: Studies in the Sociology of Culture Past and Present* (pp. 1–40). Greenwich, CT: Jai Press.

Leont'ev, A. N. (1981). *Problems of the development of the mind.* Moscow, Russia: Progress Publishers.

Lundberg, N., & Sandahl, T. I. (1998). Artifacts in work practice. In *Proceedings of IRIS 1998*. IRIS.

Lutters, W. G., & Ackerman, M. S. (2002). Achieving safety: A field study of boundary objects in aircraft technical support. In *Proceedings of the ACM Conference on Computer Supported Cooperative Work*, (pp. 266-275). New Orleans, LA: ACM Press.

Muller, M., & Carey, K. (2002). Design as a minority discipline in a software company: Toward requirements for a community of practice. In *Proceedings of CHI 2002*. Minneapolis, MN: CHI.

Nardi, B. A. (1993). *A small matter of programming: Perspectives on end user computing.* Cambridge, MA: MIT Press.

Newman, M. W., & Landay, J. A. (2000). Sitemaps, story boards, and specifications: A sketch of web site design practice. In *Proceedings of DIS 2000.* Brooklyn, NY: ACM Press.

Pahl, G., Beitz, W., Feldhusen, J., & Grote, K.-H. (2007). *Engineering design: A systematic approach* (3rd ed.). Berlin, Germany: Springer.

Schmidt, K. (2001). Classification schemes as material practice. In *Proceedings of the Second CISCPH Workshops on Cooperative Organization of Common Information Spaces*. Copenhagen, Denmark: CISCPH.

Schmidt, K., & Bannon, L. (1992). Taking CSCW seriously: Supporting articulation work. *Computer Supported Cooperative Work, 1*, 7–40. doi:10.1007/BF00752449

Schmidt, K., & Simone, C. (1996). Coordination mechanisms: Towards a conceptual foundation of CSCW systems design. *Computer Supported Cooperative Work, 5*(2-3), 155–200. doi:10.1007/BF00133655

Schmidt, K., & Wagner, I. (2002). Coordinative artefacts in architectural practice. In M. Blay-Fornarino et al. (Eds.), *Cooperative Systems Design: A Challenge of the Mobility Age: Proceedings of the Fifth International Conference on the Design of Cooperative Systems (COOP 2002),* (pp. 257-274). Amsterdam, The Netherlands: IOS Press.

Schmidt, K., & Wagner, I. (2004). Ordering systems: Coordinative practices and artifacts in architectural design and planning. *Computer Supported Cooperative Work, 13*(5-6), 349–408. doi:10.1007/s10606-004-5059-3

Sellen, A., & Harper, R. H. R. (2001). *The myth of the paperless office*. Cambridge, MA: MIT Press.

Simone, C., & Divitini, M. (1997). Ariadne: Supporting coordination through a flexible use of the knowledge on work processes. *Journal of Universal Computer Science*, *3*(8), 865–898.

Star, S. L. (1989). The structure of ill-structured solutions: Boundary objects and heterogeneous distributed problem solving. In *Distributed Artificial Intelligence* (*Vol. 2*, pp. 37–54). London, UK: Pitman.

Star, S. L., & Griesemer, J. R. (1989). Institutional ecology, 'translations' and boundary objects: Amateurs and professionals in Berkeley's museum of vertebrate zoology, 1907-1939. *Social Studies of Science*, *19*, 387–420. doi:10.1177/030631289019003001

Suchman, L. (1987). *Plans and situated actions: The problem of human-machine communication*. Cambridge, UK: Cambridge University Press.

Tellioğlu, H. (2003). Modeling coordinated work: Definition and application of the model coordinated work environment. *The Journal of Supercomputing: An International Journal of High-Performance Computer Design, Analysis, and Use*, *24*(2), 161–171.

Tellioğlu, H. (2007). *Cooperative work.* Wien, Austria: Fakultät für Informatik.

Tellioğlu, H. (2009). Model-based collaborative design in engineering. In L. Yuhua (Ed.), *Lecture Notes in Computer Science, Information Systems and Applications: Proceedings of the 6th International Conference CDVE 2009,* (vol 5738), (pp. 85-92). CDVE.

Tellioğlu, H. (2010a). Coordination 2.0: Using web-based technologies for coordination support. In *Proceedings of the International Workshop on Web Intelligence and Virtual Enterprises 2 (WIVE 2010)*. Saint-Etienne, France: WIVE.

Tellioğlu, H. (2010b). Coordination of work by using to-do lists: Practices and requirements. In *Proceedings of the International Conference on Information Technologies (InfoTech-2010)*. Elena, Bulgaria: InfoTech.

Tellioğlu, H., & Wagner, I. (2005). Work cultures in multimedia production. In *Proceedings of the ALP_IS Alpine Information Systems Seminar*, (pp. 143-161). ALP_IS.

Wagner, I. (2000). Persuasive artefacts in architectural design and planning. In *Proceedings of CoDesigning*, *2000*, 379–390.

## ENDNOTES

1. Scott Umbaugh, Textbook: Design for ECE Engineers, Ford & Coulston.
2. The name is modified.
3. The project team in Austria consisted of Hilda Tellioğlu, Ina Wagner, and Friedrich Glock.
4. Our study was conducted as part of research project "Systemic Integration of Production and Services. Case Studies in the Software and Multimedia Industries," in cooperation with the Brandenburg University of Technology Cottbus and DJI Munich (2001-2003).
5. http://mapper.eu.org/
6. The project group consisted of Hilda Tellioğlu and Ina Wagner from Vienna University of Technology as well as of Gianni Jacucci and Gian Marco Campagnolo from University of Trento.
7. The name is modified.

# Chapter 8
# Representations, Institutions, and IS Design:
## Towards a *Meth-Odos*

**Gianluigi Viscusi**
*University of Milano-Bicocca, Italy*

## ABSTRACT

*In this chapter, the authors discuss some issues emerging from the phenomenological analyses carried out by Claudio Ciborra, in particular in the Labyrinths of Information. The chapter points out that concepts such as Kairos, Drift, Bricolage, unveil a specific odos for the information systems as a discipline. In their perspective, this odos covers a meth-odos towards new opportunities offered to design by answering the provocation coming from considering information systems as infrastructures (Ge-stell). Furthermore, the authors point out that these opportunities come from a deep understanding of the philosophical background of the work of Claudio Ciborra, namely from the idea of phenomenology of Martin Heidegger, whose position refuses the idea of a subject (no matter how pure or transcendental) as the original foundation of our relationships with reality.*

## INTRODUCTION

In this chapter we discuss the issues emerging from the phenomenological analyses carried out by Claudio Ciborra, in particular in the *Labyrinths of Information* (Ciborra, 2002). Besides other contributions to the theoretical foundations of information systems as a discipline (Gregor, 2002; Jones, et al., 2008; Lyytinen, 1987), the work of Ciborra provides not only a deep analysis and understanding of the organizational issues involved in information systems design, but also an initial attempt to provide a philosophical foundation to the discipline. Furthermore, despite the

DOI: 10.4018/978-1-4666-0303-5.ch008

clear argumentations raised by Ciborra against the adoption of methodologies and the claim for an approach to the management of information systems based on concepts such as *Kairos*, *Drift*, *Bricolage,* we point out that these latter concepts unveil a specific *odos* for the information systems as a discipline in the area of organization science (Ciborra, 2002). In our perspective, this *odos* covers a *meth-odos* for the new opportunities offered by answering the provocation coming from considering information systems as infrastructure (*Ge-stell*). These opportunities arise from a deep understanding of the philosophical background of Claudio Ciborra's work, namely from Martin Heidegger idea of phenomenology, often confused with the one from his mentor Edmund Husserl, that is radically different from an ontological perspective (Heidegger, 1962).

In particular, we discuss the concept of *Ge-stell* in order to let the *Bestand* as "stock" emerge, or "standing reserve," the term Heidegger uses often. In this sense the *Ge-stell* is the provocation that leads the human being to collect what is unveiled as a "stock" or an asset, rendering the world and human being into a stockpile of raw materials (Heidegger, 1977). It is out of the scope of the chapter to provide a deep philosophical analysis of the thought of Martin Heidegger, rather we try to use a perspective closer to the concepts of Claudio Ciborra, trying to open new paths from the ones he traced.

Ciborra's choice to adopt a foundational perspective rooted in phenomenology led to the definition of relevant concepts that are peculiar to the information systems domain, such as the critical concept of *infrastructure*. We claim that this path is the most suitable one to define the specificity of information systems concepts and *meth-odoi*, beside the ones of the disciplines that concur in the design and development of information systems, such as sociology, psychology, economics, management, and engineering.

However, it is first necessary to provide a brief description of what is meant by information system. Among the various definitions proposed in the literature, we regard as sufficiently complete and useful for our purposes the one proposed by Buckingham, Hirschheim et al. (1987), for whom an information system is: "[...] a system which assembles, stores, processes and delivers information relevant to an organisation (or to a society), in such a way that the information is accessible and useful to those who wish to use it [...] An information system is a human activity (social) system which may or not involve the use of computer systems[...]." This definition highlights what is meant in general by the term *system* in the expression 'information system': the set of actors (people, objects, procedures, etc.) that interact to obtain, produce, and distribute information useful to the participants/users. Furthermore, the considered definition specifies that an information system does not necessarily have to make use of Information and Communication Technologies (ICTs). Nevertheless, today the attention is mainly on ICT adoption and use by people and organizations and mostly focused on "formalised information systems" (Avison & Fitzgerald, 1995) through formal approaches based on rules and clear structures rather than on logical-mathematical models.

In particular, an information system has the use of information as its purpose. The term *information* covering a wide range of *representations*, from unstructured or semi-structured texts, maps, images, sounds, etc. Definitions and use of information in the literature are a wide number too (Sesé, et al., 2006). Here we mention the Goguen's definition of item of information as "the interpretation of a configuration of signs for which members of some social group are accountable" (Goguen, 1997) and a General Definition of Information (GDI) by Floridi (2011) where "σ is an instance of information, understood as semantic content, if and only if: (GDI.1) σ consists of one or more data; (GDI.2) the data in σ are well-formed; (GDI.3) the well-formed data in σ are meaningful," Thus, data is considered to convey meaning only when such properties are achieved, producing a mutation into information. Information is therefore such because it is significant and comprehensible to people or groups of people, whilst data refer to the support on which those data are recorded and the language with which they are described. Information and data are therefore closely connected and stand in a systemic and formal relationship with the other actors involved in an information system. However, corresponding to this integrated view is a specialization of research consisting in studies which analyse, on the one hand, the economic-organizational aspects of information (an information system as an *organizational system*), and on the other hand, aspects more closely connected with the processing of data and information by means of technologies (*information technology system*).

This fragmentation gives rise to a separation which, in Heideggerian terms, is a forgetfulness which entails consideration of the various entities encountered in the analyses of information systems in terms of 'simple presence' or 'presence at hand' (Heidegger, 1962). To consider entities in their simple presence is to privilege a type of formal knowledge that distracts the researcher from the evidence that the world is always open to an emotionally situated understanding reducible only at the cost of a loss.

As a consequence, we can first consider information systems research area as having its place among human sciences in a sense close to the analyses carried out by Michel Foucault (1966) shown in Figure 1: following this perspective, human sciences emerge from the margin between (1) hard sciences, such as mathematics and physics, (2) the science that tries to define the human being in terms of his life (*biology*), language (*linguistics*), work (*economics*), and from (3) *philosophy* (for a detailed discussion we refer to Foucault, 1966). What is worth noting here is the *boundary nature* of human sciences using concepts and methods from the previous cited disciplines without identifying with any of them, but providing a *representation* of the issues carried out by each one of them; for example, by providing a sociological representation of the working man. Focusing on organization science, the concept of organizational structure mainly refers to a representation of what are the structural relationships between working human beings in a bounded domain (we use 'organizational structure' in a loose way as a discursive example, with no reference to a particular theoretical position).

*Figure 1. The field of the modern episteme, adapted from Foucault (1966)*

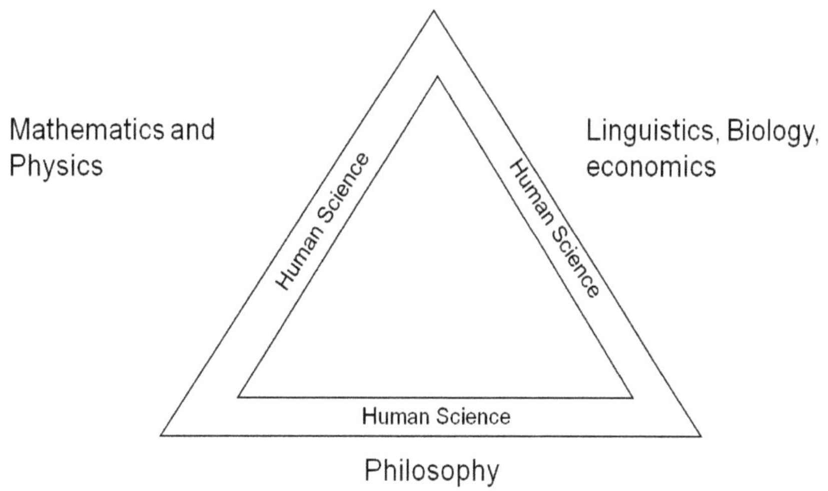

Nevertheless, in the above argumentation a relevant dimension of information systems as research area is missing: namely, *technology*.

As pointed out by Kallinikos (1995), *representation* is technology in terms of "an overall world orientation concerned with the objectification of the natural and social worlds in ways which render them amenable to calculation and mastery." Furthermore, representation is technology in terms of "an ensemble of diverse and minute techniques which embody and realize such an orientation" (Kallinikos, 1995), namely an ensemble of *representations*.

Due to these issucs, the above mentioned focus on Information and communication technology (*representation*) together with the role of modelling and formalizations (*representations*) in formal design sciences (Hevner, et al., 2004; Krogstie, et al., 2007; McKinney & Yoos, 2010), it makes information systems something different and peculiar among human sciences (as shown in Figure 2), where the concept of *representation* is the provocation for the design of information systems (see also Boland, 1999).

The main lesson from Claudio Ciborra is that focusing only on one *facet* of the *Ge-stell*, it fails to answer the challenge of systems and of information systems as a way to unveil the *stock* as something different from an asset, both at the organizational and technological level. Otherwise, on the one hand, we have sociological, organizational, economic, or technological analyses in the information systems area, but what an information systems research means remains veiled and forgotten; on the other hand, we have the design of good organizations for ideal systems or the design of complex systems for unready organizations.

At the disciplinary level, the consequence is the question about the usefulness of information systems design, both by socio-organizational

*Figure 2. Human science and representation: the place of information systems*

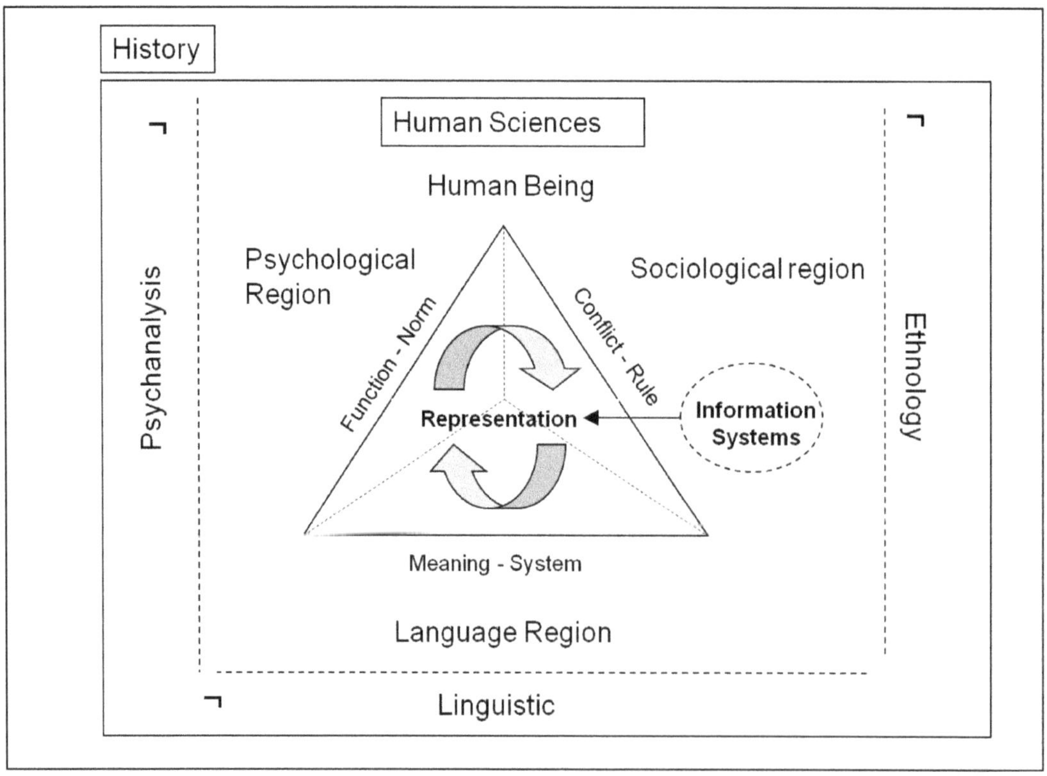

scholars and engineering scholars. Furthermore, these issues explain why the adoption of methodologies rooted in a neo-positivistic tradition fail in the information systems development (Ciborra, 2002). On the other hand, unveiling the *meth-odos* is a way of considering their positive insights in a phenomenological perspective; indeed, in our perspective *meth-odos*, is a way of answering the challenge of the infrastructure, by highlighting the idea of a path *(hodòs)* of research to cover.

The *meth-odos* exploits the literal Greek meaning of *methodos* (*meta- + hodòs)* that is "to find the way," representing new ways to be unveiled as different (*meta*) from the actual destination of the *Ge-stell*. The *meth-odos* claims for a phenomenological *praxis* of information systems, considering modelling and representation as an unveiling *poiesis*.

These issues are discussed in the following Sections in order to provide insights into organizational design frameworks (see Jacucci, 2007) towards a phenomenological approach answering to the provocation of the infrastructure. Section 2 discusses the *Ge-stell* concept by briefly analysing insights from the works of Martin Heidegger, focusing in particular on the relationship between *theory* and *praxis*. In Section 3 we discuss the issues of representation in the information systems area, focusing on conceptual models and classification schemes; these latter introduce the theme of the *archive*. In Section 4 we propose *meth-odos* as a way to challenge the *archive*. Conclusion and future work conclude the chapter.

## THE LICHTUNG

We first analyse the position of Heidegger on the technique, in order to understand the challenge of the infrastructure as *Ge-stell*. This position is in part different from the interpretation raised by Claudio Ciborra; indeed, the analysis aims to point out the issues that must be challenged in an information systems design, where infrastructure contributes to the definition of the social practices (Bowker & Star, 1999).

Against common interpretation of Heidegger thought about technique, we point out that Heidegger's perspective does not consider machinery, and the other products of the technique as per se negative (Heidegger, 1977). They depend on how the unveiling is carried out by the technique, that is a specific kind of unveiling but not the only one. The problem for Heidegger is to unveil the essence of the production technique. The modern technique is considered as an instrument for human activity, leading to consider the world as a source or a "stock" (*Bestand*) for human activity. The technique is not simply an instrument; rather it is a way to unveil, a way to answer a provocation. The unveiling is related to the complex definition of truth as *aletheia*, whose analysis is out of the scope of this chapter (Heidegger, 1962). For our aims, we point out that Heidegger considers every production (*poiesis*) not only as a way of producing something, but also as an unveiling: every production activity unveils to manifest something. The technique belongs to the *téchne* that is something poietic (in the ancient Greek terms); *téchne* refers to both manufacturing and to the arts; the main characteristic of the *téchne* is not to handle, but rather to unveil through production (Heidegger, 1977).

Nevertheless, the unveiling of the modern technique is not a *poietic* way of production, rather a provocation asking nature for energy and resources to be stocked and stored (Heidegger, 1977). The being as a *stellen* characterizes this unveiling. The *stellen* is a provocation which considers the world as a standing reserve (*Bestand*), as something that is measurable, that can be ordered and available for human activity; that is a an object available for a subject. This separation is the provocation of the *Ge-stell*, and for Heidegger it is the accomplishment of the metaphysic, considering the being as the rational representation of the entities; in the terms of *Being and Time* (Heidegger, 1962) these entities are merely *present-at-hand* and worldless, pointing out the unthinking existence that mere entities have. In a simplified perspective, the danger in the provocation of the modern technique emerges when this wordless perspective is considered by humans as the only essence of the technique, which lets nature be controlled (and in part other humans) considered as merely a stock of resources to control. This way, the *Ge-stell* is considered only as provocation and cannot be perceived as a call for other forms of unveiling found in the openness of the *Da-Sein* as *Lichtung*. The machineries are not dangerous per se; rather the closure for humans of an original form of unveiling from the call of the original openness of the *Lichtung*.

In Heidegger's *Being and Time* (Heidegger, 1962), the *Da-Sein* (literally "being-there") is always in an original relation with the world. This relation precedes every division between subjects and physical objects; the being of the things that we meet in the world is part of the *existenzial* structure of the *Da-Sein*. As already noted by Winograd and Flores (1987), Heidegger rejects both the simple "objective stance" (i.e. the objective physical world as primary reality) and the simple "subjective stance," arguing that both are intertwined and that the interpreted thing and the interpreter do not exist independently. Furthermore, the entities in the world are encountered as *ready-to-hand*, discovered through their use, and as *present-at-hand*, when entities are encountered as isolated things, with a breakdown occurring in their use (see also the interpretation from Winograd & Flores, 1987).

Taking that into account the *Da-Sein* relates itself to the others beings in different ways, because it is opened to the world and always thrown into the world and within a tradition. These different ways are the *Befindlichkeit* (i.e., affectivity), the *Verstehen* (i.e., the understanding), and the *Rede* (i.e., speaking) (Heidegger, 1962). They are co-originated and indivisible, that is an understanding always has a kind of affectivity and is articulated in a meaning expressed in a language. By referring to the Greek concept of truth as *aletheia* (non-hidden), this openess of the *Da-Sein* represents for Heidegger the background of the means for the compound of meanings of the understanding of the world; this openess of the *Da-Sein* as original openess to the truth as *aletheia*, as a non-hidden disclosure, is the *Da* of *Da-Sein* (literally the "there" of being) (Heidegger, 1997) and is called *Lichtung* by Heidegger.

The *Lichtung* is viewed as an open field of sense-making relations, the world is an "opening" that "clears" things and makes them clear, i.e., intelligible as something. The human being as *Da-Sein* is characterized by *to-be-the-open* (to be able to take-as). The facticity (*Faktizitat*) of our "cannot-not-be-the-open" is what Heidegger calls our "belonging" to the open (das *Zugehren*), but reciprocally, the open or *Da* works only as long as there is a human being. There is a necessary reciprocity between our thrown-open essence (*-sein*) and the possibility-of-sense-making (*Da-*), and this constitutes the dynamic structure of Da-sein as openess to the being. The *Lichtung* is that ontological space of *openess* to the being. The *understanding* allows *interpretation*, this latter becoming *assertion* (*Aussage*) when the entity ready-to-hand becomes *present-at-hand*, the *Who* of the *Da-Sein* becomes a subject, *Circumspection* becomes knowledge, and the discovery of the entity becomes a predicate of a judgment. Assertion in the history of metaphysics has been considered the place of truth, without considering that assertion is founded in the openness of the being-in-the world. Furthermore, Heidegger considers *theory* as the descriptive attitude aiming at the truth of the things. In the theory the world emerges as presence-at-hand and the entities are communicated through assertions. While *poiesis* is the productive attitude aimed at producing work (of art), under the *techne* disposition where the world emerges as ready to hand. The *praxis* is the way of making of the *Da-Sein*, founded in the original ways of being in the world, guided by the *phronesis*. The novelty of the Heidegger perspective is that theory is not the original orientation of the *Da-Sein*, rather it is founded in the *praxis* as the original attitude towards the openness of world. *Theory* is considered as a defective attitude with respect to *poiesis* and *praxis*. Theory considers nature as present-at-hand, as an object representation of a subject; the danger here is the provocation for humans in considering nature and the world as object of their own research, until the object itself disappears in the absence of the object (*das Gegenstandlose*) of the stock (*Bestand*). The foundation of theory in the readiness-to-hand of the being in the world and the representation of the rationalistic tradition (as called by Winograd & Flores, 1987), forget the non original existence of the concept of "subject" and "object," The entities as things present-at-hand can be ordered, measured, stocked, while forgetting their original readiness-to-hand.

## REPRESENTATIONS, INSTITUTIONS, AND IS DESIGN

The previous general and simplified discussion of Heidegger's thought allows us to understand why the *infrastructure* as *Ge-stell* is the way to uncover other ways of challenging the systems, by acting on the representations that found the infrastructure itself, bounding its openness. The concepts introduced by Claudio Ciborra fail to support the design of new systems while they are considered as separated, by focusing only on affectivity and readiness-to-hand, leaving covered the representational power that is the real provocation of the system. Furthermore, a main issue in information systems design is to introduce a *meth-odos* allowing to uncover the original openness under the provocation of the "objective" view of the systems, presently considered as originally separated from a "subjective" perspective of the organization.

The challenge of the infrastructure is to uncover the formalized representations, by grounding them in the praxis, namely as an understanding with affectivity, articulated in a meaning that is expressed in a language. To reply to the provocation of the infrastructure means to overcome the distinction between 1) a system considered as "stock" for a subject, acting on the formal and 2) institutionalized representations, making them living (Bowker & Star, 1999). Challenging the provocation of the infrastructure claims to unveil an original space that is dynamically characterized by both discursive and non discursive practices (Foucault, 1972), these latter being the 'way of making,' founding and defining their own *Rede* not as standing reserve. This original space where the practices are unveiled in their contribution to the emergence of *institutions* can be described by using the Berger and Luckmann's phenomenological approach (Berger, et al., 1967). Here, institutions emerge from a process which is characterized by the *typization* of *abitualized* actions. This process has an historical development, providing schemas of action for the participants in the institutions. The institutions cover the historical process at their basis, representing itself as present-at-hand and original. These representations are the way of unveiling of the infrastructure (as we have discussed above), inscribed in infrastructure as presence-at-hand, re-presenting the technological systems as a stock for a subject that is an asset for an organizational system. The infrastructure as an institution is inscribed in representations such as documents and schemas that must be investigated through a *meth-odos*. In the information systems area the *meth-odos* must support a change, considering documents as monuments (Foucault, 1972), unveiling from the representation, (1) *what* is represented as an institutionalized presence-at-hand, (2) *how* it has been represented, (3) and *why* it has been represented in that way (unveiling the path of construction and the related practices).

These research issues find support also in the tradition of the analytical philosophy, in particular in the complex work of Donald Davidson (2001). In a simplified way, we point out that Davidson claims that there is no supposed external reality, with no interpretation (in the terms of Heidegger the reality present-at-hand of the *Ge-stell*), a reality to which a conceptual schema is applied. For Davidson reality is a continuous process of interpretation that is expressed in a language (see above the *Assertion*), but a language (see above *Rede*) that is not a filter between a supposed reality for a given subject, rather a point of access to the openness of the being in the world. We can say that every schema is founded in a reality, and the danger arises when the schema is seen as something separated and institutionalized as the original representation (covering the object of the representation).

The challenge in information systems design is to unveil: (1) schemas as a way to social practices; (2) social practices as founding living schemas in an evolutionary process, challenging the institutionalized and disciplinary divisions between the organization and the information systems, which are considered as separated assets for producing standing reserves (*Bestand*). *Meth-odos* tries to use the Claudio Ciborra's concepts in order to trace a path for unveiling the openness within the presence-at-hand of the infrastructure as institution.

## METH-ODOS: CHALLENGING THE ARCHIVE

*Meth-odos* challenges the infrastructure as institution, by considering infrastructure as *archive* (Foucault, 1972). Following the perspective in Foucault (1972), we consider the archive as "[…]the general system of the formation and transformation of statements […]" the "[…]law of what can be said, the system that governs the appearance of statements as unique events[…]." Due to these issues, the archive encompasses both discursive and non discursive practices; due to our belonging to the archive, we need a way to uncover the archive as "the general horizon to which the description of discursive formations, the analysis of positivities, the mapping of the enunciative field belong […]"(Foucault, 1972). The infrastructure considered as an *archive* requires an archaeology of schemas and representations that raise from both discursive and non discursive practices. The term archaeology "[…]does not imply the search for a beginning; it does not relate analysis to geological excavation. It designates the general theme of a description that questions the already-said at the level of its existence: of the enunciative function that operates within it, of the discursive formation, and the general archive system to which it belongs. Archaeology describes discourses as practices specified in the element of the archive" (Foucault, 1972).

Furthermore, borrowing the concept of *simulacrum* from Gilles Deleuze (1969a, 1969b) we can define the first path of *meth-odos* as an *archaeology of schemas and representations* (considered both from the engineering and organizational perspectives) as *simulacra* of uncovering units; the archaeology must provide a detailed description of the representations and a definition of all related rules which have allowed their emergence as separated simulacra. The description of schemas and representations as not reducible simulacra in the archaeological path of *meth-odos* unveils the status of *bestand* of the schemas and of the representations used, e.g. at engineering and organizational levels as different names for a unique being, that is in our case the information systems we have to design. As previously noticed, meth-odos considers documents as monuments (Foucault, 1972), and archaeology can unveil from the representation, (1) *what* is represented as an institutionalized presence-at-hand, (2) *how* it has been represented.

Once described as not reducible *simulacra*, a *genealogy* can be carried out to uncover in a critical perspective the institutional constraints related to their adoption and use (Doolin, et al., 2002). Furthermore, genealogy has been considered in the literature as a way to escape the dualist paradigm separating humans and non-humans (Latour, 1994). In fact, using a perspective drawn from the work of Michel Foucault (1971), genealogy allows to describe both a provenance (*Herkunft*) and an emergence (*Entstehung*) for a simulacrum as an institutional fact. Genealogy can unveil from the representation (3) *why* it has been represented in that way (unveiling the path of construction and the related practices).

## CONCLUSION

In this chapter we have proposed the theoretical background and motivations for a *meth-odos* for information systems design; starting from the analyses of Claudio Ciborra we have introduced a high level description of the two main paths of *meth-odos* and the related objectives.

The proposed *meth-odos* aims to provide a preliminary support to uncover the foundation of the theory in the readiness-to-hand of the being in the world, towards the unit of *Befindlichkeit* (i.e., affectivity), *Verstehen* (i.e., the understanding), and the *Rede* (i.e., speaking) (Heidegger, 1962). Focusing on the situated and organisational oriented perspective of the information systems design, we point out that one of the main dangers for design is related to the focus on one of the simulacra which emerge from the infrastructure as *Ge-stell*. We propose to integrate *meth-odos* in state-of-the-art methodologies in order to challenge the *bestand* of schemas and representations inscribed in the technological and engineering facets of the infrastructure (in line with the analyses carried out by Bowker and Star [1999]). In particular, we propose to integrate the *archaeology of schemas* (*Krisis*) and *a genealogy of simulacra* (*Gestell*) paths in the early phases of information systems design as organizational practice; these paths must be introduced either (1) in the observation phase of social and participatory design oriented methods (Jacucci, 2007; Kensing, et al., 1998) and (2) in requirements analysis in the engineering oriented methods (Byrd, et al., 1992; Krogstie, Opdahl & Brinkkemper, 2007). In particular, the *meth-odos* integration in the early phases of systems design should support the uncovering of the opposition of organisational and engineering/technological simulacra as not reducible facets, which must be analysed and synthesized together in an evolutionary design of information systems.

## ACKNOWLEDGMENT

This chapter is a major revision and extension of an earlier version presented at WOA2009, April 29-30, 2009, Cagliari, Italy.

## REFERENCES

Avison, D. E., & Fitzgerald, G. (1995). *Information systems development: Methodologies, techniques and tools*. London, UK: McGraw-Hill International UK limited.

Berger, P. (1967). *The social construction of reality: A treatise in the sociology of knowledge*. London, UK: Penguin.

Boland, R. (1999). Accounting as a representational craft lessons for research on information systems. In Currie, W., & Galliers, B. (Eds.), *Rethinking Management Information Systems* (pp. 229–243). Oxford, UK: Oxford University Press.

Bowker, G., & Star, S. L. (1999). *Sorting things out: Classification and its consequences*. Cambridge, MA: MIT Press.

Buckingham, R. A., Hirschheim, R. A., Land, F. F., & Tully, C. J. (1987). Information systems curriculum: A basis for course design. In Buckingham, R. A., Hirschheim, R. A., Land, F. F., & Tully, C. J. (Eds.), *Information Systems Education: Recomendations and Implementation*. Cambridge, UK: Cambridge University Press.

Byrd, T. A. (1992). A synthesis of research on requirements analysis and knowledge acquisition techniques. *Management Information Systems Quarterly*, *16*(1), 117–138. doi:10.2307/249704

Ciborra, C. (2002). *The labyrinths of information- Challenging the wisdom of systems*. Oxford, UK: Oxford University Press.

Davidson, D. (2001). *Inquiries into truth and interpretation* (2nd ed.). Oxford, UK: Clarendon Press. doi:10.1093/0199246297.001.0001

Deleuze, G. (1969a). *Difference et repetition*. Paris, France: Presses Universitaires de France.

Deleuze, G. (1969b). *Logique du sens*. Paris, France: Editions de Minuit.

Doolin, B. (2002). To reveal is to critique: Actor–network theory and critical information systems research. *Journal of Information Technology*, *17*, 69–78. doi:10.1080/02683960210145986

Floridi, L. (2011). Semantic conceptions of information. In Zaltz, E. N. (Ed.), *The Stanford Encyclopedia of Philosophy*. Palo Alto, CA: Stanford University Press.

Foucault, M. (1966). *Les mots et les choses*. Paris, France: Gallimard.

Foucault, M. (1971). Nietzsche, la généalogie, l'histoire. In Bachelard, S. (Eds.), *Hommage à J. Hyppolite*. Paris, France: Presses Universitaires de France.

Foucault, M. (1972). *The archaeology of knowledge*. London, UK: Tavistock.

Goguen, J. A. (1997). Towards a social, ethical theory of information. In Bowker, G., Gasser, L., Star, L., & Turner, W. (Eds.), *Social Science Research, Technical Systems and Cooperative Work: Beyond the Great Divide* (pp. 27–56). Upper Saddle River, NJ: Erlbaum.

Gregor, S. (2002). A theory of theories in information systems. In Gregor, S., & Hart, D. (Eds.), *Information Systems Foundations: Building the Theoretical Base* (pp. 1–20). Canberra, Australia: Australian National University.

Heidegger, M. (1962). *Being and time* (Macquarrie, J., & Robinson, E., Trans.). Oxford, UK: Basil Blackwell.

Heidegger, M. (1977). *The question concerning technology and other essays* (Lovitt, W., Trans.). New York, NY: Harper & Row.

Heidegger, M. (1997). *Vom wesen der wahrheit: Zu platons höhlengleichnis und theätet*. Berlin, Germany: Klostermann.

Hevner, A. R., March, S. T., Park, J., & Ram, S. (2004). Design science in information systems research. *Management Information Systems Quarterly*, *28*(1), 75–106.

Jacucci, G. (2007). *Social practice design, pathos, improvisation, mood, and bricolage: The Mediterranean way to make place for IT?* Paper presented at the MCIS2007. Venice, Italy.

Jones, M. R. (2008). Gidden's structuration theory and information systems research. *Management Information Systems Quarterly*, *32*(1), 127–157.

Kallinikos, J. (1995). The architecture of the invisible: Technology is representation. *Organization*, *2*(1), 117–140. doi:10.1177/135050849521006

Kensing, F. (1998). Participatory design: Issues and concerns. *Computer Supported Cooperative Work*, *7*(3-4), 167–185. doi:10.1023/A:1008689307411

Krogstie, J., Opdahl, A. L., & Brinkkemper, O. S. (Eds.). (2007). *Conceptual modelling in information systems engineering*. Berlin, Germany: Springer Verlag. doi:10.1007/978-3-540-72677-7

Latour, B. (1994). On technical mediation – Philosophy, sociology, genealogy. *Common Knowledge, 3*(2), 29–64.

Lyytinen, K. (1987). A taxonomic perspective of informaiton systems development: Theoretical constructus and recommendations. In Boland, R. J., & Hirschheim, R. A. (Eds.), *Critical Issues in Information Systems Research* (pp. 3–41). New York, NY: John Wiley.

McKinney, E., & Yoos, C. (2010). Information about information: A taxonomy of views. *Management Information Systems Quarterly*, *34*(2), 329–344.

Sesé, F., Wareham, J., & Bonet, E. (2006). Words and objects in information systems development: Six paradigms of information as representation. In *Proceedings of the 14th European Conference on Information Systems*. Göteborg, Sweden: IEEE.

Winograd, T., & Flores, , (1987). *Understanding computers and cognition: A new foundation for design*. New York, NY: Addison-Wesley.

Section 3

# Phenomenology and the Social Study of Information Systems

# Chapter 9
# Studying Information Infrastructures

**Petter Nielsen**
*University of Oslo, Norway*

## ABSTRACT

*As a result of a steady increase in reach, range, and processing capabilities, information systems no longer appear as independent, but rather as integrated, parts of large scale networks. These networks offer a shared resource for information delivery and exchange to communities, which appropriate them for their respective purposes. Such information infrastructures are complex in several ways. As they are composed of a variety of different components, their openness and heterogeneity make them inherently uncontrollable; through their expansion, these various interconnected networks enter new interdependencies; while they are based on extending existing technical and social networks, they also need to develop and grow over a long period of time; and, they are developed as a distributed activity. Examples of such information infrastructures include the Internet, National Information Infrastructure (NII) initiatives and industry-wide EDI networks, as well as corporate-wide implementations of enterprise systems.*

## INTRODUCTION

The common perspective on the development or evolution of large-scale networks and information infrastructures is *network economics*. From this perspective, information infrastructures are evolving according to network effects such as increasing returns, positive feedback, network externalities, path dependency, and lock-in (for example Hanseth, 2000). Theories of network economics can be used to explain the evolving nature of information infrastructures, in particular in relation to its users (Hanseth, 2003; Hanseth & Monteiro, 1997; Shapiro & Varian, 1999). Due to heterogeneity and its character of being a network, a successful network evolves by

DOI: 10.4018/978-1-4666-0303-5.ch009

self-reinforcing mechanisms. When a network attracts new users, the value of being part of the network increases (network externalities), and it becomes even more attractive for other users to join the network (Arthur, 1994). When the user base reaches a certain threshold, the network will attract new users for enrolment almost by itself. The growth of faxes, cellular phones, and the Internet has been used to illustrate this pattern of development. However, this base of users may also introduce strong conservative forces, in that large numbers of users make it difficult to change the network, as for example illustrated by the QWERTY keyboard. Introducing a completely new network will not attract users as long as old networks provide sufficient services and have a superior number of users, thus the possibilities of introducing new and competing information infrastructures are limited.

Network economics is a macro theory, and describes the development of networks as the sum of individual actions of its users. It does at the same time say little or anything about how developers can involve themselves and influence the network as well as how these networks are interrelated with a wider context. While information systems increasingly develop as ubiquitous, heterogeneous, and pervasive networks, there is a shortage of knowledge and understanding of such development processes (Lyytinen & Yoo, 2002). There is a need to develop and strengthen our theoretical understanding of the process by which information infrastructures are built, and in particular, to open the black box of information infrastructures with respect to their malleability (Orlikowski & Iacono, 2001).

Information Systems development approaches and methodologies are based on engineering principles. While these approaches have changed their focal point from the technology to the users, and from a strict waterfall process from analysis through design and implementation to more flexible and rapid feedback and development cycles, they are still based on the traditional divide and conquer of the natural sciences. Their focus is limited to the construction of elements to support well defined and delimited user needs and how such elements can interact with other elements. Information infrastructures have been conceptualised from various perspectives (for example Branscomb & Kahin, 1996; Hanseth, 2000; McGarty, 1992; Star & Ruhleder, 1996), In this chapter, I draw upon the work related to Ciborra, Hanseth, and associates (see in particular Ciborra, et al., 2000; "Hanseth & Lyyttinen, 2010, 2010; Hanseth & Monteiro, 1997). As archetypes, McGarty (1992) has conceptualised information infrastructures as consisting of technical elements, while Star and Ruhleder (1996) have, on the contrary, conceptualised them as social structures. The conceptualisation by Ciborra, Hanseth, and associates distinguishes itself from these approaches as well as conventional software engineering by taking, as well as extending, an ensemble view (Orlikowski & Iacono, 2001) of information technology by taking into account information infrastructures' distinctive properties (socio-technical, networked, distributed, etc.). This conceptualisation also explicitly draws upon Actor-Network Theory (ANT) (e.g. Monteiro, 2000) and pictures the developments of information infrastructure as an evolutionary process which is intrinsically linked to the interplay between humans and technical components. Economic mechanisms of networks are used at the same time to describe how information infrastructures evolve and how they are strongly related to what already exists; the installed base of, for example, users and their patterns of use; design and implementation practices; standards; technical components, and investments.

This conceptualisation of information infrastructures can potentially yield several key insights. Basically, information infrastructures are understood as networks in the sense that a broad range of humans and non-human components mutually influence their development in a variety of ways. A range of factors such as people, institutions, standards, and technical components influences human actors and their acts. Thus, the acts are not independent, but constitute and produce socio-technical actor-networks (Monteiro, 2000; Walsham, 1997). As a consequence, information infrastructures do not develop due to planned and controlled actions by some developers, but rather in a process imbued with surprises, blockages, diversions, side effects and vicious circles, as well as inherent tensions between the need for universal standards and locally situated practices (Hanseth & Monteiro, 1997). It has been suggested that dealing with these challenges requires modesty—to accept and live with the complexity (Aanestad, et al., 2005). More particular, technical solutions have been suggested to interconnect incompatible networks such as, for example, the gateways linking the various protocols (e.g., SMTP and Microsoft Exchange) for e-mail (Hanseth, 2001). At the same time, flexible designs are suggested to avoid premature decisions resulting in lock-ins and obsolescence. For example, the flexibility the Internet offers has provided opportunities for a wide range of innovations and extensions over a relatively long period of time (Saltzer, et al., 1984). Strategies of cultivation and "bootstrapping" have also been suggested to expand and grow technical networks with limited content and few users (Hanseth & Aanestad, 2003). For example, the World Wide Web was "invented" by developing a minimal technical layer upon an existing network (the Internet) and by taking advantage of existing information and information needs to attract users at an early stage (Tuomi, 2002).

In this chapter, we will dwell a bit on the conceptualisation of information infrastructure, its roots, and the transition from information systems to information infrastructures. We will do this by discussing the underlying move from systems thinking to information infrastructure thinking. The aim is to point out the consequences and challenges this thinking and perspective poses to those who study large scale and complex networks. In particular, we focus on how to study and understand the balance between the autonomy of networks and the roles of its developers. By stepping down from the economic macro theories describing the development of large-scale networks, we use a more flexible approach by focusing on the developers and their privileging human agency over technical and social structures, the multiple agencies of the different developers, as well as the role of institutions and information infrastructures in shaping agency and their relationships. I argue that this *ensemble* view and concepts such as *control devolution, sinking and collage* compose a more fruitful approach to support our understanding of how information infrastructures develop, and how to take a role and influence and shape the trajectory of such processes.

## FROM SYSTEMS TO INFORMATION INFRASTRUCTURES

In the field of information systems, a key concept is *systems*. Computers are used by human beings to receive, process, store, and communicate information—the computer systems and its users comprise an information system—which supports and controls the wider organizations—yet another system. Information systems professionals work with and study these systems. We develop and maintain computer systems for others to use, and we use computer systems as our tools in doing this. This is similar to the common use of the concept systems, describing a group of elements, which are interrelated and forming a whole, such as for example the human body, the nervous system, or social, economic, and political organizational forms. This idea of a system as a group of interrelated components is reflected in how the concept has been discussed in a range of research field such as physics, chemistry, mathematics, biology, engineering, sociology, political science, and so on.

Brian Wilson (1990) has described the origin of the concept of system and what we can call modern system thinking. Even if the concept of system can be traced back to Ptolemy's model of the universe and the ancient Egyptians, what is particular to modern systems thinking is the relatively recent influence from the disciplines of engineering and biology. From the 1940s onwards, engineering methodologies were developed to cope with the need to design and optimize organizational processes that were becoming more complex and costly. This was partly a result of the general market environment becoming more competitive and thus the need to be efficient becoming more pronounced. At the same time, optimizing processes were enabled by computer systems that could to an increasing degree be used to carry out complex calculations. This should also be seen in the light of the World-War II, and how operations research in the US was developing to support the needs of the military industry. This gave rise to the development of system engineering methodologies to support the interactions between process units. Between the processes being designed and others on which it depended on, such as human beings, and between the interacting set of processes and the market environment being served. This was also a reaction against reductionism—and an emphasis on systems that are open to and interacting with their environment. Rather than reducing an entity to the properties of the parts, system engineering focused on the arrangement of and relations between the parts connecting them in a whole. *Analysts were thus trying to take into account more of the whole*—the boundary of the system being analyzed was expanding.

At the same time, and in parallel with what was happening in engineering, biologists where also concerned with studying the whole of organisms. Their thesis was that living organisms have properties that are particular to their level of complexity and which are meaningless on lower levels. This is called emergent properties or the sum is more than the parts. In his "Theory of Open Systems in Physics and Biology," the Biologist Von Bertalanffy argued that the phenomena of biology called for a new way of thinking, and that the methods of the physical sciences are inappropriate (Lilienfeld, 1988). Organic laws have to, in comparison with physical ones, require a new kind of understanding. Von Bertalanffy concluded that higher levels of organization involve new laws that are not deducible from the laws appropriate to the lower levels. Thus, modern biology was challenged by the interaction of a large but not infinite number of variables, which demanded new conceptual tools. (p. 22). What Von Bertalanffy argued was that an expansion of science is necessary to deal with the aspects that are left out in classical physics. Further, Von Bertalanffy argued that the same system concepts and principles of organization underlie different disciplines such as biology and engineering. The greatest achievement of Von Bertalanffy was to create at least a loose federation of similar concerns—linked with the concept system. His aim, arising as a result of work in different fields a high-level meta-theory of systems, expressed in mathematics, has however not happened.

To further understand how the system concept has been used in general and systems development in particular, it is useful to draw on the distinctions made by Peter Checkland. Being highly influential on information systems development approaches in his work, Checkland (1981) has termed the set of attempts in all areas of study and disciplines to explore the consequences of a holistic rather than reductionist system thinking, the "systems movement." He sees the mentioned emerging ideas in biology and in engineering as thinking in "wholes"—from reductionism to holism. It is a criticism of seeing living systems as simply machines. These ideas support the development of a new account of the "whole," and for a new "systems approach" to tackling its related challenges. Checkland saw this as being complementary to the reductionist approach embodied in the method of natural science. Checkland summarizes this system movement as follows. As we are trying to understand the world, we have found the method of science as a very successful way. We may define this method characterized by reduction, repeatability, and refutation, but science has its limitations, in particular when we move from the "restricted" (for example physics) sciences to the "unrestricted" (for example biology). Outside science, we find the method of natural science apparently weak. Complexity in general and social phenomena in particular pose difficult problems. Since science does not tackle real-life problems well, there is an incentive to examine alternative paradigms. Within natural science there is a certain level of complexity of the properties which are *emergent*, and which cannot be reduced in explanation to lower levels—that forms an alternative paradigm—that of *systems*. A paradigm concerned with *wholes* and their properties. It is holistic, but not in the sense that everything is taken into the whole—it is *wholes* and their hierarchical arrangements rather than *the* whole.

## Information Systems

To show how the system ideas have been taken up in systems development, it can be useful with an example. The idea behind the modern system concept described by Checkland is to conceptualize the world as a system, to make it accessible for us to understand it, and make us able to describe it. This is reflected in the common ideas of information systems design and implementation. Based on describing the system in various models, the system is implemented in a computer to meet the need of an organization by making their processes more efficient. It is influenced on a modern way of thought—closure, decomposition, and modelling. According to Dahlbom and Mathiassen (1993), even if system analysis as we know it today in systems development has developed largely independently of systems theory, it has strongly been influenced by its principles. It is based on principles to aid decision makers with problems of identifying, reconstructing, optimizing, and controlling a system, while taking into account multiple objectives, constraints, and resources. Also in Ian Sommerville's influential work on software engineering we can see the use of the system concept. He also explicitly describes the application of system ideas in the systems development process. According to Sommerville (2004), systems development is the activity of specifying, designing, implementing, validating, deploying, and maintaining a system as *a whole.* A system is a purposeful collection of interrelated components that work together to achieve some objective. System developers are not just concerned with the software, but also the hardware and how the system interacts with users and its environment. Thus, systems are not independent, but exist in a context. The components of the system are inextricably intermingled, and they are also hierarchical in that they include other systems. The complex relationships between the components mean that the system is more than simply the sum of its parts. These properties cannot be attributed to any specific part of the system. Sommerville describes *the reliability of the computer system* as one example of such properties. Reliability depends on the reliability of each of the system components and the relationships between the components. An even more complex example is *the usability of the computer system*, which is a property, which is not simply dependent on the system hardware and software, but also the system operators and the environment where it is used.

So, how to summarize this work on the system concept? From our common use of the system concept and the accounts of a range of authors, Charles Atkinson and Peter Checkland (1988) describe one coherent account of what they see as the basic concept of system. These ideas summarize what can be seen in the work of the significant system thinkers. *The most fundamental and general idea behind the concept of "system" is the idea of an entity constituted by connected parts.* This is the core, which finds its way into dictionary definitions. Despite the fact that there are many definitions of the system concept all assumes that system is a notion of a set of elements mutually related such that the set constitutes a whole having properties as an entity. Certain other connotations also follow: The system is separated from its surroundings, it is in an environment, its connected parts cause the system to exhibit behaviour, and the system is expected to have some long-run stability, some ability to survive in its environment. In this condensed form, system is a notion of a whole entity with boundaries, separating and isolating it from its environment in a number of ways which may include: spatially, temporally, and organizationally. Where systems are open, they are only open for input and output across clearly defined borders.

## Information Infrastructures

Compared to the vast research on the concept of system, the concept of infrastructure has had less attention. At the same time, infrastructure is like system, a common word in our vocabulary and in our natural language. A common definition is that infrastructure is the underlying installations and facilities that support our daily life and society, and examples are telecommunication networks, public institutions and so on. There is a substantial literature on infrastructure in IS related management research. This literature describes infrastructure as a strategic asset for the firm and how management can pursue and achieve strategic agility based on infrastructure investments. The IT-portfolio of a firm is founded on the IT-infrastructure. This infrastructure is a set of reliable services based on IT-components, human IT-infrastructure, shared IT-services and applications. It is usually coordinated by the IT department and used to connect different parts of the firm, and can also be used to link to suppliers, customers, and partners. As a long-term investment, the IT-infrastructure is what is common and stable across the firm, and over time. Having an infrastructure with the appropriate services in place significantly reduces time and cost to meet changes and build new systems to meet changing market conditions. At the same time, to avoid that the infrastructure becomes a liability, the IT management must ensure that the IS department knows of upcoming infrastructure needs. The management literature on infrastructure provides valuable insights in how IT-infrastructure can be a strategic asset and used actively as a means to meet change and create new business opportunities. At the same time, it adheres to a kind of "portfolio thinking," where the IT-infrastructure is seen as a *capability* composed of independent components or services which senior management can fully control, and can invest in, choose where to place, outsource if found appropriate, apply a Service Level Agreement to, and market test. From a rather different perspective of IS research, Orlikowski and Iacono describes how several IS researchers during the 1980s attempted to overcome shortcomings in what they perceived to be overly narrow views of technology. In particular, they offered alternative conceptualizations of what technology is and how it has effects. Kling and Scacchi (1982), for example, developed the concept of "web-models" of computing in contrast to what they saw as the dominant "discrete-entity" model of computing. From their perspective, information technology is more than just the tools deployed on the desktop or the factory floor. It is also the network of the commitments, additional resources such as training, skilled staff, and support services, and the development of organizational arrangements, policies, and incentives to enable the effective management and use of new technologies. This is the infrastructure that supports and is necessary for the development and use of computing in organizations. In this way, Kling and Scacchi examined the ways in which various social influences shaped how a technology was introduced into a situation, and how different user groups engaged with that technology.

As an additional perspective from the science and technology studies, Star and Ruhleder (1996) have conceptualised information infrastructures as consisting of social structures as well as technical components, in contrast to those who described them as based on technical elements only, like McGarty (1992). Based on discussing the infrastructure of a "virtual laboratory" for more than 1400 biologists, Star and Ruhleder (1996) show how an infrastructure involves a network of humans, computers and practices; its variations across different laboratories; as well as its integration with its users' local and situated contexts. As a consequence, the heterogeneity arising from the distributed and diverse nature of the components comprising the infrastructure becomes a primary challenge for the infrastructure developers to manage. Such infrastructures are not developed within some strict boundaries of a project and a organisation, but the borders of and between different information infrastructures are vague and are not always possible to predetermine (Neumann & Star, 1996). A challenge with these social structures pointed at by Star and Ruhleder is that they tend to become invisible. In certain instances, only through introducing new information technology we may discover them through breakdowns. This is also the common way in which we discover our dependence on infrastructures, such as for example when a water pipe breaks, or the introduction of information systems leads to break-downs in work processes—hence, the invisible becomes visible.

Clearly inspired by Kling and Scacchi's web model perspective and Star and Ruhleder's description of infrastructures, Ciborra, Hanseth, and associates have developed an elaborate definition of information infrastructure. Hanseth (unpublished) have made a strong argument in that the future IT-solutions, which are integrating large numbers of systems across organizational and national borders, are significantly different from traditional information systems. We need new understanding to support the development of these solutions—it should be based on a perspective of information infrastructure—and *not* systems. The reason for choosing infrastructure as a concept, and not simply talk about open or autonomous systems, is to flag that this is a radical break. Using Internet as an example of how future systems development will unfold, Hanseth argues that Internet has not been developing according to conventional IS development approaches. This move from traditional systems towards open networking is the moving force behind the development of this concept of infrastructure. The existing IS development methodologies are focused on developing single, isolated and to a large degree stand-alone systems. The development is taken care of by a project organization. The design project is assumed to have well defined start and ending times—it is an event, not an ongoing process. However, according to Hanseth, developing information infrastructures is not to develop a closed system, done by a closed project organization for a closed customer organization within a closed time frame. At the same time, the notion of system makes us believe that we through our methodologies are in complete control over the design process and accordingly that we can design an IT-solution exactly as we (and the users) want to. This is no longer the case. With this outset, information infrastructures are a contrast to information systems. Seen as an underlying IT-capability, the concept of information infrastructure can be described by certain characteristics as follows:

First, information infrastructures are *shared* and lose their value if they are being reduced or split into separate parts. Thus, their components are interdependent. For example, the next generation of telecommunication networks, or 4G, requires the existence of a physical network, handsets for the users, services as well as users and their usage. As we have seen, if one or more of these components are missing, there is no value in the rest. Second, information infrastructures are *open* in the sense that there are no limits to the number of users, stakeholders, and vendors involved, nodes in the network and other technological components, application areas, network operators and so on. This does not necessarily imply the extreme position that absolutely everything is included in every information infrastructure. However, it implies that one cannot draw borders between one infrastructure and others, and characterize them as independent, and any development project, independently of how big it is, will just cover parts of an infrastructure. The rest is already there and will be developed by others who are out of central reach and control. Thus, information infrastructures are not developed and design by one system developer and one organization. Third, information infrastructures are designed as extensions to or improvements of existing ones. Information infrastructures are thus never built from scratch related to technical components, design practices, institutions and so on—*the installed base*. As a result, information infrastructure design must recognize the installed base of existing information infrastructure, and this installed base strongly influences design activities. This point is clearly illustrated by Ciborra's critique of the management literature related to IT-infrastructure. The components and assets of information infrastructures are not easy to acquire, transfer and to dispose of. It is an illusion that management can plan and decide on how to pick or drop components of their information infrastructure. In particular, Ciborra put emphasis on how the inertia of the installed base influences how modifying, replacing or splitting parts of an information infrastructure is problematic. Further, information infrastructures as *heterogeneous* and *evolving*, properties that are consequences of the openness and related to the installed base as described. The concept of information infrastructure is influenced by a network, rather than a system perspective, in that the borders between what are inside and outside is blurred. A network will include elements of more heterogeneous character and which are more loosely coupled than what is normally the case when one talk about systems, and because of this heterogeneous and open nature, information infrastructures cannot change abrupt, but will only slowly evolve over time. They change in a small-step and incremental fashion, a process we can only influence through approaches described as cultivation. As a result of its openness and heterogeneity, its components also have to adhere to *standards* to overcome their heterogeneous nature.

This infrastructure concept from Ciborra, Hanseth, and associates distinguishes itself from the approaches by the dominating perspective in systems development, that is seeing infrastructure as a technical issue and the management literature by taking, as well as extending, an *ensemble view* (Orlikowski & Iacono, 2001) of information technology. This concept pictures the developments of information infrastructure as an evolutionary process, which is intrinsically linked to the interplay between humans and technical components. Economic mechanisms of networks are used at the same time to describe how information infrastructures evolve and how they are strongly related to what already exists; the installed base of, for example, users and their patterns of use; design and implementation practices; standards; technical components, and investments.

Where systems are defined to be open, it means that inputs and outputs can pass through their borders and interfaces. Describing information infrastructures in contrast with information systems, Hanseth (2000) describes information infrastructures as open in the sense that there are no limits to the number of users, stakeholders, and vendors involved, nodes in the network and other technological components, application areas, and so on. This does not necessarily imply the extreme position that absolutely everything is included in every information infrastructure. However, it implies that one cannot draw borders between one infrastructure and others, and characterize them as independent, and any development project, independently of how big it is, will just cover parts of an infrastructure. The rest is already there, which is the installed base, and will also be developed by others who are out of reach and out of control.

While taking complexity into account is crucial to understand information infrastructures and how they are developed, network theories and the complexity turns leave the single actor with limited if any agency and ability to make a difference. Network theory describes the action of the rational, economic actor, and is predominantly concerned with infrastructure users and how they as a part of the demand side triggers and fuels the development of infrastructures and the unfolding of battles between competing infrastructures. While such models can support our understanding of central mechanisms of infrastructure diffusion and evolution, information infrastructures are not offered by a single provider but based on a web of actors. Acknowledging this complexity brings limited value—identifying, labelling, describing, and acknowledging complexity has limited value in itself. More important, a complexity perspective shuffles control away from the single actor to the network itself, but, while centralised control obviously crumbles away from one central actor in information infrastructures, there are still ways in which control are exercised.

In the next section, I introduce and describe a case study using the information infrastructure perspective (this study has also been reported elsewhere; see e.g. Nielsen & Aanestad, 2006). Based on an ensemble view, I attribute agency and the power to make changes to actors, even if unevenly distributed and only in certain instances and to a certain extent. By focusing on the action of key actors at key points in time, I describe certain "pockets of control" where certain actors had the opportunity enact their strategy and to make a difference.

## STUDYING INFORMATION INFRASTRUCTURE: AN EXAMPLE FROM THE MOBILE PHONE CONTENT INDUSTRY IN NORWAY

This case description and discussion focuses on the strategy and the agency of certain actors in the development of an information infrastructure by focusing on three of the concepts Ciborra et al. (2000) discuss: *control devolution* as the process where a system turns into an infrastructure and centralised control vanishes; *sinking* as the nature of information infrastructures being invisible and in the background while the focus is on what it enables; and *collage* as the way in which an infrastructure is put together based on a variety of resources from different actors. Where Ciborra et al. (2000) primarily used these concepts to describe the very nature of information infrastructure as emerging through control devolution and taking a sunken nature, I use the concepts to describe how certain actors can play on these attributes to shape a certain information infrastructure configuration and nurture the unfolding of a certain collage. Each of these concepts links both to technical components such as networks, servers, and social actors, such as developers, their strategies, and social relations. These concepts embrace complexity, but they also leave the room for human agency.

The research design applied in this case traces the technical components involved, the intensions of designers, and how it is reflected in how the technology is used (Faraj, et al., 2004). The focus was on recognising the depths of interdependence of technical networks and standards, on the one hand, and the real work of politics and knowledge production on the other (Bowker & Star, 1999, p. 34). The case study is based on 39 informal and in-depth interviews with managers and system developers in 23 different organisations related to the development and operation of a platform for mobile content services: the Content Provider Access (CPA) platform. The interviews were focused on how the different actors valued the CPA platform and how they had contributed to its design and operation. While the entire set of interviews offered the required background knowledge, this analysis in particular draws upon nine of these with employees who work for one of the network operators designing and operating the CPA platform and its predecessors.

An additional four interviews occurred with the content providers who directly played a part in influencing these developments.

The CPA platform with its origin in Norway is now a common infrastructure for value added services in the mobile phone network. The CPA platform was initially designed and implemented by the two Norwegian mobile telephone network operators, NorCom and MobiNor, as two independent software platforms. Together, these platforms provide an open and public infrastructure for the provisioning of content services for mobile phones, including yellow pages, ringtones, interactive TV, news, weather information, and others. The architecture of the platform enables mobile phone subscriber to acquire services through a number of simple steps. For example, if a subscriber wishes to obtain a phone number, he or she first locates the required information for ordering this service from a variety of sources such as the Internet, in a magazine, or on a banner. This is presented in a short-code form (a phone number with only four digits) and identifies the content provider and the name or code of the item. In order to request the content, the subscriber is required to sends a SMS message from his or her mobile phone to a designated number that contains the message 'phone X' to, for example, number 1905. The SMS is processed as a text message by the message centre (SMSC) of the subscribers' mobile phone network. This recognises the short-code number 1905, with the request and the subscriber's phone number forwarded to the CPA platform. The CPA platform will forward the content of the SMS over a TCP/IP connection to the content provider associated with the number 1905, without undertaking any further processing. Upon receipt of the request, 'phone' and 'X' are recognised as the service requested with the content provider subsequently including the content and returning it to the network operator via the TCP/IP connection. The content provider is required to rate the service according to predefined rating classes ranging from NOK 1 to NOK 60. On receipt of the content, the network operator generates a Call Data Record ('CDR'), which is provided to the billing system in order for it to process the request according to the rating class. If successful, the content is delivered to the customer by the SMSC. As the subscriber is an existing customer with the network operator, there is no requirement for any further registration or to provide any personal data such as address or credit card details. The charge for the service is added to the subscriber's regular phone bill. When the subscriber pays this, revenue from the service is shared between the network operator and the content provider.

With the arrival of NorCom in 1993, MobiNor experienced competition in the mobile phone network for the first time, with the Operator eventually capturing over 30 percent of the Norwegian mobile market. In 1997, both operators launched platforms that facilitated offering exclusive content and utility-based SMS services for their respective mobile subscribers. With the exception of this closed approach to access, service acquisition almost mirrored its successor, the CPA platform. A significant difference was that that content services were charged for on a per transaction basis as standard SMS messages, regardless of any included content (NOK 3). This resulted in a limited number of content providers, as the production of content necessitated incurring some production costs and experiencing high transaction volumes in order to recoup this. Neither of these eventuated. Both of these initiatives developed on an ad hoc basis, and were largely based on the initiatives by a small number of enthusiastic employees. These unstructured approaches did not require a sound business case, but as a consequence, they also suffered from a lack of funding. This was partly addressed through joint development activities with content providers. A further difference with the earlier CPA platform was the provision of exclusive services controlled by the network operators, as noted by the founder of the platform implemented by MobiNor: "We were talking directly with the content providers, it was us that fixed the codes, so, the point was that the content providers had a relationship with us, and we fixed the rest. We were, so to say, an editor which defined which services to provide."

The services offered by the content providers were utility driven and included national news, stock quotes, weather forecasts and phone directories, and assisted the operators to develop their brand accordingly. In implementing their platforms, the primary challenge faced by the operators was to amend their billing systems. These were relatively unsophisticated at the time, reflecting the infancy of the mobile telephony sector and the priority on undertaking billing for voice services. Changes required to accommodate content driven services received a lower priority even if they were minor. In response, platforms operating outside the billing system began to emerge, as reported by the founders of the platforms implemented by MobiNor and NorCom, respectively: "When we made a new product with a new price in minutes, they used months to do the same in the old billing system y So it was just not possible to be dependent on the billing system to have the flexibility we needed. So we managed to do that by defining us out on the outside. And you can say that at that time the most flexible billing system was running on a PC in the corner of one of the developers' office.... It was not very easy to change things. We had our first version of our billing system with flat files and everything, it was a real threshing machine and when the new billing system came, we were tied on hands and feet's both by technology and possibilities. So, everything was premature deluxe. I remember we hacked everything, because we had no resources for developing." While these ad hoc efforts made the platforms operational, they did not create a successful content service market. Similarly, the network operators failed to experience economic success with their exclusive service offerings. This was caused in part by the marketing costs incurred and because the services were primarily utility based. The content providers developing services expressed frustration that the full potential of their content was not being realised, due to the pricing policy by the operators and a lacklustre service portfolio.

In the fall of 1999, a former employee of MobiNor returned to the Company after an interim period working with NTV, one of the more dynamic content providers and TV broadcasters. Aware of the issues with the platform, he began to alter MobiNor's approach to its content. At the same time, a current NTV employee who had experience working with the previous platforms was engaging in discussions with both the network operators. These centred on charging for content at more than the basic SMS price, which would result in a premium rating. The network operators did not approve of this approach to charge 5-10 NOK for content, openly ridiculing such a suggestion: "So, we felt for a long time like banging our heads to the door there, and this was not because technical issues, because that was the next thing, then one had to build that in addition. The problem was that they simply did not believe that it was possible to create revenues of from it."

As a result of external pressure and internal discussion, MobiNor initiated the development of a new and different content platform. This approach was mirrored at NorCom, and recognised that content providers were more experienced in dealing with the development of new services, marketing them, and pricing them 'correctly.' It was also recognised that content providers could associate their brands with a wider number of services than network operators, and that content providers had wider market access. Based on these factors, the decision to move away from the previous platform gained momentum, but was still hampered at managerial level. This resulted in limited investment in the implementation of the new CPA platform. As a consequence, the initial technical implementation of the CPA platform by both operators was primarily based on specifications from the previous platform. Access was granted to external content providers, but due to a lack of support and internal resources, only a limited number of individuals were made available on a day-to-day basis. Despite this, the required software was developed by the operators, complemented by the cultivation of closer relationships with content providers, as noted by one of these 'pioneers' in MobiNor: "It was a bit of entrepreneurship spirited, the project, because we had no resources assigned like you are used to in a big company. So we had to make everything ourselves, and find the resources ourselves, in a form of a project. And this resulted in, that the atmosphere, both market wise and business wise and the technical development was entrepreneur directed." The first content provider to sign an agreement with MobiNor knew little about mobile telephony, but had experience with providing non-utility services using the Teletorg platform. This followed months of rumours that the Operator had developed a new platform, and repeated effort to gain access to this by content providers. Once these providers gained access, however, many discovered that the platform was at an early stage in its evolution, and that it was still perceived as 'controversial' by some staff within MobiNor. One of the first providers to gain access iterated the haphazard manner that MobiNor approached the platform and his company's involvement: "And then they had something running and suddenly I showed up, but they had not planned to launch at this early stage. And they had to sort out, what do we do now? And that is difficult in this kind of an organisation. So, finally by being persistent, I could plug into the platform, but it was made clear that the billing could fail to function at any time and without any rights for me to claim compensation.... This project meant a lot to this particular staff member, and he did a lot that he was not allowed to by his manager. He stressed this through internally in MobiNor, in a way that he possibly would not if he were a devoted and nervous guy. So he was scolded a lot in the beginning."

The developers involved with the development of the CPA platform were also amongst its strongest promoters, and did not necessarily possess a telecommunications ethos, as observed by their peers. As the platform continued to evolve and content was added, it became evident to the network operators that the cost of its implementation and operation was marginal, while traffic continued to increase rapidly. This was a key tenet in MobiNor's strategy, as cited by one of its the middle managers: "We have not used many resources on marketing the content providers' services, which was a part of our strategy, that we initiate variety and minimise risk. Because there is many that give it a try, those who succeed we make money on, those who fail are not our problem." When launched in the spring of 2000, the CPA platform enabled content providers to deliver a wide range of services to mobile phone subscribers, and to charge them premium rates for these. Responsibility as to which services to provide and how to market and price these was now undertaken by the content providers. In contrast to previous platforms which were network-provider specific, the CPA platform established a transparent market serving the interests of both the content providers and subscribers. From providing relatively simple services such as ringtones, logos and stock quote subscriptions, the CPA platform has evolved and now provides more advanced services including as an interactive return channel for voting and discussions in TV-shows. To coordinate these events, the TV-broadcasters, content providers, and network operators continue their close coordination to develop additional services. The CPA platform also provides billing for content services delivered by MMS, WAP, Web, and positioning services. In 2004, this sector in Norway had turnover of approximately 120 million Euro, and participation from around 50 firms employing 250 people.

Before content providers were permitted to connect their own equipment to telecommunication networks, the distinction between users and vendors was clear (Branscomb & Kahin, 1996), with the incentives and the responsibility for service innovation and operation assigned to the vendors, that is the network operators. The responsibility and primary control for these still resides with the network operators, even in the case when multiple actors are connecting their equipment to the network operators' infrastructure. Within this study, this control/autonomy balance was observed in two levels. The first involved employees engaged in the development of the CPA platform within both network operators, who were granted the autonomy to pursue their personal initiatives in an entrepreneurial milieu. This autonomy emerged in part as a response to a lack of managerial interest in services, which were defined as economically marginal and appeared to be 'odd practices' and serendipitous applications of systems (Ciborra, 1994). A second level in which this balance was observed was in the development of the platform by the network operators. Recognition occurred by a number of individuals that they were not equipped with the resources, risk aversion, or the ability to innovate and create the range of services, which the market required. Through the design of the CPA platform, the network operators transferred these elements to the content providers. These companies could now exercise the flexibility they had been seeking earlier. By relinquishing control, the network operators transferred responsibility for these types of services to external participants, including services that were not congruent with their cultivated 'public image.' The CPA platform was designed, implemented, and launched in an unstructured manner that was at the periphery of strategic and managerial control. At the same time, the lack of adequate funding and resources at the outset of the project influenced the evolution of the platform's architecture. While stronger managerial control and the existence of a sound business case would not necessarily have led to a rejection of the platform, these would have impeded the process and resulted in a platform that was different to the one developed and which most likely would have been proprietary. In this particular case, the lack of managerial control was not an organisational challenge for those involved. Rather, its absence was a prerequisite for the appropriate evolution of the CPA platform that resulted in an approach of control devolution, with the CPA platform 'pushed out' of the domain of the network operators. Relinquishing control to external content providers has provided the network operators with the ability to create distance from any unfavourable, explicit, or illegal content. In addition to content providers possessing a better understanding of the market, they are also less concerned with the nature of the content they provide. This can result in expedient and often unpredictable activities, as noted by one of the founders of CPA in MobiNor: "It was primarily colourful advertising, jokes and really just nonsense. But, it appeared as what the subscribers wanted. What these guys knew was that the willingness to pay for services was high so they began at an early stage to charge NOK 10 for services. So, during a very short period of time, the turnover went far beyond the previous platform, without any investments in service development or marketing by us."

The efforts of the network operators to provide the CPA platform as an open and public business sector information infrastructure depict how information infrastructure can be designed and developed with limited effort. It also depicts the sunken nature of information infrastructure. By a strategy of withdrawal or control devolution (Nielsen & Aanestad, 2006), the previous standalone platforms were redesigned as one element within shared infrastructure. The concept of sinking depicts how this platform currently appears as a transparent infrastructure through which content providers can deliver their services. At the same time, while network operators have withdrawn in contrast to their previous platforms, content providers have taken steps forward and into focus by innovating, taking initiatives, and providing desired services: a typical example of collage that is continuously extending.

## HOW TO STUDY INFORMATION INFRASTRUCTURES

Information infrastructures as discussed here is both a real world phenomenon (e.g. the Internet) and an analytical lens. What I argue is that the information systems of today are changing as they take the nature of complex networks rather than standalone systems. With this change comes a need for new approaches to study and understand them. While theories like network economics may explain how these networks develop, it offers little if any support in understanding how developers influence their trajectory. Rather than describing these large-scale networks as autonomous and developing according to laws of economy, we must engage with (interview, discuss with and observe) the daily activities of those who actively influence.

With a complexity perspective, the world becomes more complex. Acknowledging complexity does necessarily make life or research easier, and I think it is adequate to ask whether such open infrastructures are possible to analyze and understand at all. In the words of Dahlbom and Mathiassen (1993), by taking such perspectives, don't we put ourselves in a constant turmoil? And, who knows when we will be able to deliver something more than rich descriptions? And is the contribution from such approaches mere labelling, calling, and describing something as complex without any substantial insights?

Information infrastructures are not under anyone's complete control, but that should not be taken as they are not under any form of control. They are not designed fully by one designer, but they are designed by multiple designers in space and time. Control is restricted, but control is also exercised by certain actors at certain times and by whom it is exercised is an important as well as a highly contentious matter. It develops as a negotiated process and its development has a highly political nature.

Prior research has focused on developing perspectives to better understand information infrastructures as well as how to study them (e.g. Bowker & Star, 1999; Star, 1999). These attempts are close knit with an understanding of information infrastructures as something underlying, invisible and out of sight; they are seen as being buried into practices, standards, and technologies. This invisibility of information infrastructures is deceptive, because it hides crucial social networks and practices. The way to comprehend its invisible parts is to submerge into "boring" (Star, 1999) and sometimes invisible structures, to understand how these hidden factors shape the information infrastructure and its usage. To unpack and open the black box of information infrastructures to make the underlying social networks more visible has been termed infrastructural inversion (Bowker & Star, 1999). Star (1999) describes three "tricks of the trade" for studying information infrastructures. First, she advises us to focus on the information infrastructures' "others" by identifying master narratives and how they are used to make the "others" invisible and unnamed. Master narratives are used to merge actors and actions' diverse activities into monolithic agendas, and facts are made certain and stripped of the context of their development. Star's second piece of advice is to bring the invisible work to the surface. According to Star, unnoticed and formally unrecognised work is always going on, and not taking these activities into account will likely result in non-working infrastructures. Third, Star advices us to investigate infrastructures' paradoxes. Why, for example, do small changes in routines prevent people from using new information systems? Exemplifying this paradox, Star underscores that work involving computers also involves the articulation work of bringing together computer resources, organisational routines and day-to-day tasks. Only by investigating both of these tasks can we understand why some information systems work and others do not.

To understand how information infrastructures are built and developed, we have to understand the components that compose and influence them.

This network perspective challenges us in the sense that it becomes problematic to see what is part of the network and what is not. We may focus too narrowly on the achievements of one single actor or organisation, or alternatively we might take a more universal perspective. At the same time, it is very difficult a priori to understand the nature and reach of the networks that influence its design.

The complexity turn in the Information Infrastructure stream of research has put emphasis on the forces of evolution and technologies not being under control, while the single developer is left to follow the flow. To understand these phenomena it has been argued that we need to see them as interlinked with the rest of the world in complex webs of relationships (with the social and the technical). Identifying, describing, and acknowledging this complexity does not bring much in itself rather than belittling/minimizing the role of each individual. I have in this chapter argued that only by drilling down and identifying the activities of the single developers and their "pockets of control" we can better understand how these infrastructures develops and how they can be influenced.

## REFERENCES

Atkinson, C. J., & Checkland, P. (1988). Extending the metaphor system. *Human Relations*, *41*, 709–725. doi:10.1177/001872678804101001

Bowker, G. C., & Star, S. L. (1999). *Sorting things out: Classification and its consequences*. Cambridge, MA: The MIT Press.

Branscomb, L. M., & Kahin, B. (1996). Standards processes and objectives for the national information infrastructure. In Branscomb, L. M., & Keller, J. (Eds.), *Converging Infrastructures: Intelligent Transportation Systems and the National Information Infrastructure* (pp. 3–31). Cambridge, MA: MIT Press.

Checkland, P. (1981). *Systems thinking, systems practice*. Chichester, UK: John Wiley & Sons.

Checkland, P., & Scholes, J. (1990). *Soft systems methodology in action*. Chichester, UK: John Wiley & Sons.

Ciborra, C. U. (2000). *From control to drift: The dynamics of corporate information infrastructures*. Oxford, UK: Oxford University Press.

Dahlbom, B., & Mathiassen, L. (1993). *Computers in context: The philosophy and practice of systems design*. Cambridge, MA: Blackwell Publishers.

Faraj, S. (2004). Contested artifact: Technology sensemaking, actor networks, and the shaping of the web browser. *Information Technology & People*, *17*(2), 186–209. doi:10.1108/09593840410542501

Hanseth, O. (2000). The economics of standards. In Ciborra, C. (Ed.), *From Control to Drift: The Dynamics of Corporate Information Infrastructures* (pp. 56–70). Oxford, UK: Oxford University Press.

Hanseth, O. (2001). Gateways - Just as important as standards: How the internet won the "religious war" about standards in Scandinavia. *Knowledge, Technology & Policy*, *14*(3), 71–89. doi:10.1007/s12130-001-1017-2

Hanseth, O., & Aanestad, M. (2003). Design as bootstrapping: On the evolution of ICT networks in health care. *Methods of Information in Medicine*, *42*(4), 385–391.

Hanseth, O., & Lyyttinen, K. (2010). Theorizing about the design of information infrastructures: Design kernel theories and principles. *Journal of Information Technology*, *25*(1), 1–19. doi:10.1057/jit.2009.19

Hanseth, O., & Monteiro, E. (1997). *Understanding information infrastructures*. Unpublished Manuscript. Retrieved from http://www.ifi.uio.no/~oleha/Publications/bok.pdf.

Kling, R., & Scacchi, W. (1982). The web of computing: Computer technology as social organization. *Advances in Computing, 21*, 1–90. doi:10.1016/S0065-2458(08)60567-7

Lilienfeld, R. (1988). *The rise of systems theory: An ideological analysis*. Malabar, FL: Robert E. Krieger Publishing Company.

McGarty, T. P. (1992). Alternative networking architectures: Pricing, policy, and competition. In Kahin, B. (Ed.), *Building Information Infrastructure* (pp. 218–270). New York, NY: McGraw-Hill Primis.

Monteiro, E. (2000). Actor-network theory and information infrastructure. In Ciborra, C. (Ed.), *From Control to Drift: the Dynamics of Corporate Information Infrastructures* (pp. 71–83). Oxford, UK: Oxford University Press.

Nielsen, P., & Aanestad, M. (2006). Infrastructuralisation as design strategy: A case study of a content service platform for mobile phones in Norway. *Journal of Information Technology, 21*(3), 185–194. doi:10.1057/palgrave.jit.2000064

Orlikowski, W. J. (2000). Using technology and constituting structures: A practical lens for studying technology in organizations. *Organization Science, 11*, 404–428. doi:10.1287/orsc.11.4.404.14600

Orlikowski, W. J., & Iacono, C. S. (2001). Research commentary: Desperately seeking the "IT" in IT research - A call to theorizing the IT artifact. *Information Systems Research, 12*(2), 121–134. doi:10.1287/isre.12.2.121.9700

Saltzer, J. H. (1984). End-to-end arguments in systems design. *ACM Transactions on Computer Systems, 2*(4), 277–288. doi:10.1145/357401.357402

Star, S., & Ruhleder, K. (1996). Steps toward an ecology of infrastructure: Design and access for large information spaces. *Information Systems Research, 7*, 111–134. doi:10.1287/isre.7.1.111

Star, S. L. (1999). The ethnography of infrastructure. *The American Behavioral Scientist, 43*(3), 377–391. doi:10.1177/00027649921955326

Tuomi, I. (2002). *Networks of innovation: Change and meaning in the age of the internet*. Oxford, UK: Oxford University Press.

Walsham, G. (1997). Actor-network theory and IS research: Current status and future prospects. In *Proceedings of the IFIP Working Group 8.2 Conference: Information Systems and Qualitative Research*. Philadelphia, PA: Kluwer Academic Publishers.

Wilson, B. (1990). *Systems: Concepts, methodologies, and applications*. Chichester, UK: John Wiley & Sons.

# Chapter 10
# Prioritizing Packaged Software Implementation Projects:
## The Significance of Gaps

**Nicholas J. Rowland**
*Pennsylvania State University, USA*

## ABSTRACT

*This chapter examines the dynamics of prioritizing implementation projects. Building on the notion of "fit-gap" work, this chapter emphasizes the significance of "de-prioritization" as a practical technique for managing Enterprise Resource Planning (ERP) implementation projects. "Fit-gap" is a term that resonates with current academic and professional discussions concerning the use of customization and work-arounds necessary to coax suboptimal implementations into functioning properly as the systems age. These are not idle matters given the near irreversibility of ERP projects once initiated and the reported high probability of failure following implementation. Drawn from in-depth interviews and internal documents collected from a multiyear organizational case study of ERP in an institution of higher education, this chapter reports on various uses, interpretations, and consequences of prioritization techniques used to manage implementation projects. In practice, the idea that complex software implementations can be theoretically reduced to mere gaps in fit serves to obscure the political conflict and ambiguous economic accounting that underlie committee work devoted to identifying gaps, deliberating on possible fits, and then prioritizing which gaps are fit immediately and others scheduled for fit later on. In conclusion, while fit-gap committee work is openly intended to result in fewer customizations overall, de-prioritization, as a management technique, appears to "remove without removing" agenda items from the implementation schedule. The upshot for managers: placing such decisions in purgatory delays indefinitely investments of time and finances into customizing new software to fit old policies, and all the work-arounds necessary to shore-up any lingering idiosyncrasies.*

DOI: 10.4018/978-1-4666-0303-5.ch010

## INTRODUCTION

The promise of packaged software solutions such as Enterprise Resource Planning (ERP) very much hinges on determining how to make them work on-the-ground (Wagner & Newell, 2006). The product's vendors guarantee flexible software packages (Pollock & Williams, 2008), often citing the "80/20" rule, which implies that ERP systems are roughly 80% delivered with the remaining 20% left to the discretion of local programmers (Pollock, 2005, p. 7). However, as the literature documents, malleability brings with it the burden of "fitting" work (Gasser, 1986), "work-arounds" (Pollock, 2005), and "technological adjustments" (Pfaffenberger, 1992). Therefore, it appears that implementing packaged software systems implies managing the interpersonal, financial, and temporal aspects of initial installation, ongoing maintenance, and the nearly inevitable idiosyncrasies inherited from previous system development efforts (Light, 2005; Thomas, 1994).

Building on Gasser's (1986) notion of "fitting" work and Rowland's (2008) emic term "fit-gap," this chapter examines the process and consequence of prioritizing implementation projects with special emphasis on "de-prioritization" as a practical technique for managing installation. As it happens, organizations follow elaborate formal and informal procedures in their efforts to shape the progressive implementation of ERP systems (Pollock, 2005; Wagner & Newell, 2006). The reason is simple: packaged software do not fit organizations that wish to install them (Pollock & Cornford, 2004). Therefore, a cocktail of formalized fit-gap procedures and on-the-fly work-arounds is the hallmark of packaged software implementation projects.

To manage the numerous misalignments between the organization's extant structure and packaged software off-the-shelf, which are bound to emerge during the implementation process, project managers assign key functional and technical employees to sit side-by-side on "fit-gap committees." Borrowing terminology from those charged with implementing and maintain such systems in higher education, Rowland and colleagues (Rowland, 2008; Gieryn, 2008) describe fit-gap work as a formalized process devoted to (1) identifying gaps, (2) deliberating on possible fits, and then (3) prioritizing which gaps are fit straightaway while others get de-prioritized and remain agape. Conceptually, the fit-gap idea is a valuable tool during implementation. However, the idea that complex software implementations can be theoretically reduced to mere gaps in fit serves to obscure the political conflict and ambiguous economic accounting, which are also significant elements of fit-gap work—especially in relation to de-prioritizing some gaps for fit (presumably) later on (on politics and power during implementation see Markus [1983] and Robey and Markus [1984]).

The work-around becomes a resource to shore-up lingering misfits amid the minutiae of the new machine (Gasser, 1986; Pollock, 2005). But not all gaps in fit are possible to identify despite the best efforts of fit-gap committees. Presumably, this is because committees are forced by necessity to adopt an aggregate view of the organization. This perspective brings with it a modicum of abstraction to the otherwise infinite nuance of everyday life in the organization, which is not unlike the view described by Scott (1999) in *Seeing like a State*. Returning to work-arounds, Pollock (2005) shows readers how local technicians overcome system misalignment by techno-scientific patchwork designed to mend the interface between the new system and extant practices and already established support software. Also referred to as kludges, work-arounds make the systems "work" for the moment, at least until the next round of vendor updates arrives. Pollock (2005) views work-arounds as emblematic of packaged software like ERP; ERP is a "*gray box*" precisely because a small portion of the end product is up to the discretion of users even if the bulk is delivered from the vendor (Fujimura, 1992, p. 169).

Developed from in-depth interviews and internal documentation collected from a multiyear organizational case study of ERP in an institution of higher education, the main thesis, which results from inquiry into these data, is that prioritization is a significant technique in the management of ERP implementation projects. I provide evidence of committees crafting locally justifiable strategies prioritize some gaps for fit immediately while delaying others, sometimes indefinitely. After introducing the significance of ERP for institutions of higher education, this paper reviews the literature on correcting system misalignments, and then delves into empirical support for the significance of prioritizing and de-prioritizing ERP implementation projects.

## ENTERPRISE RESOURCE PLANNING MIGRATES TO HIGHER EDUCATION

Ross (1998) called the early 21st century the "Enterprise Resource Planning Revolution." Packaged software solutions like ERP were so widespread that even software giants like IBM and Microsoft used them to manage operations (O'Leary, 2000) as did a majority of the world's multinational firms (Kumar & van Hillegersberg, 2000; Langenwalter, 1999; Norris, et al., 2000).

While American colleges and universities historically championed custom-built administrative information systems, they too have pursued ERP; a mass migration intended to make them globally competitive information institutions (Wagner & Newell, 2006, p. 307; Pollock, 2000). ERP packages are specific enough for higher education as a sector, but generic enough that the systems need to be localized to some extent during implementation (Pollock & Cornford, 2004, p. 111; Pollock, et al., 2007). Software modules mainly include human resources, student information services, and financial operations, which handle tasks essential for running a university (like payroll and promotions) but have little to do with a university's public image (like professors and promotional websites).

ERP's advantages are related. First, data storage is centralized; rather than managing numerous databases, ERP stores data in a single grid (Davenport, 1998). Second, ERP provide "real-time" data; consequently, decisions can be made with the best possible information, which improves upon previous systems wherein data were commonly outdated (Pollock, 1999). Third, data are authoritative; centralized data storage eliminates the problem of "competing" data stored in multiple databases (Swartz & Orgill, 2001, p. 21). Lastly, students, faculty, and staff are turned into "self-service" users who monitor their own relationship to the organization rather than having support staff administer those tasks (Pollock, 2005).

Ross (1998) called the early 21st century the "Enterprise Resource Planning Revolution." Packaged software solutions like ERP were so widespread that even software giants like IBM and Microsoft used them to manage operations (O'Leary, 2000) as did a majority of the world's multinational firms (Kumar & van Hillegersberg, 2000; Langenwalter, 1999; Norris, et al., 2000).

While American colleges and universities historically championed custom-built administrative information systems, they too have pursued ERP; a mass migration intended to make them globally competitive information institutions (Wagner & Newell, 2006, p. 307; Pollock, 2000). ERP packages are specific enough for higher education as a sector, but generic enough that the systems need to be localized to some extent during implementation (Pollock & Cornford, 2004, p. 111; Pollock, et al., 2007). Software modules mainly include human resources, student information services, and financial operations, which handle tasks essential for running a university (like payroll and promotions) but have little to do with a university's public image (like professors and promotional websites).

ERP's advantages are related. First, data storage is centralized; rather than managing numerous databases, ERP stores data in a single grid (Davenport, 1998). Second, ERP provide "real-time" data; consequently, decisions can be made with the best possible information, which improves upon previous systems wherein data were commonly outdated (Pollock, 1999). Third, data are authoritative; centralized data storage eliminates the problem of "competing" data stored in multiple databases (Swartz & Orgill, 2001, p. 21). Lastly, students, faculty, and staff are turned into "self-service" users who monitor their own relationship to the organization rather than having support staff administer those tasks (Pollock, 2005).

The rise of ERP is often cast as a tiresome, emotional battle: implementing organizations report project dissatisfaction and occasional user resistance; vendors appear unresponsive to user demands, their designs ever-divorced from local needs; ERP's lofty potential seems rarely realized.

However, we follow Pollock and Williams (2008) who rethink critiques of packaged software in higher education. They take issue with those that decry, for example, unresponsive vendors or predictable complaints over their ill-fitting software, and instead of leveling value-laden judgments they show that "suppliers of... packaged organisational technologies developed sophisticated strategies that *linked* them and the community of current and intended adopters" (Pollock & Williams, 2008, p. 283). That is, the authors emphasize how vendors and universities dealt with this transition in higher education; they emphasized the practical strategies and solutions to the shared problems these now-linked organizations faced, which is the perspective adopted in this chapter. Rather than blaming vendor support staff or university administrator, this chapter documents a practical solution to ongoing implementation projects.

## CURRENT EMPHASIS ON FIT AND THE SIGNIFICANCE OF GAPS

Research on localizing and modifying ERP mainly emphasizes notions of fit and alignment (Hong & Kim, 2002; Kanellis, Lycett, & Paul, 1999). This literature could be extended in fruitful ways if additional and equal analytical emphasis were placed on gaps and, in particular, competing claims concerning what constitutes a legitimate gap in the first place as well as which gaps should remain agape indefinitely. After illustrating the current emphasis on fit, this section defends the significance of gaps, and then develops the importance of prioritizing them for fit.

As a point of departure, Gasser (1986, p. 214) introduces readers to "fitting work," or the act of "changing computing or changing the structure of work to accommodate for computing misfit." Organizations can fit information systems to their work practices or vice versa. Planned alterations such as system customization constitute fitting work and so do ad-hoc readjustments to computing systems such as work-arounds. Likewise, adapting work practices, provided they are intended to accommodate or obviate socio-technical misalignment, is also "fitting work." Gasser's work, based on extensive fieldwork in ten organizations, has a kinship with what Star, Bowker, and Neumann (1997) later refer to as "convergence": "the double process by which information artifacts and social worlds are fitted to each other and come together" (see also Leonard-Barton [1988] on "mutual adaptation" and Kasvi et al. [2000] on "joint optimization").

Building on Gasser (1986), Pollock's (2005) research on work-arounds also emphasizes correcting misfits. Drawing on ethnographic data from a university's experience implementing ERP, Pollock (2005, p. 5) examines "the tendency for work-arounds ... [to be] already present in the networks that those implementing the computer system ... inhabit." The idea that local programmers will utilize work-arounds appears to be built-into current ERP systems, and the implication is that vendors appear to assume that adopting organizations are able to deploy networks already in place to localize the vendor's otherwise standardized product. Hence, the process by which local programmers fit the standardized system to local needs appears to be a central concern in the packaged software literature.

Light (2005) extends this literature in new directions. He finds that managers and consultants often frame misalignments to be either economic, functional, or technical in nature; however, Light (2005) suggests that this portrayal distracts from the power and politics that plague group-level decision-making. For example, "the implementation team ... [might] use customisation to facilitate user acceptance or as a pluralistic control mechanism" (Light, 2005, p. 616). Similar to Gasser's (1986, p. 215) point that fitting work is "a function of complex negotiations between those who control resources and those who face ... misfits," Light (2005, p. 606) argues that customization decisions reflect "the multiplicity of social groups involved in" fitting work. Light (2005, p. 606) goes a little further, however, stating:

*whilst a common reason for cutomising ERP packages might be concerned with functionality misfits ... it is necessary to examine more deeply why these might occur, and indeed, if other reasons exist.*

With the current emphasis on fitting, which Light underscores, this paper shifts analytical attention to gaps and sensitizes readers to the process by which they are prioritized for fit. Logically, gaps must be identified before they can be bridged.

What is significance about gaps? Rowland and Gieryn's (2008) research provides a hint. Snags during implementation arise from the communicative task of explicating fully the wide range of extant practices that constitute routine work rather than an organization's willingness (or not) to absorb transaction costs by merely customizing the software (on "transaction costs" see Williamson [1981] and Nooteboom [1992]).

*Each organization has its own ... culture, and its employees share tacit knowledge about their embodied practice ... [which are] impossible to retrieve and articulate precisely and unambiguously ... [thus] as the university debates about which costly customizations to buy, they are able to consider only a truncated subset of all the possible differences between their legacy system and [the packaged software solution] (Rowland & Gieryn, 2008, p. 384).*

If not all gaps are immediately explicable, then gaps that do get identified are influenced by how officials will decide to fit them and when.

An emic term from the work-worlds of those charged with maintaining ERP systems, the "fit-gap" idea refers to the fact that the implementation of innovations is a process of intermittent modification, which implicates the significance of "gaps" and "fits." When there are "gaps" between administrative practices and the design of the innovation that it plans to implement, then those gaps must be "fit" by either modifying the organization, the innovation, or both (or, conceivably, neither, in the event that it is decided that the gap remain open)—this is "fit-gap" work. As part of fit-gap work, organizations form committees to identify gaps, deliberate on possible fits, and then prioritize which gaps get fit earliest and which get fit later. Therefore, in contrast to older research on the implementation of innovations (e.g., Rogers, 1983), fit-gap work is emergent, recursive, and ongoing—it predates initial installation efforts and endures even after every identifiable practice or module is fully installed (Markus, 1983; Robey & Markus, 1984).

To be sure, work-arounds are not always the liberating or autonomy-augmenting practices they might appear to be. In a study of ERP implementation in a university setting, Kitto and Higgins (2010, p. 29) show how "the process of working around a technology can have ambiguous ... effects." This paper examines a similar charge with fit-gap work as the sensitizing concept used to understand the multiple and occasionally contradictory rationales and consequences of prioritizing gaps for fit.

## DATA COLLECTION AND ANALYSIS

Analysis centers on "Midwestern University" (MU), a large research university located in the American Midwest that has traded legacy for ERP. With 35,000 students, an annual operations budget of $1 billion, and, like most large universities, MU offers a huge array of undergraduate, graduate, and professional degree programs.

Qualitative data include in-depth interviews with senior administrators and support staff from a Midwestern research university in Fall 2005 to Summer 2006 (see Table 1). Interviews ranged in length from approximately 30 minutes to more than two hours. In addition to preparing methodological and theoretical notes, the development of on-hand memos detailing information and terminology specific to the university's administrative structure, prior legacy system, and current ERP project proved of paramount importance because senior administrators, in particular, would almost inevitably begin interviews with a string of quiz questions full of insider language (Briggs, 1986).

These interviews and internal documents have their limitations. The university under study is a relatively resource rich, large, and prestigious school. Small- and medium-sized colleges might have fundamentally different sets of problems, respond to even similar problems differently, and given the resource draining implementation process, make drastically different customization choices. Future research could compare large and small, resource rich and poor organizations to determine variation and similarity in decision-making processes and consequences for the organization and the technology. Moreover, while some interviews document work the respondents were currently engaging in, others suffer from retrospection, which is seldom exactly accurate. Secondary data are selectively drawn from to support the words of respondents whenever possible.

*Table 1. Respondent official position in the division of labor (N = 23)*

| *Senior Administrators* | *13* | *Support Staff* | *10* |
|---|---|---|---|
| Information Technology | 5 | University Technician | 6 |
| Human Resources | 1 | Administrative Assistant | 2 |
| Student Information | 3 | Financial Assistant | 1 |
| Financial Accounting | 2 | Other | 1 |
| University Officer | 1 | | |
| Other | 1 | | |

Data were analyzed according to the Straussian variant of grounded theory (Strauss, 1987, see also Weiss, 1994, p. 152). This approach is appropriate for two reasons. First, the aim of this analysis is to report on what was learned about prioritization during the fit-gap process from all respondents combined rather than emphasize particular respondents. Second, the analysis takes what was learned from concrete instances with the intention of considering the precursors and consequences of prioritizing implementation projects for universities more generally.

Rather than overtly guided by theoretical strictures, themes and the significance of events was primarily based on careful attention to the terms and concepts used by respondents and found in internal documentation. One of the most compelling reasons to conduct in-depth interviews is that *emic* terminology can be used as part of analysis, which aids the analyst to avoid imposing theoretical concepts, or *etic* terminology, from previous research. Additionally, the analyst must avoid importing outside standards, for example, that delineate "true" from "false" or "accurate" from "inaccurate" and instead accept the insider view we have been invited to view. This analysis begins by describing the exigencies of gap identification and how gaps are prioritized for fit.

### The Myth of Vanilla Implementations

Vendors promise potential clients that they can install completely standardized ERP off-the-shelf (i.e., "vanilla" implementation). However, this is misleading. In many cases, the full potential of ERP cannot be realized until it has been localized. Even so, implementation without modification comes widely recommended in scholarly publications read by respondents. For example, Swartz and Orgill (2001, p. 26) position—"the first and most important factors to control are the number of custom modifications"—or Noyes (2003, p. 54)—"the less customization, the more likely the project will be completed on time and on budget"—or Frantz, Southerland, and Johnson (2002, p. 41)—"ERP software is built according to generic best practices for functional processes, so the institution must reengineer its business processes to fit the software." According to a survey of universities, "87 percent of the study's respondents agreed that their institution's strategy was to implement software with as few customizations as possible" (Kvavik, et al., 2002, p. 54). Avoiding ERP customization means gaps in fit are solved by shifting administrative practices and other structural features of the organization. Such concerns resonated with respondents (see Table 2).

However, no university implements vanilla. Vendor promises to the contrary seem preposterous to a senior administrator in information technology:

*of course, [the vendor promises that] we provide so much in the way of functionality that you'll be able to install this vanilla, and of course that never panned out, I don't even think on a small campus, maybe some community college or something was able to do that. But, in my opinion, the only way that you could manage working with the system as delivered is if you created your university, and bought the ERP at the same time you created your university, and then you would tailor all your business practices exactly to the delivered functionality, otherwise, there's historical precedence for the way you do business.*

Hence, implementing ERP is less about installing software off-the-shelf and more about implementing systems localized through customization.

## Finding and Prioritizing Gaps

Functional and technical staff steer implementation planning committees. This is because work operations are executed differently using legacy as compared to ERP. Legacy were built according to *function* (i.e., to perform single tasks). Under legacy, employees were trained to complete one task and then, like an assembly-line, another individual would complete the subsequent task and so on. Hence, legacy design mirrored division of labor (Light, 2003). In contrast, ERP are built according to *process* (i.e., to perform a chain of tasks from start to finish). Hence, individual support staff must develop a manager's gaze over the completion of their daily tasks. By gathering employees responsible for individual functions that, in a specific order, comprised an entire process, project managers believe that inefficiencies can be identified and eliminated by implementation planning committees.

*Table 2. Report on university approaches to customization (N = 23)*

| *What is the university's stance on customization?* |
|---|
| "Go vanilla or go home" – University Technician |
| "K-I-S-S, something like Keep it Standard Stupid!" – Senior Administrator, Information Technology |
| "Avoid customization at all costs" – Senior Administrator, Financial Accounting |
| "Customizations? You avoid them, that's what you do" – Senior Administrator, Human Resources |
| "The main line sounds like this: first, you don't say customization, and second, you don't say customization, but if it is unavoidable—if the gaps are too large or a group of functional types are too entrenched, then the rare and majestic customization may be, once again, spotted in the wild" – University Officer |

*Figure 1. Implementation overview from internal documentation*

**Project Approach Update**

We on the team continue to follow the project approach laid out in April of this year; in this approach, the project was divided into several phases

1. High-level requirements definition/ planning
2. Detailed requirements definition and analysis
3. Design/ build
4. Test
5. Go-live

With the high-level requirements gathered over the summer, we are in the process of creating a master project plan in our project management tool. This plan will guide the overall project and will comprise a detailed, task-based project schedule. Each resource on the project will receive an electronic notification of his/ her tasks for the week and their expected completion date; in this way, we will be able quickly to judge the status of the entire project, as well as each task.

* Excerpt from Project Update Memo from Enterprise Resources Planning System Implementation Executive Committee.

To initiate this course of action, one university technician states:

*R: We started by working with the functional users both from central administration, core offices, and from end-user departments in order to determined current process.*

*I: Current process?*

*R: Right: what we were already doing, from a software and business perspective, before the new system. So, it was not so much desired process at the beginning [of the implementation process], but current process. The "process" part is key. We compare current process ... to the new system in hopes of finding the gaps, and we've been doing that for about seven months straight now (see also Figure 1).*

In the transition from legacy to ERP architecture, universities inventoried functions supported by their legacy systems in order to recover how those functions—spread throughout campus—combined to form entire processes. When assembled together, individuals knowledgeable about different functions produced a committee with the combined knowledge of an entire process. Where functions and process ceased to overlap, a gap was identified and once identified a potential fit would be proposed. After this, fits were prioritized; a process emphasized below.

## Priority Setting

Respondents report that politicking is part of fit-gap committee work especially related to getting your gaps identified and then prioritized. Consequently, jockeying for gap priority commences immediately. According to Joy, a senior administrator in human resources, employees voice their gaps

*They come to me and they've got their priorities and I've got mine. ... and I say to them, "Within the budget we have, three million dollars, we can't do all this!" Secondly, "we don't want to do it all now" [pause]. "We do [!], we do want*

*to do it all, because these are all good ideas, but we don't want the risk we're gonna be exposed to by trying to do a hundred-and-sixty-five extra things when we don't have enough firefighters to put out the fires, when things go wrong."*

The entire implementation project might be capsized if it were saturated with too many competing priorities. In the STS literature, design projects like ERP implementations fail when overloaded with too many priorities and too few resources. As design projects attract increasing numbers of interested parties designs become saturated with their sometimes competing priorities. Law and Callon (1992) describe how despite engineers' best efforts to accommodate rapidly accumulating sets of priorities during an aircraft design project, the resulting aircraft was so laden with design priorities that it became incapable of take-off. Universities aim to avoid a similar fate because too many customizations might capsize the entire system.

Prioritizing customizations is a survival strategy during implementation, according to Joy:

*So, we got to do about forty gaps we can fit right now because they're fundamental to when you put the system in, and we'll put the [remaining] others on our list, and besides that [laughs] I need another three million dollars if I'm gonna do these hundred and sixty-five things. I want to do the minimum possible and make it work and be successful, rather than trying to do the maximum, spend another three years building it and have all kinds of problems with it. The [oversight] committee said, "Absolutely, that makes all the sense in the world." But, my engineers were sadly disappointed. They were angry as hell at me because they did all this work [to identify gaps] ... and I told them, we weren't gonna put their recommendations into practice. I said "we're gonna do all these things, we're gonna prioritize, and put them on the list; we'll do'em eventually."*

Unsurprising in a bureaucratic setting, prioritization takes form through internal rubrics that rank gaps according to significance—high, medium, or low—and by type—policy, data conversion, process, systems integration, or security. This procedure made comparable otherwise incommensurate implementation items. However, this seemingly impartial methodology and universal criteria obscured a host of political and economic scheming hidden behind the complete cross-listing of all the gaps and fits; this distinctly sociotechnical affair is explored in the next section.

## De-Prioritizing "4 Year Promise" (4YP)

4YP was a tuition assurance program that with a few stipulations for undergraduates promised graduation in four years. As we shall see, MU took the legitimate opportunity, afforded by the implementation of new software, to discontinue an old student tuition assurance program—to some controversy.

Under 4YP, the student was provided a checklist of benchmarks and so long as the list was followed exactly, the university guaranteed that if it took the student longer than four years, then the university would pay for tuition until the student completed their degree. Upon its inception in the mid-1990s, MU reports the program was well received – so much so that universities across the American Midwest began to emulate the program. According to the *Herald Times*, Indiana University has such a program called GradPact (Wright, 1996), and according to *The Milwaukee Journal*, "University of Wisconsin System officials hope to emulate Indiana University plan that guarantees graduation in four years" (Brook, 1995). Northern Iowa University created UNIGradPact (1995), similar to MU's 4YP. And other universities copied the policy making 4YP appear like a voice of reason at a time when tuition was skyrocketing, students were taking longer to graduate, and parents were concerned about how to get their children in and then out of universities in a timely manner.

4YP's death is partially explained by the circumstances surrounding its inception. The program was initiated by MU's president, and reportedly without faculty senate input. Some informants viewed the program as "largely puff," based on "good political splash," and that "deep-down, faculty didn't care about the program because faculty didn't get any input at the outset." In the end, however, the shift to packaged software was blamed for its demise rather than faculty discontent.

And yet, 4YP was cost-effective before MU adopted ERP. MU's president branded 4YP a "darn good" program and "part of a university's promise to any student." According to one respondent, "The University did not lose a penny to the program and, really, it probably made money attracting students, well, their parents, anyways." Internal documentation confirms: not a single student that entered into a 4YP needed the additional "free" tuition after four years. "It was not always easy to run," stated a support staff member charged with administering elements of the policy, "but if I sent my son here, I'd want him to have this opportunity." Hence, 4YP, which looked so good on paper, had detractors and supporters.

Nonetheless, GradPact was retired; however, faculty senate in consultation with the president demanded the right to re-address the policy at a future time if they so pleased. This implies 4YP was not completely dead. Rather, it rested at the bottom of a long list of implementation priorities in a kind of purgatory for academic policies; it was de-prioritized, but might one day be resurrected.

Sometimes low priority customizations are just that: of limited consequence and importance. Policies and programs are also labeled "low priority" for other reasons. When gaps are identified between pre-existing policies and the new system, then members of fit-gap committees are obligated to determine how to fit that gap so that the policy could endure and at what cost. This process is the same regardless of whether or not the policy should be replaced, dismissed, or revamped—fits and their costs had to be estimated.

Once created, policies often endure long after those who created it could remember why:

*I: Do you think ... [4YP] would have endured had the university not adopted ... [ERP]?*

*R: You mean, and stay with our old policies—I can answer that: absolutely.*

However, 4YP was dismissed, and fit-gap committees documented its demise, suggesting that its infrastructure was perched on a "house of cards," as one technician concisely stated. The policy was impossible to administer without each of three on-line registration systems in operation (see Table 3).

As it became increasingly clear that not all of the functions of CWS and CRS would be retained in the transition from legacy to ERP, 4YP crumbled for lack of support—and the difference between *course* and *class* was blamed.

*Table 3. Description of 4YP support systems (N = 23)*

| |
|---|
| *Course Waitlist System* (CWS), which allowed students to waitlist for any section of a course and, as soon as a seat became available, automatically enrolled the student in the class. |
| *Course Swap System* (CSS), which worked with CWS and allowed students already enrolled in one section but who preferred another section or an altogether different course to be automatically enrolled in the new class as soon as a seat opens up and un-enrolled in their previous course. |
| *Course Reserve System* (CRS), which, if students could not get into any sections of a course they wanted or needed to graduate, got them a "saved seat" in the same course the following semester. |

Legacy operated at the course level. A single course often consisted of a set of sections (i.e., classes in ERP) that shared the same basic course number. In contrast, ERP operates at the class level. Classes could share the same course number but were treated as discrete objects. Because of the significant use of course level requirements in 4YP, especially as used in CWS and CRS, the policy could not be faithfully delivered using ERP off-the-shelf. Implementation documentation supports this conclusion:

**Description of Gap**: Students must request an "Any" (unrestricted) Waitlist if they encounter a closed course required for their program. Since [ERP], as delivered, is limited to an "only" restriction, the policy cannot be enforced.

## Impact on Business (High, Medium, Low): High

The ERP's replacement Waitlist program provided the functionality to iteratively search for dropped seats in "only" a single class rather than "any" section of a single course. "The seemingly innocuous difference between an information administration system operating at the level of course rather than class made it," according to one official, "impossible to maintain 4YP—impossible to maintain without significant modification, that is."

During legacy, "any" function in CWS placed a student in any open seat in any section under the same course number. As a policy, 4YP demanded that students comply with a pre-determined set of benchmarked courses to be taken during specific semesters. Getting students into any open seat in any of the sections under the same course number was crucial because it was definitive evidence that the student had done everything possible to adhere to the benchmarks set forth in their original graduation pact with the university. If students met these stipulations, then they were still eligible for tuition remission despite missing benchmarks. This never happened at MU.

If 4YP was going to endure the transition to ERP, then the policy's requirement for the use of unrestricted waitlisting had to be relaxed or discarded. If this policy change could not be accomplished within the bureaucracy, then 4YP would be terminated for failure of the new system to support it. Fit-gap committee records document this progression:

**Academic Policy Implications**: Waitlist adds are governed by [4YP] rules: Only "A" (no restrictions) requests with [reserved seats] may be added unless the request is for a section change (same dept/course on schedule with no dup [i.e., duplicate] course tag) or second enrollment in a variable topics course.

**Issue for Executive Committee**: If [4YP] continues to be a viable Academic Policy, then the policy governing waitlist would need to be changed to relax the mandates on "unrestricted" waitlist requests for [4YP].

## Likelihood that Team will recommend this to the Policy Change Committee (High, Medium, Low): Medium

**Trade Offs**: If this policy change is not pursued, and assuming that [on-line registration system] remains a viable service at [MU], then it would either require

1. moderate modifications to the [ERP] Waitlist functionality as well as a simple modification to carry a [4YP] attribute, to recognize and respond appropriately within registration and waitlist processing and to display/accept [4YP] compliance responses from the student or
2. a minor modification and major change to the way students request waitlist sections.

**Executive Committee decision to Proceed or Not Proceed with Recommended Policy Change Approach (Y/N)**: [MU's president] has agreed to drop [4YP] as a required policy. [Faculty Senate] has approved motion to End [4YP].

**Status**: Policy Change Pursued. Issue Closed.

**Team Action**: No Action Required.

Fit-gap committees estimated that the customization necessary to create the waitlist "any" function would cost about $50,000 initially and ongoing support would cost about $15,000 each year. However, that was only the first of four similarly price-tagged changes with similar maintenance costs, all of which were necessary to make 4YP live again as a policy at MU.

How certain were fit-gap committee members of these numbers? Sometimes estimates were "firm," according to an information technology technician, "they are estimates, of course, and estimates can be off, sure, but we've been doing mods [i.e., modifications] for a while now and have a pretty good idea how much they will cost." Looking into the future with any accuracy seemed "foolhardy," especially to more longstanding employees such as MU's registrar:

Estimates five years into the future! It makes me want to laugh. I mean, where do those numbers come from? Now I mean to be rhetorical there because the numbers are just numbers; points in the unforeseeable future that not a single one of us has any illusions will actually come true. Hell, if somebody would have told me 5 years ago that IT at [MU] was going to run us millions upon millions of dollars a year only 3 years later I would have laughed, laughed them out of my office. ... But behold! It has come true and now we're scrambling to budget this whole mess.

Another respondent was even more frank:

*... back then we were pulling those estimates out of our asses, everybody was. I mean, no one really foresaw, well, I dunno, maybe some of those consultants that had done umpteen implementations knew, but we surely didn't—I surely didn't.*

There was also an unavoidable issue of scale. Flagship universities like MU cannot operate under an economy of scale if student registration issues like administering 4YP had to be handled manually by the registrar's staff. However, that was identified as the only way to continue the policy. The final fit-gap analysis reads:

**Issue Title**: [4YP] as an Academic Policy

**Description of Issue**: [ERP] does not deliver a mechanism for tracking students enrolled in [4YP] and does not immediately enable registration/degree audit procedures.

**Impact on Business (High, Medium, Low)**: High

**Academic Policy Implications**: Issue for Executive Committee: The [4YP] policy (the 4 year graduation guarantee for [MU's main] Campus) could be retired.

**Likelihood that Team will recommend this Policy Change (High, Medium, Low)**: Low

**Trade Offs**: If this policy is not retired, then modifications medium to significant modification to [ERP] will be required to track [4YP] enrollment, enable appropriate Waitlist choices, create a [seat reservation] process, modify registration, keep an historical archive of registration activity and modify the academic advising to create Benchmarks.

**Executive Committee Decision to Proceed or Not Proceed with Recommended Policy Change Approach (Y/N)**: Executive Committee willing to change policy. [MU's president] has agreed to drop [4YP] as a required policy. [Faculty Senate] has approved motion to End [4YP].

**Status**: *Issue Closed.*

Modifications to recreate those elements of their pre-existing legacy system that supported 4YP in ERP would cost MU nearly $500,000 over 5 years. The costs, no matter how ambiguously generated and potentially inaccurate, were too great to justify an otherwise defendable policy, especially if the alternative meant going back to the sorts of paperwork ERP was designed to replaced.

Some respondents were leery of this final description. They believed that academic policies should not be designed for efficiency so much as they should be designed to foster student growth. At MU, however, they were starting to. An academic-turned-administrator stated that while "academic policy has never been driven by economic concerns, things are beginning to change." Similarly, even reluctant administrators added: "No self-respecting member of the faculty wants to see academic policies changed for sake of software." An undergraduate academic adviser active in senate asked how long before "we have stolen from us the freedom to make educational policy free from economic and technological intrusion?"

In contrast, as a seasoned official stated, "the purpose of adopting new systems was never to use them to recreate, point-for-point, the systems they replaced." Instead, organizations sought to enjoy the efficiencies created by not modifying the new system, perhaps in hopes of creating some savings on investment. In this process of selectively honoring past policies within the new system, university officials could justify discarding expensive customization projects that were not absolutely necessary for the university to operate *or* to retire policies that were not assigned cost-effective status. They could also defend their decisions to interested observers by "blaming the system"—and did.

Discarding policies for purely financial or technical reasons, rather than academic reasons, created tension among the faculty on campus, which echoes Kitto and Higgins' (2010) findings. But the policy was only de-prioritized and this quelled concerns that the policy was abandoned for "the wrong reasons" and could be re-enacted someday. Crude estimates of customization costs necessary to keep 4YP alive were substantial enough to de-prioritize the policy, which also circumvented the need for more detailed analysis. Alternative measures, which implied the work be done by hand, were untenable. Ultimately, new software afforded MU the chance to dump an old policy that many were hoping to forget, and in a politically neutral fashion—as one respondent put it, the university was able to "remove without removing."

## CONCLUSION AND DISCUSSION

Drawn from an organizational case study of ERP in higher education, this paper extended an emic term and sensitizing concept; the "fit-gap." Three steps constitute the fit-gap process, and this chapter develops their interrelation by emphasizing the third where fits for gaps are ordered hierarchically according to their perceived priority. While the term "step" does imply a linear treatment of fit-gap work, and even though other research has shown how similar processes are ongoing and recursive (Rowland, 2008), this paper emphasizes the terminology of steps. The reason is when gaps are constructed (or perceived) by committees who must then deliberate on how to fit them and when, a step-by-step process in initiated. In identifying a gap in the first place, a sequence of events begins, which must run its course and which cannot be squashed until afterwards. 4YP, the policy under fire during implementation, illustrates the phenomena in question. Once the gap is recognized between extant policy and the new system, the fit-gap sequence starts. Fits must be generated and then deliberated upon. Of all the possible fits, one goes to the queue to be prioritized by an implementation project manager. Because every fit must be prioritized, and it is this researcher's experience that the number of fits immediately available for implementation vastly outnumbers the pot of funds available to enact them, a valuable strategy for managing implementation projects is to strategically de-prioritize some fits. Moreover, internal rubrics used for fit-gap documentation help to neutralize the appearance of any political scheming or ambiguous economic accounting. In the end, a politically neutral outcome unfolds – the policy is cast into a de-prioritized purgatory of sorts with the promise that it may be reinstated someday. The interaction between implementation project managers and their fit-gap committees variously composed of functional and technical staff is significant in that it balances, in some respects, current analytical emphasis on ERP software biographies from a design standpoint (Hyysalo, 2009, p. 640; Pollock & Williams, 2008).

The upshot is that de-prioritizing implementation projects delays indefinitely the customization necessary to embed old policy in new technology and thereby obviates anticipatable economic or political costs associated with modification and local work-arounds. Given that the overall thrust of fit-gap work is to avoid customization in the first place, perhaps it is unsurprising that implementation project managers turn many potential into fewer actual gaps in fit, and see that even fewer of those gaps are actually fit.

In this context, however, emphasis only on "fit" is somewhat misleading. Scholarly literature on implementation emphasizes system fit and alignment uncritically. To balance analytical attention to fit, this paper also underscores the significance of gaps and, in particular, how they are constructed and prioritized in the context of committee work. Rowland and Gieryn (2008) maintain that only a truncated subset of gaps can be identified given the impossibility of describing work practices in enough detail to capture everyday tacit knowledge. Developing their insight, this paper showed how gaps are a resource for managers and how de-prioritizing them can serve to, as one respondent stated, "remove without removing" unwanted policy. Pollock (2005) and Kitto and Higgins (2010) sensitized readers to the significance of working-around ERP. Future research should recognize that system fitness and failure are also predicated on how gaps are locally constructed and prioritized for fit.

The practical advantage of learning to see implementation as a straightforward matter of gaps and fits is valuable. It simplifies the vast complexity of implementing ERP and helps to divide the labor necessary to do so. Nevertheless, conceptually reducing implementation to mere gaps and fits distracts from the politicking and ambiguous economic accounting used to shape ERP systems as the systems shape the processes that shape them. Ergo, the fit-gap is both a valued resource and potential liability.

Optimism surrounding ERP implementation as a strategic source of change (Sawyer & Southwick, 2002) seems rarely realized as implementing organizations report high levels of project dissatisfaction and user resistance (Cooke, Gelman, & Peterson, 2001; Scott & Wagner, 2003; Vaughan, 2005). For users on-the-ground, the rise of ERP in higher education has been a tiresome, emotional battle where the same set of chronic of issues appear recurrent (Frantz, Southerland, & Johnson, 2002; Pollock & Cornford, 2004; Pollock, Williams, & Procter, 2003). Such work ordinarily involves substantial investments of time and energy in gathering information, it always carries risks of unanticipated inefficiencies, and it inevitably disrupts workers and their habitual practices (Rowland, 2008; Soh, Kien, & Tay-Yap, 2000). But by learning from fit-gap work we might improve systems in danger of collapse and better support those already running no matter how sub-optimally.

This chapter concludes with a thought. Light (2003) sensitizes readers to the enormous value of legacy systems. Beyond vast investments of time and money, legacy systems, while cobbled-together over the years, read like a storybook of the organization's past and daily performed its many operations. Legacy operated under layers of customization encrusted like barnacles; customizations protected these systems from being disbanded, but systems were also laden and enlaced with customizations, so much so as to progressively minimize their value as the systems drift from those charged with controlling them (Ciborra, 2000). In the flight from legacy to ERP, the new systems must be customized (Pollock, 2005). This author wonders: are some adopters of packaged software initiating the very processes which are often cited as that the reason their legacy systems were in need of replacement? Encrusted with years of customization and work-arounds with few records for outsiders to effectively traverse them, legacy was abandoned for a fresh start with off-the-shelf packaged software. In the march to localized these new systems are organizations balancing these new standardized technologies and their old work practices *or* re-enacting the old ways-of-doing which got them here in the first place?

## ACKNOWLEDGMENT

The research on which this chapter is based is funded by a National Science Foundation Dissertation Improvement Grant SES #0551802 and Pennsylvania State University's Research Development Grant #276-31 AA 1001. I am also grateful to those who have commented on earlier drafts of this chapter and presentations prepared on the subject matter, especially Thomas F. Gieryn, Jan-Hendrik Passoth, Alice Robbin, Fabio Rojas, and Brian Steensland. Responsibility for the views expressed here are nevertheless my own. I would like to thank all those who contributed their time and ideas through interview for this project and, in particular, those administrators, technicians, and support staff that allowed me to rummage through old internal documents, sometimes late into the evenings. Thanks also to Thomas C. Rowland for your outstanding transcription work.

## REFERENCES

Akrich, M. (1992). The de-scription of technical objects. In Bijker, W. E., & Law, J. (Eds.), *Shaping Technology/Building Society: Studies in Sociotechnical Change* (pp. 205–224). Cambridge, MA: MIT Press.

Bowker, G. C., & Star, S. L. (1999). *Sorting things out: Classification and its consequences*. Cambridge, MA: MIT Press.

Briggs, C. L. (1986). *Learning how to ask*. Cambridge, UK: Cambridge University Press.

Brook, T. V. (1995, May 7th). UW considers guaranteeing 4-year finish. *The Milwaukee Journal*.

Cooke, D. P., Gelman, L., & Peterson, W. J. (2001). *ERP trends*. Report No. R-1292-01-RR. New York, NY: The Conference Board.

Frantz, Pollyanne S., Arthur R. Southerland, and James T. Johnson. 2002. "ERP Software: Implementation Best-Practices." *EDUCAUSE Quarterly November 4*:38-45.

Fujimura, J. (1992). Crafting science: Standardized packages, boundary objects, and "translation". In Pickering, A. (Ed.), *Science and Culture and Practice* (pp. 168–211). Chicago, IL: University of Chicago Press.

Gap Analysis. (2006). *Oxford reference online*. Retrieved from http://www.oxfordreference.com.

Gasser, L. (1986). The integration of computing and routine work. *ACM Transactions on Office Information Systems*, *4*, 257–270. doi:10.1145/214427.214429

Hong, K.-K., & Kim, Y.-G. (2002). The critical success factors for ERP implementation: An organizational fit perspective. *Information & Management*, *40*, 24–40. doi:10.1016/S0378-7206(01)00134-3

Hyysalo, S. (2009). Review: How packaged software conquers the organization: Neil Pollock and Robin Williams, software and organizations: The biography of the packaged enterprise system or how SAP conquered the world. *Social Studies of Science*, *39*, 635–642. doi:10.1177/0306312709335847

Kanellis, P., Lycett, M., & Paul, R. J. (1999). Evaluating business information systems fit: From concept to practical application. *European Journal of Information Systems*, *8*, 65–76. doi:10.1057/palgrave.ejis.3000316

Kasvi, J. J. J., Vartiainen, M., Pulkkis, A., & Nieminen, M. (2000). The role of information support systems in the joint optimization of work systems. *Human Factors and Ergonomics in Manufacturing*, *10*, 193–221. doi:10.1002/(SICI)1520-6564(200021)10:2<193::AID-HFM5>3.0.CO;2-H

Kitto, S., & Higgins, V. (2010). Working around ERPs in technological universities. *Science, Technology & Human Values*, *35*, 29–54. doi:10.1177/0162243908329535

Kumar, K., & van Hillegersberg, J. (2000). ERP experiences and evolution. *Communications of the ACM*, *43*, 23–26.

Kvavik, R. B., Katz, R. N., Beecher, K., Caruso, J., King, P., Voloudakis, J., & Williams, L.-A. (2002). *The promise and performance of enterprise systems for higher education: Educause center for applied research.* Retrieved from http://www.educause.edu.

Langenwalter, G. A. (1999). *Enterprise resources planning and beyond integrating your entire organization*. Boca Raton, FL: St. Lucie Press.

Latour, B. (1992). Where are the missing masses? The sociology of a few mundane artifacts. In Bijker, W. E., & Law, J. (Eds.), *Shaping Technology/Building Society: Studies in Sociotechnical Change* (pp. 225–258). Cambridge, MA: MIT Press.

Law, J., & Callon, M. (1992). The life and death of an aircraft: A network analysis of technical change. In Bijker, W. E., & Law, J. (Eds.), *Shaping Technology/Building Society: Studies in Sociotechnical Change* (pp. 21–52). Cambridge, MA: MIT Press.

Leonard-Barton, D. (1988). Implementation as mutual adaptation of technology and organization. *Research Policy*, *17*, 251–267. doi:10.1016/0048-7333(88)90006-6

Light, B. (2003). An alternative theory of legacy information systems. In *Proceedings of the ECIS*. ECIS.

Light, B. (2005). Going beyond "misfit" as a reason for ERP package customisation. *Computers in Industry*, *56*, 606. doi:10.1016/j.compind.2005.02.008

Mackay, H., Crane, C., Beynon-Davies, P., & Tudhope, D. (2000). Reconfiguring the user: Using rapid application development. *Social Studies of Science*, *30*, 737–759. doi:10.1177/030631200030005004

Markus, M. L. (1983). Power, politics, and MIS implementation. *Communications of the ACM*, *26*, 430–444. doi:10.1145/358141.358148

Nooteboom, B. (1992). Information technology, transaction costs and the decision to 'make or buy'. *Technology Analysis and Strategic Management*, *4*, 339. doi:10.1080/09537329208524105

Norris, G., Hurley, J. R., Dunleavy, J. R., Balls, J. D., & Hartley, K. M. (2000). *E-business and ERP: Transforming the enterprise*. Chichester, UK: John Wiley.

Noyes, J. W. (2003). The ERP dilemma: "Plain vanilla" versus customer satisfaction. *EDUCAUSE Quarterly*, *2*, 54–55.

O'Leary, D. E. (2000). *Enterprise resource planning systems: Systems, life cycle, electronic commerce, and risk*. Cambridge, MA: Cambridge University Press.

Pfaffenberger, B. (1992). Technological dramas. *Science, Technology & Human Values*, *17*, 282–312. doi:10.1177/016224399201700302

Pollock, N. (2005). When is a work-around? Conflict & negotiation in computer systems development. *Science, Technology & Human Values*, *30*, 1–19. doi:10.1177/0162243905276501

Pollock, N., & Cornford, J. (2004). ERP systems and the university as a "unique" organisation. *Information Technology & People*, *17*, 31–52. doi:10.1108/09593840410522161

Pollock, N., & Williams, R. (2008). *Software and organisations: The biography of the enterprise-wide system or how SAP conquered the world.* London, UK: Routledge.

Pollock, N., Williams, R., & D'Adderio, L. (2007). Global software and its provenance: Generification work in the production of organizational software packages. *Social Studies of Science*, *37*, 254–280. doi:10.1177/0306312706066022

Robey, D., & Markus, M. L. (1984). Rituals in information system design. *Management Information Systems Quarterly*, *8*, 5–15. doi:10.2307/249240

Rogers, E. M. (1983). *Diffusion of innovations*. New York, NY: The Free Press.

Ross, J. W. (1998). *The ERP revolution: Surviving versus thriving*. Cambridge, MA: MIT.

Rowland, N. J. (2008). *Fit-gap work: Implementation, innovation, and organizational forgetting in higher education with ERPs*. Paper presented at the Annual Meeting for the American Sociological Association. Boston, MA.

Rowland, N. J., & Gieryn, T. F. (2008). Transfer troubles: Outsourcing information technology in higher education. In Swedberg, R., & Pinch, T. J. (Eds.), *Living in a Material World: Economic Sociology Meets Science and Technology Studies* (pp. 375–391). Cambridge, MA: MIT Press.

Sawyer, S., & Southwick, R. (2002). Temporal issues in information and communication technology-enabled organizational change. *The Information Society*, *18*, 263–280. doi:10.1080/01972240290075110

Scott, J. C. (1999). *Seeing like a state: How certain schemes to improve the human condition have failed*. New Haven, CT: Yale University Press.

Scott, S. V., & Wagner, E. L. (2003). Networks, negotiations, and new times: The implementation of enterprise resource planning into an academic administration. *Information and Organization*, *13*, 285–313. doi:10.1016/S1471-7727(03)00012-5

Soh, C., Kien, S. S., & Tay-Yap, J. (2000). Cultural fits and misfits: Is ERP a universal solution? *Communications of the ACM*, *43*, 47–51. doi:10.1145/332051.332070

Star, S. L., Bowker, G. C., & Neumann, L. J. (1997). *Transparency at different levels of scale: Convergence between information artifacts and social worlds*. Retrieved 27 May 2009 from http://epl.scu.edu:16080/~gbowker/pubs.htm.

Strauss, A. L. (1987). *Qualitative analysis for social scientists*. New York, NY: Cambridge University Press. doi:10.1017/CBO9780511557842

Swartz, D., & Orgill, K. (2001). Higher education ERP: Lessons learned. *EDUCAUSE Quarterly*, *2*, 20–27.

Thomas, R. J. (1994). *What machines can't do*. Berkeley, CA: University of California Press.

University of Northern Iowa. (1995). *UNIGrad-Pact*. Retrieved 13 March 2010 from http://www.uni.edu/vpaa/gradpact.html.

Vaughan, D. (2005). *Employee resistance to enterprise resource planning technologies*. Paper presented at the Annual Meeting of the American Sociological Association. Philadelphia, PA.

Wagner, E. L., & Newell, S. (2006). Repairing ERP: Producing social order to create a working information system. *The Journal of Applied Behavioral Science*, *42*, 40–57. doi:10.1177/0021886305284290

Weiss, R. S. (1994). *Learning from strangers*. New York, NY: The Free Press.

Williamson, O. E. (1981). The economics of organization: The transaction cost approach. *American Journal of Sociology*, *87*, 548–577. doi:10.1086/227496

Wright, M. (1996). IU trustees. *Herald Times*. Retrieved 13 March 2010 from http://www.heraldtimesonline.com/stories/1996/06/10/archive.19960610.1df54f5.sto.

# Chapter 11
# The Role of Management Consultants in Long-Term ERP Customization Trajectories:
## A Case from the Italian Local Government

**Gian Marco Campagnolo**
*University of Edinburgh, UK*

## ABSTRACT

*This chapter proposes an understanding of the role of management consultants in the implementation of Enterprise Resource Planning Systems in the Public Sector. The analysis is grounded on the case of an Italian Local Government in the course of 10 years. Each period corresponds to a different key client-consultant relationship and Public Sector situation. By using situational maps, the authors have highlighted the deeply contextual reflexivity between consultants and client organizations in configuring consultants' advice over time. The purpose of the chapter is to replace "guru-istic" accounts with a more relational perspective that describes consultants as existing only in relation with their clients and the evolving Public Sector situation.*

## INTRODUCTION

New Public Management (NPM) (Hood, 1995) has become one of the most important reforms in government organizations. One of the key characteristics of New Public Management is the increased adoption of managerial and market-based principles and techniques by public-sector organizations. The adoption of Information and Communication Technologies (ICTs) has been advocated as a central aspect of the innovations undergone under the umbrella of New Public Management reforms.

DOI: 10.4018/978-1-4666-0303-5.ch011

However, government organizations may miss some of the competences that allow technology choice: public organisations find it difficult to critically assess and evaluate large IT solutions in-house. Their substantial and often business critical decisions about what may be major strategic investments (costing several millions of Euros) are carried out often for the first time and public-sector organizations often lack the expertise and experience needed for effective decision-making. For this reason, these organizations may look for expertise coming from external management consultants specializing in IT.

Trends like these have caused a significant growth of academic interest in the consultancy industry. In much of this work, consultants are seen as a distinctive group that has gained insidious power in stimulating the creation, diffusion and adoption of the latest 'best practice' fashions, and new organizational practices (Sturdy, 1997). Researchers here have examined consultants as having persuasive strategies that define clients' problems and image consultants as the possessors of problem-solving skills. Various intermediaries in the IT market have been described as "holding the ropes and setting the rules of the game" for the emergence of technological fields (Pollock & Williams, 2009).

The perspective presented in this chapter is concerned less with the way consultants organize technological fields (e.g. the technological field of ERP systems and the accounting practices supported by them) and more with the way consultancy is shaped by key negotiations and the situation.

As stated by Hislop (2002), consultants only exist in relation to their clients and, I add, they exist in relation to an extended network of actors and discourses that evolve over time.

The evolution of client-consultant relationships over time and its embeddedness in the situation is typically illustrated by configurable packages like ERP.

With ERPs, customization choices, that is the possibility to configure the system to match user requirements, are never ending. Periodic vendor system updates and new module releases require organization from time to time to renew their strategy. As a consequence, in the context of ERP implementations, client relationship with management consultants specializing in IT can be observed in its evolution over time.

By focusing on an Italian Local Government—here called 'Dante County'—this chapter provides insights into the deeply contextual reflexivity between consultants and user organizations in configuring public sector transformations over time. The case illustrates the distributed and socio-political nature of the consultancy process. First of all, during the same technological project clients replace consultants multiple times according to the evolving situation in the Local Government arena. Second, consultants expertise is defined in relation with clients: consultant advice is highly dependent upon the relevant social group they interact with within the client organization. Finally, there is not a clear boundary between 'clients' and 'consultants.' Consultants can become part of the client organization and the client organization can perform as a consultant with respect to other parties.

The chapter is structured as follows. The literature review will review the perspectives concerning the role of consultants specializing in IT in the public sector. Then the chapter will present the theoretical and methodological framework. The case study will be then presented. Discussion and conclusions will follow.

## LITERATURE REVIEW

Having somehow integrated operations within the enterprise, the suppliers of ERP software have sought to generalize it to other areas such as the public sector, which in principle are different from those for which the systems were originally conceived.

By transforming it from an application for individual productive functions (like expert systems or business process re-engineering systems) into a system to control organizational action in its entirety (Davenport, 1998), the ERP system has been considered one control system that government must possess to justify investments of public money and guarantee the reliability of their accounting procedures.

One of the major differences between private sector and public sector in Italy concerns accounting procedures. While in the private sector (the area for which the ERP systems were originally conceived) organizations utilize *ex-post* 'accruals accounting,' most Italian public sector organizations still adopt *ex-ante* budgetary accounting procedures. The 'travelling' of ERP system from private to public sector has been deeply 'flavoured' by the idea that Local Government would have adopted some sort of accrual-based accounting procedures (Anessi-Pessina, Nasi, & Steccolini, 2008).

Management consultants specializing in IT have been considered one of the main actors in this transition of ERP-supported accounting practices to Italian Local Governments.

In research on the procurement process of packaged software, the role of management consultants has been identified as an important factor in understanding the spread and spreading of such packages from private to public sector. According to Lawson and Appiaganesi (1989), consultants are professional storytellers that frame and persuade, possessing the skill-set more akin to lobbyists or salespersons than trained scientists or hired experts. Reed (1992) suggests that management consultants are a type of expert adept at managing uncertainty with esoteric knowledge or symbolic systems. These type of "guru-istic" accounts of the role of management consultants as "Swiss-army knives" of advanced capitalism can be mitigated by a more mundane analysis of long-term interactional dynamics of technology adoption in organizations. From these alternative accounts, ERP-mediated accounting practices result from a distributed socio-political process.

As observed by Pollock and Cornford (2004), the construction of similarity between public sector and the enterprise as areas of application for ERP systems does not take place within the boundaries of the organization alone. Rather, it is the result of relations that public-sector organizations have established with other institutions involved in similar technological projects. The notion of 'best practice' software (Wagner, Scott, & Galliers, 2006) is indeed fundamental for the project carried forward by the suppliers of ERP systems: after implementing the system in a single important organizations, consultants can propose the same product to other less central clients in the same sector, claiming that it incorporates the best accounting practices adopted by the 'big players.'

As emerging by the mentioned study by Pollock and Cornford (2004), decisions regarding customisation of the system in public sector organizations are often deferred or distributed. Public-sector organizations throw up so many "demands for policy" that implementation decisions can never be properly discussed among the overseeing committee, and they are "pushed down to the technical teams and then onto the system itself. In other words, change often occurred through a process of default rather than choice" (Pollock & Cornford, 2004, p. 48).

As Wagner and Scott maintain, rather than being exhaustive, ERP implementation decision-making in public-sector organizations often results from a socio-political decision process that involves a small interest group (Scott & Wagner, 2003).

In this respect, some authors (Newman & Westrup, 2005) have built the argument of how expert groups seek to influence the ERP system development by entering into strategic alliance with IT consultants, thus controlling the ERP implementation.

In this chapter, I seek to mitigate the guru-like account of the management consultant's role in shaping management processes by illustrating the distributed socio-political nature of the evolution of ERP-supported accounting practices in the public sector. I focus on the way key mediations between IT-consultants and their client changed several times during the period under scrutiny.

The example I use is related to controversies with respect to customizability of ERP systems. By focusing on an Italian Local Government—here called 'Dante County'—this chapter provides insights into the deeply contextual reflexivity between consultants and client organizations in configuring Local Government transformation over time.

## THEORETICAL FRAMEWORK

I draw upon constructivist and critical studies that emphasize the socially constructed nature of client-consultant relationship (Pozzebon & Pinsonneault, 2005). I conceive client-consultant relationships as taking place in an *extended situation* (Pollock, Williams, D'Adderio, & Grimm, 2009) which includes *(1)* a variety of other actors such as vendors, key users, operational users and other Local Governments and *(2)* where the creation of representations of ERP-supported accounting practices continues through multiple generations of product development and organizational designs, where each generation tends to build representations in relation with what has been learned by the previous one (Webster & Williams, 1993).

This chapter presents a cartographic method to locate changing client-consultant relationships based on the usage of situational maps (Clarke, 2005).

The analytical tools adopted to interpret the role of management consultants in time periods are social words/arenas maps of the public sector situation, its actors and social worlds.

With situational maps, I want to show how consultants were enlisted in negotiations on ERP customizations within the varied composition of the Local Government situation at different times.

Situational maps are a methodological extension of grounded theory (Clarke, 2005). Grounded theory has been among the leading approaches to qualitative inquiry for decades (e.g., Atkinson, Coffey, & Delamont, 2003). Initially developed by Glaser and Strauss and later by others, grounded theory focuses on systematically analysing qualitative data to elucidate key forms of action undertaken by participants in particular situations. Some more contemporary versions of grounded theory have deeply enhanced the constructionist tendencies of one of its originators, Strauss, and his social worlds and arenas framework. Instead of viewing social life as made meaningful and coherent only at the levels of the individual (micro-social) or society as a whole (macro-social), Strauss (1978a, 1978b, 1991, 1993) offered an intermediate or meso-social vision: the often hidden infrastructures through which negotiations are organized.

Situational analysis is one of these extensions (Clarke, 2005). The key concern of situational analysis is the situatedness of action (Haraway, 1988). In situational analysis, the conditions of the action are in the situation (Clarke, 2005, pp. 65-73). There is no such thing as context: the conditional elements of the situation need to be specified in the analysis of the situation itself as they are constitutive of it, not merely surrounding it or framing it or contributing to it. As part of doing situational analysis, we need maps of situations: maps that specify what is there empirically in considerable detail and from multiple angles of perception.

The maps I will use to analyse Dante County implementation choices are social worlds/arenas maps. The social worlds/arenas map has the analytic task of specifying the key social worlds. They are depicted using dotted lines to give the impression that social worlds are not fixed entities: actors can participate in more than a social world. Second, there are multiple social worlds. Each social world takes a perspective on the situation of concern. Not all social worlds will be relevant in the same way at the same time.

In each time period I will identify activities performed by actors in social worlds aimed at gaining and maintaining legitimacy of their perspective on the arena. These activities include entering into strategic alliance with other social world representatives and forming key relationships. The social world perspective will support specifying grand abstractions such as "consultants" or "clients" by grounding them within their composing social worlds. According to situational analysis, consultants can have different provenance (they can be either suppliers, consultants, or process consultants, be in-house consultants, or coming from a consultancy firm) and they can have gained experience from different public service domains from the private sector. Consultants' perspectives vary according to these social factors. To the same token, according to the analysis, the client organization is a collectivity that participates in multiple social worlds, such as the policy maker's world, the IT manager's world, the accounting manager's world, the middle-management's (e.g. key users) and the operational personnel's world, to quote a few. Second, the situation under scrutiny is populated by a variety of other organizations that, to different extents, contribute to shaping the Local Government arena (see Figure 1).

There are indeed other local organizations from different public service domains and other Local Government organizations from different regions that are engaged in similar technological projects that share participation to the above-mentioned social worlds.

## RESEARCH METHODOLOGY

The chapter presents the results of a research project conducted between April 2006 and December 2008 at an Italian Local Government—here called 'Dante County.' The data collection methods employed to gather empirical material with which to refine understanding of Dante Province's implementation strategy and create situational maps were ethnographic fieldwork, biographic interviews, participant observation, and document reading and discussion.

The field in Dante County was initially constructed by following different paths:

- the development of the accounting practice supported by the ERP system;
- the development of the ERP implementation strategy;
- the development of the people involved in ERP implementation and after.

The accounting *practice supported by the ERP system* was the management of the administrative activities related to the establishment of Dante County programs, plans, and objectives by the Council and their assignment to the administration management for fulfilment. Different Dante County organizational structures were involved in planning activities at different stages, and at each stage, they made use of the ERP system. Here the data collection method was participant observation (Van Maanen, 1998; Rosen, 1991). The observation centered on 'hot moments.' In particular, it took place during a major change in the "Long-Term Investments Plan" with subsequent evolutionary maintenance of the system (August, 2008).

The second strategy adopted to construct the field was to follow *the development of the ERP implementation*. As well as examining formal intermediaries, my strategy was to identify those intermediaries located at the interstices between organizations in informal roles and to pay closer attention to broader social relations, so as to explain the finer structure of ERP implementation and post-implementation. The key actors to interview and observe further were identified on the basis of their role in the ERP implementation and post-implementation. Starting from there, additional software modules have been identified and integrated with the initial ones, and the organizational actors connected with them as analysts, designers, programmers, management

*Figure 1. Local government arena*

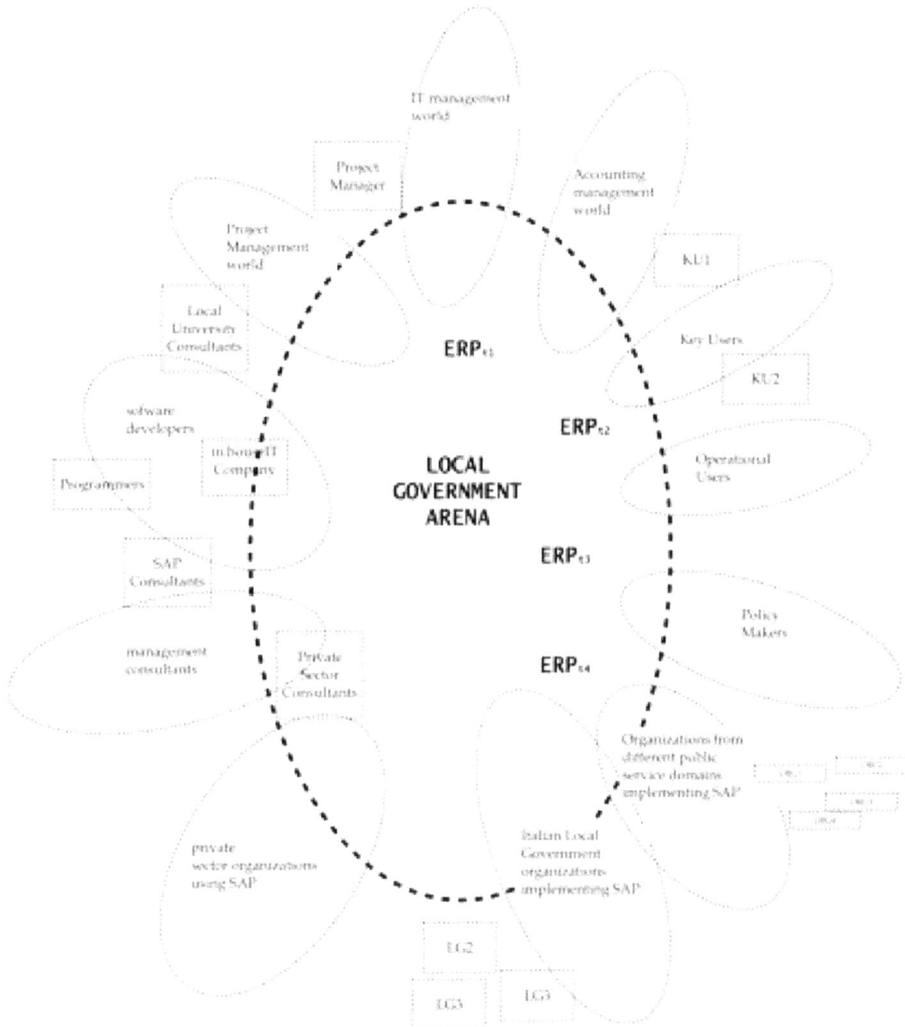


consultants, project managers, key users, content providers, and content users.

Interviews were conducted in a biographic form starting from a first invitation to analyze the professional lives of the informants concerning their connection with the technological project of ERP implementation in Dante County. The biographic interview (Portelli, 2004; Gubrium & Holstein, 2003) provides the researcher with a discourse in which the beliefs, the attitudes, the values, the representations of the biographic trajectories are expressed with a particular emotional colouring, and are inscribed within an argumentative structure that determines their sequence and manifests their connections. All the interviews were tape-recorded and verbatim transcripts were produced (Riessman, 1993). Selected interview excerpts reported in the chapter are translated by the authors (see Table 1).

The third and related strategy of field construction was to follow *the development of the people involved in ERP implementation and post-implementation* in their work settings.

*Table 1. Biographic Interviews Table*

| People involved in ERP implementation and post-implementation | |
|---|---|
| Functional Analyst of the in-house IT company | Interviewed: 5 March 2008 |
| Project manager | Interviewed: 21 November 2007; 3 March 2008. |
| key user Accounting Dept. | Interviewed: 10 January 2008 |
| key user Planning Dept. | Interviewed: 3 March 2008 |

Fieldwork has been undertaken with ethnographic methods from January to December 2008 in Dante County IT Department and occasionally in other public services organizations in the region that were undergoing ERP implementations. The fieldwork allowed the researcher to become familiar with various perspectives with respect to ERP implementation in Dante County and other connected organizations (e.g. other public service domain organizations such as the university, other research centres, the in-house IT company, ...). Long-term fieldwork also allowed identification of transformations in implementation strategy over time and connections between these transformations and the overall situation in the public sector.

The first pass through the data involved basic information sorting and inductive techniques adapted from grounded theory (Strauss & Corbin, 1998).

The empirical material covered by the observation and re-covered by biographic interviews and document analysis corresponded to a ten-year period between 1998 and 2008 and concerned the implementation of various ERP modules and applications, involving 4 different consultancy providers: Local University (LU) consultants, consultants from a consultancy firm that worked for Alitalia, SAP Public Sector consultants, and finally, a Consortium of different kind of consultants.

## CASE DESCRIPTION AND ANALYSIS

For the purpose of the analysis of management consultant relationship with the client organization, the evolution of consultants' advice with respect to ERP customization choices in Dante County is described across four different time periods, from 1998 to 2008. I have made the most influential client-consultant relationship the primary delegate for each time period, and I have organized the narrative around the ERP proposal emerging from it: each time period is distinguished by a different client-consultant relationship (see Table 2) and by a different ERP proposal.

My narrative is addressed to contribute to mitigate the "guru-like" accounts of management consultants role in multiple ways: (1) consultants exist only in relation with their clients and (2) consultancy is an interactive distributed socio-political process. First of all, clients hire and fire consultants, replace them with other consultants according to factors that go well beyond the evaluation of consultants' skills or expertise. Additionally, rather than deriving from management consultants devices and tools of knowledge commodification, consultant technical advice (i.e. about ERP implementation strategy) depend to a larger extent upon the varying nature of actors and social worlds that succeed to enter into strategic alliance with them. For this reason, the narrative will focus on key client-consultants relationships at different time periods rather than on consultants alone. Key client-consultant rela-

*Table 2. Table of key client-consultants relationship at different time periods in Dante County*

| |
|---|
| 1. Steering Committee-Local University Consultants (1998-2002) |
| 2. Key Users -Private Sector Consultants (2002-2005) |
| 3. Public Sector User Group - SAP Consultants (2005-2008) |
| 4.Policy Makers-Consortium of Consultants (2008-) |

tionships, to their turn, are not taking place in the empty space. For this reason, I also reported in detail the constellation of actors and social worlds in the Public Sector situation. Table 3 shows the different sets of actors that had a role in system implementation at different times.

## 1998-2002: Local University Consultants and the Steering Committee

This first session describes how the key relationship of Local University (LU) Consultants and the IT manager attempted to influence the initial decisions of the Steering Committee about ERP customizations in Dante County. The most relevant technical constraint with ERP systems is the amount of required customizations, that is the amount additional software that integrates the standard version in order to match the user preferences. More customizations mean more integrated accounting information (i.e. reports) for the middle management (i.e. Staff people) but at the same time more work by Operational Users. However, according to the IT management world perspective, customizations are critical because they increase the cost of implementation. On the contrary, if the system is adopted as it is, that is in its standard version, it means less sophisticated reporting functions but at the same time less work for Operational Users. At the same time, according to the IT management social world perspective, the adoption of a standard (e.g. off the shelf or 'pre-packaged' or default) version of the system costs less than a customized one.

In this first time period, LU management consultants have been hired by the Head of Dante County IT Department to manage the implementation project and provide consultancy with respect to the amount of customizations of ERP system. LU consultants would have responded to the Steering Committee, whose role was to guide the initial system implementation. The rationale of the first module implementation is described in the following way by a member of the Steering Committee:

*"Each single user, each single service was responsible for the data input and for the controls. We structured the system in a way that would solve the information requirements of the Staff Services so that they could monitor. But essentially, the system*

*Table 3. Actors involved in the implementation of the ERP system at different time periods*

| | 1. (1998-2002) | 2. (2002-2005) | 3. (2005-2007) | 4. (2008-) |
|---|---|---|---|---|
| (1) Steering Commitee | X | | | |
| (2) in-house IT Company | X | X | | |
| (3) Private Sector Consultants | | X | | |
| (4) Local University Consultants | | | | |
| (5) Consortium of Consultants | | | | X |
| (6) SAP Consultants | | | X | |
| (7) Project Manager | X | X | X | X |
| (8) Key Users | | X | X | X |
| (9) Operational Users | X | X | X | X |
| (10) Other Public Administrations implementing SAP | | | X | X |
| (11) Other Local Organizations implementing SAP | | | | X |

*was intended to serve the Operational Services users" (Member of the Steering Committee)*[1].

In this excerpt, a member of the Steering Committee presents the rationale of ERP implementation in this first period as that of supporting the perspective of the operational users' social world: the system that at that time "intended to serve the Operational Services users" was a system in which "each single user, each single Unit, was responsible for the data input and for the controls."

A different perspective on system customizations was taken by Staff Services. With the new system, accounting operations are going to be de-centralized from a central unit of specialized accountants to a larger number of under-qualified Operational Users. When de-centralization of accounting practices is taking place, the major preoccupation of Dante County Staff people is data accuracy. Data accuracy can be ensured by customizing the ERP system with the provision of additional automatic central controls on data input. However, controls have an added value for Staff people, but would have resulted in major modifications on operational users ways of working. More controls mean indeed a more restrictive and complex data input activity. In the first implementation phase, the members of the Steering Committee contrasted the Staff Service world-view on adding additional controls on the customized system because this would have resulted in a rejection of the system by Operational Users. Control requirements for monitoring purposes are defined by the Steering Committee member we interviewed as "information that does not have an added value for all," and this "should not be included" in the system.

In this context, the role of LU consultant was that of keeping the expected effort for customizations as calculated at the outset by the Steering Committee. The Steering Committee was composed of various Dante County Heads of Department (e.g. IT Dept, Accounting Dept, Human Resources Dept.). Each of them, facing the introduction of the system, required LU consultants some adaptations for matching work processes in their Unit. In 1998, nobody was implementing SAP in the Public Sector in Italy. As a consequence, LU consultants did not have a clue about what the consequences of allowing additional system customizations on an ERP system were. Rather than contrasting various Heads of Department requests, LU consultants managed the implementation project in order to host all these adaptations. As a consequence, the number of customizations had enormously risen. In 2002, an estimate of 1300 work days for customizations rose to a final 3000.

These figures have been explained by one member of the Steering Committee in the following way:

*"Reports by Local University Consultants about customizations control were used to justify customizations rather controlling them."*

In other words, having to deal as a key relationship with *all* the members of the Steering Committee, LU consultants had many different customization requests that they could not refuse. Their limited expertise about the technical or organizational consequences of the rising of customizations (given that ERP implementation in the Public Sector was at that time pioneering activity) was another factor.

Seen from the perspective of the key client-consultant relationship, while the Head of the IT Dept. and the LU consultant hired by him have been able to contrast Staff people customization requests (e.g. additional automatic controls on data input) they haven't been able to contrast those rising from other members of the Steering Committee.

## 2002-2005: The Consultant from Alitalia and the Accounting Department Key Users

Given that the amount of customizations did not match the expectations, LU consultants were fired and consultants with higher experience on ERP implementation in private organizations have been hired.

From the client side, an ERP Project Manager was appointed by the IT manager to guide the implementation process after the Steering Committee completed the system roll out. After being appointed, she was moved from the IT Department to the Accounting Department, to better control customization requirements arising especially from that side, considered by Dante County IT manager as one of the most productive requirement provider for the implementation of the ERP system.

However, the project manager perceived that the ERP implementation logic had changed from the previous period:

*"We wanted to see SAP from the point of view of the project or of the plan, not from the accounting point of view [...] but the logic had been changed." (Dante Province ERP Project Manager).*

According to the Project Manager, the change consisted of seeing the ERP system implementation from the project management social world point of view (i.e. from the point of view of the project or the plan) to the accounting management social world point of view (i.e. from the accounting point of view). This change in the underlying perspective on implementation can be better understood by looking at the changing nature of the key client-consultant relationship.

From the client side, Key Users started to have a central role in affecting/intermediating consultancy. A consequence of considering the Accounting Department one of the most productive requirement providers was indeed the appointment of most Key Users from that organizational Unit, thus contributing to changing the logic of the implementation: the system was customized in order to support more and more sector-specific reporting functions. Indeed, and contrary to what had emerged from the former period, the Key Users maintained that controls and reports "are mainly made for the Operational Structures" and should be included in the system. The following excerpt from the Planning Service Key User interview illustrates this changing perspective:

*"[controls and reports] are mainly made for the Operational Structures. With the controls, the Operational Services are confident that they cannot make mistakes. And reports allow them to have all information" (Key User Planning Service).*

It can be surprising that Staff people can have a central role in intermediating consultancy when they are appointed as Key Users. However, the strength of Key Users' role derived from the weakness of the ERP Project Manager role. The Project Manager had neither the position nor the necessary support from Private Sector management consultants to control the increasing amount of key users' customizations. The first factor was that other departments' people perceived the move of the Project Manager to the Accounting Department as a loss of impartiality and trust—especially at a time when the ERP project spread across the entire organization. Indeed, together with some additional applications in the accounting area[2], also implemented at this time was a module that serves the budget planning area.

Furthermore, each customization requiring fewer than 10 working days was not submitted to the Project Manager[3]. There was also a case when a customization took longer than 10 days, but the Key User of the proponent Unit did not report it to the Project Manager, relying instead on the support of the formerly employed LU Consultants.

Second, consultants hired in this period did not succeed in negotiating the ERP project manager view on ERP customization with Key Users. One of the consultants hired in this time period by Dante County IT Dept. to support the project manager had worked for the implementation of ERPs at Alitalia, an Italian private airline company. The Project Manager describes his contribution as follows:

*"Last year a consultant came from Alitalia. He knew everything about Alitalia. He thought he could apply the Alitalia model in the Dante County. If he had told me he had worked for the Bologna Town Council or Rome, then fair enough.... He was really way off...he was way off.... I told the consultant that public accounting is different from private accounting!" (ERP Project Manager).*

As this interview excerpt illustrates, Dante County Project Manager has been extremely critical with the Alitalia consultant contribution. Indeed, as a reaction to consultants with experience gained mainly in the private sector, Key Users presented customization as necessary because "informatics must be coherent with the provisions" (Key User interview). Thus, in order to be fully compliant with the regulations, Key Users required consultants to include in the system all possible exceptions typical of the public sector:

*"The standard module was not good because the public administration is subject to constraints concerning authorization, budget, and allocations. The private sector does not have such rigid constraints."*

Seen from the point of view of key client-consultant relationships, consultants hired in this time period, having gained experience mainly from private sector, haven't been able to support the Project Manager view on system customizations versus the accounting key users view. On the contrary, Key Users reacted to the presence of private sector consultants by raising many demands for policy and arguing that Operational users were not confident with a system coming from the private sector.

## 2005-2008: The Public Sector User Group and SAP Consultants Specializing in Public Sector

From 2005, the IT market in the Public Sector was undergoing major changes. ERP systems have been introduced in many other Italian public administrations. A Public Sector User Group has been created at national level to share ERP implementation experiences. SAP, one of the leading vendors of ERP systems in Italy, developed a "Public Sector" version of the system and started to provide consultancy targeted for SAP "Public Sector" users.

From June 2005 onwards, Dante County feasibility studies on the introduction of new ERP modules were conducted by SAP consultants specializing in Public Sector in the stead of the consultants having gained experience mainly from the Private Sector. According to the Project Manager, identification of a SAP consultant specializing in Public Sector provided an opportunity to look outside her organization to learn how other organizational contexts acted with similar projects. On looking back to the previous periods, when comparison with other Local Governments implementing ERP systems was not feasible, she said that:

*"by doing everything in-house, we also made some mistakes" (Dante County ERP PM interview).*

According to the Project Manager world perspective, with SAP consultants, feasibility studies were being completed more rapidly and the likelihood of complying with the supplier's 'best practice,' that is the standard version of the system, was increasing.

In this situation, SAP consultants specializing in Public Sector became the key intermediaries, not only between supply and use, but also, and most importantly, among different user sites. Since they were closer to the production side, SAP consultants were perceived by the Project Manager as providing more reliable feasibility studies: neither LU consultant nor the consultant that worked for Alitalia had comparable expertise on the specific topic, and they were prone to accept any requirement from Dante County Steering Committee members or accounting Key Users. As illustrated by the following interview excerpt, the perspective that the Project Manager wanted to support was instead that customizations (locally known as "personalizations") were only good for change management strategic purposes, not for functional ones:

*"Instead of encountering a rejection of the system, I can make some personalizations in order to persuade the user to come round to my way of thinking. And so I will be the rule-maker..." (Dante County ERP PM)*

Being able to travel from one user site to another this time period, consultants acted as intermediaries between different user organizations, in ways that overcame the mentioned internal struggles between competing Steering Committee members or Key Users. The stronger legitimization given by SAP consultants to the Project Manager world perspective derived from the opportunity to compare and partly reuse solutions from other Local Governments.

The Project Managers within the User Group started to exchange references to consultants they liked and talked to each about the consultancy taking place. The content of the collaboration was both re-use of the software and the comparison among the performances of consultants, in term of contracts as well as of less formal aspects of the relation. Whilst the direct re-use of software (e.g. duplication) was unlikely—hampered by the differences among regulatory frameworks—, the sharing of experiences (and even of the names of reliable consultants) started to prove valuable for Local Governments that had implemented an ERP system.

## From 2008 Up to Now: The Consortium and Other Public Service Domain Organizations

SAP consultants specializing in Public Sector were those consultants who best supported the Project Management world perspective. However, some changes occurring in the Local Government arena implied the replacement of SAP consultants working for different Local Government with a Consortium of consultants that would have worked only for Dante County. These changes were the rising cost of SAP offerings and a national regulation about the role of public-private partnerships.

Previously, SAP—the leader vendor of ERP systems to Local Governments—had offered a basic support package costing 17 percent of a customer's license fees. Since July 2008, SAP has replaced it with an Enterprise Support package that costs 22 percent. Moreover, since 2006, a new national law has dictated that in-house IT companies can only supply services in the Region where they are located.

In this situation, in the case of further ERP modules implementations, Dante County decided that consultancy had to be provided by a Consortium consisting of academic partners, local consultancy companies, and the in-house IT company. This Consortium will operate only for Public organizations in the Region.

The following interview excerpt reports that, compared with the former time period, the ERP Project Manager describes the situation as a 'return' to the 2002-2005 period, when many unnecessary customizations were performed.

*"the Consortium is more or less like the LU consultants. Just as they had said 'yes' to the Head of the Accounting Department some years ago and customized the system more than necessary.... We too will go back again: it will be the same with*

*the Consortium. If you pay and tell: 'you have to do a, b and c'.... Imagine: the analysis becomes longer! It is not that they say 'no' immediately. Especially during the analysis, they foresee all possible exceptions. But in the requirement phase it is pointless to pick up all exceptions if you know that later you will not be able to implement them."*

Given the constraints of the new national regulation on the 'in-house' model, after 2008 Dante County also decided that ERP implementations have to be extended across organizations of different public services domains in the Region. 85 educational institutions and 12 public agencies joined Dante County in the effort to implement the ERP system. An "across the board" strategy was applied with a larger number of institutions from different public services domains in the local area. As a consequence of the situation, in terms of key client-consultant relationships, the discourse on the adoption of ERP-supported accounting practices was no longer framed as exclusive Dante County's goal nor even as a goal that concerns Public Sector organizations of a single domain (i.e. Local Governments).

Rather, Dante County policy makers framed it within the administration's wider goal of "aligning a fragmented landscape of public sector organizations made up of large organizations and smaller ones with common interpretations of organizational procedures"[4].

## DISCUSSION

In order to discuss the impact of the changing nature of the key client-consultant relationship on the nature of consultancy, I choose to focus on the controversy related to system customizations, that is the amount of system adaptations that the client requires to the consultant. I want to highlight how the discourse on what a 'standard' and a 'customized' versions of the system were, and what the purposes of 'standardization' and 'customization' were, rather than deriving from consultants ability in framing and persuading customers, was a distributed and socio-political achievement. In this section, I will employ data from different time periods to discuss how consultancy about system customizations changed according to the changing key client-consultant relationship and the overall situation in the Local Government arena.

While in the period when LU consultants worked closely with the IT manager and other Steering Committee members, implementing the ERP standard version meant "giving responsibility to the Operational Services" users. This corresponded to the IT management world perspective that conceived customizations as costing too much: better was to keep the system in its standard version—without customized controls on data input—and to give Operational Users more responsibility on data accuracy. On the contrary, in the following period of accounting key users interacting with the consultant coming from Alitalia, the 'standard' ERP meant a system "coming from the private sector" that was not coherent with regulatory provisions and did not help Operational Users to be confident they did not make mistakes.

By contrast, whilst in the former period a 'customized' system was considered by the Steering Committee members to be carrying too many control requirements as "information that [does] not have an added value for all," in the latter period, according to the Key Users view, 'customizations' were important for Operational Services users to be "confident that they cannot make mistakes" since all possible controls and blocking messages were "coherent with regulatory provisions." In terms of key client-consultant relationships, in the first time period the IT manager sought to influence the implementation by entering into strategic alliance with LU consultants and the Operational users to resist additional customization requests from other Dante County Heads of Department and their staff. The result was that all Steering Committee members agreed upon the idea that the implementation of additional system customizations to improve controls by Staff members on operational users were not necessary. However, LU consultants, having not sufficient technical expertise, previous experience, and connections with other organizations implementing an ERP system, were not in the position to legitimize the IT Manager resistance to requests arising from Heads of other Dante County Departments. The result was that consultants in this time period have been conceived as *justifying* rather than controlling customizations (see Table 4).

In the second time period, when the roll out of the system was completed, responsibilities on further modules implementations were pushed down from the Steering Committee to a Project Manager and to Staff Key Users. The IT Manager hired consultants with higher experience on ERP implementation in Private Sector organizations and moved the Project Manager from the IT Department to the Accounting Department to be closer to the accounting Key Users. However, having gained experience mainly from the private sector, the new consultants did not give the Project Manager the necessary legitimization in the negotiation with Key Users about customizations. Indeed key users raised many 'demands for policy' against the standard version of the system, which they conceived as a system coming from the private sector. Consultants in this time period were considered as being *really way off.*

After 2005, owing to the above-mentioned changes in key relationships and the varying composition of the Public Sector situation, the discourses concerning system customization varied as follows. Whilst in the previous period, implementing an ERP module standard version had meant a system "coming from the private sector," with the increased exposure of various Local Governments to ERP implementation projects and with modifications for the public sector being incrementally incorporated into the supplier's package (Public Sector) verticalization, the discourse about a "standard" ERP system became that of an implementation following "best practice" criteria. By contrast, whilst in the previous period customizations were synonymous with "coherence with regulatory provisions," they now corresponded to shallow feasibility studies and an absence of change management strategies. Consultants in this time period were considered as bringing *best practices* to the Local Government arena. In this time period of key relationships between a Public Sector User Group on ERP implementation and SAP consultants specializing in the Public Sector, a 'standard' version of the system had meant a 'vertical standard' (Markus, Steinfield, & Wigand, 2006) deriving from other Local Governments which had developed a 'best practice' undertaken by a supplier serving different customers. However, when SAP consultants have been replaced by a Consortium acting only locally, the discourse on standardization changed again. Standardization meant a 'horizontal standard' across local organizations of different public sector domains (education, health care, public agencies, public administrations) with the intent to align different interpretations of the procedures. On the other side, customization efforts that in the previous period had been interpreted as synonymous with shallow feasibility studies and the absence of change management strategies were

*Table 4. Changes in the discourse of standardization and customization with respect to key client-consultant relationships (1998-2005)*

| | *Steering Committee members-LU Consultants (1998-2002)* | *Key Users-Private Sector Consultants (2002-2005)* |
|---|---|---|
| **Standard ERP** | "giving responsibility to the Operational services" | a system "coming from the private sector" |
| **Customized ERP** | carrying too many controls | "coherent with regulatory provisions" |

*Table 5. Changes in the discourses on standardization and customization with respect to key client-consultant relationships*

| | *Public Sector User Group-SAP Consultants* | *Different public service domains organizations-Consortium* |
|---|---|---|
| **Standard ERP** | "best practice" | aligning with common interpretations of organizational procedures |
| **Customized ERP** | - shallow feasibility studies<br>- absence of change management strategies | resisting isomorphic technological pressure |

now perceived by each single organization as efforts to preserve ownership over their business processes against the "technological isomorphic pressures" (Benders, Batenburg, & van der Blonk, 2006) deriving from the key relationship between Dante County and the Consortium. Dante County after 2008 has indeed turned itself from a buyer of ERP systems into a seller of consultancy about it (Rowland, 2010): having gained experience in ERP implementations, Dante County was actually participating in the Consortium consultancy activities addressed to the different local public services domains organizations more recently involved in ERP implementations (see Table 5).

## CONCLUSION

Having observed the evolution of the local government arena from the vantage point of Dante County, I provided an illustration of how many of the attributes that are taught as defining the 'consultant' as an entity playing a autonomous role in the market, in particular their ability to support strategic decisions on ERP implementations (i.e. customizations versus standardization choices), are instead the result of a relation. If we fully recognize that consultants "exist only in relation with their clients" and, as I added, they exist in relation with an extended network of actors and discourse that evolve over time, we have to be prompt to recognize that consultants' skills are not essential but enacted skills that can be observed only from a relational perspective.

In this chapter, by focussing in particular on consultancy about ERP customizations, I wanted to illustrate how the shaping of management processes and practices, organizations and technological markets cannot be reduced to singular actors' (i.e. consultants) accomplishments. Rather, these accomplishments are the outcomes of the evolution of a distributed socio-political process taking place in the situation. My proposal has been to replace grand abstractions such as "clients" and/or "consultants" and adopt a more situational perspective that focuses on key relationships between social worlds and their evolution over time. First of all, clients and consultants are collective actors that participate to different social worlds. Members of these collectives do not have homogeneous perspectives on the situation. This is evident from the analysis when I discuss the different outcomes of consultancy when it is performed in relation to different organizational actors (Steering Committee, key users, the Project Manager, Policy Makers). Secondly, as evident from all time-periods under analysis, actors (e.g. members of the clients and/or of the consultants' collectives) change their membership over time. The client organization (e.g. Dante County), taken as a singular entity, supported different perspectives on ERP implementations in different time periods, according to the actors and social worlds that succeeded to enter into strategic alliance not only with consultants and but also with other social world within the client organization and the overall Local Government arena.

What remained stabler was the discourse that the actor, acting as the representative of a social world, tried to legitimate and maintain by entering into strategic alliance with other social worlds' representatives within the client organization and beyond (Operational users, other Local Government organizations, other institutions from different public services domains in the local area). This is evident in the data from all time periods: in order to legitimize his perspective against other Heads of Departments and their staff, the IT manager formulated a discourse on ERP standardization that tried to establish an alliance with Operational Users.

The same approach has been adopted in the subsequent time period by Key Users. They formulated a discourse on ERP customization that contrasted previous discourses on ERP standardization and the perspective represented by Project Manager by establishing an alliance with Operational Users. What happened after 2005 was that new actors approached the arena of ERP implementation in Local Government organizations. In that situation, the project management and the IT management world perspectives found new alliances with similar perspectives in other Local Government organizations to contrast internal struggles between competing perspectives on system implementation. Finally, after 2008, the rising cost of SAP licences and new regulations about the 'in-house' model made Dante County policy makers find alternative arrangements. They found that in the given situation, rather than establishing alliances with other local government organizations, there was a need to establish a discourse on ERP implementation that involved other local organizations from different public service domains, and they arranged the consultancy in order to support this state of affairs.

With respect to management consultants role in particular, the statement put forward in this paper is that attributes such as consultants' 'knowledge,' 'skills,' and 'expertise' serve to the purpose of legitimizing particular discourses and social worlds perspectives within a situation that changes over time. In the first time period, LU consultants were hired to support the IT management world perspective with respect to other perspectives (e.g. the accounting management world perspective, the middle management world perspective) on ERP implementations. In the subsequent time period, Private Sector consultants were hired to support the Project Management world perspective in negotiations with Key Users. After 2005, SAP consultants specializing in Public Sector were hired to overcome these internal tensions by bringing 'best practices' from other Local Government organizations implementing the same system. Finally, the Consortium of consultants was established for the purpose of supporting the Policy Making world perspective on ERP implementations as an opportunity of alignment of procedure between local public service organizations of different domains. This does not mean that consultants have been always successful in legitimizing the social world perspective they were asked for. The main point here is that, rather than deriving from particular consultants' individual skills, consultant advice emerges at the interplay of these socio-political tensions as an adjustment of a distributed negotiation among social worlds perspectives. Consultancy takes itself part to this process as a resource for legitimization of key relationships. As such, consultancy does not emerge as a property that identifies a well-defined social world. Rather, it emerges as a property that can be enacted by different actors (or a combination of them) representing different social worlds. These actors do not have to be necessarily consultancy firms: they can be academics or even members of other similar organizations. According to this perspective, what counts for an actor to be enacted as a consultant is his/her social world membership (e.g. his/her provenance, biography, proximity with the supply side, previous client portfolio, knowledge of the local political context) with respect to the changing situation in the arena.

## REFERENCES


Anessi-Pessina, E., Nasi, G., & Steccolini, I. (2008). Accounting reforms: Determinants of local government's choices. *Financial Accountability & Management*, *24*(3), 321–342. doi:10.1111/j.1468-0408.2008.00455.x

Atkinson, P., Coffey, A., & Delamont, S. (2003). *Key themes in qualitative research: Continuities and change*. Walnut Creek, CA: Alta Mira Press/ Rowman and Littlefield.

Benders, J., Batenburg, R., & van der Blonk, H. (2006). Sticking to standards: Technical and other isomorphic pressures in deploying ERP-systems. *Information & Management*, *43*(2), 194–203. doi:10.1016/j.im.2005.06.002

Bessant, J., & Rush, H. (1995). Building bridges for innovation: The role of consultants in technology transfer. *Research Policy*, *24*, 97–114. doi:10.1016/0048-7333(93)00751-E

Brady, T., Tierney, M., & Williams, R. (1992). The commodification of industry-application software. *Industrial and Corporate Change*, *1*(3), 489–514. doi:10.1093/icc/1.3.489

Brehm, L., Heinzl, A., & Markus, M. L. (2001). Tailoring ERP systems: A spectrum of choices and their implications. In *Proceedings of the 34th Annual Hawaii International Conference of System Sciences (HICSS-34)*, (vol 8), (p. 8017). IEEE.

Clarke, A. E. (2005). *Situational analysis: Grounded theory after the postmodern turn*. Thousand Oaks, CA: SAGE.

Clausen, C., & Koch, C. (1999). The role of spaces and occasions in the transformation of information technologies: Lessons from the social shaping of IT systems for manufacturing in a Danish context. *Technology Analysis and Strategic Management*, *11*(3), 463–482. doi:10.1080/095373299107456

Czarniawska, B. (1998). *A narrative approach to organization studies*. London, UK: Sage.

Davenport, T. H. (1998). Putting the enterprise into the enterprise system. *Harvard Business Review*, *76*(4), 121–131.

Gubrium, J. F. (1988). *Analyzing field reality*. Thousand Oaks, CA: Sage.

Gubrium, J. F., & Holstein, J. A. (2003). *Postmodern interviewing*. Thousand Oaks, CA: Sage.

Haraway, D. (1988). Situated knowledges: The science question in feminism and the privilege of partial perspective. In *Her Simians, Cyborgs, and Women: The Reinvention of Nature* (pp. 183–202). New York, NY: Routledge. doi:10.2307/3178066

Hislop, D. (2002). The client role in consultancy relations during the appropriation of technological innovations. *Research Policy*, *31*(5), 657–671. doi:10.1016/S0048-7333(01)00135-4

Hood, C. (1995). The new public management in the 1980's: Variations on a theme. *Accounting, Organizations and Society*, *20*(2/3), 93–101. doi:10.1016/0361-3682(93)E0001-W

Howells, J. (2006). Intermediation and the role of intermediaries in innovation. *Research Policy*, *35*(5), 715–728. doi:10.1016/j.respol.2006.03.005

Hyysalo, S. (2006). The role of learning-by-using in the design of healthcare technologies: A case study. *The Information Society*, *22*(2), 89–100. doi:10.1080/01972240600567196

Jaeger, B., Slack, R., & Williams, R. (2000). Europe experiments with multimedia: An overview of social experiments and trials. *The Information Society*, *16*(4), 277–302. doi:10.1080/019722400457252

Knorr-Cetina, K. (1999). *Epistemic cultures: How the sciences make knowledge*. Cambridge, MA: Harvard University Press.

Kumar, K., Maheshwari, U., & Kumar, U. (2003). An investigation of critical management issues in ERP implementation: Empirical evidence from Canadian organizations. *Technovation, 23*, 793–807. doi:10.1016/S0166-4972(02)00015-9

Liang, H., Xue, Y., Boulton, W., & Byrd, T. (2004). Why western vendors don't dominate china's ERP market: Examining cases of failed ERP system implementation in China and explaining the unique circumstances. *Communications of the ACM, 47*(7), 69–72. doi:10.1145/1005817.1005824

Light, B., Holland, C., & Wills, K. (2001). ERP and best of breed: A comparative analysis. *Business Process Management Journal, 7*(3), 216–224. doi:10.1108/14637150110392683

Mabert, V., Soni, A., & Venkataramanan, M. (2001, May). Enterprise resource planning: Common myths versus evolving reality. *Business Horizons, n.d.*, 69–76. doi:10.1016/S0007-6813(01)80037-9

Mabert, V., Soni, A., & Venkataramanan, M. A. (2003). Enterprise resource planning: Managing the implementation process. *European Journal of Operational Research, 146*, 302–314. doi:10.1016/S0377-2217(02)00551-9

Marcus, G. E. (1995). Ethnography in/of the world system: The emergence of multi-sited ethnography. *Annual Review of Anthropology, 24*, 95–117. doi:10.1146/annurev.an.24.100195.000523

Markus, M. L., Steinfield, C. W., & Wigand, R. T. (2006). The evolution of vertical IS standards: Electronic interchange standards in the US home mortgage industry. *Management Information Systems Quarterly, 30*, 439–465.

Molina, A. (1989a). *The social basis of the microelectronics revolution*. Edimburgh, UK: Edinburgh University Press.

Molina, A. (1995). Sociotechnical constituencies as processes of alignment: The rise of a large-scale European information technology initiative. *Technology in Society, 17*(4), 385–412. doi:10.1016/0160-791X(95)00016-K

Newman, M., & Westrup, C. (2005). Making ERPs work: Accountants and the introduction of ERP systems. *European Journal of Information Systems, 14*(3), 258–272. doi:10.1057/palgrave.ejis.3000539

Pollock, N., & Cornford, J. (2004). ERP systems and the university as an 'unique' organization. *Information Technology & People, 17*(1), 31–52. doi:10.1108/09593840410522161

Pollock, N., & Williams, R. (2009). Constructing technology markets: How industry analysts organise technological fields. *Sprouts: Working Papers on Information Systems, 9*(10).

Pollock, N., Williams, R., D'Adderio, L., & Grimm, C. (2009). Post local forms of repair: The (extended) situation of virtualised technical support. *Information and Organization, 19*, 253–276. doi:10.1016/j.infoandorg.2009.08.001

Portelli, A. (2004). Oral history as genre. In Chamberlain, M., & Thomposn, P. (Eds.), *Narrative and Genre: Contexts and Types of Communication*. Transaction Publishers.

Riessman, C. K. (1993). *Narrative analysis*. London, UK: Sage.

Rosen, M. (1991). Coming to terms with the field: Understanding and doing organizational ethnography. *Journal of Management Studies, 28*(1), 1–24. doi:10.1111/j.1467-6486.1991.tb00268.x

Rowland, N. (2010). *Turning buyers into sellers: The spread and spreading of ERP in American higher education*. Paper presented at the Conference of the European Association for the Study of Science and Tecnhology. Trento, Italy.

Rusaw, A. C. (2007). Changing public organizations: Four approaches. *International Journal of Public Administration*, *30*(3), 347–361. doi:10.1080/01900690601117853

Scott, S. V., & Wagner, E. L. (2003). Networks, negotiations and new times: The implementation of enterprise resource planning into an academic administration. *Information and Organization*, *13*, 285–313. doi:10.1016/S1471-7727(03)00012-5

Soh, C., Kien, S. S., & Tay-Yap, J. (2000). Cultural fits and misfits: Is ERP a universal solution? *Communications of the ACM*, *43*(4), 47–51. doi:10.1145/332051.332070

Sprott, D. (2000). Enterprise resource planning: Componentizing the enterprise application packages tomorrow's customers will demand the ability to buy, reuse and build their competitive-edge solutions: Can the package vendors adapt in time? *Communications of the ACM*, *43*(4), 63–69. doi:10.1145/332051.332074

Stewart, J., & Hyysalo, S. (2008). Intermediaries, users and social learning in technological innovation. *International Journal of Innovation Management*, *12*(3), 295–325. doi:10.1142/S1363919608002035

Strauss, A. L. (1978a). A social worlds perspective. *Studies in Symbolic Interaction*, *1*, 119–128.

Strauss, A. L. (1978b). *Negotiations: Varieties, contexts, processes and social order*. San Francisco, CA: Jossey Bass.

Strauss, A. L. (1991). *Social worlds and spatial processes: An analytic perspective*. Working Paper. Berkeley, CA: University of California.

Strauss, A. L. (1993). *Continual permutation of action*. New York, NY: Aldine de Gruyter.

Strauss, A. L., & Corbin, J. (1998). *Basics of qualitative research: Techniques and procedures for developing grounded theory*. Thousand Oaks, CA: Sage.

Sturdy, A. (1997). The consultancy process: An insecure business? *Journal of Management Studies*, *34*(3), 389–413. doi:10.1111/1467-6486.00056

Van Maanen, J. (1998). *Tales of the field: On writing ethnography*. Chicago, IL: University of Chicago Press.

Wagner, E., & Newell, S. (2004). Best for whom? The tension between 'best practice' ERP packages and diverse epistemic cultures in a university context. *The Journal of Strategic Information Systems*, *13*(4), 305–328. doi:10.1016/j.jsis.2004.11.002

Wagner, E. L., Scott, S. V., & Galliers, R. D. (2006). The creation of 'best practice' software: Myth, reality and ethics. *Information and Organization*, *16*, 251–275. doi:10.1016/j.infoandorg.2006.04.001

Webster, J., & Williams, R. (1993). Mismatch and tension: Standard packages and non-standard users. In Quintas, P. (Ed.), *Social Dimensions of Systems Engineering*. Hemel Hempstead, UK: Ellis Horwood.

Wei, H., Li, C., Wang, E., Li, C., & Ju, P. (2005). Understanding misalignment and cascading change of ERP implementation: A stage view of process analysis. *European Journal of Information Systems*, *14*(4), 324–334. doi:10.1057/palgrave.ejis.3000547

Williams, R., Slack, R., & Stewart, J. (2005). *Social learning in technological innovation — Experimenting with information and communication technologies*. Cheltenham, UK: Edgar Elgar Publishing.

## ENDNOTES

1 Organizational Units in the Dante Province are divided between Staff Services and

Operational Services. Staff Services are Organizational Units that provide services to other internal units. Operational Services are organizational units that provide services to other parties (citizens, firms, etc...). The Line Services are only responsible for feeding in Long-term plans, whilst Staff Services such as Planning and Accounting merely read through them to check.

2 ("Borrowing Transactions" and "Electronic Payment")

3 The only way for the PM to control key users' decisions was via a ex-post indirect checks.

4 Published White Book on Regional IT Strategy.

# Chapter 12
# Accumulation and Erosion of User Representations or How is Situated Design Interaction Situated[1]

**Sampsa Hyysalo**
*Aalto University School of Business, Finland*

## ABSTRACT

*The design process is predominantly presented as the primordial site where creativity of highly talented people flows into material form and results in novel solutions for human concerns. Design-critical views equally place event of design on the pedestal even as they question whether designers' intuition, creativity, and aesthetic sense is enough and, particularly, whether considerations of the implications of the new technology have sufficiently permeated the process and, through that, determined its outcomes. This latter view has often been condensed into whether designers are guided by appropriate values and/or understandings in their work. There is, however, relatively little detailed research on what actually does take place in the complex and uncertain considerations wherein "design decisions" are made or whatever "design values" are followed. This chapter will contribute to this through the analysis of in-depth video-recorded ethnography of the design process of a monitoring and alarm device for the elderly. At the same time, it aims to raise questions about the centrality of the design process by examining what comprises the situated action taking place in design and examining how design interaction is situated within design-use relations.*

## INTRODUCTION

The design process is predominantly presented as the primordial site where creativity of highly talented people flows into material form and results in novel solutions for human concerns (e.g., Cross, 2000; Schon, 1983; Ulrich & Eppinger, 1995). Design-critical views equally place event of design on the pedestal even as they question whether designers' intuition, creativity, and aesthetic sense is enough and, particularly, whether considerations of the implications of the new technology have sufficiently permeated the process and, through

DOI: 10.4018/978-1-4666-0303-5.ch012

that, determined its outcomes (e.g., Papanek, 1972; Schuler & Namioka, 1993). This latter view has often been condensed into whether designers are guided by appropriate values and/or understandings in their work (Noble, 1984; Woolgar, 1991; Stewart & Williams, 2005). There is, however, relatively little detailed research on what actually does take place in the complex and uncertain considerations wherein "design decisions" are made or whatever "design values" are followed. This is what we aim to contribute in this chapter through the analysis in-depth video-recorded ethnography of the design process of a monitoring and alarm device for the elderly. At the same time, we hope to open up the question about the centrality of design process by examining what comprises the situated action taking place in design, in other words, examining how design interaction is situated within design-use relations that are temporally more long-term and spatially more distributed.

To facilitate this enquiry, we use the concept of "user representations" (Vedel, 1987; Akrich, 1995) to focus our attention on how usage is handled during design and where the ideas about future usage emerge. "User representation" and its various kin concepts link the multiple modalities a design of usages takes on prior to actual use: visions, claims, assumptions, ideas, pictures of user-practices, sketches, prototypes, the artifact wrapped for sales, and the technology entering hands of users (Hyysalo, 2004). Far from being solely an up-front 'user needs and requirements capture' process, creation of user representations can continue throughout multiple generations of product development. Recent research reveals how representations of use, users and usages tend to stem from multiple sources, and that there is considerable variation which one of these provide most actionable and adequate understandings of use in each particular case (Akrich, 1995; Oudshoorn, et al., 2004; Williams, et al., 2005; Hyysalo, 2010). But in some contrast to well known dichotomy between explicit and implicit representations (Akrich, 1995; Oudshoorn, Rommes & Stienstra, 2004), identified sources of user-representation yield at least eight major source areas with sub-categories (see Figure 1), and our research suggest that much of this range of significant inputs for designing usage is not a matter of a range of different projects but a matter of variety even within single projects (Hyysalo, 2010).

The user-representations circling around an innovation project are many. Some of these are complementary and can have mutually reinforcing relations. Others have more conflicting relation and yet other representations of use concern such different areas in the product that they tend to remain unconnected: representations related to the visual image and marketing; those resulting from engineering details of, say, reliability considerations; and representations about adequate usability tend to fall into domains of different professionals, require different expertise, and also take shape in different ways (Norman, 1999; Kotro, 2005). The fascinating thing about design is that all these considerations have to be accommodated in the limited physical and cognitive space of the product and hence their mutual relations become more urgent. Naturally, design and engineering do not stand helpless in the face of such difficult interdependencies; separation of concerns through compartmentalization, modularization, *et cetera* is one of the principal activities involved. But how this factually takes place, and how exactly are user representations raised to the fore in situated design action and interactions connected to other situations in time and space is a question that has been more assumed than established as yet.

## BIOGRAPHIES OF TECHNOLOGIES AND PRACTICES APPROACH

There are some rather pressing methodological issues involved, however, in attempts to accomplish the above research interests. Many relatively recent studies stress methodological issues that

*Figure 1. Major categories and illustrative subclasses for sources of representations of use in technology design*

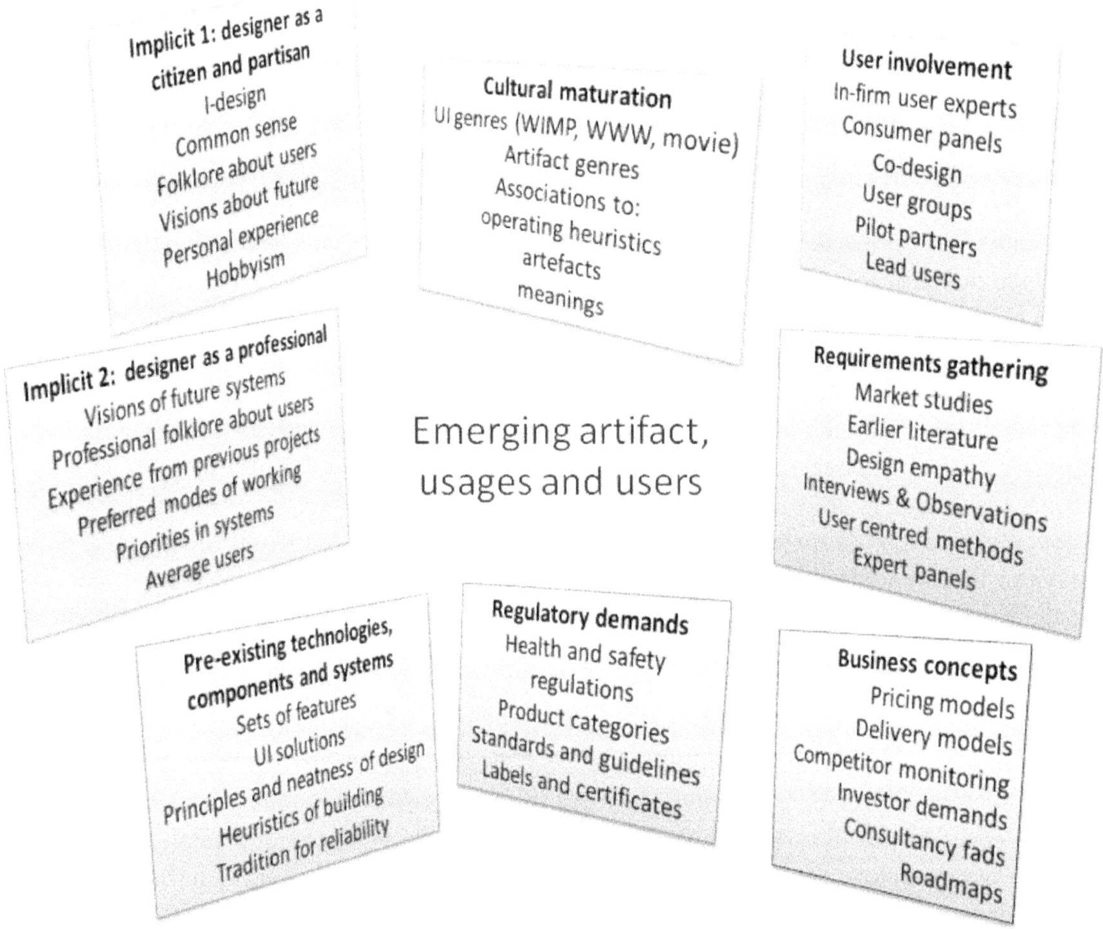


could be characterized followingly. First, the research must not fall victim to researching only a given "snap-shot" of the process be that the design process or domestication of the technology (Stewart & Williams, 2005). An adequate understanding of the shaping of a technology—or a given design situation for that matter—requires that the research is able to achieve sufficient *reach and balance* in covering the social practices that are consequential for the shaping of technology so as not to essentialize for instance design qua it begin design and use qua it assumedly being just "using" or" domestication." Second, the study should not remain confined to a single *scale* of analysis, be that a "frog's-eye view" of practitioners enmeshed in developing and using new technology or a "bird's-eye view" of policymakers and high-level analysts. Third, many studies also stress more careful attention to what is the actual *shape of technology* in different foci and loci of analysis. For instance, Zimbabwean water pumps, enterprise resource planning systems, and many health ICTs have more flexible compositions from any one particular instance and time to another than, for instance, pens, bicycles, or electricity grids (e.g., Clarke, 2005; de Laet & Mol, 2000; Mol, 2002; Sorensen & Williams, 2002; Williams, et al., 2005; Bowker & Star, 1999).

*Table 1. Methodological criteria for socio-technical change in the early 1990s (left column) and guideposts for living up to them taken on in the present volume (right column)*

| | |
|---|---|
| **Change/ continuity:** the framework should account for change as well as for continuity and stability.<br>**Symmetry:** the useful functioning of a machine should be taken as a result of socio-technical development, not its cause.<br>**Actor / Structure:** the framework should account for actor-oriented and contingent aspects of socio-technical change as well as the structurally constrained ones.<br>**Seamless web:** no a priori distinctions of what is social, technical, scientific or political as these shift within the seamless web of socio-technology. | **Reach:** the framework should not rely on or favour 'snap-shot' studies but encompass multiple times and loci where technology is shaped.<br>**Scale:** bridge birds eye view with frogs eye view.<br>**Balance**: the framework should attend to the empirical asymmetries between resources, practices, interests and make-up of different actors, yet not lead to overemphasis on technology and its development or what people do with technology.<br>**Shape of technology:** the study should not fix the ontology of technology to discreet "artefacts" or "objects" |

Taken together, these four guideposts appear not to stand in opposition to earlier programmatic articulations about how sociotechnical change should be studied—e.g., Bijker's early 1990s summary (Bijker, 1995, p. 13; see Table 1 left hand side)—but they do present a challenge in how to live up to the earlier principles in a conceptually and methodologically sound way.

Our attempts to live up to these guideposts can be called "Biographies of Technologies and (related) Practices" study (BOTP). Its core ideas can be expressed as follows:

a. Longitudinal analysis of the biography of technology from its inception to at least the early stabilization of uses.
b. Conducting this analysis and related data-gathering with multiple scales of inquiry and granularity of data: development of practices; biography of the technology and related activities; conduct of design and appropriation episodes at monthly, weekly, and minute scales. These are further linked to more obdurate patterns in socio-techno-economic change.
c. The focal points for more close-up inquiries are chosen so that they are likely to be informative with regard to broader scales of change in design–use relationship as indicated by previous studies and/or likely to reveal patterns that are of special interest for the study.
d. Combining a range of complementary data sources and methods and both historiographic and ethnographic inquiries to accomplish the above.

The research designs in this approach can be schematically portrayed by looking at the study of safety-alarm technology for elderly people that comprises the second part of this book (see Figure 2). The lines in the diagram signify research efforts targeted at different moments, sites, and scales of change in the relevant socio-technical biographies.

This diagram further reveals how this type of study operates in several timeframes of analysis within one study. These roughly correspond to three distinct timeframes on how technology projects and technological change have been (exclusively) studied in the previous research.

1. **The span of particular design and appropriation episodes.** Development of a technological innovation typically involves a series of shorter and longer design phases ranging from hours to weeks to months (Van de Ven, et al., 1999; Schon, 1983). Most studies on design interaction, design decision-making, and ethnomethodological studies of technology operate exclusively at this level.
2. **The development of a particular innovation and the organizations and people connected to it.** An innovation typically forms a biography or trajectory that lasts from five

*Figure 2. Stylized research design for a study on biography of technology on safety alarm systems for the elderly. Multiple scales of analysis were deployed on both development and use. Arrows represent research activities, circles represent shorter episodes, which the informants or the researchers regarded as particularly significant. Different bodies and granularities of data and time frames of analysis were systematically compared. (What this figure misses is the inferences and interaction that spanned both development and use as well as change at each scale).*

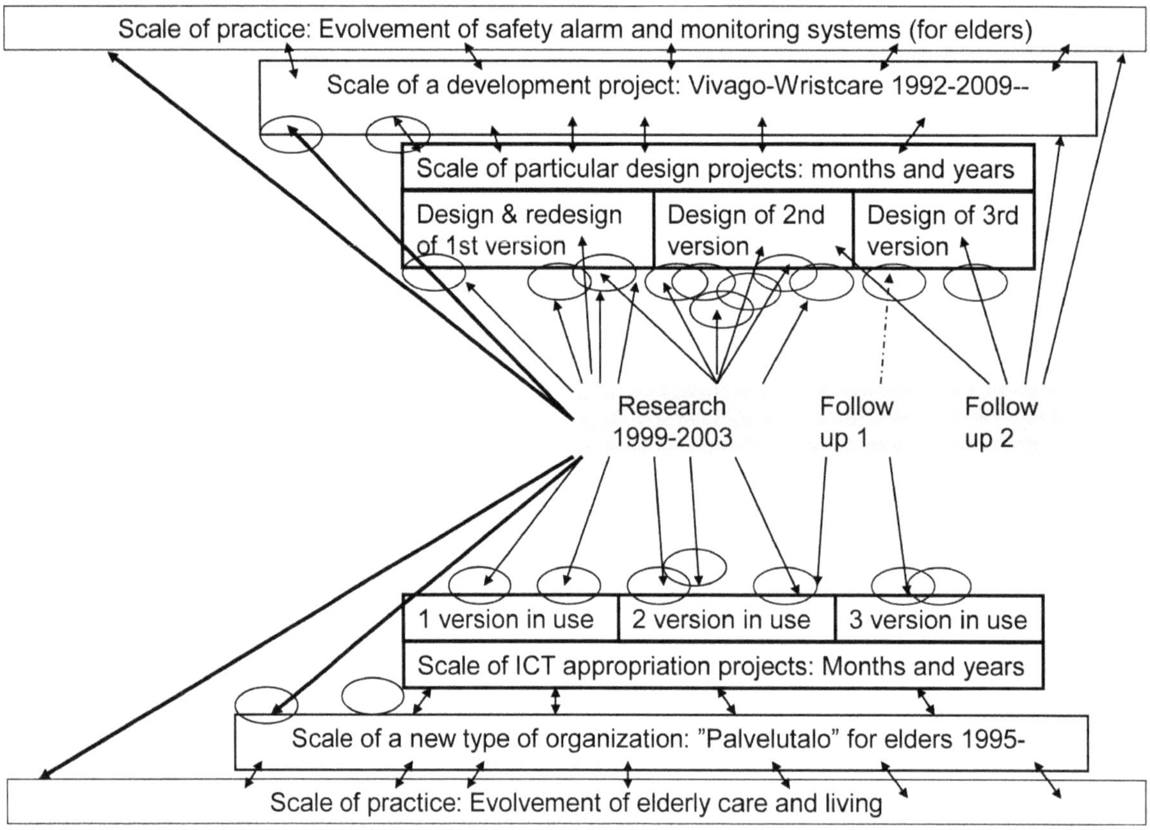


to 20 years, which has its own transitions and crises (Van de Ven, et al., 1999). These events are connected to the trajectories of its producer company (or companies) and those of its key user partners, and the professional careers and life of the individuals participating in them (Miettinen, 1998; Miettinen, et al., 2003; Hasu, 2001a).

3. **The coupling of "class of technology" and related socio-cultural practices.** The practices of both the producers and users of technology change with the passing years and decades rather than in the span of any particular event or even that of particular technology. These practices, such as elderly care or the telecommunications business, have formed relatively stable institutions and organizations of work, including a strong sense of the appropriate logic and path of development. Any new technology faces these preexisting arrangements in the organizations it is designed in (Grudin, 1993), as well as in the organizations within which it is used along with other artifacts and tasks

(cf. Clarke & Star, 2003, p. 12). Yet transitions do take place both in practices and in the way two or more practices are coupled.

4. **Long-term patterns in how design and use are organized.** Some aspects of design and use have prevailed in the industrial production of consumer goods ever since the rise of mass production (Hyysalo, 2009a). In so doing, they have been shaped, but also endured through "long-waves," significant overall reorganizations of production caused by the emergence of the upsurges of new generic technologies: electrification, motorization, and computerization (Freeman & Louçã, 2001, esp. 140-151). Attempts to characterize somewhat more rapidly changing, but still very "obdurate," socio-technical formations include industry sectors, inter-organizational fields, and socio-technical regimes (Nelson & Winter, 1982; Hoogma, et al., 2002).

The key issue here is not to regard the different time spans as somehow ontologically distinct even though studying them means employing an array of different materials and various foci of closer inquiry. An event is not pinned down to a place in the pre-ordered scheme of things but seen as simultaneously constituting and being constituted by broader patterns. Various events can be examined as evidence of these patterns, while the patterns can be constructed from a range of evidence beyond the foci and granularity of a single site under analysis.

## RESEARCHING USER REPRESENTATIONS IN DESIGN PROCESSES

Let us now focus on how usage is shaped in a design process and, in particular, how, in situated action, user-representations enter into design, how they are worked on there, and how they depart from the process of design. To study the entry and transformation of user-representations in design, we examine four interrelated "micro-biographies" of central user-representations in the design of second-generation alarm and monitoring device for elderly and disable people called Vivago, which we have analyzed with the BOTP approach as depicted in the Figure 2. These micro-biographies are inventing a new push-button solution, incorporating an LCD screen, creating auto-adjusting width of a bracelet, and that of an attempt to make the surfaces of the device anti-soiling.

The ethnography we report here spans the design of Vivago's mechanical design from September 1999 to June 2000. The process was conducted by outside industrial designers and company engineers in various specialties and a representative of the contract manufacturer. We also attended the general R&D meetings of the producer company at the time and conducted a series of interviews of the participants in the mechanical design process. In addition to field notes, over 60 hours of video recordings and close to 20 semi-structured interviews form the primary body of data for this chapter. Together with historical materials, and copies of the notes and sketches the participants made during the process, this provides a rich view of the process, even though only small fractions of it are reported in detail below.

To accommodate analysis that hangs between interaction analysis and an ethnographic account, a simplified transcription has been opted for to ease the following of the interaction-analysis-wise lengthy sequences of up to five co-present interactants that span stretches of tens of minutes in design meetings within a series of meetings spanning months. There was no space to preserve both the Finnish original and English translation in this volume, and speech in the excerpts below has been translated from Finnish by the author. Translating Finnish utterings directly into English would make technical discussion extremely hard to follow, not least because Finnish is significantly further from English in terms of vocabulary, grammar, and idioms than any Indo-European language. Consequently, words have been translated into their literal English form, and whenever differing idioms are used, meaning has been preserved here rather than verbalization. A simplified "old" conversation analysis notation (Schegloff, 2007) was then used, in which:

() marks inaudible speech or a long pause, without specifying length

// marks the beginning of overlapping speech and * the end of overlap, with the overlapping speech delivered in the line below

= symbols used in pairs, as in ==, connect talk that was broken into two lines to accommodate marking of overlapped speech

((*text in italics*)) is used to mark action and objects when speech and action fully intertwine

((text)) is used to clarify what is denoted in denotative expressions (including cases where the denotation is clear in Finnish but hard to translate adequately as well as anonymizations of the participants referred to in speech by name)

Some minor interruptions, orientation changes and gestures, and most aspects of speech delivery have not been preserved. It should hence be kept in mind that these simplifications create an overly orderly and skeletal depiction of recorded 2 – 5-person design interaction and grossly overplay verbal exchange and its orderliness. Judged from the comments to the earlier drafts of this chapter, these simplifications were, however, necessary to convey the kind of relatively lengthy sequences of interactions we shall next move into.

The micro-biographies we shall discuss below all entered design because of usage considerations. Yet their biographies invariably came to feature two other types of user-representations, ones developers recognized as affecting usage as a byproduct and ones seen to affect only something very different, even as they factually did affect the design of usage. In addition, we shall note that when these features were designed into the limited physical and cognitive space of the artifact-in-the-making, they became further linked to issues that were effectively, often purposefully, separated off from affecting usage. Hence, four types of features in relation to designing of usage become salient:

1. Ones explicitly brought to design because of usage (e.g., push-button functionality)
2. Ones recognized to have bearing on usage or users but being primarily about something else (e.g., the hold of screws in the mold)
3. Ones affecting usage but never discussed or considered from this vantage point during the design process (e.g., the layout of the circuit board inside the wrist unit)
4. Features and solutions cut off from having much (if any) bearing on use (e.g., the molding channels built into the device to allow manufacturing)

This categorization emerging from Vivago project where no particular ambition to further research on user participation was present interestingly parallels (Buscher, et al., 2009) way of stressing the type 3 and 4 issues as deeply consequential for how designing usage and choices available for envisioning uses can take place.

Furthermore, the range of issues that have bearing on use grows impressive even in the case of a small device such as Vivago. Table 2 sums up the main features that were considered as affecting usage in the design process of the second-generation Vivago. This complexity effectively waters down any attempt to describe "in full detail" how "The Usage" of even a small device like Vivago became designed, particularly if one wishes to draw connections to their sites of origin. Hence our focus on just four "mini-biographies" of features in the design process.

*Table 2. Map of features that were recognized to affect usage in the second-generation design process of Vivago-Wristcare (a "#" marks issues brought into design consideration because of usage, a "*" issues developers recognized as affecting usage or users)*

| Main feature | Main attributes; minor attributes | Interlinkages to other key features concerning use |
|---|---|---|
| Push Button # | Securing intended alarm soundings, Prevention of erroneous alarms; Tactile findability, Feel, Feedback, Look, Height, Placing, Hole and button sizes. | Cleanability, Look, Area of interface taken, Manufacturablity, Materials, Watertightness, Battery consumption, LCD-Screen etc. |
| Anti-soiling # | Dirt resistance, Hygiene, Conductivity problems, Look; seamless surfaces, no dirt gathering elements, ventilation in the arm, cleanability | Conductivity measurement, Bracelet strech, Screw placement, materials, manufacturability, color, shape, battery replacability, robustness, etc. |
| Bracelet strech # | Securing measurements, Comfort, Reducing skin irritations; Stretching when arm swells | Conductivity measurement, Anti-soiling, Robustness, Materials, Look, Feel |
| Water-tightness # | Reliability, ease of wearing (showering, dish washing, etc.), standards and requirements | Battery life-time and replacement, Robustness, Screw placement, Maintenance |
| Screws | Robustness, Price, Reliability; Screw shape, placing, number, thickness, material, retightening | Look, Feel, Anti-soiling, Conductivity contact, Price of manufacturing, Battery replacement, Robustness, Who does maintenance etc. |
| Batteries # | Life-time, Replacing, What all the device can send and display; Size, Placing, Structure, etc. | Water tightness, robustness, Maintenance, Screen, Measurements, Antenna, Sending, Feedback, etc. |
| Antenna * | Reach, Energy consumption, Usage areas; | Forwards compatibility, Battery Lifetime, Price, Battery replacement, Size, Shape, etc. |
| Cover design # | Variation of looks, Manufacturability | Water Tightness, Anti-soiling, Robustness, forwards compatibility, etc. |
| Feedback # | Feedback on pressed and automatic alarms and battery life | Tactile feel, sound, screen/LED Battery life, Price |
| Robustness * | Robustness, Sense of robustness for the user; | Materials, All structural choices, Price, etc., etc., etc. |
| Price # | Buyer acceptance, Target markets, revenue | Manufacturing, Materials, Features, etc., etc., etc. |
| Forward compatibility* | Screen, Size, Measurements | |
| Sensor plate* | Measuring, secrecy, size | Robustness, battery life, screws, size, look, etc., etc., etc. |
| Screen/LED # | Feedback, Information for user, looks, image, interactivity, new functions | Manufacturability, push-button, size, looks, electry consumption, compatibility with previous and coming electronics |
| Colors # | Plesantness, Soiling, Signalling the need of cleaning, acceptability | Push button, soiling, |
| Looks/shape # | Non-aid like, contemporary image | Price, monitoring plate, size, manufacturability, robustness, water tightness |
| Size # | Image, weight, fit to wrist, price | Push button, monitoring plate |
| Bracelent—device angle # | Fit to wrist, securing electroconductivity sensor | Sensor plate, stretch of the band, |
| In addition * issues include such as manufacturability, regulation compliance, hygiene, durability of materials… | | |

*Figure 3. Design setting: a clustering of artifacts, drawings, papers, and pens that are used to communicate and facilitate imagination. Note that even as the participants constantly talk with each other their bodily orientation and gaze rests mostly on drawings, material samples and gestures, and (most importantly) sketching by others.*

## THE SOURCES AND ACCUMULATION OF USER REPRESENTATIONS: AN ERROR-FREE ALARM BUTTON

The official kickoff of the series of design meetings took place on a sunny, crisp October morning when the head of Vivago Ltd. entered the studio of the industrial design consultancy hired to do the mechanical design. And, as is typical of design, much, if not most, of the communication took place through drawing, pointing, and gesturing accompanied by verbal descriptions (cf. Jordan & Henderson, 1994). Figure 3 gives a sense of the setting: a group of people hunched over a table filled with materials, sketches, and notebooks, as well as previous and competitors' products.

Even though much of the action takes place outside talk, verbal descriptions are important carriers of information from the accumulated knowledge about use and related future desires—and, indeed, the entering and elaboration of user-representations is where we focus with the micro-biography we trace here: that of the error-free alarm button. Its elaboration begins right at the start of the design meeting series:

*Designer 1: OK, I contacted ((the manufacturer)) yesterday and now, ((Inventor)), we have a structural idea, which we could // take as a starting point* =*

*Inventor: Yeah-h*

*Designer 1: = that is, if you do not see any reason why it would not work. So the bracelet would act as unified surface ((pointing to his drawing)) here at the top*

*Designer 1: And it is entirely possible to do it that way and the furnishing, the components, would be dropped in through the bottom*

*Inventor: Yeah*

*Designer 1: And the principle is a bit like this here ((picking up the first-generation Vivago-Wristcare—Figure 4, right-hand side—turning it around)) apart from it all being entirely unitary. No sticker, no casing, all one. Structurally this would require ((draws)) that inside this elastomer, which is this one ((draws, points)), there would be a hard plastic frame placed there in the mold. And this frame-like part could continue as cover too, only leaving open the place for the ((alarm)) button, which would then stretch. And currently it apparently has this mechanical button, which is planned ((for remaining)) there?*

*Inventor: Yes.*

*Designer 1: And this would stretch down and there would be a clear click to be felt ((upon pressing it)). And think of the advantages for keeping it clean: all one surface.*

From here, the discussion moves to hygiene, and we return to this moment as we near the end of this chapter, but we now continue to follow the push-button elaboration, which is picked up at 28 minutes, where a set of user-representations about the alarm-button is brought into consideration by the head of the company:

*Inventor: () as real-life feedback: We have gone through different push-buttons, competitors' buttons, and thought about that button. The button is going to be critical. We just had to deliver to Siilinjärvi hospital devices that do not have ((pushes the button of his own wrist device unit)) this click. Some of the people are in such poor shape that they do not have the strength to push ((even)) the 100-gram button—this one here ((on my wrist)) is a 250-gram button. Furthermore, when they seek the button, they cannot see. They search for it only by hand, somewhere there is some hole, and press that ()*

*Inventor: so let us think about it () hang on—how should I draw it? () ((draws)) that if the button is here, then in a way () I draw it ((keeps on drawing)) like this—just to give an idea—that this here is elastic ((points to top right of his figure—akin to Figure 4's top left drawing, where the arrow is later drawn to indicate "elasticity")) and then it continues somewhere from there. If this here ((keeps on drawing)) is a nub and then this an even surface, then this one would need to be—this sheds a little new light on the requirements—such that if I press here it cannot tilt sideways ((and thus not make contact))*

*Designer 1: Sure, what about if it were made into a level, then if it would form a cup, the ideas we have had // but that is really that if you press it from the side it does not. Cup form**

*Inventor: It would need to be hard =*

*Inventor: = it would need to have a larger hard surface.*

*Designer 1: Well, we can arrange that. But cup-form might have that advantage that the finger is better guided onto the bottom. In that kind of*

*Figure 4. Early drawings and first-generation Vivago-Wristcare as an artifact used to envision the push-button functioning and shape*

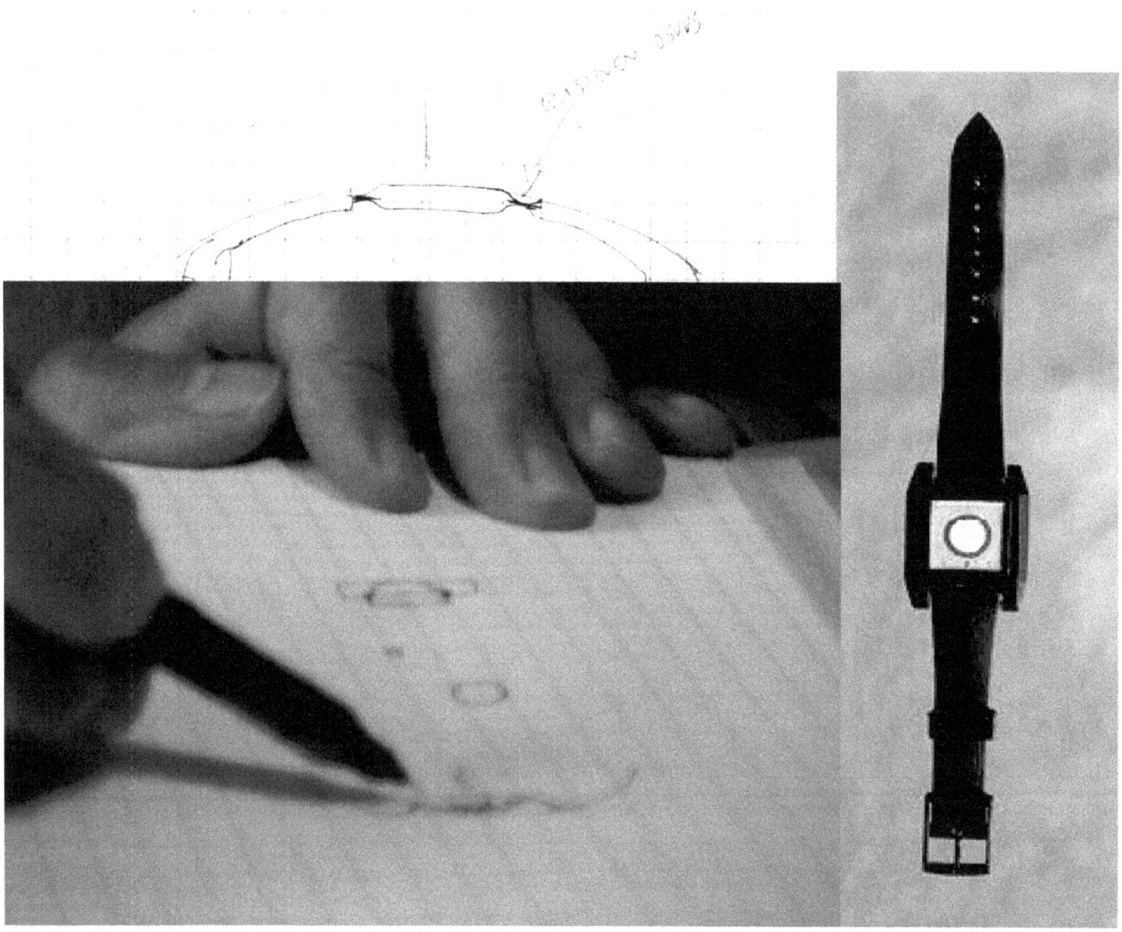

*even surface, just outlined, it's more laborious to find the center area, but it*

*Inventor: Yeah the form there =*

*Inventor: = we had Telelarm's button for comparison () no, sorry, Antenna's. Its shape is like this ((draws, a level surface and a cup)): it has the button as if finger-shaped. And they, as a matter of fact, have only a glued-in soft plastic and the pressing-hole around about there. It contains nothing else, just the soft plastic. When you press it, it too has the problem that if you press from the side, nothing happens, even as it suggests that you are in fact pushing that button. And we were just considering that the button () if you look at the prediction, not today, but prediction that people in increasingly poor condition will remain in home care and if you think about our devices // it was a total surprise for me that there are people in such a poor state in Siilinjärvi**

Let us break out of the dialogue between the developer and designer in order to elaborate on some of the interesting issues in this discussion. What we find here are the four most common ways user-representations were brought forth in this series of design meetings: comparisons to existing products and their solutions, representations that are part of company visions about the future customer (who the target customer is and how the clientele is likely to change), implications about structural solutions to be adopted in the redesign of the device, and as anecdotes about users' identity, behaviors, and needs stemming from field-use experience. Let us return to the discussion where we left it, as we next see how these four ways intertwine in the discussion.

*Inventor: And if you think our devices // it was a total surprise for me that there are people in such a poor state in Siilinjärvi**

*Designer 1: This one ((points to the first-generation device—Figure 4, right-hand side)) only has integrated plastic there. Well, that is presently () a problem in nearly all ()*

*Inventor: =hence, I was left thinking on that surprise. Our ((developers)) visited Siilinjärvi and said that the patient material was, to begin with, in the terminal stage. You know those are not even our targeted customers, but, then again, it would be sensible for us to consider the needs of these more problematic users, in ((the design of)) this button.*

*Designer 1: Yes.*

*Inventor: For that reason () it became very clear to us through that Siilinjärvi case—when doing that comparative research—that we are no worse than competitors, but no better either.*

*Designer 1: In regard to the button?*

*Inventor: Yes, in terms of just the manual button.*

*Designer 1: Sure, sure.*

*Inventor: Really, if you want to put it this way, in Siilinjärvi we noted that for people in terminal stage our automatic monitoring is no longer a working solution—alarms are coming through all the time, as they are already at such a point in their life //that* =*

*Designer 1: Yes.*

*Inventor: = It's like () even the nurses are oriented a bit like, that they just wait for when they ((the patients)) die. They no longer even plan for acute care measures. But when that person wants help, then help, of course, has to come forth.*

*Designer 1: Of course, yes*

*Inventor: and then it ((the calling for help)) has to be as easy as possible. So, in that respect I, we, back at work, pondered that if there had been the separate hard cover, it might have been beneficial to build different tactile feels for different user groups.*

Let us again interrupt the dialogue, for this point is a "natural place" to do so, because of another common facet of this design discourse; now that the concerns and requirements have been elaborated on, the discussion shifts to related requirements, since any given feature tends to be rather complexly intertwined with solutions given for other features, as illustrated in Table 2. In this case, the discussion moves first into the different releases and versions of Vivago-Wristcare, then, from there, in turn, to other topics related to these features. The ideas of how to realize the push-button are returned to an hour later, at the end of the first meeting, when the structure of the device cover is again discussed:

*Inventor: () and then it would just have a thin film, protective film on top of the ((mechanical button)), then it would be of the kind you showed, that you can press anywhere and you always score.*

This then is taken up by the designers in the second meeting two weeks later, 30 minutes from the start, initiating a more focused sequence of considerations on how to realize the push-button.

*Designer 1: OK, the organization of the push-button. It is now drawn ((points to drawing)) so as to have a small mechanical button underneath. And it would have the indentation, which would move under the finger so that it presses the button ((underneath)). The problem, then, is how to make it move so that it makes the connection in all cases. So () should the whole cup be elastic, or just an elastic rim, or what? This issue is really wide open as yet.*

*Inventor: One more thing, just a thought () is that what they use in cell phones is a sort of nudging plastic.*

*Manufacturer: Silicone button or ()*

*Inventor: Sure, like a silicone button, but not made of silica // with a nudge. So, if here is the switch element ((draws)), which can ((even)) be of the conductive plastic so it does this "dok, dok, dok" ((mimics the sound with his tongue)).* =*

*Designer 2: The whole thing can be a silicone layer, or ... there are many alternatives, it really depends on the material, if it is*

*Inventor: =for it doesn't need to be silicone, I've seen, there is one Swedish manufacturer who has done ()*

*Designer 2: Here is an example ((hands over a piece of material)).*

*Inventor: Sure. But the Swedes have done the cover not out of silicon but out of plain polycarbonate.*

The discussion now moves to details of materials and button elements for two minutes until the push-button shape is returned to.

*Inventor: Further, some criticism we got from our bunch was that the hole should be bigger here ((points to the drawing of competitor Antenna's device)) so if we are dealing with () does anyone of us have big ((hands))? ((developer 1)) at our place is a big () he has even bigger fingers than you ((designer 1))*

*Designer 1: My fingers are actually quite small*

*Inventor: So somebody just remarked that if he is to push this one, it can be that ((the finger does not fit the hole))*

*Designer 2: So it has got // to be**

*Manufacturer: Got to be bigger*

*Inventor: Yes*

*Designer 2: So there is a little bit of a dilemma there that it has to be big and easy, but then again it should not press down by itself*

At this point, there is again a shift to detail of materials for roughly two minutes which we fast-forward as it gets reiterated in the exchange that follows.

*Designer 1: How is this part ((points)) here ((manufacturer))? It has this hard plastic cover and then the elastic part. Can it be made so that there is an elastic ring somehow and the intermediate part made of hard plastic?*

*Manufacturer: Yes, it can be done*

*Designer 2: That ring has to be wide enough then, so it allows enough distance to be pressed*

*Designer 1: Mhmm*

*Manufacturer: ((looking over the table and top-down on the drawing)) I just cannot get my head around the shape of that cup that way around ((snickers))*

*Inventor: So, in principle, it would be like this, in principle, if we think that it would be here*

*Designer 1: It can be that way of course, so*

*Manufacturer: I was thinking that last time when we spoke ((draws—Figure 5, bottom left, figure in the lower part of the picture)) that if this here would be hard, then we would ((continues drawing the figure)) put an elastic part here, but the mechanics itself would be underneath. Steady enough so it produces the movement, that mechanical switch. This ((elastic part)) just gives when you press*

*Designer 1: Would this middle part ((points—same drawing)) then be elastic?*

*Manufacturer: Mmhm*

*Inventor: It can be elastic if we opt for the other switch, the bigger switch. So, if we think, this one, it has roughly a square millimeter area in the switch, but the ones we've looked into are roughly ((draws)) this size ((more than twice the area)). So then this ((elastic button)) would move in a directed channel, so that one can push wherever in the button, not just in the middle*

*Designer 2: Exactly*

*Inventor: Hence, in a way, if we look at it sideways, and—let's draw this bigger ((draws))—it has this guidance in the button, so it goes ((always down)). I don't know how the spring-structure underneath, which way it's been done, but at least one of those ((switches)) we looked at, which was a tad big for this purpose here, it was really good for one's finger () We compared these two and this one then felt just plain odd*

*Designer 2: Then it does not require ((pushing)) in the middle but you can, in a way, press its side*

*Inventor: and then this ((middle part)) could be soft. For then, what you planned as hard, it could be replaced by soft ()*

*Manufacturer: and the switch would give the mechanical movement, click, return, everything. And this // would just be* =*

*Inventor: a cap on top*

*Manufacturer: = an elastic cap on top where you can press, and thin enough that you feel that switch. Then it's really () just two places that are pressed together.*

*Designer 1: Yes, yes, so it's all one // like**

*Manufacturer: All one except from two materials*

Upon revisiting the drafts drawn by Designer 2 at the end of this meeting, the inventor exclaims, "So, in this structure we produced two solutions with one!" What happens in this episode is an inventive move from a simple button area on a level surface to one that is indented but raised to near the surface (and is therefore hard to press accidentally on surfaces, etc.), which rests on a switch element underneath that provides both "clicking" feedback upon pressing and support for structural guidance so that pressing anywhere within the finger-hole will trigger an alarm.

The episode also reveals yet another set of user-representations; using designers' own physique as compared to that of elderly users (Akrich, 1995; Kotro, 2005, 2004), to be encountered several times later with other features. Finally, the episodes also reveal how initial ideas become elaborated in drawing and physically enacting different solutions, and how some of them establish their presence in more

*Figure 5. Progression of drawings for the push-button*

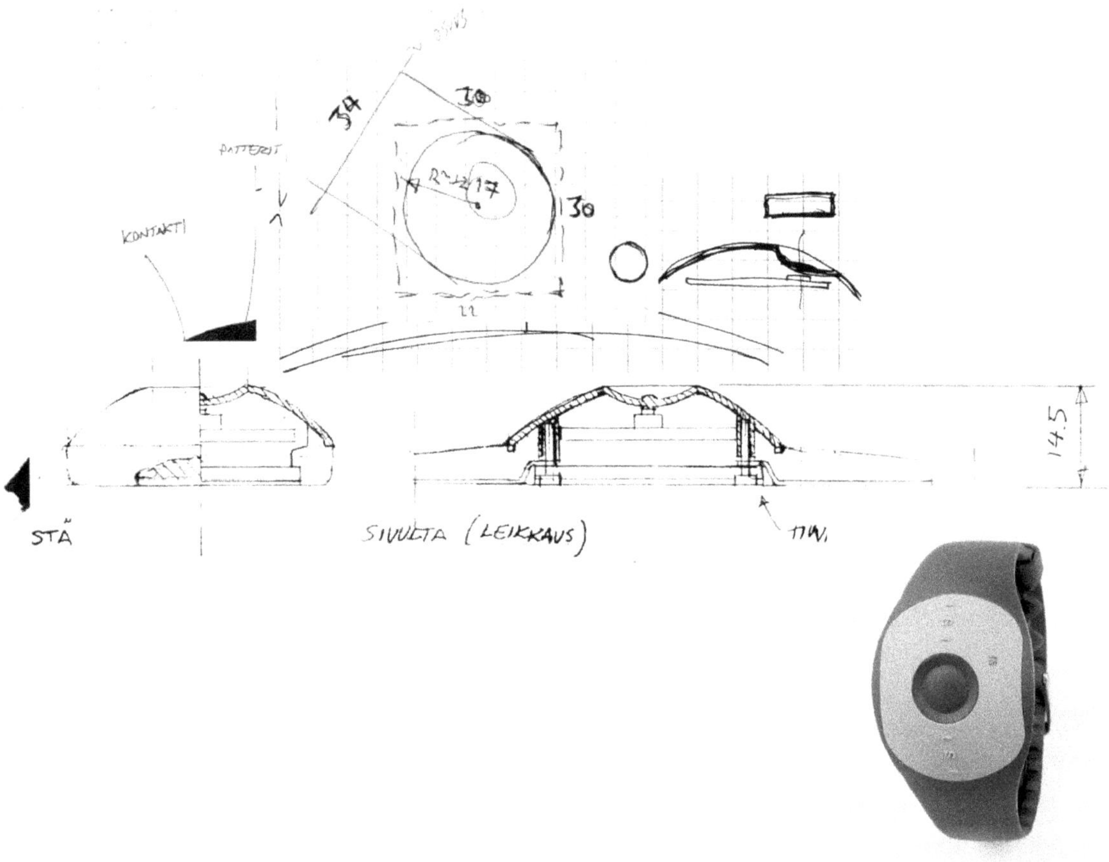


elaborate drawings that act as proxies for the future specification of the device.

Let us now rewind to near the beginning of the third design meeting, where the Inventor reviews the design drawings of Designer 2 (Figure 5, center left). The push-button solution has now been around a while, but it has been continuously challenged by other considerations, most visibly by the wish to incorporate an LCD screen into the device in its future versions. This issue is brought up in the first meeting, is reinforced in the second, and continues also after the initial eureka of the push-button and much of the third meeting is spent on the screen.

*Designer 1: So do you mean that the cover would have a ready-made place for the LCD, but when there is ((yet)) no screen there would be LEDs?*

*Inventor: For instance, so. Then there would be*

*Designer 2: ((inaudible))*

*Inventor: what?*

*Designer 2: Underneath the glass or how?*

*Inventor: Yes, this brings to mind the message when we talk about—as in your investigation ((designer 2's study on elderly users and Vivago-Wristcare design in 1995))—many wished that it would not look like an alarm // device* =*

*Designer 2: Yeah, that is*

*Inventor: = So it would have () even if it doesn't have a watch, it would look like a watch, that kind of message*

*Designer 1: Um-m, yes*

Considerations continue about what there could be in terms of LCD content and LED content and how it could be realized, and at 32 minutes:

*Designer 1: Yes. It, of course, alters a bit that basic thought. In my mind, well, if we hold on to having LEDs in the place of the screen when there is no screen, so it would // use the same spot* =*

*Inventor: Yes*

*Designer 1: = where the screen comes into for LEDs or displaying LEDs, then I think that situation needs to be considered right from the beginning. That is, build the design so that it has a clear space somewhere for it. This ((button in the center)) here is, e.g., a bit unfavorable for it has little room at many edges, but not much anywhere. So, that push-button would then need to be somehow dragged into the other corner right at the start ((as is done in the drawing in Figure 5, center right)).*

The push-button solution, designed into the middle of the device, seems now to require reiteration. Its suggested place and the related shape are now contested by a feature supported by a set of strong complementary representations of use. The 1995 inquiry on design and users (Hyysalo, 2003), to which the inventor drew attention above, had found that many users resented the "aid"-like appearance of safety phone wrist units and rather wished these would look like watches. Experience from early field trials had confirmed this user-representation, and it resonated strongly with the user-representation in the company strategy about establishing a new market among the younger, "65+" age group. Furthermore, to lure the younger users, added functionality was deemed important in the future versions, with the screen deemed vital for this, and the need to reiterate the push-button appeared likely.

But much depends also on the place a particular feature has acquired in the limited physical (and digital) space of the device. If a feature has not found its way to the progressing series of elaborated drawings—and hence not yet been elaborated in respect to the myriad of other features and constraints—it has an uphill battle to fight its way there, particularly when time pressure is high, and, consequently, the design consultancy tends to veer away from any unnecessary complications. The following interchange takes place at 27 minutes from the start of the third meeting, preceded by many questions, suggestions, and doubts about the LCD installation.

*Designer 1: So. What size would that possible screen then possibly be? If we are now to prepare for it ()*

Followed by an answer three minutes later:

*Inventor: () that which ((designer 2)) had in her design specified for a screen. It would be sufficient for appearing as a clock.*

*Designer 2: That was really small ((gestures with fingers to show how small the screen was)).*

*Inventor: It's a bit bigger. I was about to take it with me, but () ((points to the slightly enlarged gesture)) yes, it was; it was roughly that size! And it just had this shape on top. Had this *kind of // =*

*Designer 2: Round*

*Inventor: = Yes, a bit bigger. So I was // like* =*

*Designer 2: // It had**

*Inventor: = Empty space.*

*Designer 2: A little bit empty, it could be () that shape was transformed from a square. So it was about that ((draws)) size. Pretty tiny it was*

*Designer 1: Um-m, yes, exactly ()*

The client hence forces the issue onto the agenda, and so, some minutes later:

*Inventor: That button once more. Those new samples ((of switches)), as I said, have been ordered, so we can start trying this ((design)) out*

*Designer 1: Next, we create a form model and start measuring up. ((Inventor)) will give us the information on the screen: what we are to prepare in regard to space. When that is closed, we can measure that ((button)) precisely*

This appears as clear verbal consensus about taking both the push-button and the space reservation for the LCD forward, hence a call to reiterate the push-button design once more. And indeed, as we can see in Figure 5, right at the center, during the above discussion some early sketches were already drawn showing how the button could then be realized. Yet, when the fourth meeting begins, the place for the LCD is nowhere to be seen in the drawings produced by the two designers. Only a small space is reserved for a small feedback LED (see Figure 5, lower right-hand corner, for the finished device, where a small LED is visible in the top center on the right-hand side). In contrast, the push-button in its central location form has become elaborated in detail (Figure 5, bottom left and center left)—a clear expression of designers' opinion about the tradeoff and urgency of these two features. As another two weeks have passed, the client yields, and the LCD gains reality only as far as words and sketches go.

Never becoming incorporated into the actual design drawings, the screen no longer appears in further verbal considerations either. In contrast, the push-button receives much enthusiasm from all participants and stakeholders throughout spring 2000 and in the hands of the users afterwards. It becomes the most successfully realized feature in the design of the second-generation device.

Yet to understand its success, we must become aware that the conditions for its success were laid out long before. In fact, when one of the earliest safety phones, Helena, came to Finland in the early 1980s, designer 1 was commissioned to create a robust and well- functioning casing and alarm button for it[2]. The later inventor of Vivago was working with this project, and, after several events and updates, this casing was used in first-generation Vivago. And it was designer 1 who introduced designer 2 to the inventor for doing a design study of Vivago in 1995, including a literature review, expert interviews, and systematic comparison of the soiling of the alarm-button feature in all available safety phone models. In many respects, both the LCD and push-button solution discussions above reiterate the design rationales known for years—only now tied to a particular novel solution. Cast in these terms, the otherwise perhaps surprisingly swift invention and success of the new alarm button becomes more understandable. Let us examine the network of concerns and their origins that went into the push-button feature (Table 3).

There is indeed 20 years of history in fiddling with and designing the push-buttons of the devices, thinking through their weaknesses and strengths, and gaining appraisals of them from here and there. In this view, the push-button micro-biography is one of *accumulation* of social learning culminating in a new, neat design that eventually worked (even without proper testing) and during the design was able to *stabilize* against LCD-related user-representations that had strong

*Table 3. Origins for the push button solution, goals for its design and specific features realizing these*

| Push button alarm | Origins of rationales | Mission and Goals | Features |
|---|---|---|---|
| Competing against extant safety phone systems | - Buying time for R&D on automatic alarms (1992)<br>- Eliminating false alarms (M1995)<br>- Serving users in very weak condition<br>- Different pricing<br>- Less "aid" stigma | - Serving emergencies that cannot be detected by instruments<br>- Improved usability and reliability of operation<br>- Serving the segmentation of the devive for different groups of users<br>- Only manual alarm button version (= Safety phone) | - button that is hard to miss-press<br>- sheltering the button to a hole<br>- ensuring tactile finding<br>- extra-sensitive button<br>- nudge upon pressing<br>- feedback LED |
| Secure working of manual alarm function | - Manual alarms that have not come through (MS 1995, reports on SP)<br>- problems with feedback on pressing (pilots, MS 1995)<br>- Elderly to press electronics long and hard, push again if in doubt | - Achieving increased usability and reliability of alarm (see above) | - Directing the hand on device surface<br>- directing the finger to right pressing direction<br>- cap model |
| Reliability of the device | - Integrity of structure (pilots)<br>- Water resistence (devices destroyed in shower and dish washing, requirements by large telecom pilots, some country regulations) | - Maintain shock proof<br>- Maintain water proof structure<br>- Battery lifetime and replacability | - one red LED, disabled later. Vs. LCD, Vs. buzz feedback, VS. sound feedback |
| Manufacturability, steps | - Price of assembly and materials<br>- difficulty<br>- reliability | - as few assembly steps as possible<br>- modelling and molds within the limits of ease<br>- use of only tested material combinations | -Polyuretane all over, no silicone<br>- Button as a cap, nudge underneath integrated to the circuit board |
| Backwards compatibility | - Quick release of the new version<br>- Use of extant cirquit board | Button preferably in the middle, where it is in the previous model and present circuit board | - Button in the middle, fitting the circuit board<br>- Button height adjustable by the board height |
| Forwards compatiblity | - Changing cover element<br>- LCD screen reservation | - Preparing for more informational devices without new cover design<br>- Ability to produce design variations easily | - Place reservation for more information displayed on the cover<br>Vs. direction of push, Vs. button at the center |
| Stigma as aid | - To make the device appear not as an aid<br>- Avoid gathering dirt and soiling of the device<br>-Reports on elderly and technology<br>- Earlier safety phones<br>- M 1995<br>- Pilot experience and feedback | - Design elegance<br>- Round shapes "non-aid likeness"<br>- Unnoticeability<br>- Small size<br>- Changing colors<br>- Hiding the push button | -No seam cover, rounding of button hole and ascension so it allows for cleaning<br>(- make the whole cover an analog watch)<br>- Integrate LCD and digital watch<br>- Vs. large size of the device<br>- Vs. high visibility of the button |
| Target market | - From "85" to "65"<br>- Round shapes<br>- High tech looks and feel | To brand the device as high tech<br>To design device as non aid | Vs. push button and LCD next to each other on flat surface |

*Figure 6. Third-generation Vivago in 2009, second generation in 2002, first-generation device in 1998, and 1995 design sketch*

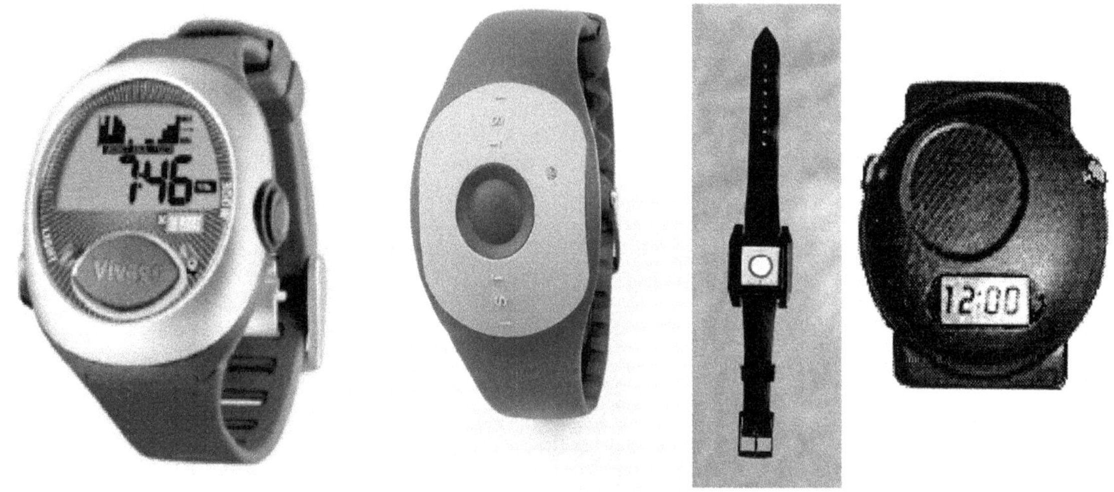


interanimated sources as well. Important issues leading to this stabilization were the tight schedule that truncated the attempts to elaborate LCD further and the fact that the push-button solution came to offer a feedback mechanism, rendering weaker any need for any kind of LED or display in the second-generation model and turning these user-representations only into a future compatibility issue.

## EROSION OF FEATURES WITHIN AND BETWEEN DESIGN PROCESSES

It would be, however, rather erroneous to assert that stabilization or rejection of technology in design would result in it being carved in stone. The LCD screen's mini-biography *in this design process* ends up as one of increasing marginalization. As noted above, it first got reduced to a feedback LED and then was later fully disabled by a software change to save on battery life. However, even a quick look at the third-generation Vivago-Wristcare probably tells all that needs to be told (Figure 6). When we compare this to the early-1995 design sketches (see Figure 6, right-hand side), the pendulum swings between different design features that are in tradeoff become apparent. The third-generation device no longer has a similarly indented–raised alarm button (perhaps producing some more accidental alarms), and, like the 1995 sketch, it capitalizes on new added functionalities and far less aid-like appearance.

Let us, however, not limit our scrutiny to how accumulation and (de)stabilization take place between different product releases. We next examine the micro-biography of the erosion of the anti-soiling characteristics of the device, which gives us further insight on interrelations between design features as well as on journeys of features during a design process.

The final micro-biography, that of the anti-soiling surfaces of the device, allows us to provide more flesh on how de-accumulation may happen. The elaboration of this feature began one minute from the kickoff of the design series, together with the push-button, and continued through the tens of iterations related to how it could be realized in practice. Let us return to the discussion we began this chapter with.

*Designer 1: So the bracelet would act as a unified surface ((pointing to his drawing)) here at the top. () and there would be a clear click to be felt ((upon pressing it)). And think of the advantages for keeping it clean: all one surface. I really got enthusiastic about the thought of "why should we have any separate cover structures?" as we discussed last time. For any and all separate covers inevitably include seams and that is always bad for cleaning*

*Inventor: () My only sort of objection here is that the device doesn't really soil from the top. It soils from the bottom*

*Designer: Yeah, well, there you have that one seam just as it presently has. That's unavoidable.*

*Inventor: Yes, but just to add here that in this way this sensor coming from here would have been deep () if you turn this, and think about it the other way around, there would be a hole and this ((sensor plate)) would come to rest against the bottom crevice. So only one small, very narrow seam would be made, no holes for the screws. The screws form a problem of their own. Their material. As we talked about last time, we need to sink them in there somehow. They cannot stay at the surface, for they react chemically even if they were of a durable material as such.*

*Designer: Sinking them in is no problem, but, of course, then the point of the screw is bad in terms of gathering dirt.*

*Inventor: Yes, that's the problem in the present design () and then our ((sensor)) plate, it cannot be taken from the surface, and you cannot sink in ((the screws)) deeper than that screw head.*

*Designer: Yes, the plate of yours, but we shall make that anew; it won't be the same plate, will it?*

*Inventor: Yes sure, but that material there won't allow it*

*Designer: Sure, but we can put more ((points to the sensor plate)) here; if we make it anew, let's make it thick enough to accommodate the sinking in of screws. ((It is)) the thickness of material there ((points)), or what determines that? What determines that thickness?*

*Inventor: It is not that // ((the)) circuit board ((the sensor plate)) can in principle be thicker. It's not up to that* =*

*Designer: so, // so this is**

*Inventor: = the circuit board material can be thicker, but that eats up the exact same amount of space from the batteries ((inside the device)) ()*

The anti-soiling presents particularly clearly how features become interlinked. Anti-soiling is now tied, on the one hand, to the screws, which have been one of the prime dirt-gathering spots in the previous design but which, as their main function, have to ensure many other things. In the course of the first six design meetings, the consideration of the screws and their various alternatives turns to considerations about compressing and locking the entire inner structure of the wrist device into place and, at that, ensuring its waterproofness, its endurance against dropping and collisions, changing of its batteries in such a reliable manner that non-specialists can do it, *et cetera*. All together, this takes more than an eighth of the total time used in the meetings, in addition to the work done outside the meetings.

On the other hand, the anti-soiling becomes elaborated against other technical issues and user-representations. In the above passage, these are the sensor plate (and through it the reliability of monitoring); the space for the batteries when it two weeks later becomes connected to battery size and battery life; and back to the integrity of the structure—that is, the number, positioning, and material of screws and other locking principles. In some of these areas, usage is implicated, and retained, as a concern; others are seen and treated as technical and structural issues from which usage is purposefully one step removed. All these connections make anti-soiling almost an omnipresent concern, rather more a design-driver than a feature. The fourth design meeting, 92 minutes from the start, is illustrative here. The topic is how to connect conductive pads to the rest of the structure.

*Developer 1: It ((the sensor plate)) has to be conical, to ensure ((points)) sealant ((structure)) here*

*Designer 1: Yeah, yeah, the idea is that it would tighten there.*

*Inventor: As such, you know, it looks good in the sense that no trail can form there that even could gather dirt. In a way, that possibility has been cut out there*

*Designer 1: You mean between the pad and the sensor?*

*Inventor: Yes, yes. In that alternative you've drawn, for the crack, it's not only the moisture but—regardless of how tiny the crack is—it can start gathering that dirt. And when you have a drop // of moisture**

*Designer 1: It is the combi // nation**

*Manufacturer: // It gathers**

*Inventor: Exactly! It gathers, yes that's clear as day. We've seen it. It's unavoidable.*

*Manufacturer: Horrible stuff. I just saw how a pulse meter that had been sent back from out there ((laughter))*

*Designer 1: Yes*

*Developer 1: Well, some of our bracelets too ((laughter))*

*Manufacturer: OK, well, let's try to find a solution, so that we can build a stop guard there ((to block dirt-and-water mixture moving into the device or forming a conductive pathway into it)). But I'm thinking that without testing you cannot say if it works.*

*Inventor: Well, can we go further here now? Well, no.*

So again we find user-representations conveyed through anecdote and comparison to existing devices: both this technology and others. Yet even as the concerns have been clearly spelled out, their order of priority remains more open to qualification. As we come to see next, a somewhat soil-gathering device can be sold and used, but one disintegrating with the first collision cannot. Let's illustrate this with the fate of the screws. The third design meeting, 55 minutes from the start:

*Inventor: () You cannot take it that far. You must () that there is the supporting structure. So if ((draws)) ((the screw support)) is here and there is plastic here, then all this is fluff—it will not hold, and even if taken all the way it won't seal or compress anything. The alternative is, as we pondered, that we have ((draws)) the polycarbonate here with a hole right here, and, and we accept that we must put the screw on top. Full stop. Then the polycarbonate acts pretty well as, as sealant*

*Designer 1: How is that screw really, protruding from the sensor surface: does it hinder monitoring, really?*

*Inventor: No, not really, if it remains in the area we are about to place it.*

*Designer 1: It will, of course, look a bit funny, but then again*

*Designer 1: So, screw here it is, either visible or if we can find low enough screw heads that we could patch it ()*

Hence the screws end up at the surface. Similar tradeoffs follow: to ensure that the conductivity measurement worked properly, diagonal entry of the pads to the device was eventually opted for, despite the crevice formed, since the "stop guard" discussed above was successfully designed such that dirt could not mess up the functioning of the device anymore. Yet the crevice as such became more accentuated when it turned out that the device had to be more watertight than expected and sealing was added. The ventilation principle and stretch solution for the bracelet introduced further channels for the dirt to cling onto. Even as the top surface indeed remained almost seamless as planned, the more difficult and more critical underside ended up disappointing the participants, as the CEO reflected more than two years after the initial design process.

*CEO: The device itself has now become conducting () and that problem is caused by the dirt () and it has caused a great number of brainteasers.*

*Interviewer: How have you solved it then?*

*CEO: We have now solved it by changing the mechanical structure of that wrist device and at the same time changing the electronics too. And it is in practice already solved.*

*Interviewer: Right. So there is going to be another round of iteration?*

*CEO: Yes, there will be. So the looks of the device are roughly the same, but when you look at the other side, now that is completely different.*

*(Interview with CEO on November 25, 2002)*

We are today unsure who came up with the term "erosion" to describe the carving of channels, nooks, and crannies in the smooth, ideal, envisioned surface in the course of the advancing design both over and underneath it. Regardless of whether it started as a joke by our informants or as our own reflection, erosion captures this chain of events well: first an accumulation leading to a clearly identified feature in the topography of the design parameters, then its gradual waning. More generally, the term "erosion of user-representation" is appealing beyond this kind of literal case to describe a process where a particular user-representation is raised to prominence and held there yet subordinated to other pressing considerations such that its original form becomes gradually and partially, for lack of a better word, eroded. Erosion is also apt metaphor, because it can be extended to cases where a given function becomes further ignored, worked around, or physically altered after the technology has been taken into use, on the scale of individual devices as well as the scale of changing generic design.

The erosion micro-biography clearly ends up in a less stable solution than, for instance, the accumulation biography of the push-button. Yet open-closed dichotomy may be too limited to be useful here. The anti-soiling surface builds on equal history of accumulated experience from previous safety phone models and their use. The repositories of this knowledge and its mediators to the Vivago project were the 1995 design study and the devices used in the 1997–1999 pilot studies, which revealed that both hygiene and monitoring reliability were affected by how the devices gathered dirt. The solutions for managing the dirt problem did become elaborated, as did the outcomes of this solution in the hands of the users, again leading to different solutions in the next design round of the device, in 2005–2007.

## DISCUSSION

Our ethnography of design underscores how this activity is highly concentrated situational engagement between people and materials producing elegant new solutions such as the novel structure of the push-button. At the same time, the micro-biographies examined in this chapter situate design as just one moment in the evolution of technology, even when examined at the feature level. The features we followed in this chapter had origins in the original Wristcare functional description, in the 1995 design study on the interface design of the Vivago-Wristcare system (see Hyysalo, 2004), in the accumulated feedback from the users of the device during 1997–1999 (see Hyysalo, 2006), or in all of these. This was also reflected in the way they came to enter the design process as issues brought forward by the inventor (who had most in contact with users at that time) or by the industrial designer who conducted the 1995 study. This design process was merely one, even if an important, period in the biographies of these features.

These micro-biographies also document (some of) the evolution of design ideas and concepts during a design process. The gradual stabilization and erosion of features indicates how attributing user-unfriendly design too straightforwardly only to *values* or poor education of designers and planners (Winner, 1980; Norman, 1988; Woolgar, 1991), appears rather daring *if the eventual artifact is used as the sole evidence*. Much happens in between the values and the outcome artifact. Design decisions affecting users negatively can result from uncertainty about, and lack of, testing of the properties of materials; from competing "good" aimed at users (the best possible functioning of the push-button vs. equipping the device with a screen to make it look less like an aid); or from mechanisms such as the erosion in the anti-soiling feature, where waterproofing and integrity of structure were directly user-related overriding concerns. This parallels but goes beyond stressing "operational readiness" in gaining advantage of usage related information in design processes (Voss, et al., 2009) in that indeed the value and possibilities to benefit from such information hinge on how the design process in general is organized, but also that even with best intentions necessary trade-offs between competing "goods" can result in erosion of one or other requirement or design feature.

In regard to design-use relation, it is noteworthy that the design process was exclusively conducted by developers, designers, and manufacturers. The only user involvement planned was user testing of the shape—and later, colors—of the new device, but this was never realized, because of time pressure. In stark contrast, users were very much present in the design discussion and a significant number of the redesigns originated from users' opinions or from problems they had voiced. These were represented in the process through the design brief's list of features and by anecdotes told by the company head and Designer 2, acting as intermediaries between users and the rest of the team. Representation of use, users, and usages by no means stopped here. Sources of user-representation visible in the above material are

- Using feedback from actual usage of the devices, brought into design mostly as anecdotes.
- Referencing one's own usage experience from actually wearing the device.
- Using oneself as reference—for instance, when the size of designers' fingers was used to determine the right size for the push-button.
- Reengineering previous push-buttons, surfaces, and buckle solutions, providing the developers with representations of how these would respond to usage.

- Business concerns, visions, and targets such as making the device appear less as an "aid" so as to make it appeal to younger users.

The last two of these sources exemplify in more detail the ways practice-bound imaginaries were discussed to offer user-representations previously (Hyysalo, 2006b). The first two owe to the devices now being in actual use, but also blend to the third one, designers using their personal experience as fellow humans as representative to the physique and behaviors of users. "I-design" can take a whole range of different ways that deserve more careful inquiry than what has been previously been done (Akrich, 1995; Oudshoorn, et al., 2004; Kotro, 2005). Indeed, there is a clear need for more thorough and encompassing "sociology of user representations" including their full variation in origins, change in and between the design processes, relative importance, stability and the various interchanges around them.

This leads us to the questions of agency in design. It is safe to say that "design is reflexive conversation with materials" (Schon, 1983) as well as to note that these materials are, in fact, heterogeneous. Indeed, much of the discussion in the above interactions is about how materials are likely to respond to particular kinds of shaping and what implications these have for users, pricing, reliability, integrity, *et cetera*. The agency in designing appears distributed to the design team and the materials and means they have available for conducting the design. Let us go further and return to the observation that ideas and solutions for each feature were pre-figured in earlier technologies and sites of usage. Nearly all alternative paths ever considered during these design episodes were equally bound to concrete events, artifacts, sites, materials, and images. Yet all the features were genuine shoots of invention: something humankind had not quite seen before. And the chain of imaginative actions (many of which were literally imaging actions) by which they came about was equally creative—devising clever solutions through analogues, creating new ways to use the available materials, and considering multiple outside-of-the-box avenues of design before choosing the eventual path to follow. Design was imaginative action, but the design imagination was unbound phantasma very rarely. Participants were quick to question the boundness of imagining, as in the case of incorporating the LCD screen (cf. Hyysalo, 2006b). Should we have resorted only to the ethnography of the second-generation Vivago-Wristcare design process, its bindings to earlier events and more long-term features of practices of safety-phone development would have remained invisible. We would have, indeed, constructed such a situated action, where the situation would have been delineated as a narrowly conceived physical and temporal setting where vocal and gestural interaction took place. A place strategic and convenient to study, yes, but in terms of the design situation at stake as we know from a more full bodied analysis of it, a rather unfortunately truncated one for the intelligibility of actions of the participants.

## REFERENCES


Akrich, M. (1995). User representations: Practices, methods and sociology. In Rip, A., Misa, T. J., & Schot, J. (Eds.), *Managing technology in society: The approach of constructive technology assessment* (pp. 167–184). London, UK: Cassel Publishers Ltd.

Büscher, M., Christensen, M., Hansen, K., Mogensen, P., & Shapiro, D. (2009). Bottom-up, top-down? Connecting software architecture design with use. In Voss, A., Hartswood, M., Procter, R., Rouncefield, M., Slack, R. S., & Büscher, M. (Eds.), *Configuring User-Designer Relations: Interdisciplinary Perspectives*. London, UK: Springer. doi:10.1007/978-1-84628-925-5_8

Cross, N. (2000). *Engineering design methods - Strategies for product design*. Chichester, UK: John Wiley & Sons.

Jordan, B., & Henderson, A. (1994). *Interaction analysis: Foundations and practice*. Palo Alto, CA: Xerox Palo Alto Research Center and Institute for Research on Learning.

Kotro, T. (2005). *Hobbyist knowing in product development - Desirable objects and passion for sports in Suunto corporation*. Academic Dissertation. Helsinki, Finland: National Consumer Research Center and University of Art and Design Helsinki.

Norman, D. (1988). *Psychologygy of everyday things*. New York, NY: Basic Books.

Norman, D. (1999). *The invisible computer: Why good products can fail, the personal computer is so complex and information appliances are the solution*. Cambridge, MA: MIT Press.

Oudshoorn, N., Rommes, E., & Stienstra, M. (2004). Configuring the user as everybody: Gender and design in information and communication technologies. *Science, Technology & Human Values*, *29*(1), 30–63. doi:10.1177/0162243903259190

Papanek, V. (1972). *Design for the real world (suom. turhaa vai tarpeellista?)*. London, UK: Thames and Hudson.

Schegloff, E. A. (2007). *Sequence organization in interaction*. Cambridge, UK: Cambridge University Press. doi:10.1017/CBO9780511791208

Schon, D. (1983). *The reflective practitioner: How professionals think in action*. New York, NY: Basic Books.

Schuler, D., & Namioka, A. (Eds.). (1993). *Participatory design: Principles and practices*. Hillsdale, NJ: Lawrence Erlbaum Associates.

Stewart, J., & Williams, R. (2005). The wrong trousers? Beyond the design fallacy: Social learning and the user. In Rohracher, H. (Ed.), *User Involvement in Innovation Processes: Strategies and Limitations from a Socio-Technical Perspective*. Munich, Germany: Profil-Verlag.

Ulrich, K. T., & Eppinger, S. D. (1995). *Product design and development*. New York, NY: McGraw-Hill, Inc.

Voss, A., Hartswood, M., Procter, R., Rouncefield, M., Slack, R. S., & Büscher, M. (Eds.). (2009). *Configuring user-designer relations--Interdisciplinary perspectives*. London, UK: Springer.

Voss, A., Procter, R., Slack, R. S., Hartswood, M., Procter, R., & Rouncefield, M. (2009). Design as and for collaboration: Making sense of and supporting practical action. In Voss, A., Hartswood, M., Procter, R., Rouncefield, M., Slack, R. S., & Büscher, M. (Eds.), *Configuring User-Designer Relations: Interdisciplinary Perspectives*. London, UK: Springer. doi:10.1007/978-1-84628-925-5_3

Williams, R., Slack, R., & Stewart, J. (2005). *Social learning in technological innovation-Experimenting with information and communication technologies*. Cheltenham, UK: Edgar Elgar.

Winner, L. (1980). Do artifacts have politics? *Deadalus*, *109*, 121–136.

Woolgar, S. (1991). Configuring the user: The case of usability trials. In Law, J. (Ed.), *A Sociology of Monsters: Essays on Power, Technology, and Domination*. London, UK: Routledge.

## ENDNOTES

1. Portions of this work have been previously published by Sampsa Hyysalo © Routledge. Used with permission.
2. Interviews with Designer 1 on October 30, 1999, and March 3, 2000.

# Chapter 13
# Strategic Ethnography and the Biography of Artefacts

**Neil Pollock**
*University of Edinburgh, UK*

**Robin Williams**
*University of Edinburgh, UK*

## ABSTRACT

*In health research and services, and in many other domains, the authors note the emergence of large-scale information systems intended for long-term use with multiple users and uses. These e-infrastructures are becoming more widespread and pervasive and, by enabling effective sharing of information and coordination of activities between diverse, dispersed groups, are expected to transform knowledge-based work. Social scientists have sought to analyse the significance of these systems and the processes by which they are created. Much current attention has been drawn to the often-problematic experience of those attempting to establish them. By contrast, this chapter is inspired by concerns about the theoretical and methodological weakness of many studies of technology and work organisation—particularly the dominance of relatively short-term, often single site studies of technology implementation. These weaknesses are particularly acute in relation to the analysis of infrastructural technologies. The authors explore the relevance to such analysis of recent developments in what they call the Biography of Artefacts (BoA) perspective—which emphasises the value of strategic ethnography: theoretically-informed, multi-site, and longitudinal studies. They seek to draw insights from a programme of empirical research into the long-term evolution of corporate e-infrastructures (reflected in current Enterprise Resource Planning systems) and review some new conceptual tools arising from recent research into e-Infrastructures (e-Is). These are particularly relevant to understanding the current and ongoing difficulties encountered in attempts to develop large-scale Health Infrastructures.*

DOI: 10.4018/978-1-4666-0303-5.ch013

## INTRODUCTION

Various writers have drawn attention to the increasing scope and scale of information systems (focusing particularly on their longevity, ever-wider application, and implications for various kinds of work). The facility that these systems offer for sharing information across the boundaries between occupational and organisational groups has been seen as transforming scientific research and knowledge-based work and services more generally. The concept of 'infrastructure' has been applied to the structures of codified information (Star & Ruhleder, 1996), and *qua* information infrastructure also to the hardware, networks and software systems via which they are increasingly stored and accessed (Monteiro & Hanseth, 1995; Hanseth, et al., 1996). The application to scientific research of powerful new computer-based tools such as 'the Grid' (in such areas as Translational research and genetic health databases) has been associated with the term e-infrastructure, e-Research or e-Science, or in the USA cyberinfrastructure (Edwards, et al., 2009)[1]. The need for such systems to cater for a wide range of current users and uses (and, given their development costs and intended longevity, potential future users and uses) makes their design and further evolution potentially challenging.

Social scientists have been drawn to address the emergence, use, and evolution of these systems—particularly in the areas of health research and health service delivery, where enormous investments have been made in the last decade, but where acute problems have also been encountered in both development and maintenance. This paper questions whether or not scholars have adequate social scientific tools for getting to grips with e-Infrastructures. It argues that we need more elaborate methodological templates and conceptual frameworks for analysing both the dynamics and the constraints surrounding these developments, for characterising the problems and factors that underpin them, and identifying how these problems may be ameliorated such that we can guide policy and practice.

The starting point for this paper is our dissatisfaction with some of the dominant analytical traditions on technology and work organisation, where we have identified shortcomings that are particularly relevant to enquiries into e-infrastructures. We have begun to develop the Biography of Artefacts (BoA) perspective to redress these theoretical and methodological weaknesses and provide more effective analytical templates to guide research and perhaps intervention.

## SOME SHORTCOMINGS OF EXISTING RESEARCH INTO TECHNOLOGY AND WORK ORGANIZATION

Research into technology and work organisation has suffered from the fragmentation of enquiry between various disciplines and schools of analysis—with their differing foci and concerns—and which have tended to be associated with different kinds of study. We draw attention particularly to the divide between a relatively small cohort of researchers (mainly from Science and Technology Studies [STS] but also from Information Systems, etc.) who have undertaken studies that encompass technology design and development (Mackay, et al., 2000) and a much larger group which has focussed more narrowly on their organisational implementation and use. Here we find a substantial body of work informed by diverse perspectives within Management Schools—including Organisation Studies, Technology Management, and Strategic Management—as well as Information Systems Research and Science and Technology Studies and what we may describe as socially-oriented computer-science, including Social Informatics and Computer Supported Cooperative Work (CSCW). Studies of organisational adoption constitute the overwhelming bulk of contemporary research into enterprise systems and other organisational technologies (ERP Research Group, 2006; Pollock & Williams, 2009); only a very small number of studies (largely from Science and Technology Studies) address technology design and use in tandem (Mackay, et al., 2000; Williams, et al., 2005; Hyysalo, 2010).

We also note the theoretical and methodological weakness of much of this research, arguably reflecting the dearth of rigorous and critical social science analysis where much of this research is undertaken (predominantly the Business School). Much of this work thus suffers from a lack of an adequate analytical framework coupled with the unreflexive adoption of rather simplistic methodologies—shortcomings that have been subject to longstanding critiques of empiricism within social sciences (Williams & Edge, 1996). Linked to this we note the predominance of short-term studies of technology adoption and in particular of company case studies.

We contend that there has been a failure within much research into technology and work to reflect upon how the choice of particular analytical framings, methodologies, tools, and research design may tend to highlight certain kinds of finding. For example, the separation between studies of technology design/development and of technology implementation/use is reflected in the circulation within many implementation studies of stereotypical accounts of technology suppliers from the perspective of the organisational user often in the early stages of implementation (e.g. stories of supplier offerings "thrown over the wall" to unhappy organisational users). Such experiences are important—they flag the gulf between necessarily generic supplier offerings and the specific practices of organisational users, and the work that must be done to adapt a new complex technology to their purposes and practices. However, this kind of account simply fails to do justice to the complex promotional and design/development strategies of successful vendors. In the next section, we examine the consequences for research findings of the prevalence of the kinds of short-term studies of technology adoption that we characterise as 'impact studies' and 'implementation studies.'

We are, of course, more than happy to acknowledge that there have also been important contributions from studies of technology implementation, and in particular from the vibrant body of more critical and analytically sophisticated work arising particularly from what we may describe as interactionist and constructivist studies within Business Studies, Information Systems Research, Social Informatics, Science and Technology Studies, CSCW, and so on. This work has, for example, offered a better understanding of the relationship between technological change and workplace activity, highlighting in particular the intricate sociotechnical processes whereby work activities such as information exchange and task coordination are achieved (Luff, et al., 2000). We recognise their valuable critique of mainstream functionalist and normative approaches within Management and Systems Development thinking. However, we have reservations about their epistemology and consequent methodological choices. In particular, we note the dominance of 'single site ethnographies' (typically studies of technology implementation) and, given the limitations of research access and funding, their limited duration.

In articulating relevant aspects of the BoA perspective below, our work seeks instead to expand the focus of research longitudinally and across different social settings and scales, addressing multiple moments and sites of innovation, and encompassing different phases of what has been described as the systems development cycle (design, selection/procurement, implementation, and use), and the multiple such cycles that constitute the product cycle for a particular artefact.

## Impact Studies

One of the first kinds of paper to emerge when a new technology is launched is what we term the 'impact study.' We find articles appearing first in the practitioner and trade journals oriented towards potential adopters. Their message is in general rather simple—geared towards the questions a potential adopter might ask: what is this technology and how can it benefit my business? Typically these are presented as a 'before and after' study, with a narrative of improvement, starting with the problems the organisation was encountering; the identification of a new technology that offered a solution, its adoption and of the benefits delivered (see for example Botta-Genoulaz, et al., 2005). These accounts are often linked to the suppliers and consultants involved; and they had a tendency to uncritically reproduce supplier claims (Webster, 1990; Fleck, et al., 1990). This was especially notable in the opening stages of the widespread historical adoption of IT within organisations when visions and promises of the benefits of new technologies were confidently articulated in advance of the often much more mixed experiences of their adoption and use. Latterly, experienced journalists and commentators have tended towards a more sceptical voice (why choose this technology amongst so many other contenders?). The case study of new technology adoption and its successful outcomes seem to provide an important currency for validating technology to potential organisational adopters (Graham, et al., 2003; Pollock & Williams, 2009). However, this kind of writing often lacks analytical distance or critical concerns. Implicitly, through their 'before and after' framing, as well as the result of any commitments of the writers, these papers often articulate an engineering or managerialist view of technology as instrumentally transforming work.

## Implementation Studies

The next wave of papers arising from practitioner and academic circles largely comprises what we describe as 'implementation studies.' These have a stronger empirical base than the impact studies, though, like the latter, they are often undertaken in the immediate aftermath of technology adoption. Such studies frequently highlight the gap between the expectations that surrounded a new technology and it immediate outcomes (for reviews of this literature in relation to Enterprise Resource Planning systems [ERP] see Esteves and Pastor [2001] and Al-Mashari [2003])[2]. This work does engage critically with supplier claims. Issues remain about how to interpret the longer-term significance of these findings. By emphasising the obstacles to technology-induced change there is risk of replacing the CAN DO messages of technology promoters with a NO CAN DO analysis and of conflating the immediate and longer term outcomes of new technology implementation (Williams, 1997a). This may produce an incomplete and misleading understanding of the consequences of technological change—consequences that may only become clear after years or decades.

Implementation studies are typically medium-term studies starting in the course of, or shortly after, the introduction of a new technology. There is thus a temporal framing—both upstream and downstream of the implementation process. The upstream framing concerns the difficulty of researching directly the 'pre-project' phase in which particular problems and the possibility of technical solutions are articulated together. The decision to adopt thus typically becomes a taken for granted feature, addressed only in hindsight—a hindsight view in which many uncertainties and shifts in perception may be suppressed by the desire to legitimate a decision already taken (McLaughlin, et al., 1999; Gerst, 2006). The fact of adoption (or at least a decision to adopt) is what has driven the selection of the cases for research. These prior 'pre-project' decisions also frame the way in which the implementation study is undertaken insofar as a number of decisions have been taken about the scope of the new technology project (including for example defining the problem to be addressed and the resort to technical solutions). In this process, we note the establishment of particular sociotechnical configurations (Sahay, et al., 2009)—emerging as a result of initial interactions amongst the complex arrays of actors potentially involved in the project—which constitute what we have described as the 'translation terrain' (Williams, et al., 2005).

The downstream framing arises from the limited duration of many studies, typically to periods of a few months or at most a couple of years. However, a body of research going back to the work of Arrow (1962) points to protracted processes of 'learning by doing' through which the efficiency and other benefits of a technology are improved by trial and error experimentation. Fleck (1988, p. 3) takes this idea further with his notion of 'innofusion' which highlights the way in which artefacts, rather than being fixed at the design stage, are transformed in their implementation and use, in the course of "the struggle to get the technology to work in useful ways at the point of application." Similar processes have been identified by other writers, using slightly differing conceptualisations described variously as appropriation (Mackay & Gillespie, 1992), domestication (Sørensen, 1996), and social learning (Sørensen, 1996; Williams, et al., 2005).

Research investigating the organisational implications of Enterprise Resource Planning (ERP) systems immediately drew attention to the implementation difficulties encountered with these complex software systems and the potential for failure. These difficulties were attributed to the 'misfit' between the standardised presumptions embodied in the packaged solutions and the specificity of user organisation practices and the consequent need either for expensive customisation or unwanted organisational adaptation (Soh, et al., 2000; van Fenema, et al., 2007). Many writers concluded from these experiences that ERP was expensive and undesirable. Such dismissive accounts leave us with a problem, however: how can we to explain the wide adoption of ERP across many different kinds of organisation worldwide? A later body of work argued that the benefits of ERP do not appear for a number of years (Mabert, et al., 2001). It was therefore necessary, they suggested, to extend the timeframe of research into the 'post-implementation' phase also described as the 'onward and upward' phase (Somers & Nelson, 2004; Wei, et al., 2005; Wu, 2008). Elsewhere, echoing some of the sentiments of this latter work, we have drawn attention to the enormous effort involved in bridging generic supplier offerings to specific organisational settings through intertwined processes of *innofusion*—in adapting these complex information systems to particular organisational settings and of *domestication*—as organisations learn to exploit the affordances of complex technologies in redeveloping their information and work practices (Pollock & Williams, 2009).

## Design Oriented Studies

An important stream of studies of workplace settings has been directed towards improving the design of computer-based systems. Much of this work has emerged from the socially-oriented computing perspective (including, for example, Social Informatics and CSCW). This work has emphasised the double value of workplace ethnographies, *directly* as a source of detailed empirical information about particular organisations, activities that can be a resource for system design and also *more generally* as a way of achieving a more adequate understanding of how work is constituted and achieved. The rich empirical picture of workplace activities that can be achieved by ethnographic research is envisaged as helping overcome the difficulties encountered with traditional methods of 'requirements capture' that only engage with the formal descriptions of how work tasks are supposed to be undertaken. By drawing attention to the range of informal procedures through which work goals are carried out, including dealing with frequent 'abnormal instances,' such studies were seen as providing the information required for designing tools and systems that could better support the ways in which work activities are actually performed

(Plowman, et al., 1995; Luff, et al., 2000). However, this goal has proved somewhat elusive for a number of reasons (Ackerman, 2000; Schmidt, 2000; Stewart & Williams, 2005). In particular, those involved in design stressed the difficulties of packaging sociological understandings into a form that could inform workplace analysis and design (Dourish, 2001; Fitzpatrick, 2003). The social scientists involved conversely raised two sets of questions. The first revolved around what kind of empirical investigation was needed to acquire an adequate understanding of work settings. The developers' need for timely information about potential users and uses, and the prohibitively high costs of protracted labour intensive ethnographic research, prompted suggestions for the adoption of 'quick and dirty' ethnographies, which could yield information better targeted to designers' needs and in more manageable volumes (Hughes, et al., 1994; Anderson, 2000; Martin & Sommerville, 2006). Such methods were, however, widely criticised as potentially generating misleadingly simplified accounts of the social settings being studied compared to the more robust understandings available from extended ethnography[3]. The second set of questions concerned the role and responsibility of social scientists. Social scientists observed that the knowledge they generated could not simply be packaged into a form that others could use. Moreover, sociologists and anthropologists were generally unhappy about being cast into an instrumental role of providing information that others (designers, managers) would use (and might abuse!). They sought instead to retain some accountability and control over how their findings would be utilised. Rather than aligning with managerial or developer objectives, they saw the need to make their interventions accountable to various stakeholders—including the various groups of workers who might be affected by a new development (Suchman, 2002). Difficulties in determining how their knowledge would be applied, and also in developing effective communications with design specialists led some to suggest that the gap between organisation members, ethnographers and design specialists should be collapsed in a process of 'co-realisation' (Hartswood, et al., 2002; Voss, 2006).

## The Link Between Types of Study and Research Findings

This brief review has given some illustrations for our suggestion that particular types of study are closely associated with certain kinds of findings. This association is not intentional—indeed it has gone almost totally unrecognised by scholars who will typically be involved almost exclusively in one or other type of study and with their particular research perspective (Law, 2004). It may be no coincidence that the insights we report here come to the fore as we reflect upon an extended programme of research that has occupied us over two decades.

At the outset of this period, one of the authors conducted a set of case studies of new technology adoption in the workplace. This included two firms in the same area (the West Midlands in the UK) adopting the same technology (Computer Numerical Control Machine tools) to produce similar products (aerospace components) but with radically differing forms of work organisation (operator programming versus technician programming) (Williams & Steward, 1985). These kinds of short-term case study comparisons, perhaps inevitably, draw attention to the scope for variability between cases and thus for *choice* (Bessant, 1983) surrounding technology and work organisation, shaped by local contingency including the micropolitical strategies of organisation players.

In contrast historical studies of technology and work, comparing workplaces over a long timeframe—perhaps contrasting these workplaces studied in the 1980s with their counterparts today—would immediately draw the observer's attention to the substantial concerted changes over time in how work is done. We would be struck by similarities in, say, office or factory work settings today and striking differences from their counterparts 20 years previously. Social scientists seeking to explain these concerted changes might be drawn to look for institutionalist explanations (for example in terms of what DiMaggio and Powell [1983] describe as isomorphic pressures due to the influence of prevalent managerial programmes or technologies [Tingling & Parent, 2002]), the globalisation of the economy and the influence of technological change.

## Narrative Bias

Particular kinds of study have characteristic framings, it transpires, in terms of the duration of study, in the tradeoff between breadth and depth of empirical enquiry and in the temporal and societal boundaries around what is addressed. In addition the various schools of analysis and types of study seem also to be associated with certain narrative repertoires. What makes a 'good study' is linked to its ability to touch upon the core concerns of a school of enquiry including ideas about the key problems needing to be addressed and possible solutions. We have used the concept of *narrative bias* to invite more reflection upon how particular schools of analysis often come with classic stories attached; they are associated with dramatic repertoires; heroes and pitfalls, problem diagnoses, dilemmas, and solutions (Stewart & Williams, 2005). Consider, for example, the contrasting narrative structures and concerns of two important strands of analysis within STS, which are of central relevance to understanding the emergence of certain kinds of expert knowledge and technological infrastructures. These are the Sociology of Classification and studies of the socioeconomic shaping of inter-operability standards.

The Sociology of Classification (as exemplified by Star and Ruhleder [1996] and Bowker and Star [1999]) emphasises the ability of new classification systems to impose order and in this way to prioritise one worldview over others (Timmermans & Berg, 1997; Berg & Timmermans, 2000; Hardstone, et al., 2006). In contrast, socioeconomic studies of the development, adoption, and implications of the interoperability standards which are at the heart of the emergence of Inter-Organisational Network Systems and other e-infrastructures emphasise the difficulties of achieving common standards—in terms both of reconciling the competing needs of different stakeholders in initially agreeing standards and, once particular local (regional or sectoral) standards have been implemented, the difficulties of migrating towards later generic standards (Graham, et al., 1995; Graham, et al., 1996; Williams, 1997b; Ciborra, et al., 2001). These traditions have each theorised, and, perhaps unconsciously, given particular emphasis to one aspect of infrastructures. We argue that an adequate account needs to develop a balanced theorisation that for example addresses both the immediate barriers to standardisation efforts and the possibility and significance of such alignments (Timmermans & Berg, 1997; Williams, 1997b; Hardstone, et al., 2006). Moreover, these necessarily incomplete alignments of meaning and practice, can become 'infrastructural,' crystallised into standards and classifications, embedded in software and organisational routines and can in turn become components in the further innovation of practices and systems developments (as Berg and Timmermans [2000] have noted in relation to systems geared towards health professional work).

# THE NEED FOR NEW KINDS OF RESEARCH PERSPECTIVES

Our broader thesis here is that various critical engagements with what we may call 'the modernist project' have produced a pattern of dichotomised debates and partial accounts. One of the most influential here has been the 'localist' and 'interactionist' turn, which has emerged in reaction against 'universalist' claims that powerful new technical capabilities provide solutions that can readily be applied and bring improvements in many diverse organisations in different sectoral and regional settings, and across different departmental structures and work groups. We note the growth of ethnographic accounts of technology in the workplace, inspired by ethnomethodology (Button, 1993) and in particular by the influential work of Suchman (1987). These 'interactionist accounts,' which attend to how social relationships are constituted and maintained through social interaction, emphasise the improvisatory role of local actors in sustaining organisational practices. The consequent diversity of work practices and uniqueness of organisational settings mean that standardised technical solutions often fail or can only succeed as a result of the active repair and workarounds by users (see Pollock, 2005). These issues have been primarily addressed through rich ethnographic studies, often of particular places of work. Such research has provided an inspiring and rich intellectual context for our research. However, we are becoming frustrated with their embedded epistemologies and characteristic research designs (Pollock & Williams, 2009). Their privileging of the local is frustrating, particularly in a context of increasingly pervasive information system adoption, which mean that local interactions are increasingly bound up with technology development processes and organisational practices elsewhere (Suchman, 2002; Kallinikos, 2004a; Ellingsen & Monteiro, 2003). We have, in addition, had long-term reservations about the limited ability of such research (that typically involves short-term ethnographic case-studies of particular work groups or organisations) to theorise longer-term processes and the influence of the broader historical and institutional context which pattern and structure these local processes (Williams & Edge, 1996). The question this paper seeks to pose is whether we can go beyond these simple analytical lenses to generate richer and more effective accounts of complex sociotechnical phenomena such as the evolution of e-infrastructures?

### A Framework for Addressing the Evolution of e-Infrastructures

To this end, we need an analytical framework that can account for both stability and change in sociotechnical relations and how these can be addressed over different time spans. Thus, we need concepts that can encompass both:

- the short-term dynamics surrounding the selection, implementation and embedding of new technologies encompassing incremental changes and also the continuity of existing social relations. Here, Webster (1990) has observed that where there was a collision between existing work practices and the presumptions built-in to technology, it is the former which often seem to be more immediately durable, and
- the longer-term evolution of work practices and technologies in which we may simultaneously see both:
    - gradual alignment around generic and specific standards, technical infrastructures and other crystallised social relations; and,
    - new dynamism and differentiation.

The Biography of Artefacts (BoA) perspective (Pollock & Williams, 2009; Hyysalo, 2010) emerges from our attempts, with others, to undertake such an analysis in relation to one of the most successful current organisational e-infrastructures—the ERP systems that have become widely adopted in private and public sector organisations.

## THE BIOGRAPHY OF ARTEFACTS PERSPECTIVE

The BoA perspective emerged in the course of protracted efforts amongst a small community of scholars to produce a deeper theoretical and empirical account of the relationship between technology and organisation. Our early research had demonstrated the need to move beyond episodic (short-term single site) studies of settings of technology design or its organisational implementation/use and instead to address the evolution of workplace technologies over multiple cycles of design and implementation. Brady et al. (1992) suggested that packaged software artefacts had biographies—highlighting how bespoke applications were sold-on to other firms in the same (financial services) sector and formed the basis for commodified solutions. Building on earlier research which showed that new workplace technologies often emerged through the enhancement of existing applications (Fleck, et al.,1990), Williams (1997a) further applied the concept of biographies to analysing the historical evolution of integrated information systems like Computer-Aided Production Management (CAPM). When supplier offerings were implemented, they inevitably had to be adapted to fit the technical and operational circumstances of adopting organisations. This process often threw up further useful innovations that could feed into future technology supply. Workplace technologies thus evolved through successive cycles of technical development and industrial implementation and use, a 'spiral of innovation,' oscillating between moments of development, implementation and use. These short-term cycles were phases in a longer-term biography; and longitudinal studies showed how the CAPM and Manufacturing Resource Planning (MRP II) systems of the 1980s and 1990s, widely seen as the precursors for today's ERP systems, themselves emerged from stock and production control systems developed in the 1960s in vehicle and aerospace sectors (Fleck, et al.,1990).

## Research Designs to Capture Biographies of Infrastructural Artefacts

Following on from this, a number of studies explored the subsequent emergence of ERP and its wide uptake. These provided a rich conceptual and empirical base for articulating and testing the BoA perspective (see for example Clausen & Williams, 1997). In developing this perspective, we sought to exploit the richness of ethnographic study and its open-ness to contingency in processes and outcomes. However, we also wanted to redress weaknesses of the naïve research designs frequently adopted under the dominant action-centred traditions (for example, the relatively short-term, single site ethnographies favoured by many interactionist analysts)[4].

Our articulation of the BoA perspective reflects our concern to engage more coherently with the ways in which longer term history and the broader context shape innovation processes and outcomes. Koch (2003, 2007) pointed out that we should move away from a focus on particular innovation moments, organisations, and artefacts, and perhaps think of complex organisational technologies like ERP systems as a 'community,' embracing user organisations and other intermediaries as well as suppliers, and as 'heterogeneous assemblages' encompassing software artefacts, suppliers, organisational users, consultants, work practices and visions of organisational improvement.

## The Need for Strategic Ethnography

Very different kinds of research design are needed to address this. We have proposed (Pollock & Williams, 2009) what we describe as 'strategic ethnography'—whereby the choice of research settings and the scope of studies is informed by provisional theoretical/empirical understandings of the locales in which new technologies are being shaped[5] as well as by the specific research concerns and issues under examination (though we note that research design choices are also necessarily influenced by opportunism and pragmatic exigencies). Thus, we concur with proposals for multi-sited ethnography to provide more robust, contexted understandings of complex objects (Marcus, 1995; Hine, 2007). In relation to technological change, we propose that research should engage with the multiple locales and moments of innovation (encompassing, for example, technology design, procurement, implementation, maintenance) (see also Kallinikos, 2004b).

In contrast to the short-termism of much research (conditioned by the typical duration of research project funding), studies need to tackle extended timeframes, through more complex temporal designs including longitudinal studies, follow-up studies and long-term historical investigations. Working specifically with various health ICTs, Hyysalo's (2004, 2010) work has offered a more systematic theorisation of the multiple histories that intersect in every episode in the biography of an artefact. His framework (building upon Hutchins, 1995) ranges from the short timeframes of conducting particular tasks to the long-timeframes of the development of wider institutional practices. In our work, we approached this challenge in a number of ways. These included studying mature products and those at an early stage in their product lifecycle; combining historical and contemporary study; and, by undertaking several research projects over a period of more than 10 years. These multiple temporalities equally encompass both immediate contexts of interaction and the more generalised contexts in which innovation unfolds. Here we have focused upon the emergence and evolution of technological fields, through interactions between suppliers, organisational users and various kinds of experts and consultants (and especially noting the emergence of organisations such as the industry analyst Gartner that have come to play a key role in drawing boundaries around and distinctions within technological fields like ERP [Pollock & Williams, 2007]).

These research design considerations suggest a radical rethinking of the model of how research should be designed and undertaken. Detailed studies of particular innovation moments (e.g. of design/development, implementation/use, post-implementation support), though valuable, need to be supplemented by studies at the interstices between organisations (e.g. of marketing/procurement and the construction of markets). Research needs to be extended spatially to see how technological fields and markets are constituted and organised as well as temporally to address the longer-term evolution of the technology and its associated organisational practices (e.g. including further implementations and feedback into future technology supply and the extension of a product into other markets). This is no small feat. An effective account of the biography of an artefact perhaps needs to be a team project at least and perhaps seen as the outcome of a research programme amongst a community of enquiry. We can explore these issues through our case study of the evolution of ERP systems and their precursors (Pollock & Williams, 2009).

## THE BIOGRAPHY OF ENTERPRISE RESOURCE PLANNING (ERP) SYSTEMS

### Packaged Enterprise Systems: An Impossible Project?

Most of the top UK, European, and US corporations and a growing range of public sector organisations have adopted packaged ERP systems. Sold for example by firms such as SAP, these are integrated solutions, which handle a vast array of different kinds of transactions right across the organisation. Their success presents something of a challenge to the dominant analytical fashion across a range of traditions (including socially oriented Computer Science, CSCW, Information Systems Research, Technology, and Organisation Studies) which, as already noted, informed by ethnomethodological and interactionist accounts, share an emphasis upon:

- the uniqueness of particular organisational settings and practices (Hartswood, et al., 2002);

- the consequent gulf between standard generic enterprise software solutions and these practices (Soh, et al., 2000; Walsham, 2001; Taylor & Virgili, 2008).

Packaged ERP implementation is accordingly often portrayed in these accounts as risky and imposing high costs (as well as the high purchase price for these enormously complex software systems, costs include the significant expense and effort of adapting software to user needs or of adopting unwanted organisational change to align to the templates built-in to the packaged solutions) and uncertain in its benefits.

Arguably, the frameworks adopted in the bulk of the literature are thus not well-equipped to explain the phenomenal success of standardised ERP solutions. The success of the ERP suppliers did not, however, correspond to the modernist dream of a new e-infrastructure conceived as a universal solution that can somehow anticipate and cater for the entire range of organisational contexts and practices. We have shown that SAP (and other ERP package suppliers) did *conquer the world*. However ERP's success was achieved through an arduous process of selectively accommodating and sorting user requirements in close relationship with existing and prospective organisational users to produce generic solutions for market segments. And ERP expanded into other markets one segment at a time, through the gradual extension of their offerings from the manufacturing sector to chemicals, financial services, and latterly public services etc.. This process of building product markets and extending into other segments was characterised by frequent setbacks failures and reversals as the supplier grappled with difficult and contradictory contingencies. Our work explored the detailed techniques and strategies developed by vendors to manage these tensions and their attendant trade-offs (Pollock & Williams, 2009).

## Mapping the Biography of Artefacts

We were, for example, able to document and explain the extension of packaged enterprise solutions through a series of historical and contemporary ethnographic studies that addressed different moments in the life cycle of an ERP package (technology design, procurement, implementation, maintenance). Thus we were able to explore how, in their design and development decisions, suppliers of standard ERP packages were able to build viable 'bridges' to diverse organisational users by various kinds of 'generification work' (Pollock, et al., 2007). Generification involved a complex set of interactions and alignment efforts between the developer and its user community. Our studies revealed a number of linked strategies deployed by suppliers to manage this process including segmenting the market, enrolling selected user organisations as development sites and as members of 'user groups,' and subsequently by sorting, aligning and prioritising user requirements. For example, when a major ERP supplier moved into the Higher Education sector, we saw how certain organisational users sought to gain influence over the development of the package by establishing themselves as pilot sites. We also observed how a major University (Big Civic) experienced a certain loss of control as the supplier expanded the array of pilot users around which it was developing its products. Later, the supplier enlisted members of its user community to help them identify homologies of practice that would be implemented in their new product. This was not so much a search for identical ways of doing things between their different customers but was an attempt to establish limited spreads of practice that could be handled in similar ways by their necessarily generic software package. Through what we described as 'process alignment' work, they established equivalences between disparate organisations around a manageable level of diversity, which their package would cater for in its generic functionality. Other kinds of requirement were in contrast rejected as 'organisationally specific.' These design choices reflected the operation of a complex 'political economy' as the supplier established boundaries around the market for its product and segmented the market. Thus, product enhancement proposals were assessed against the standing of the customer, its representativeness and importance. Equally, the supplier selected organisations as pilot sites on the grounds of their size, reputation, and salience in the sector and their willingness to invest in the pilot implementation and thus their potential as visible exemplars to others in sector. There was a hierarchy of users. At the other end of the spectrum, were the "transactional users" which the supplier treated on a more strictly commodified basis—offering to install additional functionality only if they paid for it. Our study revealed the array of techniques that suppliers had developed to manage this process—to align and organise its relationship with their user markets and achieve effective closure around product features.

A parallel study explored the homologous challenges faced by a software supplier at a much earlier stage in the product cycle. The newly established supplier seeking to convert its new organisational software into a packaged solution had to reconcile escalating levels of customer diversity as it sold its product on to other organisations with differing needs and practices. Its initial strategy of 'accumulating functionality' to cater for the additional requirements of each new customer rapidly gave way to an attempt to sort and regulate these requirements in order to keep the product development and maintenance process manageable.

These generification strategies do not constitute universal principles for the design of software solutions[6]. Instead they represent the lessons learnt by suppliers and the techniques they developed to manage a set of broadly homologous trade-offs which probably confront all suppliers of complex packaged solutions. Crucially, these dilemmas change over the product life cycle—in the early stages as an application was recycled to other organisational users and developed as a packaged solution, and again when a packaged product designed for one environment was re-worked to allow it to travel to other sectors and segments.

We have also been able to chart, through this and previous studies, the historical roots and evolution of ERP. These could be traced directly back to 1960s stock control systems and 1970s Materials Requirements Planning (MRP) systems, and their progressive extension to Manufacturing Resource Planning (MRP II) systems in the 1980s, through (in some accounts) Computer Aided Production Management in the late 1980s, to the 1990s emergence of ERP. Reflecting upon this longer-term history also draws our attention to the changing fabric of the sociotechnical terrain. Thus, MRP emerged from discussion amongst production managers about optimising work scheduling—who initially developed their own applications. Later these become bound up with packaged software solutions with extended functionality across the whole organisation (Clark & Staunton, 1989; Webster, 1991). And whilst accounts of these early stages stress the role of key individuals involved—the 'three gurus': George Plossl, Joe Orlicky, and Ollie Wight—the final stages of the story are populated by a web of various kinds of organisation as the field becomes more institutionalised—not just managers in user organisations but also software suppliers, consultants, and industry analysts who help shape and guide users through the ERP market. To address these we need analytical tools that can encompass this changing register and the relative salience of different kinds of individual or collective action across different periods and contexts, rather than from the outset privilege action or structure.

## TOWARDS A BIOGRAPHY OF e-INFRASTRUCTURES

### What Can We Learn from Looking at Different (e)Infrastructures?

We have elaborated our BoA framework through studies over recent decades charting the unfolding of ERP and some other major organisational technologies at multiple sites of design, procurement, and implementation. Our findings have been supplemented by those of an informal community of researchers also active in these areas (see for example, Clausen & Williams, 1997) and united by a common interest in 'the Social Shaping of Technology' (MacKenzie & Wajcman, 1985; Williams & Edge, 1996). Our goal was to provide an approach that could be relevant to other long-lived complex technologies. The framework seems particularly relevant for addressing the emergence and evolution of technology infrastructures.

Indeed ERP is arguably an e-Infrastructure. Though in much of the literature, the design and implementation of ERP has been analysed as an organisational *IT application*, Hanseth and Braa (1998) suggested that ERP should instead be analysed as an organisational *Information Infrastructure*, noting that it is subject to the same dilemmas, for example over alignment of multiple stakeholders over extended periods, that such infrastructures face (Monteiro & Hanseth, 1995; Hanseth & Braa, 1998; Ciborra, et al., 2001). ERP is often implemented across multiple sites within large branch-plant organisational structures (Markus, et al., 2000; van Fenema, et al., 2007). Indeed one reason cited for the early success of SAP's R/2 system was its wide adoption by European multinationals, which valued its ability to handle multiple currencies and languages (Pollock & Williams, 2009). The success of packaged enterprise solutions may in part reside in the way they provide a template for the corporate Information Infrastructure, offering a solution to otherwise intractable issues of system standardisation and information harmonisation across multi-branch enterprises. In this sense, ERP can be seen to provide an architecture for intra-organisational communication across large heterogeneous organisations, as well as specific application solutions. In addition, Kallinikos (2004a) notes that ERP packages reproduce and reinforce prevalent corporate organisation structures. As a result, ERP and its embedded business processes and presumptions become taken for granted. This discussion strongly parallels discussions in the sociology of classification of how particular categories become infrastructural (Star & Ruhleder, 1996).

Our own work has also noted the convergence between two hitherto somewhat separate technology trajectories concerned, respectively with intra-organisational integration (underpinning today's enterprise systems) and inter-organisational information networks and systems. Our studies observed the extension of ERP to support inter-organisational exchange (e.g. through Customer Relationship Management systems). With the growth of e-Business technologies, whereby organisations collaborate on procurement and supply-chain integrations (Gerst, 2006), ERP systems are now coming to be seen as parts, albeit large components, of wider e-Infrastructures.

As Monteiro and Hanseth (1995) point out, (information) infrastructures differ one from another. Analysis must attend to their specific technical and organisational features. Some of the factors that led us to develop the BoA perspective (in the case of organisational technologies) are particularly relevant to emerging e-infrastructures (e-Is). They are large-scale pervasive initiatives that must accommodate a large number and diverse range of users and uses. Linked to this, substantial investments must be made in their development—and (in order to secure the benefits in this investment) in their subsequent maintenance and enhancement. They thus emerge and evolve over long timeframes, and their development and redevelopment must consequently find ways to cater for users and uses not currently anticipated.

These features—which underpin our decision to describe them as e-infrastructures rather than as computer systems (Edwards, et al., 2009)—throw up a number of theoretical and practical issues. To address these, we propose a programme of research to explicate 'the biographies of e-infrastructures.' The BoA perspective suggests the advantages of multi-local and multi-temporal studies. However, the need for multi-local analysis is not just a question of addressing an ever-greater number of sites. Rather it is a question of which locales and moments to address. We need instead specific sensitising concepts to guide both the empirical research design and its analysis, but do we have a set of specific concepts for analysing e-Is, beyond these somewhat generic constructs?

## SOME RECENT CONCEPTS FOR ANALYSING e-INFRASTRUCTURES

Recent social scientific research on e-Is provides some important pointers. Though not specifically informed by the BoA perspective this work provides valuable examples of the kind of analysis needed to create a biography of e-Is and, more importantly, throws up valuable novel ways of conceptualising these biographies.

e-Is exhibit some common features with computerised systems in general. For example, with integrated organisational applications, issues arise regarding the alignment of differing departments and functional groups within the organisation. In addition to the consequent need to accommodate potentially diverging 'user requirements' across different parts of the organisation, we note the need to reconcile current and emergent uses and users (Williams, et al., 2005; Johannessen & Ellingsen, 2009).

At the same time there are also some distinctive features of e-Is which underpin our decision to distinguish them from conventional information systems. These include the particularly large-scale of e-Is and their (intended) longevity. Scale here is a question of both the number of users and the range of users/uses. This means that they cannot typically be developed or procured as integrated solutions but require extended periods of investment for their initial development/procurement and subsequent enhancement and evolution. In recognition of this, Edwards et al. (2007) suggest that it may be more helpful to apply a metaphor of *growing* rather than designing or building e-Is. These features mean that e-Is result from large-scale, protracted investments, and tend also to have very long timescales. They are typically erected on the foundations of earlier systems development and implementation work (Chae & Poole, 2005). Edwards et al. (2007) have described how earlier, more small-scale/localised e-Is come into contact with others based around different purposes, standards or classification systems. Difficulties in aligning the entrenched differences between these local systems generate pressures of competition or accommodation between systems that may be resolved through the creation of *gateways,* which allow multiple divergent systems to interoperate. The tensions and discrepancies between these local systems may in due course generate pressures leading to periodic adjustments and redevelopments to accommodate changing internal and external circumstances (Ribes & Finholt, 2009).

e-I development and evolution thus involves simultaneous work on many fronts. For example, e-I design needs to serve as a bridge forwards towards future anticipated users/uses. At the same time, e-I implementation involves building workable bridges between necessarily generic features of the e-infrastructure and the particular locales of use. This calls into question the traditional privileging of systems design in much of the early Science and Technology Studies and social oriented computing literature, and its treatment as a discrete, prior episode in isolation from implementation. e-I development is about evolving user practices as well as system design efforts—a process which Pipek and Wulf (2009) describe as *infrastructuring*. Hepsø et al. (2009) make a similar point in describing this interplay between top down and bottom up efforts as *ecologies*. Actors involved must make choices about the architecture of an e-I: choices about boundaries and alliances; about which black boxes to open and which to leave shut. These *configurational politics* (Sahay, et al., 2009) shape the techno-political landscape (or 'translation terrain') for the subsequent (ongoing, more or less overt) conflicts and negotiation to which the evolution of e-Is is subject. Ribes and Finholt (2009) note that those trying to initiate, promote, and grow e-Is need to integrate short-term tactics within longer-term strategies to institutionalise roles and organisations. These are also issues about bridging multiple temporal scales—which Ribes and Finholt (2009) describe, after Braudel (1949), as 'The Long Now' (*longue durée*). Efforts to grow infrastructures are thus deeply paradoxical. For example, e-I design strives to cater for all purposes including those not yet envisaged—but, given the unpredictability of these activities and goals, these anticipatory efforts necessarily fail. Today's imagined future-proof systems rapidly become tomorrow's legacy systems.

## Tensions and Contradictions: Theorising the Dynamics of e-Infrastructure Evolution

Given this state of affairs, can we find a way to theorise the dynamics of e-I evolution? We have argued earlier about the need for frameworks that can give due attention to both stability and dynamism and that will capture the ways in which particular contingencies and reconfigurations will open up different pathways and generate different outcomes (Jørgensen & Sørensen, 1999; Williams, et al., 2004). Can we, in particular, find a mode of analysis that will avoid the already noted pitfalls of dichotomised modes of analysis (e.g. between Can-Do and No-Can-Do interpretations)? For example, are there concepts that can help us capture both the short-term resistance to aligning around particular classifications and standards and the longer-term alignment that may arise as these once contested components become taken-for-granted and infrastructural?

Ciborra et al. (2001) have addressed the contradictory requirements surrounding the development of increasingly complex and wide-reaching inter-organisational systems—and the pressures thereby arising both to increase the number and range of users and to increase the scope of activities that are supported. They describe a shift from "Control to Drift," as attempts to reconcile the increasingly diverse and complex array of stakeholder requirements and uses inevitably derail developer intentions. This pessimistic formulation runs the risk of offering a fatalistic account of the necessity of failure. In contrast, Gerst (2006), in her study of the emergence and evolution of a large e-I (a portal for automotive industry component procurement), shows how the senior managers involved deployed considerable reserves of past experience in anticipating and alleviating the social and technical problems that might emerge. Though there is no 'correct solution' that would resolve the competing exigencies the decision-makers are grappling with (for example increasing integration within the organisation is in competition with another exigency—of aligning the systems of multiple vehicle builders), the outcome is what she describes as 'managed drift.' A growing body of analysis has begun to address such tensions ad contradictions surrounding e-I and other complex systems development (Hanseth, et al., 2006; Gerst, 2006; Ure, et al., 2009).

The analysis of tensions and contradictions provides an alternative to the dichotomisation in modes of analysis we have previously described, between for example modernist and critical accounts that, respectively, promote and dismiss the claims of technology induced restructuring of organisational practice. Instead, this work highlights incremental and partial changes, which may contribute to thoroughgoing transformations over extended periods of time. Those seeking to steer such developments with only incomplete understanding (and given the inevitability of changes in system goals, tools and context that surround their long-term evolution, such complete knowledge is only reliably available with hindsight). For actors needing to make decisions under such circumstances these concepts draw attention to the difficult trade-offs that must be made between competing exigencies, as well as the reflexive efforts of players involved to anticipate and accommodate these and achieve 'do-able' solutions.

## APPLYING THESE CONCEPTS TO ANALYSING HEALTH INFRASTRUCTURES

How can we apply these concepts to understanding the evolution of e-Is in Health?

Findings from other application domains and sectors are relevant to Health e-Is. Indeed ERP systems have been successfully ported across from manufacturing and private service to public services including Higher Education and Local Authorities (see for example Pollock & Cornford, 2004; Pollock, et al., 2007), and have been proposed as a solution for health service organisations as a means of overcoming long-term weaknesses in their utilisation of IT to achieve better integration of information within the organisation. More generally, we see a shift towards supply of health information systems as Commercial-Off-The-Shelf solutions (Martin, et al., 2007a; Johannessen & Ellingsen, 2009; Johnson, 2009; Johnson, 2011) in pursuit of anticipated cost savings and dependability advantages.

At the same time, the integration challenge in health services outstrips pretty-well all other commercial and public sector organisations in terms of both scale and complexity. Health services in developed economies represent some of the largest and most complex organisational structures of all time. For example, the UK National Health Service (NHS) is one of the largest employers in the world with around one million staff, spread across some two thousand hospitals as well as the widely dispersed primary care system with some 40,000 general practitioners[7]. Major IT investments have been proposed for more than a decade as a way of tackling these organisational challenges, responding to political pressures to improve efficiency and quality of service.

Currently emerging Health e-Infrastructures (HeIs) thus represent a rather different socio-technical terrain to the ERP systems previously described, which are integrated solutions promising sharing of information between the range of functions across a single organisation. Today's Health e-Infrastructures are geared towards inter-organisational integration at the interface between the multiple differing organisations involved in health service delivery and, for governance and research purposes, across the entire health service (comprising a huge array of hospitals, clinics and general practices) as well as intra-organisational information integration (Norris, 2002). They seek to offer improvements in health service delivery through more effective sharing of information between specialised functions in secondary care and especially between secondary care and the highly dispersed primary care system[8]. Moreover, in public health systems, demands for better management of resources create powerful additional political incentives to link systems together—for instance by linking in local systems to centralised databases to allow health professionals across the service to access to Electronic Patient Records for their patients[9].

The creation and evolution of Health e-Is, perhaps above all, exemplify first, the need for an analytical shift from a 'discrete systems' to an 'infrastructural' perspective and second, the need to explore long-term biographies rather than isolated moments and sites of change. One feature is the long duration of information systems in health care and especially of the information that they contain (patient records need to last a lifetime for healthcare purposes; perhaps even longer for research purposes). Though acknowledging that HeIs present both a larger scale and a higher level of complexity than most enterprise level solutions in other sectors, both face similar issues—for example around how to cater for the diversity of user requirements and activities—and the consequent need for suppliers to pursue generification strategies in designing packaged solutions

(Johannessen & Ellingsen, 2009; Hyysalo, 2010). These challenges have underpinned some of the, widely publicised, implementation difficulties often encountered with HeIs.

Health care work does of course exhibit a number of features that distinguish it from manufacturing activities and even from other public services like local government and higher education settings we addressed in our ERP studies. We note, in particular, the centrality of expert judgement in clinical diagnosis and treatment, the ongoing elaboration of expert knowledge and practices, reflected in strong professional autonomy and a high degree of professional and departmental specialisation[10]. These circumstances may resist the kinds of alignment and formalisation that underpin the adoption of the kinds of standardised packaged solution that have been seen with ERP and associated ideas of a single central information kernel. However, it would be unhelpful and misleading to portray health services as wholly exceptional[11]. International classification systems have been successfully introduced into health care (Bowker & Star, 1999). In addition, we see the widespread resort to packaged solutions in health service information systems procurement, and, though the barriers to alignment in knowledge codification and practice are great, we must remember that information systems integration needs to be analysed as a cumulative process: local efforts in realising health information systems provide the foundations for subsequent extensions in the spread and scope of health e-Is (though for the reasons outlined above they may be unfolding at a slower rate than in other sectors)[12]. In short, the biographies of the evolution of e-Is in health services will have distinctive differences from other sectors. They will also encounter similar, indeed homologous, challenges. The BoA perspective and its application to analysing e-Is specifically seeks to improve understanding, providing tools for comparative analysis of the evolution of e-Is in their differing settings.

## UK Health Infrastructures: Connecting for Health (NPfIT)

The UK presents a particular challenge for researchers seeking to characterise the evolution of health e-Is given the huge investments currently being made by the UK NHS in the National Programme for Information Technology (NPfIT; subsequently renamed Connecting for Health)[13]. This major programme, launched in 2002 with a projected £7bn budget, was the largest non-military procurement in the world. By 2008, it was already 4 years late and costs had risen to £12.7 billions (National Audit Office, 2008). Substantial savings, estimated at more than £860 million, have been secured through centralised procurement of software through a limited set of Commercial-Off-The-Shelf solutions (Johnson, 2011).

Some dissatisfaction has been expressed from the outset within the health service with NPfIT and with the component applications and services of the system installed to date. These have flagged usability and reliability problems, and frustrations that the new systems installed did not always match the level of functionality of the old systems they displaced. This has been associated with an ongoing and occasionally intense debate, which has involved a number of leading figures in the study of computing, about the philosophy and architecture of this 'System of systems.' An early longitudinal study has flagged the competing institutional exigencies surrounding this development (for example, between patient choice, professional autonomy, efficiency) which may impede it fulfilling its goals (Currie & Guah, 2007). Packaged supply of solutions has thrown up a number of issues. In particular, the initial NPfIT specifications did not anticipate the extent of demands for additional functionality by adopting NHS organisations, which continued to the later stages of the programme. The consequent escalation of development costs for the suppliers threatened their profitability. This led to restructuring of the supply system (withdrawal of suppliers and mergers of weaker players) which had been established to avoid monopoly in provision, with most Strategic Health Authorities choosing between two major configurations (Cerner/BT's Millennium system, and Lorenzo from CSC and IBA Health) (Johnson, 2011).

A number of commentators have expressed concern about the centralised architecture of the NPfIT system, and in particular its central Data Spine, a central national database where summary patient records are stored. This centralisation has been seen as amplifying the dependability and privacy risks of failures (Randell, 2007; Peltu, et al., 2008). Different architectures for integration, it was argued, could dissipate these risks. Thus, Randell (2007, p. 222) argued for "Evolutionary Acquisition: Specifying, Implementing, Deploying, and Evaluating a Sequence of Ever-More Complete IT Systems." This kind of 'socio-technical' approach has been espoused as also offering greater scope for adapting systems to the needs of local users (Eason, 2007). These writings thus foreground many of the same issues that we have seen surround e-I development. These discussions also draw attention to some of the distinctive features of the NPfIT, which distinguish it from previous technical change programmes in the UK and from developments elsewhere. Indeed, comparison between different national patterns of health technology acquisition (see for example Ellingsen & Monteiro, 2003; Currie & Guah, 2007; Johnson, 2009; Hyysalo, 2010) highlight many differences in the way that the vision of e-health has been pursued and the overarching influence of the particular diverse governance, financial and institutional settings of national health care systems which stamp their mark on e-I development.

To date our own research has managed to engage with only small strips of these complex developments (see for example Bunduchi, et al., 2005, 2006; Anderson, et al., 2006; Hardstone, et al., 2006). The enormous scale of these developments, the diversity of specific computer systems being linked together, their coupling with changes in the organisation of health care delivery and its management all present profound challenges to researchers pursuing an adequate account of these unfolding developments. The various specific studies we have been involved in provide interesting pointers to the current large-scale, rapid evolution of the UK health infrastructure, but can only be seen as initial survey points on a much larger map. Some longitudinal analyses of NPfIT have been undertaken (for example, Currie & Guah, 2007). However, in the UK setting, any particular local study seems dwarfed by the overarching scale and scope of developments. Capturing these adequately presents very different analytical challenges than many preceding organisational technologies. This is not an argument against the application of a BoA perspective, but instead underpins the necessity of a broader analytical perspective, and tools to allow individual studies to be integrated within a bigger picture.

## CONCLUSION

The starting point for this chapter was our critique of some of the prevalent models adopted for social scientific research into technology design and implementation. We have noted here our frustration with the research designs and epistemologies not only of impact studies driven by supply side perspectives, but with more critical studies, often based on ethnographic methods, and informed by more sophisticated frameworks such as Actor Network Theory, social interactionism, and ethnomethodology. We criticised the naïve methodologies often adopted—for example the popularity of single site ethnographies of technology adoption—as well as the segmentation of research between technology design and implementation that results from both disciplinary divisions and the practicalities of short-term research. These had produced an inadequate and polarised account of technological and organisational change. There has been a marked failure to reflect upon the ways in which particular forms of research design may be associated with particular findings.

To overcome these shortcomings, we proposed the *Biography of Artefacts* (BoA) framework and have further refined it as an outcome of a long-term programme of collaborative enquiry into the emergence of Enterprise Resource Planning (ERP) and its predecessors. This perspective is underpinned, we argued, by the need to address technologies at different moments in the systems development life cycle (design, implementation, use, and further enhancement) and their broader product cycle (encompassing the shift from emerging to mature products and their subsequent extension and evolution). This is not a criticism of ethnographic methods which have shown themselves to be particularly helpful in understanding richness of local contexts and the detailed social interactions involved in technological innovation. However rather than unreflexively selecting particular sites of study (typically sites of technology adoption) we proposed instead *strategic ethnography*, guided by a provisional understanding of the moments, locales and nexuses in which artefacts and attendant practices and knowledges were being created, exchanged, traded and validated. Our intellectual journey, occasioned by a study of ERP emergence and evolution, seemed particularly relevant to our growing interest in e-Infrastructures.

The first generations of information systems, localised around particular tasks and professional/organisational structures, have in turn provided the foundations for the progressive extension of information sharing across departmental and functional boundaries, resulting in the emergence of the large-scale information systems with multiple users and uses, which we have described as e-infrastructures. These developments call into question both the technology design/management frameworks deployed by practitioners and the analytical tools of social scientists studying them. For example, previously separate research traditions have addressed, respectively, the intra-organisational extension of systems and the emergence of inter-organisational networks/systems. The analysis of e-Is that we have begun to develop here has drawn insights from both traditions.

The huge sunk investments in data and systems in e-Is mean they are long-lived and have to cater for extensions to the activities supported and unanticipated changes in technical environments and social purposes to which they are directed. An exploration of e-Is thus requires an evolutionary analytical framework, that can address multiple temporal scales. The BoA perspective seeks to provide analytical tools for such long-term exploration and is itself informed by an analysis of the complex and protracted interplay between actors and intermediaries in technology supply, implementation, and use that we have analysed under the rubric of *social learning* in technological innovation (Williams, et al., 2005).

These considerations suggest the need to reflect upon the mission of socially oriented computer science to attend to the 'highly flexible, nuanced, and contextualized' character of collaborative human activities and develop systems develop systems that will support these effectively (Ackerman, 2000, p. 197). Without rejecting this mission, our argument supports a shift in perspective away from its traditional focus on the design from scratch of discrete systems (informed by workplace studies). This shift specifically suggests:

1. an altered view of the system design and development process not as *ab initio* design, but as a process of configuration (Fleck, 1988; Stewart & Williams, 2005)—also described by Büscher et al. (2001) as 'bricolage': the skilful selection, combination and reworking of an array of already existing components (hardware, software, classification systems, organisational processes) and their integration with new components; and,
2. an altered view of how and where to engage with organisational users. Thus, workplace studies might usefully go beyond studying the *status quo ante* and instead address the evolution of technology and work over time. In particular, at-

tention may perhaps most productively be focused on how organisation members engage with newly implemented technologies, since, as Fleck (1988) observes, the 'implementation arena' constitutes a key site for innovation in which organisational users engage with and explore the affordances of new technology systems and where new practices are elaborated—a space in which social learning is particularly intense and where new innovation opportunities are thrown up that may enhance future application development. Moreover, these locales bring analysts (and suppliers for that matter) into in contact with actual users in real settings of everyday use, yielding more detailed and robust understandings of emerging user requirements. This can arguably generate information with more immediate implications for system enhancement and evolution.

3. a broader view of development activities and thus of spaces for analysis and for intervention. Recognising the diverse kinds of activities which are salient in system development and evolution, takes us out with the traditional sequential presumptions of when and where design takes place, to include, for example, system configuration, integration, and testing activities (Martin, et al., 2006). For example, issues about hitherto unrevealed diversity in the practices and requirements of user organisations frequently surfaced late in the procurement process, in the course of package implementation and initial use. As a result, negotiations about how far these can be met often took the form of conflicts over the contract with the supplier (Anderson, et al., 2006; Martin, et al., 2007a)[14]. Attention has been drawn to the importance of *integration work* in relation both to the adoption of COTS solutions (Martin, et al., 2007) and in the development of e-Is (Ellingsen & Monteiro, 2003; Johannessen & Ellingsen, 2009)—involving not just technical integration of components but efforts to integrate practices across organisational divides and to link them with applications. As Martin et al. (2007, p. 56) succinctly state: "The 'design problem' becomes concerned not so much with the simple creation of new technical artefacts or the 'computerization' and replacement of work practices as with the effective integration of computer systems with existing and developing localized work practices."

A further corollary is that socially oriented computer-science may need to rethink its received attitude towards the spread of ERP and other packaged solutions that have come to be the major way in which organisation-wide solutions are acquired. In contrast to the emphasis of ethnomethodological and interactionist accounts on the 'unique adequacy' of organisational processes—and thus also the need for specificity of designed systems and of methods of creating them (Voss, 2006)—our studies have suggested that packaged solutions have prevailed because they offer workable 'bridges' between generic organisational/informational templates and the particularities of diverse organisation practices. The prevalence of packaged ERP solutions cannot be simply attributed (as some have suggested) to the success of powerful vendors in marketing their products to naïve organisations (though issues of supplier provenance [past reputation; future prospects] do figure in procurement decisions [Pollock & Williams, 2007]). Their adoption has, moreover, not been without cost and effort and includes a number of widely publicised instances of failure. Nevertheless, large numbers of organisations have adopted them. We suggest that standardised enterprise solutions have been adopted because they provide an information and IT architecture around which organisational information systems can be constructed and do this (stakeholders presumably believe) more cheaply and more robustly than by building such systems from scratch. This is not to suggest the 'victory' of standardisation over local diversity—standardisation of information systems and practices across different groups and organisations is always incomplete and has paradoxical and unintended outcomes (Anderson, et al., 2006; Ellingsen & Monteiro,

2003)[15]. It is important, however, to acknowledge, as well as barriers to standardisation, the gradual sedimentation of certain standardised procedures, codifications, technologies—which provide the foundation upon which further differentiation and elaboration (and, in due course, further integration and standardisation) efforts take place.

As a result, socially oriented computing need to pay attention to the longer-term evolution of technologies and practices. Here the BoA perspective offers not only methodological guidance but also analytical cues regarding the variable dynamics of innovation in differing contexts (for example at different stages of maturation of a technological field). This bears also upon the opportunities for intervention (Stewart & Williams, 2005). These in turn are strongly influenced by the location of a particular technology offering or technical field in the 'historical arc'[16] of its development. For example, in the early stages of the evolution of an information infrastructure, development processes may be relatively open, and particular organisational groups may seek to articulate their purposes through new system development. At later stages, however, the 'translation terrain' (Williams, et al., 2005) on which further development takes place is more complex—already densely populated by past investments and players. These on the one hand provide resources around which new systems will be installed. However, on the other hand, as a result of these commitments and alignments, many technical and social options may be effectively closed off (Sahay, et al., 2009). As a result, in these later stages of information system/e-I evolution, further development opportunities may be primarily for local reconfigurations within a broader already formed information infrastructure. There will of course be periods in which a more radical restructuring of information systems and their underlying architectures prevail, but even here, we suggest, we are likely to find increasing resort to large-scale packaged solutions, given the enormous cost and dependability advantages of recycling code.

These issues are likely to become increasingly salient in relation to health information systems and e-Is. Thus, Johannessen and Ellingsen (2009) have noted the different challenges in the supply of health e-Is in a case in which an information system, designed for one customer, is transferred to other contexts and subsequently to a larger market. The generification strategies adopted by the integration supplier in this study bear striking resemblances to the generification strategies articulated over time by our ERP suppliers.

This is to suggest that we are moving to a situation in which work activities are conducted through extended heterogeneous e-Is rather than discrete systems. In this context, developments are of necessity inevitably partial interventions within the broader, longer-term evolution of e-Is. We are not the first to make this observation, of course. We note important contributions from others, based on work which is frequently directed at the health domain. These draw our attention, for example, to the emergence of efforts directed towards *integration* between existing services, rather than discrete systems design (Ellingsen & Monteiro, 2003; Martin, et al., 2006; Martin, et al., 2007; Johannessen & Ellingsen, 2009). Moreover in these contexts, some of the issues around e-I emergence are salient such as balancing multiple local and central requirements, and the consequently complex and paradoxical (sometimes unintended/undesired) outcomes of such developments (Ellingsen & Monteiro, 2003; Winthereik & Vikkelsø, 2005; Anderson, et al., 2006; Martin, et al., 2007a; Johannessen & Ellingsen, 2009).

e-Is are becoming more widespread, pervasive and salient in health and other sectors. There is, accordingly, a growing number of studies of particular instances of emerging e-Is—not least in Health services. We have proposed a systematic programme of enquiry into e-Is, especially HeIs (and their biographies). Comparative studies between different kinds of e-I may offer important insights (Edwards, et al., 2009)—and there has been a recent conceptual flowering of work in this field. Some of this work, as revealed by the brief review above, has thrown up valuable new conceptual tools—which can contribute to the development of analysis of the biography of e-Is[17]. However, we are still in the early stages in articulating this framework. We hope this brief paper has been able to provide pointers and tools for the intellectual journey to come.

## REFERENCES


Ackerman, M. S. (2000). The intellectual challenge of CSCW: The gap between social requirements and technical feasibility. *Human-Computer Interaction*, *15*(2-3), 179–203. doi:10.1207/S15327051HCI1523_5

Al-Mashari, M. (2003). Enterprise resource planning (ERP) systems: A research agenda. *Industrial Management & Data Systems*, *102*(3), 165–170. doi:10.1108/02635570210421354

Anderson, B. (2000). Where the rubber hits the road: notes on the deployment problem in workplace studies. In Luff, P., Hindmarsh, J., & Heath, C. (Eds.), *Workplace Studies: Recovering Work Practice and Informing System Design*. Cambridge, UK: Cambridge University Press. doi:10.1017/CBO9780511628122.011

Anderson, S., Hardstone, G., Procter, R., & Williams, R. (2006). Down in the (data)base(ment): Supporting configuration in organisational information systems. In Ackerman, M. S., Halverson, C. A., Erickson, T., & Kellogg, W. A. (Eds.), *Resources, Co-Evolution and Artifacts: Theory in CSCW* (pp. 221–254). Dordrecht, The Netherlands: Kluwer. doi:10.1007/978-1-84628-901-9_9

Arrow, K. J. (1962). The economic implications of learning by doing. *The Review of Economic Studies*, *29*, 155–173. doi:10.2307/2295952

Berg, M., & Timmermans, S. (2000). Orders and their others: On the constitution of universalities in medical work. *Configurations*, *8*(1), 31–61.

Bessant, J. (1983). Management and manufacturing innovation: The case of information technology. In Winch, G. (Ed.), *Information Technology in Manufacturing Processes* (pp. 14–30). London, UK: Rossendale.

Botta-Genoulaz, V., Millet, P. A., & Grabot, B. (2005). A survey on the recent literature on ERP systems. *Computers in Industry*, *56*, 510–522. doi:10.1016/j.compind.2005.02.004

Bowker, G. C., & Star, S. L. (1999). *Sorting things out: Classification and its consequences*. Cambridge, MA: MIT Press.

Brady, T., Tierney, M., & Williams, R. (1992). The commodification of industry-application software. *Industrial and Corporate Change*, *1*(3), 489–514. doi:10.1093/icc/1.3.489

Bragato, L., & Jacobs, K. (2003). Care pathways: The road to better health services? *Journal of Health Organization and Management*, *17*(3), 164–180. doi:10.1108/14777260310480721

Braudel, F. (1949). *The Mediterranean and the Mediterranean world in the Age of Philip II* (*Vol. 1*). London, UK: Collins.

Bunduchi, R., Graham, I., Smart, A., & Williams, R. (2005). The tensions shaping the emergence of standard bodies: The case of a national health informatics standard standards body. *Prometheus*, *23*(2), 149–166. doi:10.1080/08109020500085544

Bunduchi, R., Williams, R., Graham, I., & Smart, A. (2006). XML based clinical data standardisation in the national health service Scotland. *Informatics in Primary Care*, *14*(4), 227–234.

Büscher, M., Gill, S., Mogensen, P., & Shapiro, D. (2001). Landscapes of practice: Bricolage as a method for situated design. *Computer Supported Cooperative Work*, *10*(1), 1–28. doi:10.1023/A:1011293210539

Button, G. (1993). *Technology in working order: Studies of work, interaction, and technology*. New York, NY: Routledge.

Callon, M., & Latour, B. (1981). Unscrewing the big leviathan: How actors macrostructure reality and how sociologists help them to do so. In Knorr-Cetina, K. D., & Cicourel, A. V. (Eds.), *Advances in Social Theory and Methodology: Toward an Integration of Micro- and Macro-Sociologies*. Boston, MA: Routledge and Kegan Paul.

Chae, B., & Poole, M. S. (2005). The surface of emergence in systems development: Agency, institutions, and large-scale information systems. *European Journal of Information Systems*, *14*(1), 19–36. doi:10.1057/palgrave.ejis.3000519

Ciborra, C. U., Braa, K., Cordella, A., Dahlbom, B., Failla, A., Hanseth, O., & Simon, K. A. (2001). *From control to drift: The dynamics of corporate information infrastructures*. Oxford, UK: Oxford University Press.

Clark, P., & Staunton, N. (1989). *Innovation in technology and organisation*. London, UK: Routledge.

Clausen, C., & Williams, R. (Eds.). (1997). The social shaping of computer-aided production management and computer integrated manufacture. In *Proceedings of International Conference, COST A4*. Luxembourg: European Commission DGXIII.

Currie, W. L., & Guah, M. W. (2007). Conflicting institutional logics: A national programme for IT in the organisational field of healthcare. *Journal of Information Technology*, *22*(3), 235–247. doi:10.1057/palgrave.jit.2000102

DiMaggio, P. J., & Powell, W. W. (1983). The iron cage revisited: Institutional isomorphism and collective rationality in organisational fields. *American Sociological Review*, *48*(2), 147–160. doi:10.2307/2095101

Eason, K. E. (2007). Local sociotechnical system development in the NHS national programme for information technology. *Journal of Information Technology*, *22*(3), 257–264. doi:10.1057/palgrave.jit.2000101

Edwards, P. N., Bowker, G. C., Jackson, S. J., & Williams, R. (2009). Introduction: An agenda for infrastructure studies. *Journal of the Association for Information Systems*, *10*(5), 155–173. Retrieved from http://aisel.aisnet.org/jais/vol10/iss5/

Edwards, P. N., Jackson, S. J., Bowker, G. C., & Knobel, C. P. (2007). *Understanding infrastructure: Dynamics, tensions, and design*. Ann Arbor, MI: Deep Blue.

Ellingsen, G., & Monteiro, E. (2003). A patchwork planet integration and cooperation in hospitals. *Computer Supported Cooperative Work*, *12*(1), 71–95. doi:10.1023/A:1022469522932

ERP Research Group. (2006). ERP systems and related issues bibliography: 600 articles. *Victoria University*. Retrieved from http://www.business.vu.edu.au/sap/Research.html#ERP_Systems_and_Related_Issues_Bibliogr.

Esteves, J., & Pastor, J. (2001). Enterprise resource planning systems research: An annotated bibliography. *Communications of AIS*, *7*(8), 1–52.

Fitzpatrick, G. (Ed.). (2003). *The locales framework: Understanding and designing for wicked problems*. Dordrecht, The Netherlands: Kluwer Academic Publishers.

Fleck, J. (1988). Innofusion or diffusation? *The nature of technological development in robotics*. Working Paper. Edinburgh, UK: The University of Edinburgh.

Fleck, J., Webster, J., & Williams, R. (1990). The dynamics of IT implementation: A reassessment of paradigms and trajectories of development. *Futures*, *22*, 618–640. doi:10.1016/0016-3287(90)90131-Z

Gerst, M. (2006). *The socio-technical dynamics of ICT innovation: A social shaping analysis of portals*. Unpublished PhD Thesis. Edinburgh, UK: University of Edinburgh.

Graham, I., Lloyd, A., Slack, R., & Williams, R. (2003). *Uptake of BPR*. PRECEPT Working Paper. Lyngby, Denmark: Danish Technical University.

Graham, I., Spinardi, G., & Williams, R. (1996). Diversity in the emergence of electronic commerce. *Journal of Information Technology*, *11*(1), 161–172. doi:10.1080/026839696345360

Graham, I., Spinardi, G., Williams, R., & Webster, J. (1995). The dynamics of EDI standards development. *Technology Analysis and Strategic Management*, *7*(1), 3–20. doi:10.1080/09537329508524192

Hanseth, O., & Braa, K. (1998). Technology as traitor: Emergent SAP infrastructure in a global organization. In *Proceedings of the International Conference on Information Systems,* (pp. 188-196). Helsinki, Finland: Association for Information Systems.

Hanseth, O., Jacucci, E., Grisot, M., & Aanestad, M. (2006). Reflexive standardization: Side effects and complexity in standard-making. *Management Information Systems Quarterly*, *30*, 563–581.

Hanseth, O., Monteiro, E., & Hatling, M. (1996). Developing information infrastructure standards: The tension between standardization and flexibility. *Science, Technology & Human Values*, *21*(4), 407–426. doi:10.1177/016224399602100402

Hardstone, G., D'Adderio, L., & Williams, R. (2006). Standardization, trust and dependability. In Clarke, K., Hardstone, G., Rouncefield, M., & Sommerville, I. (Eds.), *Trust in Technology: A Socio-Technical Perspective* (pp. 69–103). Dordrecht, The Netherlands: Springer. doi:10.1007/1-4020-4258-2_4

Hartswood, M., Procter, R., Slack, R., Voss, A., Büscher, M., Rouncefield, M., & Rouchy, R. (2002). Co-realisation: Towards a principled synthesis of ethnomethodology and participatory design. *Scandinavian Journal of Information Systems*, *14*(2), 9–30.

Hepsø, V., Monteiro, E., & Rolland, K. H. (2009). Ecologies of e-infrastructures. *Journal of the Association for Information Systems*, *10*(5), 430–446. Retrieved from http://aisel.aisnet.org/jais/vol10/iss5/

Hine, C. (2007). Multi-sited ethnography as a middle range methodology for contemporary STS. *Science, Technology & Human Values*, *32*(6), 652–671. doi:10.1177/0162243907303598

Hughes, J. A., King, V., Rodden, T., & Andersen, H. (1994). Moving out of the control room: Ethnography in system design. In *Proceedings of the 1994 ACM Conference on Computer Supported Cooperative Work (CSCW 1994),* (pp. 429-438). Chapel Hill, NC: Association of Computing Machinery.

Hutchins, E. (1995). *Cognition in the wild*. Cambridge, MA: The MIT Press.

Hyysalo, S. (2004). *Uses of innovation: Wristcare in the practices of engineers and elderly*. Academic Dissertation. Helsinki, Finland: University of Helsinki.

Hyysalo, S. (2010). *Health technology development and use*. New York, NY: Routledge.

Johannessen, L. K., & Ellingsen, G. (2009). Integration and generification--Agile software development in the healthcare market. *Computer Supported Cooperative Work*, *18*(5-6), 607–634. doi:10.1007/s10606-009-9097-8

Johnson, C. W. (2009). Politics and patient safety don't mix: Understanding the failure of large-scale software procurement for healthcare systems. In P. Casely & C. W. Johnson (Eds.), Proceedings, *Fourth IET Systems Safety Conference*. London, UK: IET Conference Publications.

Johnson, C. W. (2011). *Case studies in the failure of healthcare information systems*. Retrieved from http://www.dcs.gla.ac.uk/~johnson/papers/AHRQ/case_study.pdf.

Jørgensen, U., & Sørensen, O. (1999). Arenas of development: A space populated by actor-worlds, artefacts, and surprises. *Technology Analysis and Strategic Management*, *11*(3), 409–429. doi:10.1080/09537329910743 8

Kallinikos, J. (2004a). Farewell to constructivism: Technology and context-embedded action. In Avgerou, C., Ciborra, C., & Land, F. (Eds.), *The Social Study of Information and Communication Technology: Innovation, Actors, and Contexts*. Oxford, UK: Oxford University Press.

Kallinikos, J. (2004b). Deconstructing information packages: Organizational and behavioural implications of ERP systems. *Information Technology & People*, *17*(1), 8–30. doi:10.1108/09593840410522152

Koch, C. (2003). 'ERP-software packages: Between mass production communities and intraorganisational political processes. In Preece, D., & Laurila, J. (Eds.), *Technological Change and Organizational Action* (pp. 56–76). London, UK: Routledge.

Koch, C. (2007). ERP: A moving target. *International Journal of Business Information Systems*, *2*(4), 426–443. doi:10.1504/IJBIS.2007.012544

Latour, B. (1987). *Science in action*. Cambridge, MA: Harvard University Press.

Law, J. (2004). *After method: Mess in social science research*. London, UK: Routledge.

Luff, P., Hindmarsh, J., Heath, C., & Hinds, P. J. (Eds.). (2000). *Workplace studies: Recovering work practice and informing system design*. Cambridge, UK: Cambridge University Press. doi:10.1017/CBO9780511628122

Mabert, V., Soni, A., & Venkataramanan, M. (2001, May). Enterprise resource planning: Common myths versus evolving reality. *Business Horizons*, *n.d.*, 69–76. doi:10.1016/S0007-6813(01)80037-9

Mackay, H., Carne, H., Beynon-Davies, P., & Tudhope, D. (2000). Reconfiguring the user: Using rapid application development. *Social Studies of Science*, *30*(5), 737–757. doi:10.1177/030631200030005004

Mackay, H., & Gillespie, G. (1992). Extending the social shaping of technology approach: Ideology and appropriation. *Social Studies of Science*, *22*, 685–716. doi:10.1177/030631292022004006

MacKenzie, D., & Wajcman, J. (Eds.). (1985). *The social shaping of technology: How the refrigerator got its hum*. Milton Keynes, UK: Open University Press.

Marcus, G. E. (1995). Ethnography in/of the world system: The emergence of multi-sited ethnography. *Annual Review of Anthropology*, *24*, 95–117. doi:10.1146/annurev.an.24.100195.000523

Markus, M. L., Tanis, C., & van Fenema, P. C. (2000). Enterprise resource planning: Multisite ERP implementations. *Communications of the ACM*, *43*(4), 42–46. doi:10.1145/332051.332068

Martin, D., Hartswood, M., Slack, M., & Voss, A. (2006). Achieving dependability in the configuration, integration and testing of healthcare technologies. *Computer Supported Cooperative Work*, *15*(5-6), 467–499. doi:10.1007/s10606-006-9032-1

Martin, D., Mariani, J., & Rouncefield, M. (2007). Managing integration work in an NHS electronic patient record (EPR) project. *Health Informatics Journal*, *13*(1), 47–56. doi:10.1177/1460458207073648

Martin, D., Procter, R., Mariani, J., & Rouncefield, M. (2007a). *Working the contract*. Paper presented at OZCHI 2007: Australian Computer-Human Interaction Conference. Adelaide, Australia.

Martin, D., & Sommerville, I. (2006). Ethnography and the social structure of work. In Besnard, D., Gacek, C., & Jones, C. B. (Eds.), *Structure for Dependability: Computer-Based Systems from an Interdisciplinary Perspective* (pp. 169–188). London, UK: Springer. doi:10.1007/1-84628-111-3_9

McLaughlin, J., Rosen, P., Skinner, D., & Webster, A. (1999). *Valuing technology: Organisations, culture and change*. London, UK: Routledge. doi:10.4324/9780203019870

Monteiro, E., & Hanseth, O. (1995). Social shaping of information infrastructure: On being specific about the technology. In Orlikowski, W., Walsham, G., Jones, M. R., & DeGross, J. I. (Eds.), *Information Technology and Changes in Organisational Work* (pp. 325–343). London, UK: Chapman & Hall.

National Audit Office. (2008). *The national programme for IT in the NHS: Progress since 2006 National Audit Office Value for Money Report, HC 484-I Session 2007-2008: 16 May 2008*. London, UK: National Audit Office.

Norris, A. C. (2002). Current trends and challenges in health informatics. *Health Informatics Journal*, *8*, 205–213. doi:10.1177/146045820200800407

Peltu, M., Eason, K., & Clegg, C. (2008). *How a sociotechnical approach can help NPfIT deliver better NHS patient care*. London, UK: Bayswater Institute. Retrieved from http://www.bayswaterinst.org/downloads/Sociotech%20Group%20NPfIT%20report%20May08.pdf.

Pipek, V., & Wulf, V. (2009). Infrastructuring: Toward an integrated perspective on the design and use of information technology. *Journal of the Association for Information Systems*, *10*(5). Retrieved from http://aisel.aisnet.org/jais/vol10/iss5/

Plowman, L., Rogers, Y., & Ramage, M. (1995). What are workplace studies for? In H. Marmolin, Y. Sundblad, & K. Schmidt (Eds.), *Proceedings of the Fourth European Conference on Computer-Supported Cooperative Work*, (pp. 309-324). Stockholm, Sweden: Kluwer Academic Publishers.

Pollock, N. (2005). When is a work-around? Conflict and struggle in computer systems development. *Science, Technology & Human Values*, *30*(4), 496–514. doi:10.1177/0162243905276501

Pollock, N., & Cornford, J. (2004). ERP systems and the university as an 'unique' organisation. *Information Technology & People*, *17*(1), 31–52. doi:10.1108/09593840410522161

Pollock, N., & Williams, R. (2007). Technology choice and its performance: Towards a sociology of software package procurement. *Information and Organization*, *17*(3), 131–160. doi:10.1016/j.infoandorg.2007.05.001

Pollock, N., & Williams, R. (2009). *Software and organisations: The biography of the enterprise-wide system or how SAP conquered the world?* London, UK: Routledge.

Pollock, N., Williams, R., & D'Adderio, L. (2007). Global software and its provenance: Generification work in the production of organizational software packages. *Social Studies of Science*, *37*(2), 254–280. doi:10.1177/0306312706066022

Randell, B. (2007). A computer scientist's reactions to NPfIT. *Journal of Information Technology*, *22*, 222–234. doi:10.1057/palgrave.jit.2000106

Ribes, D., & Finholt, T. A. (2009). The long now of technology infrastructure: Articulating tensions in development. *Journal of the Association for Information Systems*, *10*(5), 375–398. Retrieved from http://aisel.aisnet.org/jais/vol10/iss5/

Sahay, S., Monteiro, E., & Aanestad, M. (2009). Configurable politics and asymmetric integration: Health e-infrastructures in India. *Journal of the Association for Information Systems*, *10*(5), 399–414. Retrieved from http://aisel.aisnet.org/jais/vol10/iss5/

Schmidt, K. (2000). The critical role of workplace studies in CSCW. In Luff, P., Hindmarsh, J., Heath, C., & Hinds, P. J. (Eds.), *Workplace Studies: Recovering Work Practice and Informing System Design* (pp. 141–149). Cambridge, UK: Cambridge University Press. doi:10.1017/CBO9780511628122.007

Soh, C., Kien, S., & Tay-Yap, J. (2000). Cultural fits and misfits: Is ERP a universal solution? *Communications of the ACM, 43*(4), 47–51. doi:10.1145/332051.332070

Somers, T., & Nelson, K. (2004). A taxonomy of players and activities across the ERP project life cycle. *Information & Management, 41*, 257–278. doi:10.1016/S0378-7206(03)00023-5

Sørensen, K., & Levold, N. (1992). Tacit networks, heterogeneous engineers, and embodied technology. *Science, Technology & Human Values, 17*(1), 13–35. doi:10.1177/016224399201700102

Star, S. L., & Ruhleder, K. (1996). Steps towards an ecology of infrastructure: Design and access for large information spaces. *Information Systems Research, 7*(1), 111–134. doi:10.1287/isre.7.1.111

Stewart, J., & Williams, R. (2005). The wrong trousers? Beyond the design fallacy: Social learning and the user. In Howcroft, D., & Trauth, E. M. (Eds.), *Handbook of Critical Information Systems Research: Theory and Application* (pp. 195–223). Cheltenham, UK: Edward Elgar.

Suchman, L. (2002). Located accountabilities in technology production. *Scandinavian Journal of Information Systems, 14*(2), 91–105.

Suchman, L. A. (1987). *Plans and situation actions: The problem of human-machine communication*. Cambridge, UK: Cambridge University Press.

Taylor, J. R., & Virgili, S. (2008). Why ERPs disappoint: The importance of getting the organisational text right. In Grabot, B., Mayère, A., & Baze, I. (Eds.), *ERP Systems and Organisational Change: A Socio-Technical Insight* (pp. 59–84). London, UK: Springer. doi:10.1007/978-1-84800-183-1_5

Timmermans, S., & Berg, M. (1997). Standardization in action: Achieving universalism and localization in medical protocols. *Social Studies of Science, 27*, 273–305. doi:10.1177/030631297027002003

Tingling, P., & Parent, M. (2002). Mimetic isomorphism and technology evaluation: Does imitation transcend judgement. *Journal of the Association for Information Systems, 3*(5), 113–143.

Turner, W., Bowker, G., Gasser, L., & Zacklad, M. (2006). Information infrastructures for distributed collective practices. *Computer Supported Cooperative Work, 15*, 93–110. doi:10.1007/s10606-006-9014-3

Ure, J., Procter, R. N., Lin, Y., Hartswood, M., Anderson, S., & Lloyd, S. (2009). The development of data infrastructures for ehealth: A socio-technical perspective. *Journal of the Association for Information Systems, 10*(5), 415–419. Retrieved from http://aisel.aisnet.org/jais/vol10/iss5/

van Fenema, P. C., Koppius, O. R., & van Baalen, P. J. (2007). Implementing packaged enterprise software in multi-site firms: Intensification of organizing and learning. *European Journal of Information Systems, 16*, 584–598. doi:10.1057/palgrave.ejis.3000708

Voss, A. (2006). *Corealisation: A radical respecification of the working division of labour in systems development*. Unpublished PhD Thesis. Edinburgh, UK: University of Edinburgh.

Walsham, G. (2001). *Making a world of difference: IT in a global context*. Chichester, UK: Wiley.

Webster, J. (1990). *Office automation: The labour process and women's work in Britain*. Hemel Hempstead, UK: Harvester Wheatsheaf.

Webster, J. (1991). Advanced manufacturing technologies: Work organisation and social relations crystallised. In Law, J. (Ed.), *A Sociology of Monsters: Essays on Power, Technology, and Domination* (pp. 192–222). London, UK: Routledge.

Wei, H., Li, C., Wang, E., Li, C., & Ju, P. (2005). Understanding misalignment and cascading change of ERP implementation: A stage view of process analysis. *European Journal of Information Systems, 14*(4), 324–334. doi:10.1057/palgrave.ejis.3000547

Williams, R. (1997a). Universal solutions or local contingencies? Tensions and contradictions in the mutual shaping of technology and work organisation. In McLoughlin, I., & Mason, D. (Eds.), *Innovation Organizational Change and Technology* (pp. 170–185). London, UK: International Thompson Business Press.

Williams, R. (1997b). The social shaping of a failed technology: Mismatch and tension in the supply and use of computer-aided production management. In Clausen, C., & Williams, R. (Eds.), *The Social Shaping of Computer-Aided Production Management and Computer Integrated Manufacture* (*Vol. 5*, pp. 109–130). Luxembourg: European Commission DGXIII.

Williams, R., Bunduchi, R., Graham, I., Pollock, N., Procter, R., & Voss, A. (2004). Understanding the evolution of standards: Alignment and reconfiguration in standards development and implementation arenas. In *Proceedings of the 4S/EASST Conference*. Paris, France: 4S/EASST.

Williams, R., & Edge, D. (1996). The social shaping of technology. *Research Policy*, *25*(6), 865–899. doi:10.1016/0048-7333(96)00885-2

Williams, R., & Steward, F. (1985). *The role of the parties concerned in the introduction of new technology - Case studies on the implementation of collective agreements on new technology in Britain*. Dunblin, Ireland: European Foundation for the Improvement of Living and Working Conditions.

Williams, R. A., Stewart, J. K., & Slack, R. (2005). *Social learning in technological innovation: Experimenting with information and communication technologies*. Upper Saddle River, NJ: Edward Elgar.

Winthereik, B. R., & Vikkelsø, S. (2005). ICT and integrated care: Some dilemmas of standardising inter-organisational communication. *Computer Supported Cooperative Work*, *14*, 43–67. doi:10.1007/s10606-004-6442-9

Wu, J. (2008). *Critical success factors for ERP system implementation. Research and Practical Issues of Enterprise Information Systems II* (pp. 739–745). Boston, MA: Springer.

## ENDNOTES

1 Though we use the term e-infrastructure we note that these various terms (cyberinfrastructure, information infrastructure etc) are used in inconsistent and overlapping ways.

2 We find it interesting that there are very few CSCW studies of ERP implementation. (Indeed, the area of packaged software appears rather strangely to have been mostly ignored by this group of scholars). The few studies we did find, however, exemplified similar aspects of the more general information systems literature in highlighting the gap between expectations and outcomes. Taylor and Virgili (2008, p. 68), for instance, write: '...it then became possible to conceptualise the gap between current modes of working, and those that SAP envisioned. As this process transpired, however, the complexity of the SAP technology was also beginning to reveal itself. How to reconcile accepted practice and new system now became less a simple matter of identifying discrepancies and correcting them than it did of finding a way to deal with the intractable realities of practice either by modifying the technology, or abandoning the practice—or both. This was not exactly the way the development process had been envisioned. It was more complex—considerably more.'

3 Indeed classical ethnography, which involves immersive acquaintance with what is involved with being 'a member of the tribe,' might in many ways escape our critique of short-term, local case studies.

4 Our critique of the naïve empiricism of localist studies (especially those inspired by eth-

nomethodology) also applies to the simplistic methodological nostrums of Actor-Network Theory (ANT) as exemplified by Latour's (1987) much cited exhortation to 'follow the actor.' This begs the obvious question, "which actors should we follow?" (see Sørensen & Levold, 1992). What is at stake here is an orientation to theory. ANT Theorists have argued against the resort to the existing body of social science findings on the grounds that these may offer misleading generalisations and presume that the world of tomorrow will simply repeat patterns entrenched in current structures. Similarly, ANT writers suggest, social science methodologies embed potentially misleading presumptions about what are the key factors at play—and in this way may unintentionally constitute reality, for example by naturalising the power imputed to existing institutions (Callon & Latour, 1981; Law, 2004). However their claims to be able to apply 'naturalistic' observation methods, unencumbered by theoretical commitments, fall foul of well-established critiques of *empiricism,* and not least of failing to address taken-for-granted social relations (and they have been criticised for overlooking entrenched asymmetries for example of class and gender). We argue instead that the existing research base should be treated as provisionally or partially valid background knowledge—that we can use to *inform* research design choices and interpretation rather than to *prejudge* outcomes (Pollock & Williams, 2009).

5 See previous endnote

6 There will be sharp differences for example between the exigencies surrounding the development of different kinds of software development—for example between complex organisational technologies discussed here and discrete or component technologies such as personal productivity tools (Brady, et al., 1992).

7 Data from NHS Information Centre http://www.ic.nhs.uk/statistics-and-data-collections/workforce sampled 10 March 2010.

8 And many current visions of health service reform envisage the extension and integration of services, and the IT systems that support them, between health and social care.

9 The centralised architecture proposed in England for an electronic Summary Care Record that will be available to NHS staff involved in a patient's care, anywhere in the country, has proved rather controversial, in a context in which some have expressed deep concern about the confidentiality of Electronic Patient Records. The potential sensitivity of information stored on Health e-Is, as well as the clinical importance of data integrity, underpin the strict regulation of access and exchange of health information, and has a profound influence on system design and architecture and the procurement and implementation process. The competing exigencies—for example the stringent data protection requirements for personal health data coupled with the frequent need to share such data for risk abatement and effective care (Norris, 2002)—create additional contradictions to the development of health e-Is.

10 Personal correspondence with Sampsa Hyysalo 18th March 2010.

11 Indeed there are evident similarities between the visions and models of change between healthcare and other industrial sectors, including the notions of information integration as well as a process orientation to technology enabled restructuring (Norris, 2002; Bragato & Jacobs, 2003).

12 A wide range of other factors have been advanced to help explain the particular difficulties that seem to be associated with information technology adoption in health services. As well as their complexity and large scale, these include organisational

fragmentation and the difficulties integrating political reform with the often rather longer timeframes for large-scale procurement, implementation and evolution of health e-Is (Johnson, 2009). As a result, despite an extremely long history of IT adoption in health services, dating back to the 1960s and before, we do not always find the *cumulative* development of IT capabilities and infrastructures. At the same time, our colleague Mark Hartswood has drawn our attention to an implication of our analysis (personal communication, 30th March 2010) that the emergence of Health e-Is is inevitably going to be a 'long road': a mixed bag of failure and partial success in the development and maturation of generic technologies and systems that support complex organisational work over a number of decades.

13 See http://www.connectingforhealth.nhs.uk/. The description that follows addresses the unfolding of NPfIT in its major area of England. Other parts of the UK have followed somewhat differing approaches.

14 As we have noted elsewhere (Pollock & Williams, 2009, p. 66), market provision of complex organisational solutions frequently encounter 'incomplete contracting' issues. Though effective monitoring of the contract calls for strict prior specification of customer requirements which can be embedded in the contract and their fulfilment policed, in practice the user organisation only has an imperfect initial understanding of its own requirements, which invariably evolve as the system becomes implemented, partly in reaction to the capabilities of the package.

15 For example, as Anderson et al. (2006) observe, when classification systems become embedded in organisational software solutions, deficiencies in agreeing classification systems are transposed into a categorisation problem – of getting idiosyncratic events to conform to standard categories.

16 Personal correspondence with Mark Hartswood 30 March 2010.

17 Indeed these tools developed for analysing the evolution of e-Is can also contribute to analyses of the biographies of other large-scale, complex and evolving technologies.

Section 4

# Phenomenology, Organizational Politics, and Organizational Design

# Chapter 14
# The Horizons of Experience: The Limits of Rational Thought upon Irrational Phenomena

**Tony Hines**
*Manchester Metropolitan University, UK*

## ABSTRACT

*Social researchers consciously and intentionally set out to understand the life-world (Lebenswelt). It is different than knowing the inanimate world of objects. Social study is systematic and as rigorous as natural science but focuses on human experience. The social world we inhabit cradles all human experience. It is the context for meaning, for all being and becoming. Life is a temporal stream of experience that, if we are to understand it, needs to be kept at bay (bracketing it). In doing so, it is as Husserl said at the 'horizon of experience,' pre-theoretical. It is only through critical reflection that we understand our lived experience in relation to others, present and past. This chapter reflects on Husserl's notion of 'horizon of experience' to intentionally understand the limits of rational thought on irrational objects. It offers insights through these reflections and possibilities for researching information systems applying phenomenology.*

## INTRODUCTION

*We shall not cease from exploration*
*And the end of all our exploring*
*Will be to arrive where we started*
*And know the place for the first time*
*T.S. Eliot (1969, pp. 329-330)*

DOI: 10.4018/978-1-4666-0303-5.ch014

It often appears to social researchers that we begin to explore ontological and epistemological questions by drawing on our historical biographies and philosophical repositories to make sense of who we are, only to find that after we have done so, we arrive at the place we started from. Our intention is to understand our subjective selves in relation to our objective world. This somewhat crude duality distinguishing between subject and object situates our arguments of existence in space and time in relation to agents and structures, individualism and holism, reason and unreason, the variant and invariant nature of things. However, what is different about this becoming, is that we begin to understand ourselves (who we are), and where we are, for the first time. Thus, we make sense of everyday lives through the lived experience. This is a point made well by T.S. Eliot in the extract above from his poem 'Little Gidding.' It is the reflexive interest in knowing ourselves and our place in the world that concerns each of us in our attempts to understand the phenomenon of human existence. To understand the phenomenon of human existence through human endeavour and human organization we address our questions to philosophy, science, and sociology.

The phenomenon appearing within my horizon placed under scrutiny here is phenomenology itself. In this quest I want to particularly focus attention upon meanings attributed to the notion of 'horizon' and its correlates: epoché, intentionality, objects of consciousness, and meaning-making. The motivation is to demonstrate that a phenomenological approach offers researchers' a repertoire of possibilities with potential to understand information systems. Conceptualising this phenomenological approach draws inspiration from the past and traces the correlates in other strands of interpretive social research.

Hegel's phenomenology of spirit is foundational to phenomenology. Adorno said in relation to phenomenology that Husserl "discovered the North Pole for the second time"(Hyppolite, 1974, p. vii). This is somewhat ungenerous given Husserl's immense contribution and yet, it confirms Hegel's foundational dialectic phenomenology. Hegel's approach in phenomenology had much in common with his own German idealism. Characteristics of phenomenology are also shared with 'radical empiricism' discussed by William James and his discussions of a 'world of pure experience.' However, phenomenology in its motivation for Husserl and in its early conception owed a debt to the descriptive psychology of consciousness of Brentano (Moran, 2000). There are three distinct developments of approach influential to my own understanding: 1) Hegel's phenomenology of spirit setting an agenda of dialectic phenomenology; 2) Husserl and his followers, particularly Schütz, Merleau-Ponty, and Scheler, amongst others, developing transcendental phenomenology; and 3) Heidegger's focus on being and time leading to the development of *Dasein* drawing a line under Cartesian duality. It is from these voices of the past, and from, past experience as well as from contemporary social researchers and my own lived experience [Erfahrung] that I draw inspiration for this chapter examining the horizons of experience.

It is worth prefacing the unfolding discussion with a caveat that phenomenology is not a unified doctrine but rather that each of the foundational writers brought their distinct interpretation to the meaning of phenomenology. It is a movement, a method, and a methodology offering understanding and meta-theoretical insights into our everyday lived experiences. This chapter sketches the ways our everyday life experience shape action and thought, through a symbiotic process called, nominally, living, or human-being. It begins by examining reason and perception before examining what Husserl and others said in relation to horizons of experience and how this has shaped phenomenology. It traces later developments releasing phenomenology from its Cartesian chains of cognition, conceptualising the essence of phenomenology as a movement and a means of studying social situations. Thus, allowing everyday experience to contribute understanding without recourse to Cartesian dualisms [cogito ergo sum] and recognises that the subject-object duality is of greater concern. The embodied mind becomes a part of lived experience rather than separated from it, contrary to the way Descartes puts it describing his *res cogitans* (mindful thinking) as primary to his secondary *res extensa* (bodily doing).

What we are able to know given the nature of reality (ontology) and how we know, given methodological consideration of how knowledge is constructed (epistemology) is emphasised by Ryle in his statement about 'knowing what' and 'knowing how' (Ryle, 1949). This is a distinction taken from the earliest writings of Plato and Aristotle in their discussions of *phronesis* (practical knowledge or wisdom) and *techne* (application of knowledge). In their schema other forms of knowledge included episteme (scientia in Latin—pure knowledge), *sophia* (theoretical wisdom), *nous* (activity of the mind—intelligence and it also means intention) and *politike* (political science). For Plato there is an essence of form that exists hidden behind the phenomena observed in empirical inquiry of the world of objects (Plato, 1941). Knowledge of forms is *a priori* to experience located in rational thought. This is foundational to structuralism. For Aristotle the crafting of knowledge is in its application and this is not its essence but rather comes *a posterioiri* through experience (Aristotle, 1976). Experience of the life-world gives a context to its meaning both for individuals and social groups where there is shared culture that informs agent actions in the world (agency). Let us begin by considering reason and perception in relation to ontology and epistemology. The primary question addressed is: what can we know about phenomena? Secondary questions relate to: how do we come to know and what type of knowing is it? The first question is ontological and the second is epistemological.

## RATIONALITY AND THE HUMAN PSYCHE (REASON AND PERCEPTION)

Husserl, locates his development of phenomenology tracing it to the rationality of Descartes expressed most clearly in the thought experiment of the *cogito* argument (Husserl, 1950). His later transcendental phenomenology owes more to Kant's rationality in his critique of 'pure reason' a reference to Plato's legacy of reason (Gardner, 1999). Transcendental phenomenology shares with Kant's transcendental subjectivity the notion that the natural world is a realm of objects "existing as the correlates of acts of consciousness" (Gardner, 1999, p. 343) and is close to Kant's empirical world of appearance. As Gardner points out this fact is obscured by the later practitioners of phenomenology in their efforts to distance themselves from this Kantian and neo-Kantian rationality, particularly Heidegger, Merleau-Ponty, and Sartre.

Kant made a careful distinction between phenomena and noumena in his 'critique of pure reason' (Kant, 1900). Phenomena were experienced through human senses and endowed with meaning through subjective mindful sensibility whereas noumena are objects, 'things in themselves.' While reason is subject to the mind, the 'things in themselves,' the objects, have no meaning other than that assigned by the subject through reason. The conception of 'pure reason' is an acknowledgement that reason is according to Plato the highest form of knowledge. It is not contaminated by the non-pure empirical world in which we live and therefore, is not subject to 'noise' from that world. Abstract thought enabled clarity and existence of higher forms of knowledge [Erlebnis]. Of course, purity in any sense of meaning is an ideal rather than an actuality and so it was with Plato and other idealist philosophers in their search for purity of reason, which Kant makes clear in his critique. This abstract reason is ontologically individualist in conception. It is not possible to know in detail the reasoning of other minds (Wittgenstein, 1998). Nor is it possible to know the perception of other minds apart from what is declared by the individual through action: deeds and words. This is something Wittgenstein draws from Kant's reasoning that to perceive is not to know(Kant, 1900).

It is perhaps worth noting here that this distinction between the known (object) and the knower (subject) is at the centre of many epistemic arguments in scientific debate. As an example, it is cause for concern when we attempt to understand the lived experience of others, since researchers step into other worlds and view new horizons to make sense of them. In order to make sense of others experience (new horizons) we need to develop shared understandings, which for the researcher become a part of their own experience. It is a cultural understanding.

William James's radical empiricism had much in common with neo Kantians, but an important point of departure is that consciousness has a duality not of mind and matter in the Cartesian sense, but rather it is a duality of subject and object and it is the logical correlation of 'content' in an experience (James, 1912). Nevertheless, for James the deeper meanings of reality were only found in perceptual experience. In this construction of radical empiricism, James introduces the notion that the connections made by the subject in relation to the objects of experience are a part of that experience. It is in the connections that meanings are made. It offers a bridge between experience and understanding without the Cartesian dualism. In his conception, James gets closer to explaining radical empiricism as phenomenology.

The focus on experience emphasises that phenomenology studies data from everyday life. The movement was critical of natural science and abstract theory moving it away from the everyday human experience. *'Zu den Sachen selbst'* is the essence of the phenomenological approach examining the things themselves. For some phenomenology was perceived as similar to the British empiricist movement and there are indeed similarities with Hume's psychology of human nature. However, the emphasis on the lived experience distances phenomenology from Locke's 'tabula rasa,' although some consider the epoché has similarities. Phenomenology conceives that experience is rather formed intentionally with purpose, with interpretation shaping relations between objects and subjects of the experience within our horizon.

Phenomenology has roots located in the idealism of Plato, the posterior analytics of Aristotle, Bacon, and Hume's empiricism, Cartesian rationality, Kant's critique of pure reason, Hegel's dialectic phenomenology, Marx's false consciousness, superstructures and substructures, Weber's social action, Brentano's psychology, Husserl's method of reductionism, Schütz social relations, Merleau-Ponty's perception, and Heidegger's *Dasein*. Popper opined "The much debated question whether it is a new invention, as pure phenomenologists think, or perhaps a version of Cartesianism or Hegelianism, can easily be decided; it is a version of Aristotlianism,"(Popper, 1945, pp. 18-19). However, this is to overlook the essence of idealism from Plato and the contributions listed. Table 1 summarises some key contributions to the development of phenomenology although the list is by no means exhaustive. For example, I have left out the legacy from Plato, Aristotle, Descartes, and Kant choosing to begin the table with Weber, Windleband, and Dilthey since they laid the path of human understanding with the development of interpretive sociology, distinguishing between natural and social science, and the history of relations. Brentano was ploughing his own furrow separately developing descriptive psychology and it was from this Husserl drew his motivation to begin phenomenology as we now understand it. Husserl's project was to build a bridge between positivist science and human interpretivism although clearly phenomenology has become a much more subjective epistemology with leanings to idiographic method recognising individual agency of human nature through intentionality.

Phenomenology fits within the interpretive sociological traditions. The following statement supports this claim and puts it thus: "The task of interpretive sociology is to understand and interpret social action" according to Weber (Schütz, 1967, p. 15). There are three axioms accepted by Schütz in Husserl's phenomenology taken from Weber: firstly, that social science should be value free; secondly, Weber's methodological indi-

*Table 1. Key developments in establishing phenomenology*

| The writers | Their script | Importance of contribution |
|---|---|---|
| Weber 1864-1920 | Social action – interpretive sociology | Understanding social action in its historical context. Action directed at significant others and recognition of subjective meaning. Value free social science. Human rationality in social action. Interpretive understanding to arrive at causal explanations. Ideal types. |
| Windelband 1848-1915 | Distinction between idiographic and nomothetic studies | The distinction laid the foundations of difference between social and natural science in its aims. |
| Dilthey 1833-1911 | Distinction between natural and human science | Human science is about understanding human life and its history of relations. Influential thinker on hermeneutic phenomenology. |
| Brentano 1838-1917 | Descriptive psychology and intentionality | Mental phenomena intentionally directed. inner consciousness distinguished from inner observation |
| Husserl 1859-1938 | Acknowledged founder of phenomenology | Subjectivity, perception, reduction and the epoché, eidetic seeing, self-conscious, presuppositionless science, noetic-noematic structure of experience, the horizon, transcendental turn, intersubjectivity |
| Scheler 1874-1928 | Man's place in the cosmos<br>Questioned Husserl about phenomenology being a method<br>It was an attitude of spiritual seeing of objects that otherwise remain hidden. He studied medicine, philosophy and sociology, the latter under Simmel and Dilthey. | Contemporary of Husserl knew Brentano through association at the University of Munich where he taught. Heidegger said all philosophers of the century owed a debt to Scheler. Value-based [werstein] metaphysics referring to the lived-body. Distinction between societal presence (*Dasein*) and the nature of ideas (Sosein). |
| Heidegger 1889-1976 | Sein und zeit | Being and time, transformation of phenomenology fusing hermeneutics with phenomenology, *Dasein*. |
| Gadamer 1900-2002 | Philosophical hermeneutics | Importance of language and meaning |
| Ricoeur 1913-2005 | Hermeneutic anthropology | Anthropology of the "capable human being." Hermeneutic suspicion and contributions to the linguistic turn. Self, identity, and change over time. Memory, history, and forgetting. |
| Levinas 1906-1995 | Phenomenology of alterity [otherness] | Role of philosophy, defence of subjectivity |
| Arendt 1906-1975 | Phenomenology of the public sphere | The human condition |
| Sartre 1905-1980 | Passionate description | Intellectual discursive, philosophy, transcendence of ego, phenomenology of imagining, ontology of being and nothingness, the meaning of other. The view from nowhere, a commentary intentionally directed at scientific method. |
| Merleau-Ponty 1908-1961 | Phenomenology of perception | Philosophy of embodiment, phenomenology of origins, structure of behaviour, critique of reductionism |
| Derrida 1930-2004 | From phenomenology to deconstruction | The end of philosophy, deconstruction and the linguistic turn, différance, offered critiques of Husserl, Heidegger, amongst others |

vidualism and the notion of ideal types; and thirdly, that of social action with social being defined in its relationships to which we attach subjective meaning through interpretation. Weber claimed that 'ideal types' were central to theory construction helping to characterise relationships and offered the example of rational commodity markets under free market conditions in an exchange economy helping to understand the particular complexities (Weber, 1949). For Weber, ideal types are a means to an end and not the end itself.

Phenomenology's philosophical roots are located in German idealism with the nature of reality located in the "spirit or idea" rather than residing in any sense data of empiricism. Objects are irrational until a subject is able to make sense of them. It is through understanding we come to know these objects. It is a tradition that offers an alternative to positivism and empiricism. The commonality of the interpretive paradigm is the attempt to understand the world from "the point of view of the actors directly involved in the social process"(Burrell & Morgan, 1979, p. 227). The everyday language of the participants in the social process is an important part of developing understanding. Kant's rationality explained that to understand any sense data there is a necessary *'a priori'* knowledge required. Such knowledge is independent of any external reality appearing in the sense data of empirical experience. Thus, we witness in Kant a return to the *a priori* 'pure reason' of Plato. This follows the zeitgeist of German idealism with its mindful intuition. There was clearly a period of intellectual fermentation taking place towards the end of the 19th century and the beginning of the 20th century that saw these ideas develop further. As Burrell and Morgan note, phenomenology was "subject to diverse influences" such as the romanticism of Goethe and Schiller as well as Hegel's dominant and sometimes dogmatic philosophy (Burrell & Morgan, 1979, p. 227). During this period there was a growing interest in the subjective and irrational with Freud, Weber, and Husserl each focussing their intention at understanding the epistemology of Kant's critique of pure reason (Hughes, 1958). These arguments return to the very nature of science and the epistemological question how knowledge is acquired and to the metaphysical problem of what we can know given the reality of human being [i.e. the ontological question]. Debates raised questions about the differences and similarities between natural science [Naturwissenschaften], cultural science [Geistesswissenschaften], and intuition and holism as opposed to hitherto atomistic analysis [Gestalten]. It is from this cauldron of concern with philosophy, science, and sociology that Husserl began to work out his ideas in relation to phenomenology. He drew upon the interests of his teacher Brentano to elaborate his ideas of consciousness and intentionality in relation to phenomenology. As a mathematician, Husserl had an affinity with science and his own notion of 'horizon' is probably drawn from his understanding of Euclidian geometry. It is to this notion of horizon that we now turn our gaze.

## MEANINGS ATTRIBUTED TO HORIZON

In this section, I want to explore the meanings attributed to 'horizon,' which is foundational to understanding phenomenology. In doing this I turn to the 'texts themselves,' thus, drawing attention to the words of others that have informed my own horizon of understanding. "Every experience has its own horizon" was Husserl's phrase limiting experience of the object by the subject (Husserl, 1973, p. 32). Kant had previously written that in "...respect of objects the horizon is either historical or rational." In respect of the subject "... the horizon is either the universal and absolute horizon, or a particular and conditioned (private) horizon"(Kant, 1988, p. 46). For Kant, the private horizon is determined by "empirical conditions" and "special considerations, e.g. age, sex, position, way of life, and the like" (p. 46). Kant argued that every mind had its own horizon. This argument sits well with psychology. Thus, rational thought is bounded and limits what we experience. What we are unable to know is beyond our horizon. This needs some further consideration applying spatial reasoning to gain further understanding of why the notion of horizon is so important to Husserl and other phenomenologists.

*In whatever way we may be* ***conscious*** *of the world as universal horizon, as coherent universe of existing objects, we, each "I-the-man" and all of us together, belong to the world as living with one* *another in the world; and the world is our world, valid for our consciousness as existing precisely through this living together. (Husserl, 1970, p. 108)*

In examining this sentence we note the world as a universal horizon, in other words, this is the limit for all objects existing in that world, the *I* and the *us*. It is a world we are conscious of in our minds and through experience of living together. It is also evident from Husserl that the subject *I* is part of the world as much as the collective *us* and all become objects in that world.

Schütz, is close to Husserl in discussing the horizon of his life-world in relation to his biography and nature as we observe in the following text extract:

*I find myself in my everyday life within a world not of my own making. I know this fact, and this knowledge itself belongs to my biographical situation. There is, first, my knowledge that Nature transcends the reality of my everyday life both in time and in space. In time, the world of Nature existed before my birth and will continue to exist after my death.. . I know, furthermore, that in a similar way the social world transcends the reality of my everyday life. I was born into a preorganized social world which will survive me, a world shared from the outset with fellow-men who are organized in groups, a world which has its particular open horizons in time, in space, and also what sociologists call social distance. (Alfred Schütz, 1962, pp. 329-330)*

Clearly, this text shares with Husserl, notions of a pre-existing world that *I* and *fellow-men* share through lived experience and observes that both nature and society transcends the reality of his everyday life. It is a holistic statement of the nature of reality connecting the individual to a system that transcend his personal existence. Nevertheless, it may be argued that this is also a Durkheimian 'social fact' that the everyday world of experience has a reality *suis generis.*

Merleau-Ponty refers to horizon as "vision brought into being from somewhere without being enclosed in its perspective"(Merleau-Ponty, 1958, p. 178). He continues seeing objects peripherally located in the visual field brought into being by concentrating on it. One becomes 'anchored' in the object by the 'gaze' closing up the landscape and opening up the object.

*The inner horizon of an object cannot become an object without the surrounding objects becoming a horizon... The horizon, then is what guarantees the identity of the object throughout the exploration; it is the correlative of the inspecting power which my gaze retains over the objects which it has just surveyed, and which it already has over the fresh details which it is about to discover. No distinct memory and no explicit conjecture could fill this role: they would give only a probable synthesis, whereas my perception presents itself as actual. The object-horizon structure, or perspective, is no obstacle to me when I want to see the object: for just as it is the means whereby objects are distinguished from each other, it is also the means whereby they are disclosed. To see is to enter a universe of beings which display themselves, and they would not do this if they could be hidden behind each other, or behind me. In other words to look at an object is to inhabit it, and from this habitation to group all things in terms of the aspect which they present to it... Thus every object is the mirror of all others. (Merleau-Ponty, 1958, pp. 178-179)*

In this passage from Merleau-Ponty, we observe Heidegger's notions of dwelling equating with habitation and *Dasein* through disclosure. Merleau-Ponty continues arguing that what has been noted here in relation to the spatial could equally apply to the temporal domain. The notion of horizon is both spatial and temporal.

Heidegger makes clear reference to past horizons, present horizons, and future horizons in his conversation (Heidegger, 1966). In the introduction to *Being and Time* Heidegger states "Our provisional aim is the Interpretation [capitalized in original] of time as the possible horizon for any understanding whatsoever of Being" [capitalized in original] (Heidegger, 1962, p. 1). The translators for the English edition remark that Heidegger's 'horizon' "provides limits for certain intellectual activities performed 'within' it." It is an existential phenomenology that concerns Heidegger. Thus, Heidegger emphasises the temporal dimension of being there [*Da-sein*] in his horizon. Heidegger is less concerned with a spatial horizon in the sense of what constitutes it, as Husserl was. Heidegger was reclaiming the 'autonomy' of time incorrectly conceptualised in his view as spatial by Aristotle through to Bergson. Heidegger's temporality was ontological rather than ontical.

Ricoeur refers to a "horizon of ignorance" perceptual limits to our understanding the world (Ricoeur, 1991, p. 269). This draws attention to the capability of the phenomenologist in making sense of what is in the horizon a point overlooked by many other commentators. There are important differences between the approach of Ricoeur and Gadamer in working through and developing their own ideas of hermeneutic phenomenology, which hinge around distanciation. Ricoeur thinks that Gadamer is somewhat hasty in closing down distances between objects in the horizon of experience that if they were given more attention would allow for understanding between the horizon of experience and the horizon of meaning. This is important because we "both belong to and participate in our historical situation"(Kaplan, 2008, p. 200). Kaplan puts it thus taken from Ricoeur, "self understanding must take a detour through the critique of ideology, informed by the matter of the text"(Kaplan, 2008, p. 201). Alienating distanciation is a priority for Ricoeur in creating meaning from the horizon.

Gadamer's conception of horizon is captured in the following text:

> *Every finite presentation has its limitations. We define the concept of "situation" by saying that it represents a standpoint that limits the possibility of vision. Hence, an essential part of the concept of situation is the concept of "Horizon." The horizon is the range of vision that includes everything that can be seen from a particular vantage point.... A person who has no horizon is a man who does not see far enough and hence overvalues what is nearest to him. Contrariwise, to have an horizon means not to be limited to what is nearest, but to be able to see beyond it.... The working out of the Hermeneutical situation means the achievement of the right horizon of enquiry for the questions evoked by the encounter with tradition. (Gadamer, 1988, p. 269)*

Gadamer in developing his hermeneutic stance towards horizon refers to multiple horizons and the necessity for fusion. Thus, the hermeneutic horizon is a "fusion of horizons" (Alvesson & Sköldberg, 2009, p. 120). Researchers make sense of other 'meaning-fields.' For example, researchers in information systems enter the different worlds in which meaning is constructed between participants through their lived experience. In these situations, the researcher is the sense-maker attempting to understand their world, a world, which is limited by the researcher's 'horizon' as they enter these different meaning-fields. The primary task of the sociologist remains as Weber put it to understand the meaning for the actor, not the observer. This is a flexible horizon that can and does change (Gadamer, 1988). The individual exists within their spatio-temporal place in their own meaning field and this pre-exists the present. The individual is a product of past meanings and each one carries with them their preconceptions, which are as Husserl noted a barrier to understanding. Heidegger, cited by Alvesson and Sköldberg presents his own version of the hermeneutic circle which they put thus:- "to understand presupposes preunderstanding, but at the same time preunderstanding is an obstacle to understanding" (Alvesson & Sköldberg, 2009, p. 120).

## THE CORRELATES OF HORIZON

The correlates of horizon are epoché (or bracketing), intentionality, and objects of consciousness.

Unlike the objects of science, social researchers study 'thinking things' [res cogitans]. These are not inanimate objects: materials and machines. Studying patterns of behaviour for thinking things is different because they reason and act or act and reason. To be human entails subjectivity, which gives rise to inconsistency. Inconsistency we sometimes equate with irrationality. Nevertheless, it is important to recognise that we sometimes hold subjective attitudes with good reason. Hence, subjectivity is rational. If subjectivity is without reason, it leads to *doxa* [opinion without evidence] and *dogma* [belief without reason]. In scholarship and research, it is unacceptable to form opinions making claims unsupported by logic and evidence, or founded on irrational belief, for example prejudice. This is why epoché (bracketing) is central to phenomenology and phenomenological reduction allowing understanding to emerge.

Husserl uses the term noumena to mean the object of thought or perception. The word has its roots in the Gk. Nous (mind). Noetic content is of the mind. For example, when we observe what we see is represented in the mind, whereas, noumena are of the mind phenomena, objects of the world outside the mind. Husserl was careful to draw his distinction so as not to simply represent objects in the mind but rather to make sense of these objectives outside the mind. Phenomena the word originates from the Gk. *Phainómenon* meaning "that which appears." Phenomenology has become a method applied through systematic reflection to uncover the conscious experience and to locate the properties and structures of its essence. Consciousness is always a consciousness about something, an object. This is what Brentano, Husserl, Schütz, and other phenomenologists refer to as the '*intentional object.*' In his contemplation of consciousness, Husserl foreshadowed many of the debates that later philosophy of the mind would address. It is a very subjective position, ontologically individualist with a stream of consciousness constituting objectivities. It has been noted by a number of commentators how thin the line is between phenomenology and solipsism giving rise to charges of psychologism and idealism (Crotty, 1998; Scruton, 2004). However, this is to overemphasise a unity of natural scientific method and underestimate what it means to be human studying social phenomena.

Husserl's early conception of phenomenology was heavily criticised as psychologism[1] by Gottlob Frege verging on solipsism[2], something his later work addressed and the works of his followers were aware of and attempted to avoid. It has also been criticised by philosophers claiming it to be eliminativist because of bracketing but this is to misunderstand the sociology of everyday lived experience, which takes so much for granted, and which any social scientist needs to set aside for *verisimilitude*. Heidegger, for example was heavily influenced as a student of Husserl's phenomenology and Hegel's dialectic phenomenology and ploughed his furrow returning to "things in themselves" developing his ideas of unconcealment in *Dasein*. Husserl himself made a distinction between: meaning, intention, and objects of consciousness; and physical acts and experience. While logic can deal with what is meant, it does not deal with acts of meaning. Objects of consciousness are phenomena, what Heidegger refers to as the things themselves. The physical acts are experience, which through reflection can be transformed to phenomena. Such phenomena are no longer real experience acts but reflections of them or put differently, objects of further acts (Honderich, 1999).

Supposing I observe an object that is 'T' shaped with the bar of the T being steel and the length of the T being wood. I see this for the first time and someone tells me, it is a hammer. I now know it by name. They also tell me it is a tool. Therefore, I now know two properties for this object: the name and that it is a tool, its category. However,

I need to ask about tool since I have never heard this term before. I am told again that there are all kinds of tool and that it is a term for useful things to do something. I am given further examples of the category of 'tool'; spoons for eating, cups for drinking and pens for writing. I now have a category by which I can file and locate the hammer in my mind as a tool but I still do not know its use. So I can now ask the question: what is a hammer for? I receive the answer it is for striking nails into wood with a purpose of fixing the wood to another piece of wood or another object. Thus, I know the intention of the object, its purpose of existence. The properties and structure of its essence have been disclosed. I now know the hammer by name (nominal), I also know it is a tool (category) and I understand what it is used for (essence). I have yet to perform the physical act, to experience or observe how it is used in practice.

Husserl refers to "The Galilean natural science, mathematical-physical nature is objective-true nature; it is this nature that is supposed to manifest itself in the merely subjective appearances." (Husserl, 1970, p. 221). He goes on to state that conceptual method of idealisation is fundamental to 'exact' science, "It is not the experienced nature of the life-world" (p. 221). The exact science method leads to "inventing 'exact' theories and formulae and also of reapplying them within the praxis which takes place in the world of actual experience" (p221). Thus, for Husserl, here are Plato's forms idealised as exact theories abstract in nature to be applied and tested in practice. He contrasts the life-world as the "outer experience" [to distinguish it from psychology—"the inner experience"] subjective in nature. He distinguishes the *mens* and the *corpus* in this discussion and emphasises a parallelization process of Cartesian duality. A further example is the way Husserl considers Brentano's notion of intentionality. "The first thing we must do here is overcome the naiveté which makes the conscious life, in and through which the world is what it is for us—as the universe of actual and possible experience—into a real property of man, real in the same sense as his corporeity,"(Husserl, 1970, p. 233). Brentano's intentionality is similar to Aristotle's "final cause" the teleology in 'Posterior Analytics.' It is a rational argument of causation. Merleau-Ponty developed a notion of making sense recognising that there is a basic form of intentionality which appeared as a symbiotic relationship between perception and the perceiver's environment that he argued is a dialectic process (Merleau-Ponty, 1958). This is a return to Hegel's conception of dialectic phenomenology.

Heidegger emphasises being and time: existence in temporal context is important to understanding things themselves, phenomena. Consider a musical thought experiment whereby I hear one note 'c' what am I to make of it. I hear 'd' but it does not mean anything until I hear a sequence of notes that I recognise as a tune. However, supposing I hear 'c' in isolation at a certain time t1 and the next note 'e' is heard at time t2, I may not relate these two notes because of the time distance in between. Therefore, I cannot make sense of melody. Time is important in relation to the existence of these notes. It is only when I hear a sequence with notes before 'd' and notes after 'd' that I begin to recognise a pattern which as I reflect upon them in relational space begin to make sense to me as a tune. If for example, I now give you the notes before d and after d in full sequence (c,d,e,d,e,d,c,a,g,e) and play the notes in 4/4 time you will experience a pattern of a melody you may recognise. Thus, in Husserl's notion of phenomenology on reflection I transform these physical acts [the musical notes] into phenomena recognising patterns [sequences in time]. In essence, there is a rhythm of everyday life. Out of the monotony of rhythm we understand, relating time and space.

## UNDERSTANDING: PHENOMENA APPEARING WITHIN OUR HORIZON

Consider another context examining organizational change, I need to enter the 'life-world' of actors who are experiencing change. In this, I need to bring their spatio-temporal relations into my horizon as I attempt to understand their understanding of the phenomenon of change. The gaze is focused on the actors and their actions through their interrelations. I am attempting to see the 'natural life' of their existence in a flow of time. I need to understand the context in which their change takes place and the things that brought the change they are experiencing into their horizon. Only by doing so can I begin to understand their social construction of change through critical reflection. Current meanings attributed to organizational change emerge within my horizon and are critically examined removing the 'taken for granted' interpretations given. There is a reinterpretation of experience to generate new meaning and subsequent understanding. In doing this, their phenomenon of change becomes an object of consciousness for the subjects (people in the organization) focusing their attention to understand change and change in their context becomes an object of consciousness for me where the subjects also become objects for me as researcher. A recent study examining organizational change this way examined a local government organization undergoing change and what it meant for the employees as new technology (SAP) was introduced with the managerial intention to improve efficiency. The SAP Enterprise System reduced organizational effectiveness and employees became suspicious of surveillance coming to see the new system as such with the unintended consequence of managerial action being to reduce both organizational effectiveness and efficiency. For the researchers understanding the phenomenon of change in this context brought panopticism into the horizon of experience (Kayas, McLean, Hines, & Wright, 2008).

Husserl describes everyday existence as "original and naïve, prior to critical reflection" (Husserl, 1970, p. 281). For Husserl, waking life is always directed towards some end or means, to the relevant and irrelevant, to the private or public, to the routine or new. Thus, for Husserl it is original life experienced before any critical reflection, as it is presented to consciousness that is the way to study phenomena. Inductive reasoning applied to the horizon of experience, which is a dynamic flow of action in a pre-given world.

Husserl's notion of horizon is unthematic [outer world] that is thematizable [inner world] (Welton, 2000). We experience the world and then conceptualize it to understand our experience. Objects are presented to our consciousness. Habermas is cited by Welton commenting on Husserl's reflection of horizon and says that "this produces an open horizon of possible objects that admits of a manifold of different types of objectivities that can be grasped only descriptively" (Welton, 2000, p. 31) Welton states that open subjectivity is bound to the specificity of the fields whose horizon it provides. Horizons are constitutive but they are also transformed through continuing experience. Husserl refers to multiple horizons; nevertheless, this does not mean there are multiple worlds but rather multiple possibilities within one world. It recognizes complexities in the way we constitute horizons for example "historical specification of one world viewed as possibility" (Welton, 2000, p. 47).

## CONCEPTUALIZATION

It becomes clear from these discussions that there are differences in conception of phenomenology brought into horizon here with at least five moments that have emerged since Husserl's original interpretation. These strands emphasise a particular aspect and are:

1. Realist phenomenology (from *Logical Investigations,* 1901) placing emphasis on description of universal *essence*.
2. Transcendental phenomenology (from *Ideas*, 1913) made distinction between the act of consciousness (noeisis) and phenomena at which it is directed (noemata)

3. Constitutive phenomenology prioritizing objects brought into *consciousness*.
4. Existential phenomenology prioritizing *human-being in its temporality*.
5. Hermeneutical phenomenology emphasizing interpretation of *lived experience*.

Each of the types of phenomenology identified has roots in Husserl's original conception through his philosophical argument. Although identified as particular types it is perhaps important to draw attention to priorities and emphases rather than difference since each of them draws inspiration from the same source. Nevertheless, there are epistemic differences that have been presented in discussion. For example, apart from Husserl's 'bridging between positivism, empiricism, and subjectivism, Schütz and Scheler are most closely associated with realistic phenomenology. The constitutive strand relates closely to Husserl's intentionality arguments and is linked with Merleau-Ponty, Sartre, and Arendt. The existential and Hermeneutical strands being most closely associated with Heidegger and Simone de Beauvoir for the former and Gadamer along with Ricoeur in the latter.

There are characteristics shared by each of them identified in Figure 1, in particular the concepts and the praxis. There are also differences in relation to the way each of the different strands has given meaning to them, which I have sketched within the paper in the discussions of horizon and its correlates. Commonalities emphasised are lived experience by returning to things themselves. In doing so, we are required to bracket the natural attitude removing prejudice through preconceptions we give to meanings of things we see in our horizon. It is the pre-theoretical world we examine and which is brought to the attention of our consciousness. The essence of phenomenology as Husserl outlined it with structural description is employed mainly by the realist, transcendental and constitutive strands whereas the existential and hermeneutic strands emphasise textural descriptions of meaningful experience. The former structural description emphasising the invariant meanings of experience or at least giving more weight to them. Still from the root, we see the branches that grew on the same tree of knowledge.

The following quote comes from Bourdieu and I include it within my horizon because it captures *habitas,* which owes much to Husserl's formation of phenomenology. Within this short passage as it is presented we see clearly an example of constitutive phenomenology.

*Knowledge of the social world has to take into account a practical knowledge of this world which pre-exists it and which it must not fail to include in its object, although, as a first stage, this knowledge has to be constituted against the partial and interested representations provided by practical knowledge. To speak of habitus [my emphasis] is to include in the object the knowledge, which the agents, who are part of the object, have of the object, and the contribution this knowledge makes to the reality of the object. However, it is not only a matter of putting back into the real world that one is endeavouring to know, a knowledge of the real world that contributes to its reality (and also to the force it exerts). It means conferring on this knowledge a genuinely constitutive power, the very power it is denied when, in the name of an objectivist conception of objectivity, one makes common knowledge or theoretical knowledge a mere reflection of the real world. (Bourdieu, 1984, p. 467)*

Cartesian duality is radically revised with the subject-object dualism receiving more prominence in discussions. Scepticism is dissolved when objects are brought to consciousness and subjected to critical reflection surfacing assumptions founded in empirical knowledge and action, as they are perceived. Husserl moved Brentano's notion of 'intentionality' away from the latter's psychological interpretation and was careful to speak of objects brought into the horizon rather

*Figure 1. Conceptualising phenomenology*

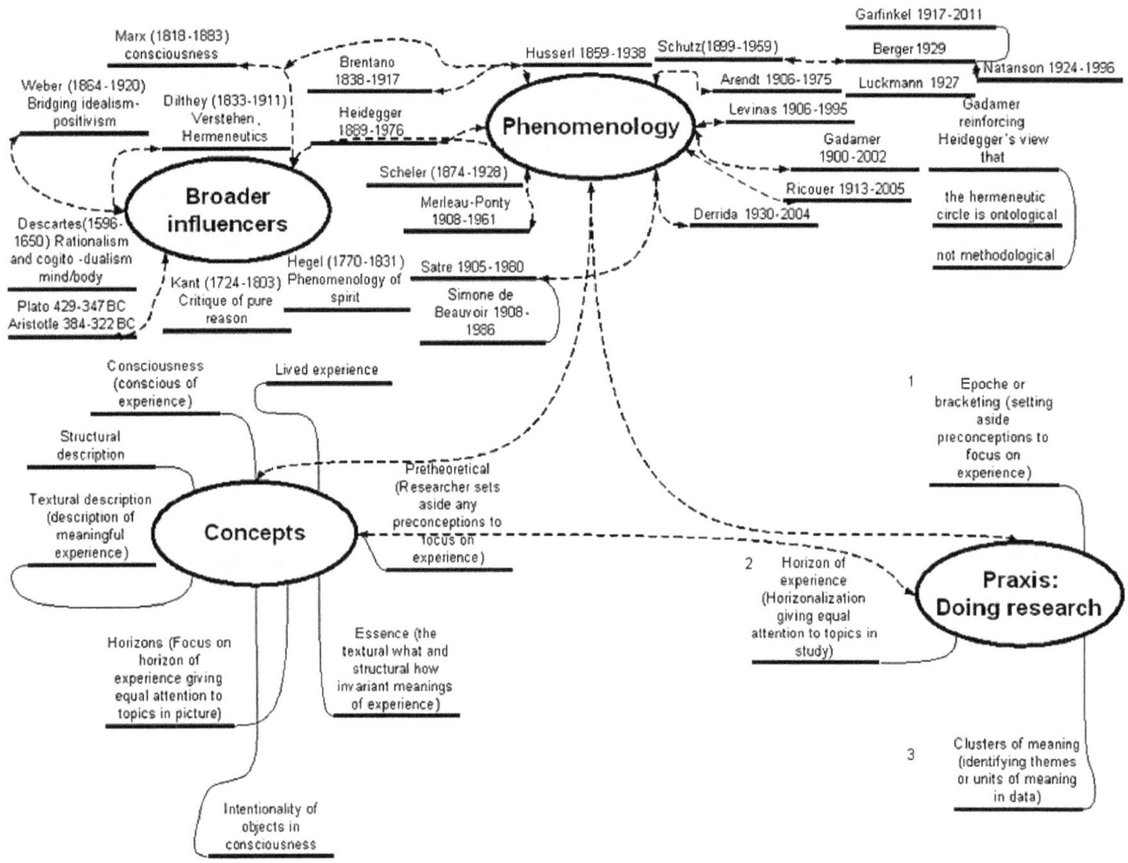


than represented in mind. In the spatio-temporal horizon, there are relations between subjects and objects and the duality dissolves as we bring all objects of study to our conscious conceptualization of the world we see.

Phenomenology is clearly a human method construing meaning to understand and in so doing, retaining a scientific approach but it is not coloured by science or preconception. It is necessary to remove all furniture of the mind to allow intuition to form. "Our minds are cluttered with beliefs about contingent and the inessential; we can approach essences therefore, only if we bracket beliefs" (Scruton, 2004, p. 11). It allows attention to be directed at the objects of study. It is perhaps worth adding it is not just the clutter but misconception, error, doxa, and dogma colouring beliefs, which, if we allow them to, may obscure the view. Secondly, by giving equal attention to objects within the horizon of experience we may be able to develop clusters of meaning identifying themes and units of meaning in the data.

This *lebenswelt* of phenomenology has in its spatio-temporal relations been conceptualised in Figure 1.

## UNDERSTANDING RATIONAL SUBJECTS AND IRRATIONAL OBJECTS

It is perhaps important at this juncture to return to the word understanding and bring it into sharper focus [the thing itself] so as not to simply take it for granted. Understanding in phenomenology comes from the German *verstehen* which has no less than two meanings in this context: Firstly, there is Weber's 'primary task' of the sociologist to understand the meaning the act has for the actors themselves (Natanson, 1963) and secondly there is the sociologist making interpretation of these meanings through the socially constructed realities of actors (Berger & Luckmann, 1966). Strikingly, Gadamer said that all understanding is to understand differently (Bernstein, 1983). It is essential to illuminate how we achieve understanding and whose understanding it is. This relates back to the earlier point in the introduction to the chapter about who exactly is making sense of phenomena, which has been a source of epistemological confusion and debate.

The Cartesian wedge driven between mind and body in the seventeenth century caused a rupture in the way we think about reason as something different to bodily experience and particularly emotion. Developments in contemporary neuroscience have begun to put the embodied mind with its emotion as the other part to rationality back into equilibrium (Ramachandran, 2003). Neuroscience and philosophy dialogues are offering new understandings of these long lived arguments (Bennett, Dennett, Hacker, Searle, & Robinson, 2007). It is a category error thinking that mind exists independently of body (Damasio, 1995). Accepting that rationality and the previously non-rational are parts of the whole system of corporeal human being. No longer are we merely 'rational puppets' a term coined by Schütz and portrayed in the Cartesian *cogito* argument that preceded the dualism of mind and body. Only by invoking God as transcendent did Descartes avoid the charge of solipsism that hindered Husserl. A ghost only laid to rest after Heidegger moved phenomenology above philosophy and psychology. For Heidegger, it was not simply appearance but a return to 'the things themselves' whereby phenomena show themselves to consciousness in the original Greek sense. The question for Heidegger was the meaning of being [sein]. Da-sein[3] is in essence the "thing which understands being" (Scruton, 2002, p. 270).

Phenomenology considers the irrational object is only irrational until the subject is able to make sense of its relation to other objects in time and space. We exist in a continuous flow of experience and what went before is sometimes ignored and we only see what appears to us in the present. In such circumstances, we have understanding of the present but do not sense any wider pattern relating existing phenomena to the contextual flow. If we repeatedly existed without recognising this interconnectivity through time, it is tantamount to learning without remembering. This may help our bracketing but it narrows to a spatial horizon and ignores the temporal horizon. The notion of where the horizon is drawn in applications of phenomenology is critical to understanding the lived experience. Let us turn attention towards phenomenology and its descendants[4], hermeneutics and ethnomethodology as possible methods to study information systems.

## POSSIBILITIES FOR RESEARCHING INFORMATION SYSTEMS

Early systems theorists concerned themselves mainly with technical problems (Ashby, 1956; von Bertalanffy, 1950). Although some recognised the importance of social relations (Katz & Kahn, 1969) and the psychology of decision-making and rationality (Simon, 1969). The work of the Tavistock Institute examining socio-technical systems also moved systems thinking towards critical insights into social relations in technical systems (Loveridge, Willman, & Deery, 2007). These human studies although often classified as 'action research' share much in common with a phenomenological study. Phenomenological studies were taken up firstly in systems thinking which is outlined in a 30 year retrospective by Checkland (2000) and more specifically in information systems (Boland, 1983, 1985, 1991). Understanding computers and cognition has been studied in different guises over time (Button, Coulter, Lee, & Sharrock, 1995; Winograd & Flores, 1986) including addressing the Turing test of what computers can and cannot do (Dreyfus, 1982, 1992). In recent years Information Systems (IS) have evolved from a positivist, primarily statistical discipline to become more pluralist in chosen methods (Chen & Hirschheim, 2004; Mingers & Willcocks, 2004). This has allowed the discipline to examine a broader range of problems embedded in their social relations. There has also been greater concern to question assumptions, evaluate and justify approaches to research (Introna & Ilharco, 2004; Introna & Whitley, 1997; Introna & Whittaker, 2002; Mingers, 2003). Lee's early paper considering positivist and interpretivist methods including phenomenology pioneered this evaluative approach in IS and is worth revisiting given the current emphasis on mixed method in social research (Lee, 1991). A set of principles for evaluating IS research adopting interpretive methods drawn from anthropology, phenomenology and hermeneutics is useful for researchers considering phenomenology, hermeneutic phenomenology and ethnomethodology (Klein & Myers, 1999). Let us now draw these discussions to a close reassessing the contribution phenomenology can make considering the horizon through experience in the concluding section.

## CONCLUSION

In summary, Husserl set out his idea on phenomenological method, which he said, is descriptive being distinct from psychological description. The method requires the researcher to suspend or bracket what Husserl referred to as 'natural attitude,' which we now refer to as preconceptions. Viewing this outlook, we observe the world from a 'natural standpoint' where the world "spread out in space endlessly, and in time becoming and become, without end." We bring this world in and out of focus temporally adjusting the horizon. It is a world that Husserl said is constantly present. It is not a world of facts in any Durkheimian sense of a world stuffed full of factual content waiting for discovery. It is rather a world constituted by values and if I were to focus on one aspect of this world, I am still aware of the totality of the world ever present beyond my immediate horizon. My experience of the world is as Husserl says "a piece of pure description prior to all theories" (Husserl, 1983, p. 56). This is the way we both inhabit and relate to a world existing before we enter, and as we live in and experience it. It is the context in which we exist. The contents of this world are objects of study for scientists adopting a natural standpoint. We suspend evaluations bracketing or setting aside our preconceptions simply to experience the world as it appears to us. This bracketing or epoch is what Husserl refers to as phenomenological reduction. The description of what remains after bracketing is structures that constitute possibilities in mental experience becoming visible in consciousness. For Husserl, this form of reduction allows one to discover one's own transcendental ego. Husserl argued it was existence of ego that is pure consciousness distinct from psychic self of interest to psychologists. He later revised this notion arguing it correlated to the world and was not absolute in relation to all other objects in the world. This latter position is not one that all phenomenologists would subscribe to. There are different views of phenomenology particularly from those identified with Husserl's early position and those that came later, Merleau-Ponty and Heidegger, for example. They reach different conclusions in relation to this.

Summarising in the most general terms, there are three stages in the process of understanding following a phenomenological method. Epoché or bracketing is the first step in a process of data reduction as the researcher sets to focus upon the current experience of participants setting aside any preconceptions. This is referred to as pre-theoretical or unthematic (Husserl, 1970). 'Horizon-alization' is the second step whereby the researcher focuses on all aspects of the experience deemed relevant to the topic studied and it is given equal attention in critical examination. This in effect is the implementation of Husserl's notion of a horizon of experience being central to studying phenomena. The third step is to identify themes from strands of data that are clustered into meaningful units for analysis (Moustakas, 1994). When the stages are complete, we understand.

Natanson captures the essence of phenomenology as follows:

*Phenomenology is a presuppositionless philosophy which holds consciousness to be the matrix of all phenomena, considers phenomena to be objects of intentional acts, and treats them as essences, demands its own method, concerns itself with prepredictative experience, offers itself as the foundation of science, and comprises a philosophy of the life world, a defence of reason, and ultimately a critique of philosophy (Natanson, 1973, p. 19).*

Phenomenology questions common sense with its accepting attitude of everyday life and the realms of natural science, and as such, phenomenology "transcends the natural attitude of daily life," to render it an object of philosophical examination by describing the essence of its structure, (Burrell & Morgan, 1979, p. 233; Natanson, 1973, p. 3). Burrell and Morgan further comment "that there is an objective external world which exists in space and time and is real for all men…subjected to thorough scrutiny," (Burrell & Morgan, 1979, p. 233). Scientific presuppositions being reduced to 'implicit metaphysical commitments' with the external objective world an 'artefact of consciousness' brought into everyday life as 'intentional acts.' Bracketing, consciousness, and intentionality are central to Husserl's analytical approach generating knowledge by studying the essence and structures of experience. It is necessarily a subjective position, ontologically individualist with a stream of consciousness constituting objectivities.

It is important to understand the world and the spirit of the age from which that world appeared if you want to trace the interconnectedness of ideas emerging from a particular world-view. Phenomenology has itself been subjected to such scrutiny in this contribution. If we firstly consider the temporal horizon, we note it drew its source from Plato, Descartes, and Kant's rationality during the Englightenment. The spatial horizon can be traced to traditions of continental philosophy particularly centred on Germany and Austria. If we take a historical perspective, the primacy of scientific thought with its distancing of subject from object of study, focusing on method and data with discussions detached from the everyday language and framed in abstraction would perhaps be the picture we paint. Materialism and Newtonian physics had shown the way to enrich human existence (Gribbin, 2003). At the start of the twentieth century Einstein was actively pursuing his theories of science culminating in his relativity theory in 1913 (Einstein, 1954). Research of the Englightenment was nomothetic in nature searching for universal laws observed through physical movements and patterns of behaviour (Bortolotti, 2008). This is the intellectual climate experienced by Husserl, Heidegger, Merleau-Ponty, and Schütz, amongst others, as they developed their respective projects. I say projects because they were just that and should not be seen as a continuum since they really were not. What phenomenologists share is recognition that human experience is subjective, embedded in social relations, emphasising agency over structure, with description primary to analysis emphasising particulars of spatio-temporal relation to the whole, and interpretation over measurement. Italian historian Benedetto Croce said that "All history is contemporary history" meaning we write about the past from the perspective of the present (Evans, 1997, p. 30). This is inevitably what I have written in this chapter and therefore, it is an interpretation, my understanding. Understanding, which is intentionally captured through description in the flow of time, from existing relational objects, suspending judgements until critical reflection and personal reflexivity is exercised in the process. From here, it is but a short step through logical inference to 'thematize' meanings through interpretation giving equal attention to 'the things themselves' through inner anticipation and dwelling. It is as a consequence evident from this study of phenomenology that the horizon is raised in the process of human understanding.

## REFERENCES


Alvesson, M., & Sköldberg, K. (2009). *Reflexive methodology: New vistas for qualitative research*. London, UK: Sage.

Aristotle,. (1976). *The nicomachean ethics* (2nd ed.). London, UK: Penguin Classics.

Ashby, W. R. (1956). *An introduction to cybernetics*. London, UK: Methuen.

Bennett, M., Dennett, D. C., Hacker, P., Searle, J. R., & Robinson, D. (2007). *Neuroscience & philosophy: Brain, mind and language*. New York, NY: Columbia University Press.

Berger, P., & Luckmann, T. (1966). *The social construction of reality: A treatise in the sociology of knowledge*. London, UK: Penguin.

Bernstein, R. J. (1983). *Beyond objectivism and relativism: Science, hermeneutics and praxis*. Philadelphia, PA: University of Pennsylvania Press.

Boland, R. J. (1983). The in-formation of information systems. In Boland, R. J., & Hirschheim, R. A. (Eds.), *Critical Issues in Information Systems Research* (pp. 363–394). New York, NY: John Wiley & Sons.

Boland, R. J. (1985). Phenomenology: A preferred approach to research on information systems. In Mumford, E., Hirschheim, R. A., Fitgerald, G., & Wood-Harper, A. T. (Eds.), *Research Methods on Information Systems* (pp. 193–201). Amsterdam, The Netherlands: North Holland.

Boland, R. J. (1991). Information system use as a hermeneutic process. In Nissen, H. E., Klein, H. K., & Hirschheim, R. A. (Eds.), *Information Systems Research: Contemporary Approaches and Emergent Conditions* (pp. 439–464). Amsterdam, The Netherlands: North Holland.

Bortolotti, L. (2008). *An introduction to the philosophy of science*. Cambridge, UK: Polity.

Bourdieu, P. (1984). *Distinctions: A social critique of the judgment of taste* (Nice, R., Trans.). Boston, MA: Harvard University Press.

Burrell, G., & Morgan, G. (1979). *Sociological paradigms and organizational analysis*. London, UK: Heinemann Educational.

Button, G., Coulter, J., Lee, J. R. E., & Sharrock, W. (1995). *Computers, minds and conduct*. Cambridge, UK: Polity.

Checkland, P. (2000). Soft systems methodology: A thirty year retrospective. *Systems Research and Behavioral Science*, *17*(1), 11. doi:10.1002/1099-1743(200011)17:1+<::AID-SRES374>3.0.CO;2-O

Chen, W., & Hirschheim, R. (2004). A paradigmatic and methodological examination of information systems research from 1991 to 2001. *Information Systems Journal*, *14*, 197–235. doi:10.1111/j.1365-2575.2004.00173.x

Crotty, M. (1998). *The foundations of social research - Meaning and perspective in the research process*. London, UK: Sage.

Damasio, A. R. (1995). *Descartes' error: Emotion, reason, and the human brain*. London, UK: Harper.

Dreyfus, H. (1982). *What compuers can't do: The limits of artificial reason*. New York, NY: Harper & Row.

Dreyfus, H. (1992). *What computers still can't do: A critique of artifical reason*. Cambridge, MA: MIT Press.

Einstein, A. (1954). *Relativity*. London, UK: Routledge.

Eliot, T. S. (1969). Four quartets: Little gidding. In *The Complete Poems & Plays*. London, UK: Faber & Faber.

Evans, R. J. (1997). *In defence of history*. London, UK: Granta.

Gadamer, H.-G. (1988). *Truth and method*. New York, NY: Crossroad.

Gardner, S. (1999). *Routledge philosophy guidebook to Kant and the critique of pure reason*. London, UK: Routledge.

Gribbin, J. (2003). *Science: A history 1543-2001*. London, UK: Penguin.

Heidegger, M. (1962). *Being and time* (Macquarrie, J., & Robinson, E., Trans.). Oxford, UK: Blackwell.

Heidegger, M. (1966). Conversation on a country path about thinking. In *Discourse on Thinking* (Anderson, J. M., & Freund, E. H., Trans.). New York, NY: Harper and Row.

Honderich, T. (Ed.). (1999). *The philosophers: Introducing great western thinkers*. Oxford, UK: Oxford University Press.

Hughes, H. S. (1958). *Consciousness and society*. New York, NY: Alfred Knopf.

Husserl, E. (1950). *Cartesian meditations: An introduction to phenomenology* (Cairns, D., Trans.). The Hague, The Netherlands: Martinus Nijhoff.

Husserl, E. (1970). *The crisis of European sciences and transcendental phenomenology: An introduction to phenomenological philosophy* (Carr, D., Trans.). Evanston, IL: Northwestern University Press.

Husserl, E. (1973). *Experience and judgement* (1st ed.). (Churchill, S., Trans.). Evanston, IL: Northwestern.

Husserl, E. (1983). *Ideas pertaining to a pure phenomenology and to a phenomenological philoshophy, first book* (Kersten, F., Trans.). Dordrecht, The Netherlands: Kluwer.

Hyppolite, J. (1974). *Genesis and structure of Hegel's "phenomenology of spirit"* (Cherniak, S., & Heckman, J., Trans.). Evanston, IL: Northwestern University Press.

Introna, L. D., & Ilharco, F. M. (2004). Phenomenology, screens and the world: A journey with Husserl and Heidegger into phenomenology. In Mingers, J., & Wilcocks, L. (Eds.), *Social Theory and Philosophy for Information Systems*. Chichester, UK: Wiley.

Introna, L. D., & Whitley, E. A. (1997). Against method-ism. *Information Technology & People*, *10*(1), 31–45. doi:10.1108/09593849710166147

Introna, L. D., & Whittaker, L. (2002). The phenomenology of information systems evaluation: Overcoming the subject object dualism. In Wynn, E. H., Myers, M. D., & DeGross, J. I. (Eds.), *Global Organizational Discourse about Information Technology* (pp. 155–175). Dordrecht, The Netherlands: Kluwer Academic Publishers.

James, W. (1912). *Essays in radical empiricism*. New York, NY: Longmans, Green, & Co.

Kant, I. (1900). *Critique of pure reason* (1st ed.). (Meiklejohn, J. M. D., Trans.). London, UK: Colonial Press.

Kant, I. (1988). *Logic* (1st ed.). (Hartzman, R. S., & Schwarz, W., Trans.). New York, NY: Dover.

Kaplan, D. (2008). *Reading Ricoeur*. New York, NY: SUNY.

Katz, D., & Kahn, R. L. (1969). Common characteristics of open systems. In Emery, F. E. (Ed.), *Systems Thinking* (pp. 86–104). Harmondsworth, UK: Penguin.

Kayas, O. G., McLean, R., Hines, T., & Wright, G. H. (2008). The panoptic gaze: Analysing the interaction between enterprise resource planning technology and organisational culture. *International Journal of Information Management*, *28*, 446–452. doi:10.1016/j.ijinfomgt.2008.08.005

Klein, H. K., & Myers, M. D. (1999). A set of principles for conducting and evaluating interpretive field studies in information systems. *Management Information Systems Quarterly*, *23*(1), 67–94. doi:10.2307/249410

Lee, A. S. (1991). Integrating positivist and interpretive approaches to organizational research. *Organization Science*, *2*(4), 342–365. doi:10.1287/orsc.2.4.342

Loveridge, R., Willman, P., & Deery, S. (2007). 60 years of human relations. *Human Relations*, *60*(12), 1873–1888. doi:10.1177/0018726707084917

Merleau-Ponty, M. (1958). *Phenomenology of perception* (Smith, C., Trans.). London, UK: Routledge.

Mingers, J. (2003). A classification of the philosophical assumptions of management science methods. *The Journal of the Operational Research Society*, *54*, 559–570. doi:10.1057/palgrave.jors.2601436

Mingers, J., & Willcocks, L. (Eds.). (2004). *Social theory and philosophy for information systems*. Chichester, UK: Wiley.

Moran, D. (2000). *Introduction to phenomenology*. London, UK: Routledge.

Moustakas, C. (1994). *Phenomenological research methods*. Thousand Oaks, CA: Sage.

Natanson, M. (1963). A study in philosophy and the social sciences. In Natanson, M. (Ed.), *Philosophy of the Social Sciences* (pp. 271–285). New York, NY: Random House.

Natanson, M. (1973). *Edmund Husserl: Philosopher of infinite tasks*. Evantson, IL: Northwestern University.

Plato,. (1941). *The republic of Plato* (Cornford, F. M., Trans.). Oxford, UK: Oxford University Press.

Popper, K. R. (1945). The open society and it's enemies: *Vol. 2. Hegel and Marx*. London, UK: Routledge.

Ramachandran, V. S. (2003). *The emerging mind: BBC Reith lectures 2003*. London, UK: Profile Books.

Ricoeur, P. (1991). *From text to action*. Evanston, IL: Northwestern University Press.

Ryle, G. (1949). *The concept of mind*. London, UK: Hutchinson.

Schütz, A. (1962). *Collected papers: The probelm of social reality* (*Vol. 1*). The Hague, The Netherlands: Martinus Nijhoff.

Schütz, A. (1967). *The phenomenolgy of the social world* (Walsh, G., & Lehnert, F., Trans.). Evanston, IL: Northwestern University Press.

Scruton, R. (2002). *A short history of modern philosophy*. Oxon, UK: Routledge.

Scruton, R. (2004). *Modern philosophy*. London, UK: Pimlico.

Simon, H. A. (1969). Rational choice and the structure of the environment. In Emery, F. E. (Ed.), *Systems Thinking* (pp. 214–229). Harmondsworth, UK: Penguin. doi:10.1037/h0042769

von Bertalanffy, L. (1950). The theory of open systems in physics and biology. In Emery, F. E. (Ed.), *Systems Thinking* (pp. 70–85). Harmondsworth, UK: Penguin. doi:10.1126/science.111.2872.23

Weber, M. (1949). *The methodology of the social sciences*. New York, NY: Free Press.

Welton, D. (2000). *The other Husserl: The horizons of transcendental phenomenology*. Bloomington, IN: Indiana University Press.

Winograd, T., & Flores, F. (1986). *Understanding computers and cognition: A new foundation for design*. Reading, MA: Addison Wesley.

Wittgenstein, L. (1998). *Philosophical investigations* (2nd ed.). (Anscombe, G. E. M., Trans.). Oxford, UK: Blackwell Publishers.

## KEY TERMS AND DEFINITIONS

**Consciousness:** Subject awareness as opposed to unconsciousness. Being conscious of objects is always intentional (see intentionality).

***Dasein:*** Unconcealment or disclosure encaptulates Heidegger's conception of *Dasein*. It is existence, being in the world.

**Enlightenment:** An age of reason. Its genesis is dated to the year of publication of Newton's *Philosophiae Naturalis Principia Mathematica 1687*. The end is less clear although postmodernists argue it has ended.

**Epistemology:** Study of knowledge

**Epoché:** (or Bracketing): The setting aside of any preconceptions as far as is humanly possible to understand the experience.

**Hermeneutics:** (meaning-making): Meaning-making which requires interpretation. Hermes was the messenger from the Gods (Gk.) which is where the word hermeneutics originates from. See sensemaking.

**Horizon:** A term used by Husserl and other phenomenologists whereby objects/subjects (data) are given equal attention when they are brought into the horizon. The intention is to understand phenomena through a process of 'horizonalization.'

**Intentionality:** A term coined by Brentano acknowledging that we may only understand, if we experience the outward nature of appearance balanced against the inner meaning, memory or image we attach to the object. Phenomenologists refer to *intentional objects* of consciousness.

**Irrational:** The opposite of rational. Ideas and actions which are difficult to make sense of, understand and explain.

**Object(s):** Things, which in themselves have no meaning until subjects bring them into their horizon. Interpreting the meaning of objects that appear to us or that we experience requires reason and logic to understand. Meaning, is usually understood through relations in space and time.

**Ontology:** Study of the nature of reality (world-views).

**Phenomenology:** Study of phenomena. Phenomena are objects of the world. Phenomena can be studied. Kant made the distinction between phenomena (the things in themselves) and nouemena (the things of the mind). Phenomena appear and can be experienced.

**Rational (Rationality):** Ideas and actions arrived at through powers of reason or logic. The Greeks refer to *logos* as logic.

**Sensemaking:** and sensemaker: Understanding achieved through the senses (sight, smell, touch and hearing). Sensemaking is a form of meaning-making requiring interpretation.

**Subject(s):** In this context it usually means people (actors), those possessing powers of reason.

## ENDNOTES

1 Psychologism is in the philosophy of logic the idea that logic is based upon laws of thought.

2 Solipsism is the idea that only one's own mind exists and this criticism is something that Descartes managed to sidestep by employing God as a third party in his cogito argument in Meditations. Most people would accept there are of course other minds but as Wittgenstein said we may not know them.

3 Da-sein is deliberately hyphenated to disclose the prefix Da (from) to the verb (being) usually Dasein.

4 Husserl's student Schütz escaped German nationalism in 1939, emigrating to the USA, and was responsible for developing a new generation of sociologists in New York, including Garfinkel (ethnomethodology), Berger, Luckmann (social construction of knowledge), and Natanson (Husserl's biographer); each carried on the tradition of phenomenology, some spawning new branches.

# Chapter 15
# Social Practice Design

**Gianni Jacucci**
*University of Trento, Italy*

**Gian Marco Campagnolo**
*University of Edinburgh, UK*

## ABSTRACT

*This chapter introduces an action research methodology called Social Practice Design and its application in the case of a small-medium enterprise production management innovation process. Social Practice Design includes three main phases: observation and the production of how questions; analysis and the production of visions of solutions; synthesis and the production of action plans or SPD paths. Social Practice Design is proposed as an extension of Participatory Design to the implementation phase of Information Systems Development. The research team in Social Practice Design includes three modes of knowing: the ethnographer, the counsellor, and the action researcher modes of knowing.*

## INTRODUCTION

It has been long recognized that the development of information systems (Information Systems Develpment, ISD) is constitutively inevitably inseparably socio-technical. Yet the design practices of information systems still seem to neglect the social dimension of technology and especially the problem deriving from technology adoption in organizations. The Participatory Design (Bødker, et al., 2004), which is the most advanced method currently in ISD, incorporates users in the design of information systems through intense participation in the design process, with the aim to set up an artifact that is used effectively in practice. Compared to traditional approaches, the Participatory Design (PD) recognizes the incompleteness of the strategic alignment between business processes and technology, and therefore finds it necessary an in depth examination of work practices, to be conducted with ethnographic methods, able to better support the development of a vision for change. However, the focus of participatory design is basically limited to the management of the technical dimension of the problem, ne-

DOI: 10.4018/978-1-4666-0303-5.ch015

glecting the fact that not only the introduction of a technology but the same application of PD for the introduction of a technology involves a modification of socially established practices.

The *Social Practice Design*—i.e. design of social practices—is an approach that aims to fill this gap, in order to restore the necessary balance between the two dimensions of the problem of design (specifically, information systems design)—the technical and the social dimensions—emphasizing the need to support the organizational change impied by not only the introduction of new technologies but also by the introduction of new participative methods. Social Practice Design does this with the ultimate objective of ensuring that the potential benefit of not-taken-for-granted participatory methodologies offered to support the introduction of organizational technologies can be understood and realized.

The term 'social practice design' is an oxymoron, because we all acknowledge that the social practices emerge by social construction and not through by design, but the choice of this name is an intentional and motivated provocation. On the one hand, it serves to underscore the intentionality of the creative or constructive action: the proactive, deliberate, and planned ingredient of the process of organizational innovation. On the other hand, we choose to name the method social practice design to underscore that the role the word design has in the technological discourse has to be paralleled with equal centrality in the methodological discourse regarding social practices: organizational changes brought about by new participatory methods aren't less relevant than those implied by technological change.

This chapter intends to illustrate the constituent elements of Social Practice Design through a concise guide to its application in an electronics Small / Medium Size Enterprise (SME), granted by an European STREP project called MAPPER (Model-based Adaptive Product and Process Engineering, IST-016527).

## SOCIAL PRACTICE DESIGN

Social Practice Design (SPD) is an innovative approach to organizational change in the context of the development of information systems. Phenomenology and counselling (Rogers, 1951) are its two distinctive features. SPD offers a way to design sustainable interaction between people, computers and work, through the activation of staff and management people in processes of co-construction. It can be considered an extension of the Participatory Design to the implementation phase of information systems. Its object is the design and introduction of new participatory activities, or new ways to accomplish tasks by humans, in order to 'make room' (Ehn, 2006) for technology, addressing various issues related to the social deployment and use of technology in organizations. SPD is a form of action research or action-oriented research (Bødker, et al., 2004). Although the last two decades research on Participatory Design has developed activities to support decision making at every stage of the design process, the phase that has remained the most neglected is the implementation phase of system development, that is when users need to integrate technology within their own activities, in many cases having to change the way they work. Another aspect that Participatory Design has weakly addressed is the organizational aspect: namely, what organizational measures can be imagined and experimented by/with the users, in order to reflexively support the expression of the potential of the introduction of participatory methodologies in the development of technology.

Within the domain of social studies of information systems, Ciborra (2002) has strongly supported the implementation of an organizational vision on technology development, saying that the unveiling of the organizational forms of the real world requires a different analytical approach than that of business research:

*"The gap between theoretical, what ex post explanations and models can deliver, and the actual garbage-can style of managerial choice is considered to be a fact of life by practitioners, and an unavoidable result of the limitations of modeling any approach by scholars" (Ciborra, 2002, p. 174).*

Weick and Quinn (1999) speak of "episodic change" as what you should consider when adaptation—such as it can happen in the case of the introduction of technology—is slow to occur. Episodic change can be dramatic, as recognized by Lewin:

*"To break open the shell of complacency and self-righteousness it is sometimes necessary to bring about deliberately and emotional stir-up" (Lewin, 1951; quoted in Marshak, 1993, p. 400).*

Referring to the four distinct types of change by Prochaska et al. (1992), Weick and Quinn also pointed out that the episodic process of change is not linear but follows a spiral. They also argue that organizational change requires a change in the system of meanings: it is about using a different vocabulary, alternative schemes of communication, build opportunities for coordination and care.

Social Practice Design developed from these discussions, and from the recognition of the epistemological assumption that we can learn something of the real world only by trying to change it (Lewin, 1946). SPD creates a space for the design of social practices and for social design of organizational practices in the application of Participatory Design techniques in organizations. During the implementation of a technology, there are often problems and concerns that transcend its operations as technically understood, and, by being organizational in nature, invoke solutions that often only marginally relate to the technical capabilities of the system. Social Practice Design can thus be placed within that group of perspectives who look to technology as 'inscription' (Latour, 1986), considering the reflection on social practices the key to a good implementation. Similarly, in the Social Practice Design perspective, we recognize that the process of hierarchical decomposition embedded in technology and methods development, takes place through the mobilization and consolidation of networks of people, objects, contexts, data and information. The view that Social Practice Design inherits from the actor network genre is therefore that the complexity reduction made possible by the development of information systems and their 'hard' achievements in actual facts is dependent upon the formation of networks of social practices, those practice Social Practice Design is about to analyse. The engineering decomposition and subsequent managerial perspectives on information systems development are nothing but the result of networks of practices that create values, goals, objectives, constraints, changes, assemblies of tasks and errors. Engineers and managers ability to sort out and disassemble these networks of practice as methods and techniques is dependent upon additional classification work, which in turn depends on the mobilization, the articulation, and consolidation of further networks of social practices.

In order to emphasize the obstinacy of practices with respect to management methods and engineering decomposition techniques, SPD echoes the theories of organizational change that emphasize improvisation as a characteristic of successful companies that "have well-defined managerial responsibilities and clear project priorities and then let the design process being highly flexible, characterized by improvisation and continuous change" (Weick & Quinn, 1999, p. 371). Orlikowski (1996), in her study on changes in a system for tracking incidents, found that work practices were based on continuous improvisation that resulted in a restructuring of those same practices. Moorman and Miner (1997) argue that 'the more an act is improvised, the narrower is the gap between composition and performance, design and production, planning and implementation." In an excerpt from Labyrinths of Information (Ciborra, 2002), Ciborra emphasizes that improvisation is just one element of the shift to the more radical shift to the idea of an entirely phenomenological design:

*"I suggest that the information systems field, with views of the ITS rational knowledge, decision making, strategy, and orderly development systems, is based on a narrow model of rational, ideal actors. [...] By focusing on the mundane and the existential, I want to transition to contribute towards the field ... passion and improvisation, bricolage and moods, emotions and workaday chores; existence and procedures will become integral to systems design and use, casting new shadows and lights on the unfolding world of technology (ITS deployment and management in organizations and societies)" (Ciborra, 2002, p. 9).*

Social Practice Design reflects this vision, serving as a design approach where the researcher sets the stage for the ethnographic work, and through it he defines the key points around which to organize the process of social practice change connected to the introduction of new technology in the organization. In this process, participants analyze, co-construct, and perform in ways closer to improvisation than to rationality.

Specifically, the SPD can be described as a series of steps or phases that invoke multiple perspectives and that, far from unfurling strictly in accordance with a predetermined sequence, can be repeated recursively:

## Observation (Output: How Questions)

1. Some initial conversations are carried up to open the process of interaction and to understand their objectives and perceived problems;
2. thereafter, concepts are generated that reveal the existing problems through a period of ethnographic observation and analysis of data from multiple perspectives;
3. within the research staff, the issues raised during observation are consolidated into questions about how to overcome them ("how questions"). The process of co-construction of the how questions also involves feedbacks from the personnel involved, thus generating awareness and shared meanings, giving initiation to the change process.

During the observation phase, the focus is on: the complexity of the work; the flexibility of people in ordering their work processes and to adapt to the needs of the situation as they arise; the necessity for people to have an overview of the process and the state of advancement of their work. During this phase under scrutiny will be: the collaboration practices; the use of artefacts and their combination; the strategies to align work across boundaries of various kinds (physical, organizational, geographical, cultural). Particular attention is paid to the key objects, the artefacts in use, the role of standards and procedures, the use of physical space to make work visible to others and the location of physical objects and documents. The observations are carried out through the use of equipment for audio and video recording of group meetings and individual workstations. At each stage of observation, there are three people involved, in order to simultaneously observe, note, and record. Fieldwork notes are read, compared and discussed within twelve hours from the observation in a group comprising the ethnographer, the counsellor and the researcher (action researcher). Workgroup also involves discussions on the summary of the ethnographic fieldwork, where excepts of audio-video recordings are exposed to the different background and specific knowledge of those involved (the ethnographer, the a counsellor, the action researcher). Emerging assumptions on 'how questions' are critically evaluated. A refined report on field study that summarizes the results of the group work is send out to the members of the organization involved in the project. The feedback by the personnel on the report on field study represent the conclusion of the first phase (observation phase) of the SPD method.

The activity that takes place during the research staff group meetings for the analysis of ethnographic notes to draft the initial 'how questions' is an interpretation activity aiming at making sense from observational details (Walsham, 2004) through the use of "personal knowledge" (Polanyi, 1966). A word of caution is necessary here: letting emerge critical aspects is much easier than describing things that work good. What works well is invisible to the eye and to the consciousness. For this reason, the design of social practices proceeds from what does not work, from what can be improved.

## Analysis (Output: Visions of Solutions)

1. relevant theoretical principles within the social study of information systems domain are identified with respect to how questions;
2. *visions of solutions* are conceived and generated by confronting theories with the 'how questions' deriving from observation. Visions of solutions may include instructions to people, changes in work processes, training modules, working proposals of supporting technologies, or mockups;
3. the proposed visions of solution are modified and then consolidated through repeated meetings with the involved members of the organization.

The analysis phase is based on the encounter between how questions and relevant principles arising from scholarly case studies within the various disciplinary domains within the area of social studies of information systems (HCI, CSCW, Participatory Design, and others). Given the mixed nature of the research staff involved, the logic leading to the production of visions of solutions is neither strictly inductive nor deductive. Rather, tacit sense making skills of the research staff involved may refer in the first instance to theoretical principles as well as directly to proposals of visions of solution. There is no necessity to proceed from the theoretical stage to the proposal of solution. Based on the agreed-upon visions of solutions, the counsellor provides an agenda for meeting the members of the organization, together with prioritized issues. The purpose of the meetings taking place in this phase is to make sure that, according to the personnel, the proposed visions of solutions meet the needs that emerged during the observation phase.

## Synthesis (Output: Action Plans or *SPD Paths*)

1. The combination of visions of solutions and the results of fieldwork generate strategies and action plans (such as workshops for the co-construction of social practices, training modules for managers who want to plan and manage change);
2. the action plan is submitted to the personnel involved who co-construct it and put it into practice in interaction with the research group;
3. the results of the action plan is then evaluated and the possibility of repeating the whole process is discussed.

In the early stages of the generation of the *how questions*, *visions of the solutions* and *action plans* that characterize the SPD method echoes the usual triad of the scientific paradigm: observation, analysis, and synthesis. In addition, Social Practice Design merges three distinct disciplinary practices:

- the practice of the ethnographic researcher, able to identify and describe the social practices of the actors involved in the change and innovation process, and conduct a qualitative assessment on the sustainability of change and innovation;

• the counselling practice, to identify stakeholders and to establish a state of hospitality to innovation, making explicit the methods in use, and, on that basis, undertake a process of co-construction of innovation in which all parties are involved and learn;

• the practice of the Action researcher, which ensures the realization and acceptance of innovation by supporting organizational change and by creating *paths* that users can follow.

## An Application: Social Practice Design

This section describes how Social Practice Design has been applied in the case of a European research project of the Sixth Framework Programme STREP (IST project MAPPER). The project relies on enterprise modeling to promote the adoption by small and medium enterprise of ICT for cooperation in the context of global manufacturing. The project has given us the opportunity to apply the social practice design method in different contexts of use, and to involve people from different organizations to identify their problems and rethink organizational constraints.

The enterprise modeling methodology is state of the art in different kinds of scientific and technological domains such as knowledge management, organizational design, computer science (especially in software engineering and computer simulation), architecture, and industrial design. There are several different modeling languages and notations accepted in enterprise modeling such as the Universal Modeling Language (UML)—to model objects, activities, states, and architectures—, Petri nets—to model the flow—, ER notation—to model ontologies, the data structure, etc. In addition to modeling languages, enterprise modeling adopts a series of modeling methodologies or approaches, accompanied by their own modeling notation (IDEF, MERISE, * POP). These approaches, languages, and notations have different objectives and can be used for different purposes. In the industrial case analyzed here, the approach used is called POP*. The main objective of the MAPPER project was to enable fast and flexible manufacturing processes through reusable methodologies, infrastructures, and services for participatory design in networked manufacturing enterprise, demonstrating its practical benefits and scientific value in three different industrial cases.

VCP is one of the three industrial cases of the MAPPER project. VCP is a company manufacturing electronic components. In VCP we had the opportunity to observe modeling sessions and production in seven different workplaces. We have also participated in several meetings—a meeting with managers, a marketing meeting, a meeting to handle a 'crisis,' and a Skype conference call with a customer and a reseller of VCP products in the U.S. The VCP engineers work in project teams co-located (four or five people per room) and cooperate with a number of external partners. The engineers at VCP are involved in 4 different types of activities (provision of engineering services to other companies, production of different types of virtual components, provision of pre- and post-sales support services).

VCP is a company characterized by a strong engineering culture with widespread usage of sophisticated knowledge management tools (long e-mails, distributed target platform, video conferencing systems at low cost) in the context of globally distributed outsourcing. The company has achieved, over the past 15 years, a position that allowed the exploration of new limits: the lack of a culture of business management, due to the lack of a direct market access, and de-concentration of best skills and expertise of personnel from the mission critical areas of business (such as customer relationship management, development of new products) that undermine their marketing strategy.

Coinciding with a major effort for the survival, VCP is in danger of failing to identify in a precise core business activities, due to the opportunistic scattering of production efforts in multiple directions without correlation. All these factors indicate the need for better management strategy of the production processes, where the lack of resources often causes the waste of valuable intellectual resources and the delayed adoption of a platform for customer management.

The VCP personnel is cooperative and team-oriented. The solving-problem meetings between developers, though called by the CEO, seem self-managed. These meetings proceed in an efficient and democratic way. The CEO expects the solution will emerge from the initiative of the developers. Participants proceed independently without interference except in the final phase, in which the assignment of roles and tasks to roles takes place. Procedural documents did not appear to be part of the organizational culture of the company. If we exclude the documents describing the software development tasks, management procedures are not described in documents.

## The Observation of Modeling Sessions (Observation)

A modeling session in MAPPER is a gathering of people to produce a model. It is a collaborative action taking place during the requirements engineering phase. Modelers assign to the people who participate in a modeling session prescribed roles. These roles are: use case manager, domain expert (the user the finished model), modeling expert and coach (who facilitates the modeling session). The result of a modeling session is usually a model. The production of the model is based on a methodology for modeling client organization's processes and sub-processes. The objective of the methodology is to create a model of the client organization to support, in this case, the formation of a production management culture. The modeling, in VCP, was seen as a cultural facilitator for collaboration among VCP as a producer of virtual components and a partner who manufactures the physical circuit. During the modeling session analyzed, the VCP model realized by the VCP engineers involved in the project MAPPER has been analysed and corrected by the modeler. Subsequently, a model of the physical circuit production process in the partner organization has been realised, when the engineers of VCP were co-present. The objective of this session was to increase the understanding the production process of the partner organization by VCP engineers in order to find points of collaboration (see Figure 1).

There are several reasons to apply the MAPPER modeling approach to establish a production management culture. First of all, according to the proposed approach (POP*) any organization process can be decomposed in five dimensions: process, organization, product, decision, and infrastructure. Secondly, the AKM approach (Active Knowledge Modeling; see Lillehagen, et al., 2002) can be used to create desktop interfaces for managing work processes. Third, these interfaces can be further customized through a cooperative modeling process. Finally, user interfaces can be re-created and re-configured based on the same models without the need for additional programming work. The model is useful to identify structural problems within the organization, such as the lack of correlations between different roles or processes, boundaries in data sharing, lack of shared understanding with respect to roles, objectives, responsibilities, and deadlines. The models can be used as artefacts to establish shared collaboration processes between the actors involved.

Together with these advantages, some existing problem areas have emerged in the creation and use of models in the organizational context for the management of production processes. First, although the models are rich representations of individual organizational aspects, is not always easy to access them. Some users may not have the necessary application tools (and in the correct updated version) to view models or modify them. This factor can inhibit the participation of some users to the modeling activities. An effect of modeling is that the object-of-design (say, the modeling of a process) becomes invisible to

*Figure 1. Modeling*

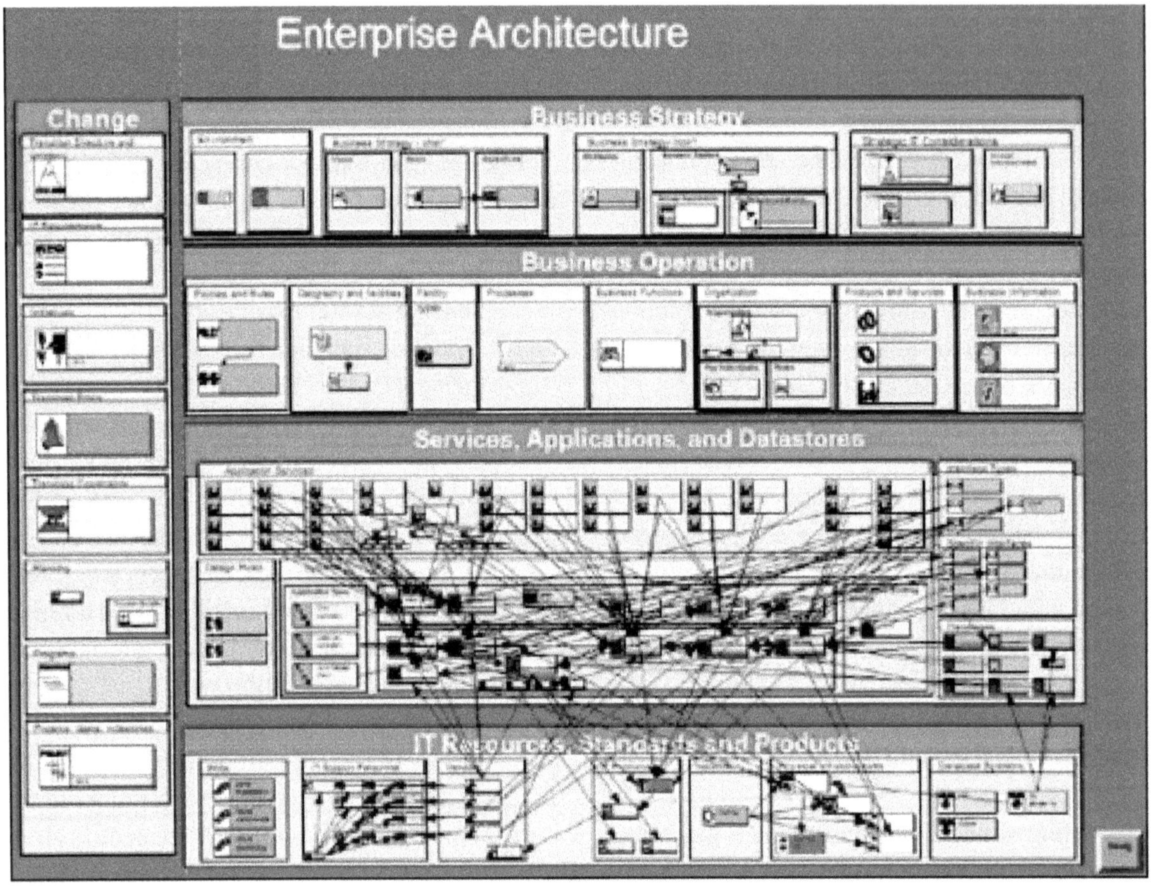


those who do not have access to the model. This factor can inhibit participation by those excluded from access. Further, if modeling is adopted by a company for the representation of organisational aspects, there might happen a 'model-everything' tendency: working practices, social relations, informal exchanges. This is problematic because the models are limited and it is not currently possible to model informal exchanges among team members, especially when these are generated by contingent and unforeseen activities, highly dependent on the situation. Finally, models offer the possibility of limited representations (mainly containers, boxes, and arrows). In most cases, modeling means creating workflows. It is well known, however, that workflows do not represent all working practices. The workflows are created a priori, through a top-down process: they are highly structured and logically ordered. As already mentioned, there are contingencies and work practices that cannot be represented as workflows. In the next section, we describe how we have moved from identification of problem areas related to modeling to their existing consolidation in more structured 'how questions.'

## How Questions on the Modeling of Manufacturing Processes (Analsysis)

Observations of modeling sections in VCP provided a detailed description of the modeling process. Such descriptions have led to the identification of some qualitative aspects of modeling as a practice and of models as a representation. These aspects have been captured in the following 'how questions':

- How to deal with the granularity of modeling? How to decide where to begin to model and what to exclude from the modeling process, leaving it to other formats (documents, informal interactions)? Despite some models arguing the opposite, it makes no sense to model every aspect of the working context. Complex and interconnected processes, organizational structures, and products will certainly need simplified representations to make them accessible, and this may be the role of models. Less desirable is instead to model the exchanges between the actors ad-hoc cooperative and informal existing social interactions. These interactions are different from time to time, depending on the specific people involved, in many cases unique. In addition, if the model contains every aspect of the organization (finance, strategy, organization, etc.), it may become a sensitive document.
- How is it possible to establish a standard that makes models flexible and interoperable? There are several modeling approaches, depending on the modeling tools and approaches. If an organization has already established its own language, it is difficult to change. The merging of models created with different notations is a complex task and requires the re-modeling of all models written in a different language.
- How to transform procedural documents in models? Documents are different from models. Transforming documents into models requires the definition of relationships and dependencies that are not spelled out in documents. The document is usually a 'live' artefact that continues to grow. It is not always clear, in a document, which has changed it, what were the changes and when they actually took place. Rather than transforming documents into model, it might be more appropriate for the time being just to create a link in the model pointing to an external document, thus creating a composition of different types of artifacts (the model and the document).
- When is it appropriate to transfer the company knowledge in a model? When this transfer does not make sense? Many organizations keep their knowledge in procedural documents. Some of these documents are static; others are more similar to working documents that are constantly updated. The models are often created with the express intent to update the entire organizational structure in the advent of any change. This is not always a wanted effect. That is why it is important to think in advance about the consequences of the migration from a static to a dynamic vision of company knowledge.
- How to manage access rights when the necessary modeling skills are external to the company? When managing documents in organizations, there are some actors that have full control over them, while others may have read only access. This makes of models and documents key artefacts, not only because they may contain sensitive information but also because they incorporate specific access rights. Normally, access rights are defined by document administrators. Since models are often administered by external experts, access rights will be determined by them, potentially creating interference with the allocation criteria within the organization.
- How to define the criteria for data exchange between the modeling system and other applications? One of the benefits of having models for the management of production processes is the ease of replication of knowledge in different organizational contexts.

This makes the models a central resource with respect to other applications currently in use. The models cannot however be the only application in an organization. Especially in the early stages of adoption, the model must operate in the presence of other applications. The definition of principles through which models exchange data to/from other systems (SAP, CAD) is crucial.

- How to keep track of the learning generated during the modeling process? All modeling systems provide an interface that simplifies software development and reduces it complexity of use. Modelers and domain experts agree that the learning generated during the modeling sessions is often forgotten once the model is produced.

With these 'how questions' in mind, we had conversations with the CEO and middle management of VCP, with the aim of co-construct visions of solutions, with the aim of commenting and give priority to the proposed 'how questions.' These conversations aimed also at promoting their awareness on the issues involved at at collecting their views on visions of solutions. This is done before the exposure of visions of solution developed by the research team. The analysis phase involved, beyond the CEO, also two middle management groups: a group from the technical area and a group from the area of marketing. It has been the counsellor's responsibility to enable in the people involved an appreciative attitude, oriented to the *pars construens*. Our primary mission in the workshops in the VCP was to make people involved the authors of organizational change, in their quest toward a production management culture through the use of models. This perspective was clearly contrasting with the vision of top management, in favour of the procurement of outside expertise to solve business problems.

The synthesis phase of Social Practice Design is described in the following section. In that phase, aiming at the production of visions of solutions, SPD researchers proposed groupwork sessions, based on the adoption of an analogical version of modeling, called the 'organizational kit game.'

Two groups have been involved in the sessions: a first group of four managers of the production line, including VCP product development manager and a second group, formed by the new Director of Marketing, a marketing specialist and two designers.

## The Organizational Kit Game (Synthesis)

Participants were involved in an 'exercise,' whose reference to the 'how questions' and its objective—to take a step forward compared to the problems that emerged during the analysis—were made explicit. Involvement has been generated by using a state of the art participatory design technique, the 'design games,' in particular through the game of the organisational toolkit. The kit consists of: wooden building blocks of different shape and colour (some featuring anthropomorphic characters), paper, stickers, and coloured pencils.

The organizational kit game (Figure 2) was introduced to address the issue of how to acquire a production management culture through the use of modeling. The participants were given instructions to use the building blocks to create a shared representation of their organization as they see it today. The words used were: "use these constructions to represent your organization in a creative way. You have a few characters to do this. You can give them names. Also you have building blocks to represent places. These places may be within this building or elsewhere. The circles are artefacts: cars, documents, or anything." In a second time, they were asked to represent their organization constraints. Participants were assigned cards containing

*Figure 2. The organizational kit game*

images that symbolized a variety of aspects. Finally, participants were asked to think about what they wanted to change of the organization they represented and to 'implement' these changes with the organizational kit resources.

At the beginning of each session, the research team introduced the participants to the objectives of the workshop and the SPD method, focusing on the aspect of creating visions, and to design games as a method of inspiring activities to help support participants to generate ideas, concentrating on 'how questions.' The researchers also reminded participants the set of 'how questions' and presented the workshop as having the objective to test the organizational kit game on the participants and to support them in creating their visions of solution.

## The Vision of the Organization from the Point of View of Production

The first group consisted of four production line managers and the VCP1 product development manager—as it will become apparent, VCP has two quarters, which we will call VCP1 and VCP2. Previously, VCP1 had a flattened hierarchy: it was the stage of the formalization of the tasks in production management, where teams were formed around the various production lines. Since they started naming product managers on the basis of the projects ("This is a dynamic position"), line managers have had to plan work, supervise and manage critical situations by themselves. The line manager is also responsible for marketing, promotional materials and so on—although "the engineers are not interested in, clients are not willing to meet the ambitions of the designers."

The 'how question' on which this group has helped has been that of how to acquire a production management culture through the use of models. The activity developed according to three steps: represent your organization, identify the constraints, re-designs. The group immediately began to build a representation of their own organization, completing the task quickly. The focus was on the VCP1 quarter and its president (one of two), while the VCP2 site was considered marginally.

The discussion then focused on the need to improve communication between the two quarters and, in order to represent that, one of the line managers moved the character who was responsible for the development in a location between the two locations, "we have to analyze the real problems in the product lifecycle... and improve communication with VCP2, point towards a coordination of the demands deriving from the production line to VCP2, someone in VCP2 has to coordinate and prioritize." The group then quickly found the stickers representing the problems and constraints of the organization, agreeing on a set of problems (see Figures 3 and 4).

## The Marketing Vision of the Organization

Marketing and production are not well connected and cooperative in VCP1. One consequence of this is that marketing find it difficult to express a marketing perspective from the way product

*Figure 3. The President's most pressing problem – lack of resources*

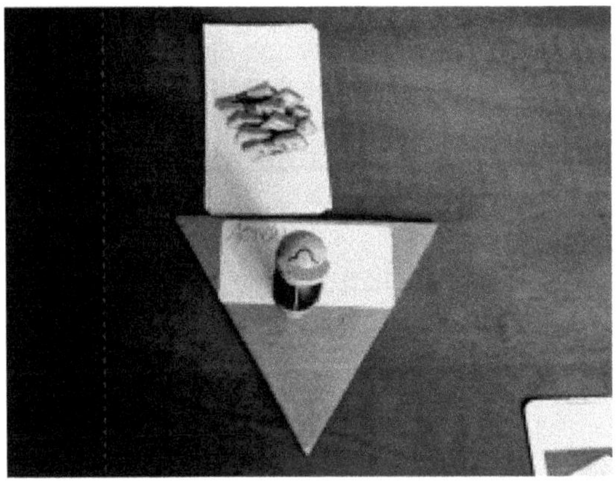

- The president is described as facing a mountain of problems
- There is much confusion on the production line and waste of resources: "someone has to coordinate priorities among production lines"
- Faster decisions: "sometimes we have ideas on how to solve the problems but the decision is not taken quickly enough because the president has to deal with everything"
- The responsibilities are not dear: "who is in charge"; "a more rigid structure is needed."

*Figure 4. Other arguments: incentives system, human resources allocation, planning*

- Improve the system of incentives: "after a training period in VCP, the staff moves to other companies," looking for more interesting and better paid work.
- There is no rigid allocation of human resources on production lines: "it is difficult to plan something without a planned budget."
- Product managers do not keep track of scheduling, because there are too many changes in the production cycle.
- We need a list of repetitive tasks, more detailed targets for each product for each quarter.

information is conveyed to them from production. The website contains information sheets, written from a purely technical point of view. The details, the main asset—low-cost customizations—is not expressed in a way that someone who is not expert can understand. The website contains a general description of the orientation of the production process to customers, and important aspects of configurability are hidden in the technical details. Recently, a design engineer has been appointed marketing manager after having taken a marketing course. Here it is how he describes the organization:

*"This line is the IP Core Department in VCP2 and this is the same thing in VCP1. This is the development department in VCP1, led by M. This is the department of embedded systems and software in VCP2, which SF is responsible for. These two departments, perhaps sub-departments, making the total of the electronics department. This is the marketing and sales in VCP2, and this is the one in VCP1. Here we have the sales and marketing manager M, and here we have the sales manager S. Here we have the VCP1 department of administration which I is responsible for. And this is the administration and finance in VCP2 and the VCP2 Chief Financial Officer, Chief Administrative Officer Mrs. J. This in the middle is of our main product, which is the '"IP core," the virtual component. This has influence on the whole business. Here we have the IT department. The IT department is in the middle because the people who work there works for both VCP1 VCP2."*

While the first group was more focused on his own work, the second had a more diffused attention for the two sites (and VCP1 VCP2) and their relationships. The next step was to identify constraints. The group identified a major problem in the failure to communicate with top management: "This Aladdin's lamp (one of the kit's cards), illustrates some problems that come from the boss. His expectation is that we have to know what he thinks about the different things. He thinks that we must do the way he thinks without knowing what he thinks ... that just means that he did not properly explained himself. We do not know what he thinks and he thinks that we should know ... It should explain himself better ... He gives orders, but does not ask if we have understood them correctly. He does not know how we have understood them. There is no feedback expected from our side." A map showing a road sign is explained as symbolizing that "the company is in permanent construction," and the group thinks that this is due to a lack of real management skills. Another card (which symbolizes a cable), is used to illustrate the lack of communication between the various units of the company, marketing and development, the VCP1 department of electronics and VCP2, between the top management and the managerial levels below. This group has several ideas how to drive change too. According to participants, managerial skills have to be increased, so as to be able to re-design products and penetrate new market segments. According to them, the company needs restructuring. It needs to be divided into two entities, extracting some activities on the virtual components that are currently based in VCP2 and transferring them to VCP1. In addition, the marketing and sales department should only be in one place (VCP1). The discussion on managerial skills was then concluded by stating that "management problems are not just our boss problems. They are lower levels groups, managers of the production line, and others problems. We should all have these skills. And we should do just that and not perform other operational functions."

## DISCUSSION

The relevance of the 'how question' on how to improve production management skills through the use of modeling has been confirmed by the work group on the organizational tool kit. Both groups—the marketing and the product development ones—fundamentally agree in the analysis of VCP problems and have many ideas on how to change the organization. Design game techniques contributed to the creation of an open, non-judgmental context for the discussion of organizational issues. The organizational game kit was also interpreted as the proposal of a tangible approach to modeling, which can solve some problems raised in the 'how question' by acting as an active support to the creation of a production management culture.

## The Non-Judgmental Environment

The staff of VCP is rarely found in the situation to be encouraged to have an open debate without censorship, in full recognition of their skills. Usually their conversations remain private. Design games have supported participants in shaping their comments, to create a mutual understanding, and to go beyond the ideas already discussed privately between them. The cards and the building blocks that are served as part of the game were used as 'thinking tools': we observed the participants select, move, label these elements while giving shape to their ideas. The playful element has enabled participants to get involved, explore options, entering the realm of imagination. The design game also favoured the establishment of a position of trust and confidence among participants: they not only closed the door, but also locked it during the session.

## The Organizational Kit as a Tangible Modeling Tool

The organizational kit game was immediately appropriated by the participants as a modeling tool. The participants were delighted: "This is as a modeling process should be." Comparing this approach to the modeling system used in the MAPPER project, they said that this had all the requirements of a good modeling system. The organizational kit turned out to be:

- **Object-oriented:** the possibility of using different types of objects (building blocks of different shape and colour that represent people, activities, objects, places, stickers to name building blocks, cards with images) and to improvise using post-it notes, freehand drawing, cut out of additional forms in the cartoon.
- **Mobile and portable:** the kit can be transported in a small rectangular box. It is reusable: the interpretive flexibility of wooden building blocks allows reuse and customization. It is also three-dimensional: the participants can put a character in a higher position simply by elevating it on a pile of buildings. New combinations of elements can be invented on the fly and easily interpreted.
- **It does not contain sensitive information and it does not retain changes:** it is a simple construction kit that does not save information nor can serve data sharing as a software application does. In addition, changes made on the position of the buildings do not have direct implications on all the others.
- **it helps preserve the modeling process experience:** when audio and video recorded, the physical movements and exchanges made by participants during usage of the kit can easily be reconstructed, along with their reflections-on-practice (Schon, 1983);
- **it supports the creation of visual scenes:** building blocks and cards featuring images help visualize complex ideas and relationships, inspirational objects provide implicit visualization modes;
- **it contains an activating and playful element:** participants used cards with pictures and building blocks such as thinking tools. Ideas are enacted rather than just talked about. Implications and alternatives can be experienced as building blocks moves. There is a strong element of exploration, and space for the unusual;

- **it invites participation:** it corresponds to the idea of user-driven modeling, it supports cooperation. The items scattered on the table invited participants to become part of the modeling session. The context makes clear the scale of the map and the need for cooperation for the representation of complex situations.
- **It is user-oriented and environmentally friendly:** The kit is easy to understand, easy to carry, easy to extend. It's simple and made of recyclable material.

## CONCLUSION

The chapter intended to illustrate the constituent elements of Social Practice Design through the description of one of its applications oriented guide to the acquisition of a production management culture through the use of enterprise modeling tools by a company producing electronic components (VCP). In this case, the observation phase covered a range of qualitative aspects concerning modeling sessions. The analysis phase has consisted in the consolidation of some 'how questions' on how to acquire a production management culture through the use of enterprise modeling. In the synthesis phase of the SPD process, VCP people were involved in the creation of their visions of solution through a design game based on a tangible modeling tool. The phenomenological view on information systems development from which the Social Practice Design takes its shape suggests that the engineering breakdown approach and subsequent managerial perspective on information systems development methodologies are embedded in networks of practices that create values and goals. The possibility to break down these networks of practice in methods and techniques depends on the mobilization, the articulation and consolidation of social practices. The way in which the constitutive elements of the Social Practice Design were shown (observation, analysis, and synthesis) corresponds to the phenomenological awareness that not only the introduction of a technology (in this case modeling), but the same application of methodologies oriented to its introduction involves a modification of established practices. The main aspect of the Social Practice Design is thus the attention paid to making explicit to the users the methods adopted by the researchers in all their phases: in the observation phase, ethnography is preceded by some initial conversations with users on the method; in the analysis phase, the knowledge of the counselor seeks to establish a state of hospitality to innovation (consisting of the technology—in this case modeling—but also of method) making explicit the methods in use; at the beginning of each synthesis phase workshop, the organizational kit game has been introduced not only as a support for the creation of visions of solution but also, reflexively, as part of the SPD path and as such subjected to the evaluation of the participants. As an extension of participatory methodologies to the implementation phase of technology development, aiming at the generation of strategies and action plans, Social Practice Design is also characterized by the absence of an established order or even the necessity for the intervention aiming to 'make room' technology to have a deadline. The synthesis phase culminates with the evaluation of the intervention and the possibility of a reiteration of the whole process.

## REFERENCES

Bødker, K., Kensing, F., & Simonsen, G. (2004). *Participatory IT design*. Cambridge, MA: MIT Press.

Ciborra, C. (2000). *From control to drift: The dynamics of corporate information infrastructures*. Oxford, UK: Oxford University Press.

Ciborra, C. (2002). *The labyrinths of information*. Oxford, UK: Oxford University Press.

Ehn, P. (2006). *Invited talk*. Retrieved from http://www.mcis2006.org.

Latour, B. (1986). Visualization and cognition: Thinking with eyes and hands. *Knowledge in Society*, *n.d.*, 6.

Lewin, K. (1946). Action research and minority problems. *The Journal of Social Issues*, *2*, 34–46. doi:10.1111/j.1540-4560.1946.tb02295.x

Marshak, R. J. (1993). Lewin meets Confucius: A review of the OD model of change. *The Journal of Applied Behavioral Science*, *29*, 393–415. doi:10.1177/0021886393294002

Moorman, C., & Miner, A. S. (1997). The impact of organizational memory on new product performance and creativity. *JMR, Journal of Marketing Research*, *34*, 91–106. doi:10.2307/3152067

Orlikowski, W. J. (1996). Improvising organizational transformation overtime: A situated change perspective. *Information Systems Research*, *7*(1), 63–92. doi:10.1287/isre.7.1.63

Polanyi, M. (1966). *The tacit dimension*. London, UK: Routledge & Kegan Paul.

Prochaska, J. O., Di Clemente, C. C., & Norcross, J. C. (1992). In search of how people change: Applications to addictive behaviors. *The American Psychologist*, *47*, 1102–1114. doi:10.1037/0003-066X.47.9.1102

Rogers, C. R. (1951). *Client-centred therapy*. New York, NY: Constable.

Schon, D. A. (1983). *The reflective practitioner*. New York, NY: Harper Collins.

Walsham, G. (2004). Knowledge management systems: Action and representation. In *Proceedings of ALOIS 2004*. ALOIS.

Weick, K., & Quinn, R. (1999). Organizational change and development. *Annual Review of Psychology*, *50*, 361–386. doi:10.1146/annurev.psych.50.1.361

# Chapter 16
# IS-Related Organizational Change and the Necessity of Techno-Organizational Co-Design(-In-Use):
## An Experience with Ethnomethodologically Oriented Ethnography

**Chiara Bassetti**
*University of Bologna, Italy*

## ABSTRACT

*This chapter considers some aspects of an ethnomethodologically oriented ethnography that has been carried out in a medical Emergency Response Centre (ERC) before, during, and after an IS-related organizational change. After a description of the everyday work in the ERC and its larger social arena, the authors discuss the main changes and the users group's resistance that mediated the new technologies' transformative potential: the rejection of abandoning 'old' cooperative work practices, and the emergence of an innovative one, with its own condition of appropriateness, applicability, and accountability. Finally, starting from the evidence that solutions to problems emerging in a field must be coherent with the endogenous organization of activities of that field, with the configuration of inter-actions that actually sets up that context, the authors discuss the necessity of co-design(-in-use), and the possibilities provided by ethnomethodological ethnography as a tool for action research in IT design and techno-organizational change management.*

DOI: 10.4018/978-1-4666-0303-5.ch016

## INTRODUCTION

The chapter considers some aspects of an ethnomethodologically oriented ethnography (e.g. Crabtree, Nichols, O'Brien, Rouncefield, & Twidale, 2000; Fele, 2008; Sharrock & Randall, 2004) that have been carried out in a medical emergency dispatch "centre of coordination" (Suchman, 1993, 1997) before, during, and after an IS-related organizational change.

Emergency Response Centres constitute stressful and time-pressing, hi-tech, and information-rich environments, in which communication, on the one hand, and information management (gathering, use, transmission, archiving, etc.), on the other hand, are the primary work activities—and are largely technologically-assisted activities. It is a complex and changing, situated configuration of people, artifacts, knowledge, information, activities, and practices that constitutes the organization itself. ERCs, therefore, can be considered as socio-technical systems (e.g. Avgerou, Ciborra, & Land, 2004; Button & Sharrock, 1998; Suchman, 2002; Suchman, Trigg, & Blomberg, 2002).

The process of design, introduction, implementation, and use of new Information and Communication Technologies (ICTs), indeed, is not simply a technological one, but also a social one, and always entails related changes in the organization of activities and work practices, changes that need to be addressed, managed and *designed.*

It is not that, accordingly to the "technological imperative" (Markus & Robey, 1988), IT "impact" in a univocal way on individuals' and organizations' acting: as we shall see, (groups of) people appropriate, make sense, and use artifacts in various emergent and dynamic ways. It is neither that, accordingly to the "organizational imperative" instead, IT designers and/or organizational managers have "almost unlimited choice over technological options and almost unlimited control over the consequences" (Markus & Robey, 1988, p. 587), so that they could exploit IT, in a predictable way, as a tool for solving problems and satisfying needs of the organization. As I shall try to show, innovation is never a merely institutional process, it is never just about standardized procedures and formal routines; on the contrary, it is also an interactional process, it is also about situated practices and improvisation.

Looking at IS-related organizational change processes from a socio-technical perspective, therefore, means considering not only technology itself, but also and especially technology-in-use (Suchman, 2002); not only the formal properties of facilities and artifacts, but also their actual usability in context (Adler & Winograd, 1992). The latter consists of the new technologies' appropriateness and usefulness with respect to the actual organization of working activities in the time and space specific context in which specific (groups of) people shall use them alongside pre-existent tools and ICTs. Usability consists of technologies' coherence with everyday work practices and the logic(s) underlying such practices.

Furthermore, a socio-technical approach to techno-organizational change entails taking into consideration the whole set of individual and collective actors—designers, (future) users, managers, etc.—as well as the different ways in which they frame, appropriate, and make sense of the everyday, technologically-mediated work practices, on the one hand, and, on the other, of the change itself.

It is thus necessary, first of all, to analyze the everyday work practices in and through which participants produce, recognize, and co-ordinate their technologically oriented activities (Heath & Luff, 2000), the way in which they orient themselves towards each other's (verbal and non-verbal) conduct, and "constitute the occasioned sense and significance of features of the setting such as objects, artifacts and the like" (Heath, Hindmarsh, & Luff, 2010, p. 87). As I shall contend, an ethnomethodologically oriented ethnography, with its focus on the minute details of (naturally occurring) action and interaction, on order and routine as practical accomplishment, on "actions-as-constitutive-of-their-settings and settings-as-constitutive-of-their-actions" (Heritage, 1984, p. 308), could be extremely helpful in such a task.

The 30 months long ethnography has been carryied out at the medical emergency response centre of the Province of Trento (Italy): "Trentino Emergenza 118." Data include: field notes, informal interviews, photographs, screen captures and other documentary materials collected on the field; audio-recordings (about 39 hours) of the phone calls and video-recordings (about 56 hours) of the everyday work in the centre, then transcribed and analyzed accordingly to the principles and procedures of conversation analysis (Jefferson, 1984; Heritage, 1984) and video-based research (Heath, Hindmarsh, & Luff, 2010).

The research was not initially designed as the study of a techno-organizational change, but of the collaborative and cooperative work practices enacted everyday by the operators, the purpose being to analyze the organization of emergency assistance as a situated social practice connected with the use of technology/ies. The process of innovation started when the ethnography was going on since about 18 months; the unexpected opportunity of studying such a process brought the fieldwork to be prolonged.

Although, as we shall see, the research process, and the ethnographer carrying it out, have obviously affected the IS-related change process, the research remained scientific in its aim, and was not intended as an intervention. Nevertheless, I believe that such a research experience may contribute in answering one of the questions that this book is posing: Whether and to what extent organizational "design" connected to information system adoption and use is possible? As I shall try to show, it is not only possible, but also necessary.

After a brief description of the organization of the everyday work in the ERC, and the more general social arena in which such a work is embedded, I shall discuss the more relevant IS-related changes, their design and implementation, and their implications on everyday work practices, as well as on information and knowledge management. I shall then describe the users group's *resistance* that mediated the transformative potential of the new technologies (Hakken, 1999): from refusing to give up to the "old" work practices of coordination and cooperation, to creating an inNovATIVE practice, accompanied by its own conditions of appropriateness, as well as its own (ethno)methods of application and recognition (or accountability). I shall try to outline the emergence of an innovative, situated organization of the activities in the field; of a new work practice, collaboratively designed and enacted by field's inhabitants (or "natives"). Starting from this, I will finally draw some reflections on Action Research (AR), trying to enlarge ethnomethodologically oriented ethnography potential for IS-related organizational design both behind and beyond the role of organizational problem finder (Jacucci, 2007; Jacucci, Tellioglu, & Wagner, 2007) or problem solver (Akdere, 2003; French & Bell, 1999; cf. also Kemmis, 2001).

## SCENARIO

### Everyday Work in the ERC: Communication in Action

The medical emergency response centre of Trento is basically composed of four rooms, called boxes, and equipped for receiving and processing emergency calls: aligned along a single corridor, boxes are separated by glass walls, in order to allow the operators to see what is going on in the box to side and communicate from a box to the other through gesture, mimicry, bodily conduct, gaze and other non-verbal semiotic resources—or "systems of signs" (Goodwin, 2000)[1]. The first box, called "room chief box" is occupied by the ERC's coordinator, who is usually present at the centre from 8 AM to 6 PM; the second box, called "taxi box," is active monday to friday from 7 AM to 9 PM and manages non-urgent transports (hospital transfer, discharge, etc.); the third and fourth boxes, active 24/7 and respectively called "province box" and "city box," are instead fully operative for managing the emergency calls, and differ for territorial jurisdiction.

In each of the latter boxes two operators work in close cooperation and alternate themselves, during the work shift[2], between the roles of call-taker, assigned to phone calls management, and dispatcher, responsible for radio communication with first aid units. This division of work activities and responsibilities, however, is not entrapping nor established once and for all. It can happen, for instance, that the two colleagues deal in parallel with two different calls, even if they mostly enact complex forms of coordination and cooperation in order to manage in parallel diverse aspects of a same emergency (for example, in case of road accident, dispatching ambulances, giving first aid instruction to the caller, inform the police, etc.). Nevertheless, it is important to highlight that the engagement of the call-taker role implies also the task and the responsibility to make all the necessary decisions concerning the incoming calls and related emergencies. For this reason—and other ones connected to the change process that shall be clear later—I will define the above mentioned roles, respectively, operator A and operator B.

Emergency centre operators communicate: a) via telephone with the callers, having the immediate task to gather crucial information in order to determine the accident, the place and the patient's or patients' conditions; b) via radio with first aid units, an informative, quick and largely unidirectional communication; and c) in co-presence with colleagues, an internal parallel communication aimed at the management of the external emergent one with callers and rescue units. Incoming communications, in fact, continuously and unexpectedly intersect the ongoing flow of activities in the ERC (Fele, 2005); operators, therefore, must situatedly and interactionally coordinate their task(s) with others' ones, as well as with emergent communications, by collaboratively establishing priorities and assigning/dividing tasks moment by moment.

All these communicative work practices are produced and coordinated in real time by the participants, through talk, mediated or not; bodily conduct, especially in its kinesics, proxemics and gestural components; and the use of tools, technological artifacts and the like. This situated coordination, that allows participants maintaining a dynamic monitoring of the ongoing situation in which they insert their own flexible contribute, finds help in another IT, the automated one which is the computer system. Operators exploit the latter in order to: manage incoming and outgoing, radio and phone communication; select and monitor medical vehicles; locate the accident on a digital map; record information; and get real-time awareness of the colleagues' activities.

## Social Arena: Spaces and Places, Actors and Roles, Actions and Interactions

In order to briefly describe the social arena in which the techno-organizational change took place, I shall exploit a multi-level map (Figure 1). The first level is the cartographic one concerning physical spaces, while the second level is that of the chromatic areas, which represent those that have become places, the zones of work and "jurisdiction," so to speak, of each group of actors. Actors, whose name is written slantwise in the map, and who represent the third level, are grouped on the basis of institutional professional roles, but also of the material dislocation of bodies and work activities, and the consequent perception of actors themselves in terms of membership and belonging. The last level (dotted lines) concerns the material and interactional trajectories of each group of actors, and thus the circulation of knowledge and practices.

The centre of coordination "Trentino Emergenza 118" is located in front of the Hospital; the access is placed in front of the First Aid building's

*Figure 1. Multilevel map of the ERC compound*

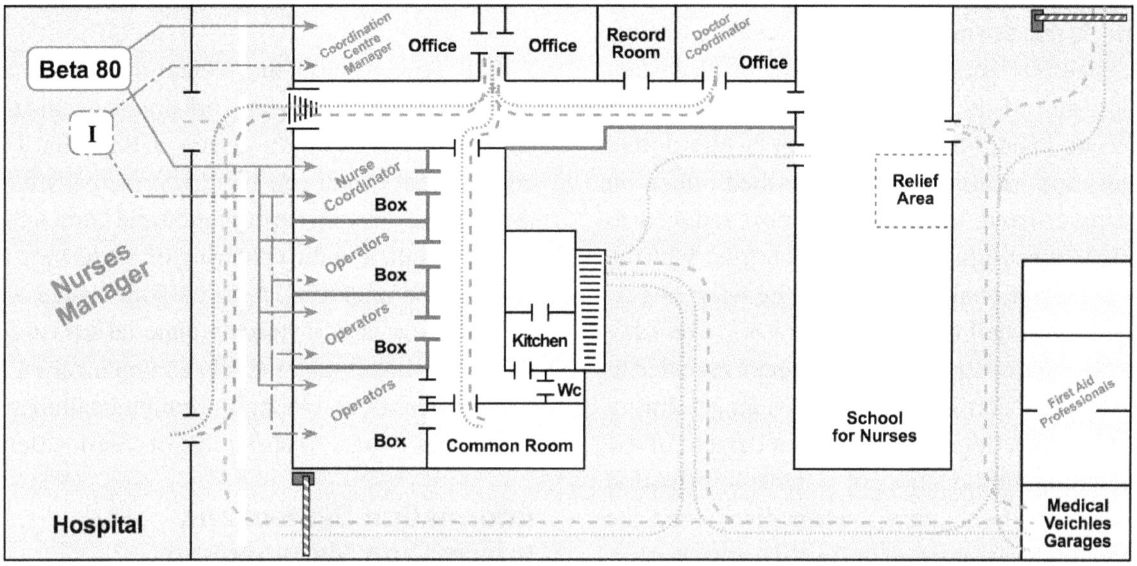


entrance that is reserved for ambulances (up on the left in the map). Inside the hospital there is the office of the Nurses Manager, which is responsible for all the nurses of the Provincial Agency for Medical Services, ERC's operators included.

Once entered the centre, one finds him/herself in front of a long corridor, which brings to the adjacent building of the professional school for nurses, where is also located the relief area that is used by ERC operators and first aid unit members too. The latter generally stay in/near the garages, ready to "exit" for rescue, but sometimes they enter the ERC—using the fire stairs on the backyard—in order to entry data concerning closed events on the computer of the common room. On the left of the corridor there are the offices of the "Trentino Emergenza 118"'s managers: the manager of the whole coordination centre and his secretary, both doctors, and the Doctor Coordinator of the ERC. Along the corridor one finds also the record room, with a LAN-connected Racal recorder that keeps trace on DDS2 tape of all incoming and outgoing, phone and radio communications.

On the right of the corridor, instead, there is the entrance to the actual ERC, where nurse operators work. The door is always closed and one can access the centre only after identification trough an interphone connected to a closed circuit camera. Personnel too, when off duty, must follow this procedure, although some of the above-mentioned managers, having the keys, do not have to. Along the narrow corridor of the ERC there are the boxes of the operators and the Nurse Coordinator. Down the corridor, on the left, one accesses the common area, which consists of a bathroom, a kitchen and a common room, equipped with one computer and some shelves where each operator can put her/his personal effects, and where his/her mail can be deposited. These areas are frequented by the operators, especially at the end/beginning of the work shift; by first aid units members; and, just in case of meetings, by the managers, which otherwise restrict themselves to walk along the corridor and look through the glass doors.

We thus reach the fourth level of the map, that of the material, interactional and cultural trajectories of each group of actors; that, therefore, of the knowledge, discourses and practices circulation. What I am interested in underlining, here, is that the interactions between the various managerial figures (of the hospital: black dashed lines; and of the ERC: black dotted lines), on the one hand, and the operators (white dotted lines) and paramedics (white dashed lines), on the other hand, is minimal: almost reduced to formal meetings. Think, for example, that the re-organizational hypothesis—heavily affecting, as we shall better see, operators' everyday practices of cooperation—have been revealed to me by the Coordination Centre Manager during our first meeting after the announcement of the upcoming introduction of a new information system. On the contrary, as I later discovered, the operators were not at all informed on that regard, and get to know about thanks to my questions on the topic.

Last actors of the arena, in fact, are myself-the-ethnographer (I) and the information system designers group (*Beta80*). Both have been in touch with managers as well as operators. However, it is important to highlight that the designers, differently from me, have interacted with operators only at a second stage: for the major part of the design process, in fact, they have discussed just with the managers; the interactions with those that would have then been the IS actual users have been restricted to some meetings, held when the system was already almost completely designed, in order to illustrate its functioning and evaluate the necessity of minimal modifications that the user could have requested. The consequences of such a choice—totally indifferent to the social organization of work activities which was going on in the context of application and use of the new information system—revealed their limits soon enough: we shall see how the inhabitants of the socio-technical system, left aside during the design phase, have been able to put themselves back at the centre during the phase of actual implementation and use.

## TECHNO-ORGANIZATIONAL CHANGE AND *RESISTANCE*

I shall now discuss the techno-organizational change, its implication for work practices, and the reactions, both explicit and tacit, of the users. The changes that I shall consider are principally two: the first one concerns information and knowledge management, and the emerging of a new kind of information and, thus, potential knowledge; the second one regards instead the material infrastructure of the boxes, and the related organization of the practices of cooperation, communication and tasks/roles' situated distribution (or coordination).

### Information System and Information Management

The main innovation of the new information system regards information management and consists in the shift from the "accident form" (Figure 2), opened and filled by the operator in case of aid intervention and pre-formatted to contain all the necessary information, to the "call form," which automatically opens itself for each incoming call and must be filled by the call-taker—in real time, according to designers' and managers' intentions—with information regarding the caller and the reason of the call[3]. The operators, in case of an actual emergency, have then to fill an "aid form," with information about the accident, its seriousness, etc.; a "mission form" for each vehicle chosen and dispatched for that aid; and a "patient form," including vital statistics and medical information, for each rescued person (see Figure 3).

A first implication of such a change consists of the workload increasing for the ERC professionals, whose time—a basic and fundamental resource in emergency contexts—comes reduced. As brought out by the individual interviews concerning the new IS and carried out with the operators before its implementation (which took place in May 2005), most of the operators were in disagreement with the upcoming organization of work activities, and

*Figure 2. Capture of the accident form's main screen of the "old" IS*

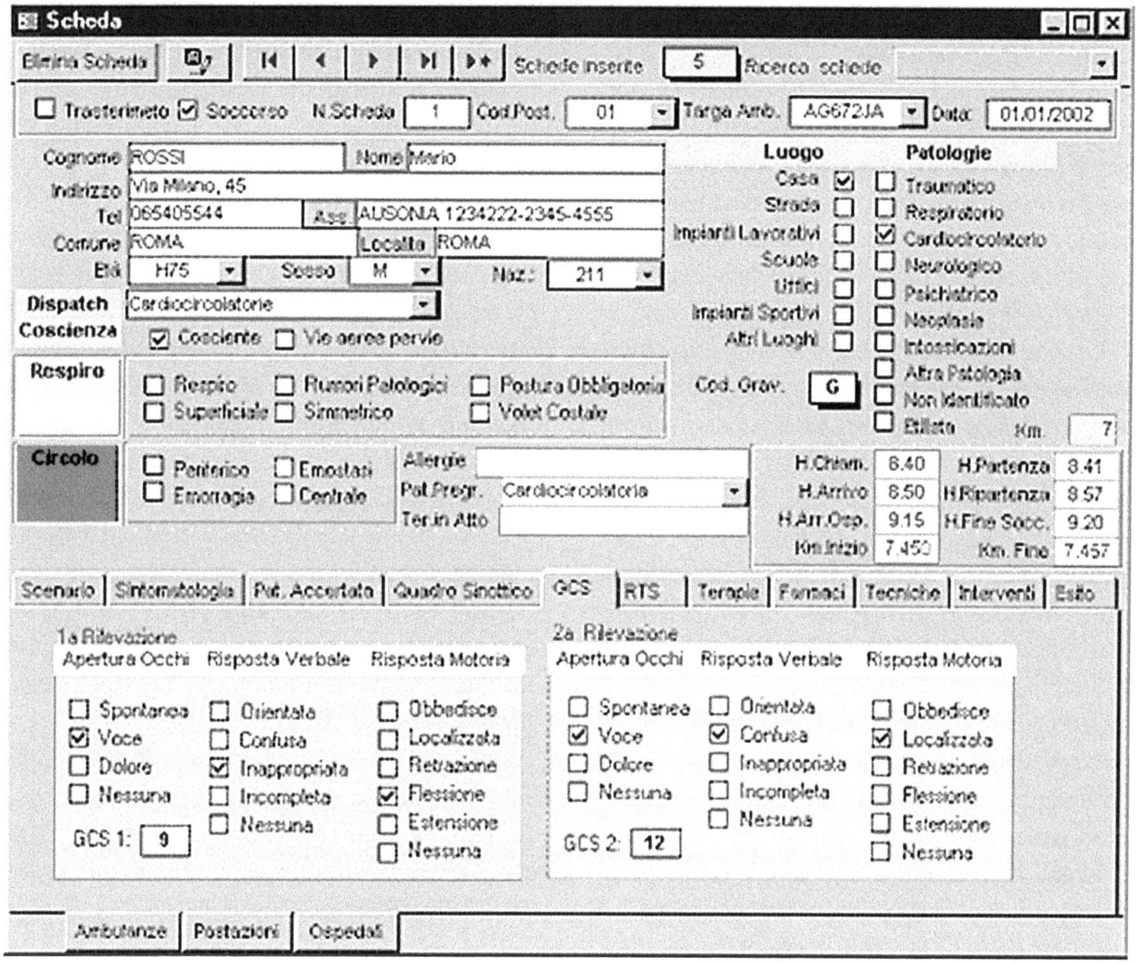


*Figure 3. Scheme of the data entry forms of the new IS*

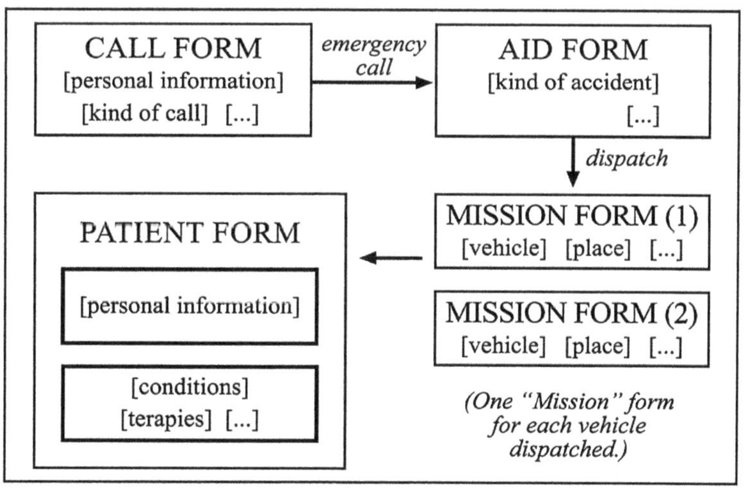

immediately started to plan personal alternatives. See for instance the following excerpt.

*No, but I won't give a toss! That is, probably I'll not entry some useless calls. Many calls too, if I haven't time, if I'm very busy... Out of experience, when there's a lot of chaos, in order to protect yourself, you must deal only with real urgencies, and leave aside less important things, such as "event open," "event closed," number of kilometres. Nobody can incriminate you because you haven't wrote down the kilometres or, in the worst hypothesis, because you've skipped a scheduled transfer, but if you let somebody die... [Ma. 03-02-2005]*

Another consequence of the technological change consists of the emergence of a new kind of information. With the new IS, in fact, collected information does no more regard just the various characteristics of the accidents, and what is possible to derive from that, but also information someway organizational: how many calls in a determined period of time, of which kind and in which percentage, which is the everyday workload of an operator, and so on. While such information could be useful for the managers, its collection complicates the operators' everyday work. This kind of information, furthermore, is never available for operators (whose work is now both more demanding and surveilled): in fact, it comes collected, managed and archived by the information system in an automated manner, and it can then be recovered, dealt with and turned into knowledge only by members of the managerial group. The prospective organizational changes that could have been decided—but have not been, at least during the ethnographic observation—on the basis of such a knowledge would have reached the professionals unexpectedly. In her study on IT for a home assistance service, Sara Eriksen (2002) presents a similar case and talks of *over-empashized economic accountability*; she then claims for the necessity to support local visibility and reflexivity in action, instead of managerial intervention, control, and surveillance.

Moreover, the new system, by itself, stimulates the abandonment of the use of paper supports. The reason is twofold: first of all, the amount of data that need to be entered has increased, so that it has become difficult to find the time for collecting information on paper supports and then entering them on the computer; secondarily, the new IS is connected to a cartographic software, so that, by entering in the call form the accident's place, the latter comes visualized in real time on a second monitor, which also indicate the coordinates. Nevertheless, as brought to light by informal interviews, operators were not at all disposed to abandon the *pen and pencil* system. In a work of many, and socially very recognizable, responsibilities, a work where the temporal pressure is as high as the stake, completely relying on an automated technology produces insecurity. See, for instance, the following excerpts.

*It will be necessary, however, to hold paper sheets in case of a problem with the computer. [...] I feel myself surer with some paper sheets on my desk. [R. 22-02-2005]*

*It will be difficult anyway to take note of all the information and interview details on the computer, some data will be lost. [F. 20-01-2005]*

Moreover, apart from risk perception and evaluation that clearly differ for managers, designers, and operators, the *pen and pencil* system is the protagonist of various communicative and cooperative practices among colleagues. Such a tool, in fact, helps operators to coordinate their work without talking, to "talk to the room" (Artman & Wærn, 1999), and to more easily insert their mutual communication in the midst of the incoming, inevitably verbal, ones. This constitutes an example of artifact-based coordination of cooperative work. Consider the following fieldnotes excerpts.

*During the phone interview [operator A] attracts [operator B]'s attention, by touching his shoulder, and asks him to dispatch, by showing the sheet on which [operator A] has written down the address. [Operator B] dispatches, while [operator A] continues giving first aid instruction on the phone. [Fieldnotes: 02-18-2005, 3.09 PM]*

*Emergency call coming from Trento-city, which should therefore have been taken by the box4 (territorially competent), but [operator A, box4] is engaged in another call and [operator B, box4] has taken a break. Therefore [operator A, box3] takes the call. Few moments later [operator A, box4] enters box3 and asks: "Was it ours?." Without saying a word, [Operator A, box3] gives her a sheet with notes regarding the call. [Operator A, box4] takes it and comes back in her box for dispatching. [Fieldnotes: 02-02-05, 5.07 PM]*

The last issue regards the communication between ERC operators and rescue units. With the new information system, such a communication undergoes a slowdown: by entering any kind of call on the system, in fact, the numeric code of the emergency calls gains some extra digit, so that its verbal transmission via radio requires more time. Although it could seem a trifle, as it has certainly done in the eyes of managers and designers, it is not at all about a detail; on the contrary, in a context where communication is almost the only resource and time is a fatal issue, it is something that operators have immediately noticed, pointed out, and discussed in various occasions.

## Infrastructure and Cooperative Work Practices

The second main change concerns the infrastructure of the emergency boxes and the organization of the cooperation between the work-shift colleagues in each box. More precisely, it consists of the introduction of two identical workstations in each box. With the "old" information system, each one was equipped with:

- a workstation with double monitor, one for data entry, the other one for the cartographer,
- and a PC Client, connected to the switchboard, the LAN, a digital phone, and an headset[4], both used by operator A;
- a second PC Client with single monitor, used for data entry by operator B, as well as, in the majority of cases,
- the radio-set[5] and
- the "traditional" phone (in reserve, so to speak, just in case of computer system failure[6]).

With the new system, instead, each box has two workstations with double monitor (one for calls management and data entry, the other one for the cartographer), two headsets, and two radio-sets, so that each operator could potentially—conditional is needed, as we shall see—work by her/himself, without cooperating or dividing tasks and responsibilities with the work-shift colleague.

A first consequence of the above-mentioned change concerns, once again, information and knowledge management. With the old system, data regarding the accident could be inserted by both operator A and operator B (usually, during moments of relative calm). The latter, moreover, came anyway to acquaintance of that information since he had listened to the colleague on the phone, since the colleague had informed him in order to allow the dispatch or a consultation, and/or since he had read the paper notes taken by operator A during the call. See also the following fieldnotes excerpt.

*Before leaving for lunch break, [operator A] checks with [operator B] the last unclosed events, dispatched vehicles, patients' form, etc. The two colleagues, in fact, keep each other informed. They use term such as "updating" and "briefing."*

*Moreover, as soon as possible, they discuss and choose together how many and which vehicles to dispatch. [Fieldnotes: 05-01-2005, 12.28 PM]*

The new information system, instead—with the breathtaking increase of the data to insert, and with the possibility, for each operator, of taking calls and dispatching at the same time of the colleague—decreases the opportunities of achieving an adequate degree of mutual awareness, and discourages communication between colleagues, thus diminishing the cooperation degree. Therefore, it is about a technology, which fails in one of his first and more important tasks, that of supporting users in acting in a way which is compatible with the configuration of activities that constitute the context of use.

Hence, a second, fundamental implication of the IS-related change regards the organization and coordination of work activities in each emergency box. The operators had their opinions in this respect too: they strongly disagreed with the incoming organizational change, and complained about not having been consulted by those decision makers that, differently than them, had no experience of the everyday work in the ERC.

*An operator cannot work alone. [Managers] don't work in the boxes and don't know, but alone you can't... in real time... It's better to work in couple in the box, helping each other and discussing. I mean, two brains are better than one. [R. 05-01-2005]*

*You work less quickly when you're alone. And, anyway, you can't discuss. While the possibility to discuss with the colleague for making decisions is fundamental. [Mi. 16-02-2005]*

Like many studies have already shown (e.g. Button & Harper, 1996; Crabtree, Nichols, O'Brien, Rouncefield, & Twidale, 2000; Dourish & Button, 1998; Eriksen, 2002; Suchman, 1987), the lack of dialogue between different groups of actors, and thus between the diverse occupational communities (Van Maanen & Barley, 1984) involved in the techno-organizational innovation process, as well as the lack of attention for the context of use and the actual practices of the users, may cause various problems and certainly increases the risk of non-use and/or under-use (Markus, 2000) of new technologies. In the case at hand, the operators, aware of the logic at the basis of the practices that they everyday enact in order to organize and coordinate work activities—a logic which constitutes the occupational culture that they themselves (re)produce—did not change their mind after the ICT reorganization in the boxes.

On the contrary, emergency operators basically continued to work in couple—something that was still possible thanks to the material environment of the boxes, which remained reciprocally separated by glass walls—and carried out two different and mutually interchangeable practices. The first one consisted of simply behaving like if nothing, from this point of view, had changed (non-use of new technologies). The other one consisted instead of deciding who was the operator A for every half of the work shift—hitherto, nothing new—but considering such a decision as not strictly binding. By playing and joking on the idea of what we can call the *cowboy operator*—i.e. the first who takes the call "gains" the role of operator A for *that* call—the operators moved the first steps in this direction (under-use of new technologies).

The "choice" of the one or the other practice was largely dependent upon the kind of relation existing between the two work-shift colleagues. One operator, for instance, could adopt the first cooperative practice, the "traditional" way to organize and coordinate activities, with some colleagues, while the second practice, the new, "joking" way to work together, with other colleagues, with whom s/he was much more familiar, close, or intimate. Furthermore, whatever the relation was, the cooperation inside the box was a taken-for-granted: nobody tried to work by him/herself, not even among the little group of new IS fans, and nobody was in search of confirmation about the colleagues' agreement on such a conduct. The various couples of box colleagues collaboratively found, *each* time, their way to co-operate and coordinate work activities. They did so situatedly and mostly tacitly, by grounding bodily conduct and visual orientation within the context of the ongoing situation and the phenomenal field (Merleau-Ponty, 1942, 1945; Garfinkel, 2002); by interacting in, with, and tanks to the material environment and the object, artifacts, and bodies that populated it.

The distribution of roles and tasks, therefore, became not only situated, but also potentially contingent and, especially, negotiable. A inNovA-TIVE practice, accompanied by its conditions of appropriateness, as well as its (ethno)methods of enactment and accountability, emerged in the field.

## An inNovATIVE Organization of the Activities

With this wordplay I would like to underline two issues. First of all, the way in which the operators have organized their own work activities is innovative. On the one hand, it includes a new practice, and, in a certain way, one that innovates the institutional innovation itself (an actual revolution, literary speaking). In fact, notwithstanding what the new technological infrastructure was intended to support, roles still exists in the box, but, thanks to that same infrastructure, they have become contingent, and more easily switchable. On the other hand, this new organization of the activities is innovative in its introducing the possibility of a situated, interactional choice regarding institutional work procedures, tasks' and roles' division, and the contextual use of artifacts and technologies. If the "revolutionary" practice of the *cowboy operator* is now possible, what is perhaps even more interesting is the permanence—as an alternative, each time negotiable option—of the "traditional" practice.

Secondarily, it is important to bear in mind that this new organization of the activities is endogenous to the phenomenal field (or "native"). It has been collaboratively created and repeatedly enacted (product an reproduct) by participants; it emerged in the field through a sort of collaborative improvisation—not so different from what happens for some artistic activities (e.g. Sawyer, 2000), or in ordinary conversation—and has then gradually been established. Indeed, it is through the social practices through which the activities of a determined context are accomplished everyday that procedures and routines emerge (Hughes, King, Rodden, & Andersen, 1994). "People are able to accomplish collectively what they could not do individually. The design that produces this complex mixture tends to be emergent and visible only after the fact" (Weick, 2001, p. 58).

Apart from the by now banal observations that one could put forward in criticizing the choices—almost completely indifferent to real users' point of view, actual work practices, and new technologies' context of use—that has brought to the failure of the new techno-organizational system as for the managers' intentions, I would like to underline, once again, that collaboration and cooperation in a (working) group are phenomena that lay before and beyond formal distribution of duties and responsibilities, and emerge from the situated interaction of the members of an occupational community who share a mostly practical, tacit, and embodied corpus of knowledge and know-how.

# TECHNO-ORGANIZATIONAL CO-DESIGN(-IN-USE) AND ACTION RESEARCH

As I mentioned, the ethnomethodologically oriented ethnography on which this contribution is based was not intended as an action research. Nevertheless, I think it teaches us something important in order to foster IT co-design in organization and co-construct organizational solutions to (IS-related) change.

I shall not review, here, the large action research literature (Dick, 2004, 2006, 2009; Nielsen & Svensson, 2006; Susman & Evered, 1978), nor its IS sub-section (Baskerville & Myers, 2004; Baskerville & Wood-Harper, 1998; Kock, 2007; Kock & Lau, 2001; Lau, 1997), nor, finally, its organizational (change/development) one (Berg & Eikeland, 2008; Coghlan, Rashford, 2006; Yu, 2004). However, it is important to acknowledge the variety of AR approaches, and the different relevance that they accord to cooperation and participation. Maurer and Githens (2010) have recently distinguished between conventional, critical and dialogic AR. They contend that, whereas the first one seeks to solve predetermined problems and fulfill predetermined goals, the second one—which includes Participatory Action Research (PAR)—is mainly intended to create emancipatory knowledge and to "free individuals and organizations from false assumptions, ideological distortions, and social distortions" (Maurer & Githens, 2010, p. 278). Dialogic AR, finally, while sharing with the second one "a commitment to matters of everyday life" and a "critical reflection upon current practices" (Maurer & Githens, 2010), is intended to cultivate practical or hermeneutic knowledge through intercultural dialogue. While Maurer and Githens conclude by proposing dialogic AR to be used in conjunction with conventional one, the approach that I shall outline points towards a combination between critical and dialogic AR.

Finally, it should be mentioned that issues of cooperative design and participation—in one word, co-design—constitute the cornerstone of another approach to computer design: namely, Participatory Design (PD) (Ehn, 1993; Kensing & Blomberg, 1998; Kyng & Mathiassen, 1997; Mumford, 1983; Schuler & Namioka, 1993). PD regards the user as the real expert (instead of assigning such a role to the designer or the researcher); looks at technologies in their context; considers the users' sensemaking activity about technology as at least as important as its technical features; and takes a critical position between workers and managers (both potential users of the IS) in favour of the former. Given the centrality accorded to the context of use (the workplace), as well as the active involvement of the user(s), it follows that designing technology should be not only a cooperative activity, but also a practical (instead of abstract) one, a one concomitant to use itself—i.e. design-in-use.

Starting from the need for techno-organizational co-design(-in-use), which are the objectives orienting the proposed approach to action research in IS and organizational studies? And which is the role of the action researcher and her/his "expert" knowledge? These are the issues that I shall consider in the following sections.

First, I shall briefly summarize what I believe is the lesson of the research experience that I have presented here, and focus, in particular, on the aspects one needs to pay attention to for conducting effective AR for/in techno-organizational change. I shall underline how a phenomenological point of view and, more specifically, an Ethnomethdological (Garfinkel, 1967, 2002) and Conversation-Analytical (EM/CA) approach, with its focus on situated interaction and practical, collaborative accomplishment of order and routine, can help in making those aspects emerge from the tacit "taken-for-grantedness" of organizational everyday life.

On the basis of such an approach, I shall then propose some implications and guidelines for future action research, with particular reference to the role of the researcher and her/his "expert" knowledge, know-how, and gaze, as well as the actions that s/he sh/could carry out, and the values and objectives guiding the research process. In this respect, I shall finally underline some differences between the proposed approach and those considering the research(er) as organizational problem finder/solver.

## Phenomenal Field, Situated Interaction, and the Logic(s) of Practice(s)

In terms of IT design and innovation processes, the acknowledgement of the socio-technical nature of the organizational setting points to the fact that, in Lucy Suchman's (1997, p. 56-57) words, the design of any one aspect of a particular working order "potentially implicates an open horizon of other aspects," therefore "one needs to understand that aspect's relationship to the extended system of activities and technologies of which it is a part." Solutions to problems and issues emerging in a field, in fact, must be coherent with the endogenous organization of activities of that field; they must be proper and suitable to the configuration of actions and interactions setting up that specific context. When it is not so, resistance to IS finds one of its fundamental conditions of existence (cf. Markus, 1984).

This means, firstly, that an IS-related change process is never merely technical or technological, rather always organizational too. A workplace is "a complex but habitual field of equipment and action, involving intimate relations of technology and practice, body and person, place, and activity" (Suchman, 1996, p. 36). From this point of view, it is important to bear in mind that organizational (change) design is a necessity, and should be strictly intertwined with technological (change) design and, more specifically, IS design. What needs to be (re)designed, in other words, are a socio-technical system and a techno-organizational change process? IT design, therefore, cannot avoid a parallel, careful analysis of the phenomenal field. Designing an entire IS on the basis of the managers' more or less formal description of work and the working order, and present it to the future users when modifications cannot be else than incidental, is nonsense.

Secondarily, and consequently, the coherence bind entails the need for starting from who inhabit the field, since participants embody in their acting the (ongoing) logic of organization of the (situated) context itself. This holds not only for (action) researchers, but also for IT designers and organizational decision makers. Workplace "natives," as the above presented case demonstrates, "are the ones best qualified"—for keeping with Suchman's (1997, p. 57) discourse—"to carry forward and fine tune the design of the workplaces they inhabit. Design in use is always a process of improvisation (...) The role of professional design"—being the technological one of IT designers, or the organizational and managerial one of decision makers and policy makers, or, finally, the social one of the researcher—"is to lay the groundwork for that process."

As we have seen, furthermore, such improvisational process is collaborative in nature, takes place in and through situated interaction, and often in a tacit way. This means that we have not just to start from the field inhabitants, but from the group(s) of inhabitants and their mutual interaction. The first meaning of the expression co-design(-in-use) points exactly to the fact that the appropriation of new technologies by actual users—the inhabitants of the phenomenal field—is a social, interactional, and collaborative process. A second meaning that such an expression is intended to underline concerns the active engagement of the research subjects/participants, and refers to a set of principles and objectives common to PD and AR—particularly, PAR (Lau, 1997; Yu, 2004). The field inhabitants, in this case, usually encompass a larger spectrum: not only those that I have defined workplace "natives" (e.g. operators), but also the other inhabitants of the organizational setting and the larger social arena (IT designers, managers, decision makers, etc.). I shall come back on this issue in the next section.

If innovation must be coherent with the dynamic but habitual organization of activities and related supporting technologies in the phenomenal field in which participants act every day, a third task of the researcher is then that of taking into consideration, observe, and analyse in detail the material, practical, even carnal aspects of the way in which participants coordinate, carry out, and accomplish, in and through interaction, the technologically-assisted activities at hand. The idea of getting rid of paper supports, for instance, immediately appears ridiculous if one stops looking into the *minutie* of action-in-interaction. Order and routine constitute a situated, collective accomplishment, with a clear improvisational dimension; nevertheless, any situated working group develop an habitual way for collectively organizing work and collectively accomplishing every day the routine and the habitual order themselves. Designing innovation means, once again, designing the groundwork for the cooperative and emergent accomplishment of new order(s) and routine(s).

Finally, it is therefore necessary to acknowledge the relevance of everyday work practices, and, more specifically, of those of communication and cooperation. A successful coordination of work activities, in fact, heavily rests on the dynamic circulation of information and knowledge, on the one hand, and, on the other hand, on the opportunities for ongoing, situated cooperation and negotiation. IT can support or constrain both; IT design—or, better, techno-organizational design—should be committed to the first aim[7], to which the researcher sh/could largely contribute.

From this point of view, ethnometodologically oriented ethnography enables to catch, in its minute details, the native, endogenous organization of activities in a field, so allowing the action researcher a) to be not a misfit, nor clumsy, so to speak, and b) to bring to light the situated meaningfulness of practices and the tacit logic underlying them. Such a logic, indeed, needs to be made explicit to the other actors of the techno-organizational (re)design process.

## Implications for Action Research

The first objective of AR for techno-organizational design and innovation process management consists of the coherence bind in respect to the local—practical and interactional—organization of the activities. From this point of view, the action researcher should both remember and foster the awareness that workplace natives are the best possible innovators of their own work environment and working order. Practices, in fact, are inseparable from the context in which they take place as well as the culture in which they are embedded.

A further, complementary issue should be taken into account. What two advocates of "dialogic action reserach" have been recently referring to in arguing that "participants, professional researchers, and local stakeholders represent a variety of value orientations and commitments that influence both the courses of action and the goals of a project" (Maurer & Githens, 2010, p. 282) can be easily defined as a variety of occupational cultures. This is an evidence that we cannot ignore when stating that action research should (help in) orient(ing) and/or design(ing) techno-organizational change along a direction which is desired by—or desirable for—the members of the organization. Members, in fact, belong to different groups; each of them has a specific idea and representation of the (work in the) organization and its desirable future. These ideas, representations, and desires are often conflicting, or at least diverging; at the same time, different groups are equipped with different resources, power, and decision-making opportunities.

From the point of view of the action researcher, this entails, first of all, an analytical task. Before any attempt to foster what, starting from the concept of occupational culture, or membership culture, could be defined intercultural dialogue—something akin to dialogic AR (cf. Maurer & Githens, 2010)—and even before bringing out the tacit, underlying logic of workplace natives' practices—something somehow akin to critical AR (Maurer & Githens, 2010; cf. also Fenwick, 2005; Kemmis, 2001) and PD—one should, first and foremost, look at the interactions in/between different groups of organizational members, and analyze them in detail.

As mentioned above, a third, fundamental objective of the action researcher consists of the deep understanding of everyday practices of communication, cooperation, and, more generally, work, till up the unveiling of their underlying logic. Such an analytical endeavour, furthermore, should be complemented with a more "interventionist" one: it is necessary to foster the awareness, in both IT designers and organizational (and/or political) decision makers, of the relevance of such practices and their logic. The latter, in fact, are the main tools through which the work activities are coordinated and interactionally accomplished. If, on the one hand, it is about to take a critical stance and give voice, so to speak, to who usually get the work done but lack in power and decision-making opportunities, on the other hand, it is also about constructing the groundwork for actual dialogue between different groups of actors and, thus, for the discussion (and partial relativization) of the respective cultures and logics, and the potential modification—i.e., the innovation—of determinate practices.

Within this frame, looking at the material, interactional, and cultural trajectories undertaken by the various participants, and the related circulation of information, discourses, knowledge, and practices within the socio-technical system and the larger social arena, constitutes a quintessential prerequisite for action researcher's effective intervention. Mapping such trajectories and analysing the ways in which different groups of actors communicate and interact constitutes a goal that ethnography, and particularly ethnomethodologically oriented one, is well-equipped for reaching.

The ethnomethodologically oriented ethnographer brings thus his/her particular knowledge and analytical gaze—that is, her/his occupational culture and practices—into the socio-technical system. S/he brings a gaze on occupational cultures and the interaction between different (group of) actors; on the material, experiential, and interactional aspects of technologically assisted work; on the situated practices of coordination, communication, and cooperation taking place in the phenomenal field, their meaningfulness and tacit logic.

That of the action researcher is not a better gaze than other actors' one, but it is different. The ethnomethodologically oriented researcher, in fact, is not only trained in the observation and analysis of the above mentioned aspects; s/he is also positioned outside the more or less conflicting relationships between different (groups of) actors, and the previous (hi)story, in a manner of speaking, of their reciprocal inter-actions.

The researcher, therefore, should basically act:

- as an agent of visibility and recognizability (of the local organization of activities and work practices, and their logic);
- as an agent of mutual awareness and intercultural dialogue (between different occupational cultures);
- as an agent of legitimization (of the point of view of some actors—primary, but not exclusively, the ones more lacking in power—in front of the others, starting from his/her both neutral and "expert" social position).

As for organizational members' satisfaction, I believe that action research should consider their self-reflexive evaluation in terms of a) the feasibility of tasks' accomplishment; b) the degree of responsibility and consequent stress, and their management; c) the degree and quality of 1) cooperation, 2) communication, and 3) knowledge circulation—i.e. the primary social tools of any working order.

The whole intervention, however, should be as less interventional as possible.[8] I mean, for instance, that actual and prolonged presence on the field—i.e. participant observation—should be more important than periodical formal meetings. More than proposing solutions for change, the action researcher should "be there"; attend and participate as much as possible to the everyday activities of the considered setting; and disperse her/his questions, comments, indications, and explanations within the different occasions of such a situation. In this respect, the approach that I am proposing significantly differs from what Mårtensson and Lee (2004) call "dialogical action research"[9].

The purpose of a "low-interventional intervention" is twofold: on the one hand, it allows situated learning (Lave & Wenger, 1991) and gradual sedimentation for all the participants (the researcher included); on the other hand, it links more or less general observations and issues with concrete, practical, situated examples grounded in the lived experience of research participants. Equipped with additional knowledge, understanding, and awareness of themselves and the relevant others, as well as the ways in which routine is collaboratively and improvisationally achieved, the participants would then be able both to find—or, better, identify—and solve—or, collectively negotiate and deal with—techno-organizational problems on their own. They would be in better conditions for innovating their work practices and instruments, for designing the change process. "The setting," in Richard Baskerville's (1999, p. 17) words, would be "free to self-reorganize."

It is from this point of view that I regard ethnography—and, more particularly, ethnomethodologically oriented one—as a method for AR that goes both behind and beyond problem-finding and problem-solving. According to the proposed approach, the role of the researcher does not consist of informing managers and/or designers about the problems and needs of the (future) users or the organization, nor of proposing specific and precise organizational and/or technical solutions. Alongside to the analytical endeavour, it consists, instead, of orienting members' attention, being a vehicle for knowledge circulation, and thus fostering techno-organizational co-design(-in-use) and the co-construction of solutions for/to change.

## CONCLUSION

In this chapter, I look at IS-related organizational change from a socio-technical perspective. On the one hand, therefore, the focus is on the new technologies' usability in context, and especially on the technology in use—or, we could say with Heath and Luff (2000), technology in action. On the other hand, adopting such a perspective means taking into consideration the whole set of individual and collective actors (designers, managers, and future users), and the different ways in which they make sense of tools and artifacts, of the everyday technologically-mediated work practices, and of the techno-organizational change process itself.

This implies, firstly, to start the analysis by focusing on the situated work practices in and through which participants carry out, accomplish, mutually recognize, and co-ordinate their everyday activities. As the case of the Trento ERC shows, in fact, when new IT artifacts are designed and adopted in a manner completely inconsistent with real users' point of view, actual work practices, the endogenous organization of activities, and the specific configuration of interactions that actually sets up the specific context of use, then the new technology/ies could turn out to be non/under—or at least differently—used.

Secondly, and consequently, it is necessary to analyze in detail the interactions in and between different groups of organizational members, or different occupational communities, and then to make mutually intelligible and accountable the tacit logics of their respective practices. From this point of view, the action researcher should work as an agent of visibility and recognizability, as well as of mutual awareness and intercultural dialogue.

Finally, in order to foster technologies' usability, the proposed approach entails the need for what I have called co-design(-in-use) of techno-organizational change. Besides the "failure" side of the coin, the Trento ERC case also shows how an inNovATIVE practice of use had emerged and developed in the field through a sort of collaborative improvisation, and, therefore, how the "workplace natives" constitute the best innovators of their own working order—which is precisely what needs to be (re)designed during an IS-related change. The action researcher, in this respect, represents also an agent of legitimization.

Ultimately, co-design(-in-use) should be a process of collective improvisation on the basis of the endogenous and habitual practices and organization of activities in a workplace. As a consequence, the role of professional design—being the technological one of IT designers, the organizational one of managers, or the social one of action researchers—should be to lay the groundwork for that process.

It is my contention that ethnomethodologically oriented ethnography constitutes a particularly suitable and helpful methodology to accomplish the above-mentioned tasks. In particular, a phenomenological point of view, and more specifically an EM/CA approach, equip the researcher with a way to look at, and (inter)act in, the world that may contribute to the conduction of effective action research projects for/in techno-organizational change[10].

## ACKNOWLEDGMENT

I am grateful to the anonymous reviewers and Maurizio Teli for their advice on earlier drafts of this chapter. A special thanks goes to the operators and managers of "Trentino Emergenza 118" who made the research possible. The research participated to the project *Traffic Control as Work*, funded by the Italian Ministry of University and Scientific Research (MIUR, 2001), and to the project *MEDiCA—Medical Emergency Dispatch Centre Analysis*, funded by the European Commission (Marie Curie Outgoing International Fellowship: # MOIF-CT-2004-002884).

## REFERENCES

Adler, P. S., & Winograd, T. A. (Eds.). (1992). *Usability: Turning technologies into tools*. New York, NY: Oxford University Press.

Akdere, M. (2003). Action research paradigm in the field of training and development. *Journal of European Industrial Training*, *27*(8), 413–422. doi:10.1108/03090590310498559

Artman, H., & Wærn, Y. (1999). Distributed cognition in an emergency co-ordination centre. *Cognition Technology and Work*, *1*, 237–246. doi:10.1007/s101110050020

Avgerou, C., Ciborra, C., & Land, F. F. (Eds.). (2004). *The social study of information and communications technology: Innovation, actors and context*. Oxford, UK: Oxford University Press.

Baskerville, R. (1999). Investigating information systems with action research. *Communications of AIS*, *2*, 19.

Baskerville, R., & Myers, M. D. (2004). Making IS research relevant to practice – Foreword. *Management Information Systems Quarterly*, *28*(3), 329–335.

Baskerville, R., & Wood-Harper, A. T. (1998). Diversity in information systems action research methods. *European Journal of Information Systems*, *7*(2), 90–107. doi:10.1057/palgrave.ejis.3000298

Berg, A. M., & Eikeland, O. (Eds.). (2008). *Action research and organization theory*. Frankfurt, Germany: Peter Lang.

Button, G., & Harper, R. (1996). The relevance of "work-practice" for design. *Computer Supported Cooperative Work*, *4*, 263–280. doi:10.1007/BF01846695

Button, G., & Sharrock, W. (1998). The organizational accountability of technological work. *Social Studies of Science*, *28*(1), 73–102. doi:10.1177/030631298028001003

Coghlan, D., & Rashford, N. S. (2006). *Organizational change and strategy: An interlevel dynamics approach*. New York, NY: Routledge.

Crabtree, A., Nichols, D. M., O'Brien, J., Rouncefield, M., & Twidale, M. B. (2000). Ethnomethodological informed ethnography and information system design. *Journal of the American Society for Information Science American Society for Information Science*, *51*(7), 666–682. doi:10.1002/(SICI)1097-4571(2000)51:7<666::AID-ASI8>3.0.CO;2-5

Dick, B. (2004). Action research literature: Themes and trends. *Action Research*, *2*(4), 425–444. doi:10.1177/1476750304047985

Dick, B. (2006). Action research literature 2004-2006: Themes and trends. *Action Research*, *4*(4), 439–458. doi:10.1177/1476750306070105

Dick, B. (2009). Action research literature 2006-2008: Themes and trends. *Action Research*, *7*(4), 423–441. doi:10.1177/1476750309350701

Dourish, P., & Button, G. (1998). On "technomethodology": Foundational relationships between ethnomethodology and system design. *Human-Computer Interaction*, *13*(4), 395–432. doi:10.1207/s15327051hci1304_2

Ehn, P. (1993). Scandinavian deign: On participation and skill. In Schuler, D., & Namioka, A. (Eds.), *Participatory Design: Principles and Practices* (pp. 41–78). Hillsdale, NJ: Lawrence Erlbaum Associates.

Eriksen, S. (2002). Designing for accountability. In O. W. Bertelsen, S. Bødker, & K. Kutti (Eds.), *Nordichi 2002 – Proceedings of the Second Nordic Conference on Human-Computer Interaction,* (pp. 177-186). Aarhus, Denmark: Nordichi.

Fele, G. (2005). *Interaction and collaborative work in a medical emergency dispatch centre*. Paper presented at the 100th Meeting of the American Sociological Association. Philadelphia, PA.

Fele, G. (2008). In praise of thin descriptions: Ethnography, ethnomethodology and the study of practice. In *Proceedings of ALPIS 2008 – Alpine Ski Seminar on Information Systems,* (pp. 51-58). Carisolo, Italy: University of Trento Press.

Fenwick, T. (2005). Conception of critical HROD: Dilemmas for theory and practice. *Human Resource Development International*, *8*(2), 225–238. doi:10.1080/13678860500100541

French, W., & Bell, C. (1999). *Organization development: Behavioural science interventions for organization improvement*. Upper Saddle River, NJ: Prentice Hall.

Garfinkel, H. (1967). *Studies in ethnomethodology*. Engelwood Cliffs, NJ: Prentice Hall.

Garfinkel, H. (2002). *Ethnomethodology's Program: Working out Durkheim's aphorism*. Lanham, MD: Roman & Littlefield.

Gartner, J., & Wagner, I. (1996). Mapping actors and agendas: Political frameworks of system design and participation. *Human-Computer Interaction*, *11*(3), 187–214. doi:10.1207/s15327051hci1103_1

Goodwin, C. (2000). Action and embodiment within situated human interaction. *Journal of Pragmatics*, *32*, 1489–1522. doi:10.1016/S0378-2166(99)00096-X

Hakken, D. (1999). *Cyborgs@Cyberspace?: An ethnographer looks to the future*. New York, NY: Routeledge.

Harper, R. H. R., & Hughes, J. A. (1992). What a f ing system: Send 'em all to the same place and then expect us to stop 'em hitting: Making technology work in air traffic control. In Button, G. (Ed.), *Technology in Working Order: Studies of Work, Interaction and Technology* (pp. 127–144). London, UK: Routeledge.

Heath, C., Hindmarsh, J., & Luff, P. (2010). *Video in qualitative analysis*. London, UK: Sage.

Heath, C., & Luff, P. (1996). Convergent activities: Line control and passenger information on the London underground. In Engeström, Y., & Middleton, D. (Eds.), *Cognition and Communication at Work* (pp. 96–129). Cambridge, UK: Cambridge University Press.

Heath, C., & Luff, P. (2000). *Technology in action*. Cambridge, UK: Cambridge University Press. doi:10.1017/CBO9780511489839

Heritage, J. (1984). *Garfinkel and ethnomethodology*. Cambridge, UK: Polity Press.

Hughes, J. A., King, V., Rodden, T., & Andersen, H. (1994). Moving out from the control room: Ethnography in system design. In *Proceedings of the 1994 ACM Conference on Computer Supported Cooperative Work,* (pp. 429-439). New York, NY: ACM Press.

Jacucci, G. (2007). Social practice design (SPD), pathos, improvisation, mood, and bricolage: The Mediterranean way to make place for IT? In I. Benbasat, M. Limayem, D. Te'eni, G. Jacucci, & E. Monod (Eds.), *Proceedings of the Eighth Mediterranean Conference on Information Systems,* (pp. 89-103). Venice, Italy: University of Trento Press.

Jacucci, G., Tellioglu, H., & Wagner, I. (2007). Design games as a part of social practice design: A case of employees elaborating on organizational problems. In I. Benbasat, M. Limayem, D. Te'eni, G. Jacucci, & E. Monod (Eds.), *Proceedings of the Eighth Mediterranean Conference on Information Systems,* (vol 1), (pp. 141-155). Venice, Italy: University of Trento Press.

Jefferson, G. (1984). Transcription notation. In Atkinson, J., & Heritage, J. (Eds.), *Structures of Social Interaction* (pp. ix–xvi). New York, NY: Cambridge University Press.

Kemmis, S. (2001). Exploring the relevance of critical theory for action research: Emancipatory action research in the footsteps of Jurgen Habermas. In Reason, P., & Bradbury, H. (Eds.), *Handbook of Action Research: Participative Inquiry and Practice* (pp. 91–102). London, UK: Sage.

Kensing, F., & Blomberg, J. (1998). Participatory design: Issues and concerns. *Computer Supported Cooperative Work, 7*, 167–185. doi:10.1023/A:1008689307411

Kock, N. (Ed.). (2007). *Information systems action research: An applied view of emerging concepts and methods*. New York, NY: Springer.

Kock, N., & Lau, F. (2001). Information systems action research: Serving two demanding masters. *Information Technology & People*, *4*(1), 6–11.

Kyng, M., & Mathiassen, L. (Eds.). (1997). *Computers and Design in Context*. Cambridge, MA: MIT press.

Lau, F. (1997). A review on the use of action research in information systems studies. In Lee, J., Liebenau, J., & DeGross, J. (Eds.), *Information Systems and Qualitative Research* (pp. 31–68). London, UK: Chapman & Hall.

Lave, J., & Wenger, E. (1991). *Situated learning: Legitimate peripheral participation*. Cambridge, UK: Cambridge University Press.

Markus, M. L. (1984). *Systems in organizations: Bugs and features*. Boston, MA: Pitman.

Markus, M. L. (2000). Toward an integrative theory of risk control. In Baskerville, R., Stage, J., & DeGross, J. I. (Eds.), *Organizational and Social Perspectives on Information Technology* (pp. 167–178). Boston, MA: Kluwer Academic Publishers.

Markus, M. L., & Robey, D. (1988). Information technology and organizational change: Causal structure in theory and research. *Management Science*, *34*(5), 583–598. doi:10.1287/mnsc.34.5.583

Mårtensson, P., & Lee, A. S. (2004). Dialogical action research at omega corporation. *Management Information Systems Quarterly*, *28*(3), 507–536.

Maurer, M., & Githens, R. P. (2010). Toward a reframing of action research for human resource and organization development: Moving beyond problem solving and toward dialogue. *Action Research*, *8*(3), 267–292. doi:10.1177/1476750309351361

Merleau-Ponty, M. (1942). *La structure du comportement*. Paris, France: Presses Universitaires de France.

Merleau-Ponty, M. (1945). *Phénoménologie de la perception*. Paris, France: Gallimard.

Mumford, E. (1983). Participative systems design: Practice and theory. *Journal of Occupational Behaviour*, *4*(1), 47–57.

Nielsen, K. A., & Svensson, L. G. (Eds.). (2006). *Action research and interactive research: Beyond practice and theory*. Maastricht, Germany: Shaker.

Normark, M. (2002). *Using technology for real-time coordination of work: A study of work and artefacts in the everyday activities at SOS alarm*. Stockholm, Sweden: Kungl Tekniska Högskolan.

Pettersson, M., Randall, D., & Helgeson, B. (2004). Ambiguities, awareness and economy: A study of emergency service work. *Computer Supported Cooperative Work*, *13*, 125–154. doi:10.1023/B:COSU.0000045707.37815.d1

Sawyer, R. K. (2000). Improvisational cultures: Collaborative emergence and creativity in improvisation: Commentary on lead article by Howard Becker. *The Etiquette of Improvisation. Mind, Culture, and Activity*, *7*(3), 180–185. doi:10.1207/S15327884MCA0703_05

Schuler, D., & Namioka, A. (Eds.). (1993). *Participatory design: Principles and practices*. Hillsdale, NJ: Lawrence Erlbaum Associates.

Sharrock, W., & Randall, D. (2004). Ethnography, ethnomethodology and the problem of generalisation in design. *European Journal of Information Systems*, *13*(3), 186–194. doi:10.1057/palgrave.ejis.3000502

Suchman, L. (1987). *Plans and situated actions: The problem of human-machine communication*. New York, NY: Cambridge University Press.

Suchman, L. (1993). Technologies of accountability: On lizards and airplanes. In Button, G. (Ed.), *Technology in Working Order: Studies of Work, Interaction, and Technology* (pp. 113–126). New York, NY: Routeledge.

Suchman, L. (1996). Constituting shared workplaces. In Engstrom, Y., & Middleton, D. (Eds.), *Cognition and Communication at Work* (pp. 35–60). New York, NY: Cambridge University Press.

Suchman, L. (1997). Centres of coordination: A case and some themes. In Reisnick, L. B., Saljo, R., Pontecorvo, C., & Burge, B. (Eds.), *Discourse, Tools, and Reasoning: Essays on Situated Cognition* (pp. 41–62). Berlin, Germany: Springer-Verlag.

Suchman, L. (2002). Practice-based design of information systems: Notes from the hyperdeveloped world. *The Information Society*, *8*(2), 139–144. doi:10.1080/01972240290075066

Suchman, L., Blomberg, J., Orr, J., & Trigg, R. (1999). Reconstructing technologies as social practice. *The American Behavioral Scientist*, *43*, 392–408. doi:10.1177/00027649921955335

Suchman, L., Trigg, R., & Blomberg, J. (2002). Working artefacts: Ethnomethods of the prototype. *The British Journal of Sociology*, *53*, 163–179. doi:10.1080/00071310220133287

Susman, G., & Evered, R. (1978). An assessment of the scientific merits of action research. *Administrative Science Quarterly*, *23*(4), 582–603. doi:10.2307/2392581

Van Maanen, J., & Barley, S. R. (1984). Occupational communities: Culture and control in organizations. *Research in Organizational Behavior*, *6*, 287–365.

Weick, K. E. (2001). *Making sense of the organization*. Oxford, UK: Blackwell Publishers.

Whalen, J., & Zimmerman, D. H. (2005). Working a call: Multiparty management and interactional infrastructure in calls for help. *Pragmatics and Beyond New Series*, *143*, 309–345.

## KEY TERMS AND DEFINITIONS

**Co-design(-in-use):** is a situated and interactional process of collective and collaborative improvisation carried out on the basis of the endogenous and habitual practices and organization of activities in a workplace. It is not just about workplace natives' experience, but also their collective know-ledge/how, which is both exploited and increased during the process.

**Cooperative Work Practices:** are the practices through which members of a workplace carry out every day the activities at hand by coordinating their actions(-in-interaction) with the other members through mediated or non-mediated talk, gesture, bodily conduct, and the use of objects, artifacts, etc.

**Endogenous Organization:** refers to the ways in which local members practically and interactionally coordinate their work activities. In so doing, they both rely on and everyday collectively accomplish a common (and often taken-for-granted) definition of the situation, thus achieving situated order and routine.

**ERCs (Emergency Response Centres):** are stressful and time-pressing, hi-tech and information-rich environments, in which communication, on the one hand, and information management, on the other hand, are the primary work activities – and are largely technologically-assisted activities.

**Ethnomethodologically Oriented Ethnography:** is a methodology that employs diverse qualitative techniques – among which extended participant observation, informal interviews and videoanalysis are prevalent – and directs attention towards a) giving voice to the inhabitants of the field and their culture by focusing on their everyday practices, discourses and experiences – a traditional ethnographic mission – and b) analysing the (mostly tacit) ways in which members situatedly and interactionally carry out everyday activities in a manner accountable for the other members, thus collectively accomplishing order – the ethnomethodological perspective.

**ICTs (Information and Communication Technologies):** are an ensemble of (often interconnected) devices and applications intended to support information management/sharing and mediated communication.

**Information and Knowledge Management:** entails acquiring, storing, organizing, retrieving, and manipulating information. These activities, on the basis of (different) culture-dependent knowing practices, produce and reproduce (different) knowledge, which can be more or less shared. Issues at stake include: a) who produce b) what kind of information and c) who (instead) can access and use information d) in order to do what e/f) with/to whom.

**Phenomenal Field:** consists of the perceived reality and lived experience of the (body-)subject in relation to the world that s/he inhabits together with the co-present bodies and objects. The elements of the phenomenal field, in other words, are constituted by the perception, orientation, and focused attention of the subject in the intersubjective context.

**Socio-Technical Perspective:** takes into consideration the social embeddedness of technologically-assisted activities and technology-occasioned changes; directs attention towards work practices and organizational culture(s); and regards workplaces as contexts that incorporate as a whole technology and practice, bodies and tools, place and activity.

**Techno-Organizational Change:** refers to organizational change strictly intertwined with technological change, in order to foster new technologies' usability and enhance actual work practices(' innovation). In the context of modern and contemporary organizations, with their increasing necessity to manage information, we can talk of IS-related organizational change.

## ENDNOTES

1 In order to communicate from one box to the other, an interphono exists, but it comes rarely used to such an aim. A similar phenomenon has been observed also in a Swedish ERC (Pettersson, Randall, & Helgeson, 2004, p. 149). The interphono, connected to a closed circuit camera, comes instead generally used in order to communicate with the garages behind the ERC.

2 There are three work shifts for the 24/7 active boxes: 7AM-2PM, 2PM-9PM and 9PM-7AM.

3 The ERC, in fact, does not only receive emergency calls, but also those coming from diverse hospital wards; newspaper editorial offices; institutional agencies such as police, fire department, etc.; other ERCs; or elders in search of some company. Sometimes, they even receive calls that are then revealed as phone jokes.

4 In order to answer a call one had, and still has, just to press the space-bar; the call is then transferred to the associated digital phone—which, after one ring, transfers the communication on the headset—and disappears from the monitor of the other computers.

5 The system anyway allowed operator A, by pressing a pedal, to use the microphone for radio communication while keeping listening to the phone call through the headphones.

6 Like *pen and pencil system*, this is a way for managing the risk associated with automated technologies.

7 In the case of the Trento ERC, the new technologies ended up in "unintendedly" support dynamic, contingent, and situatedly negotiable forms of cooperation.

8 Richard Baskerville (1999, p. 16) distinguishes between "directive" and "non-directive" interventions.

9 Although being close to PD's critical and democratic values, the proposed approach also differs from participatory design: the former assigns to the researcher a less active role in co-design(-in-use) activities than the latter. This is a matter that would require larger space and further reflection.

10 I acknowledge, though, that such a methodology, with its analytical requirements and low interventional character, takes a large amount of time.

# Chapter 17
# A Drifting Service Development: Applying Sociotechnical Design in an Ambient assisted living Project

**Kai-Uwe Loser**
*Ruhr-University Bochum, Germany*

**Alexander Nolte**
*Ruhr-University Bochum, Germany*

**Michael Prilla**
*Ruhr-University Bochum, Germany*

**Rainer Skrotzki**
*Ruhr-University Bochum, Germany*

**Thomas Herrmann**
*Ruhr-University Bochum, Germany*

## ABSTRACT

*Drift is one of the visible phenomena observed in an ambient assisted living project. In this project, services for elderly people were developed that would be ordered using a digital pen technology. In this project, the method of the Sociotechnical Walkthrough (STWT) for an integrated development was applied to clarify technology usage, technical aspects, and the work processes. This approach was combined with several other methods to form a multi-facetted sociotechnical design approach. During the course of the project several shifts in perspectives, breakdowns of understanding, and negotiations could be observed. This chapter describes how using this approach of sociotechnical design facilitated the identification of drift phenomena and its processing in service design. The authors observations also clearly show the limitations of up-front process planning for complex environments such as service processes.*

DOI: 10.4018/978-1-4666-0303-5.ch017

## INTRODUCTION

Todays' design practice has already been influenced and shaped by the philosophical background and theoretical concepts presented in this book. Methods such as prototyping or end-user development are well known techniques in this context, especially when non-designers are empowered to design, but today's collaborative infrastructures are more complex than tools merely used to support a personal task. Cooperative work needs to be designed as well and is per se abstract, which makes it hard to apply well-known hands-on design methods. The design process is a dialectical reflection process, in which the artifact under design as well as its understanding is possibly in conflict and drifts alongside. Breakdowns on each side—technology concepts as well as people's interaction—regularly happen during cooperative design work and in all stages of the development—from first concepts to already tuned practice.

Ciborra (1996) mentioned several possible causes of breakdowns when IT is used:

- deficits of user interfaces and other technical problems,
- lack of knowledge about the modality of use,
- a critical mass of needed users is missing,
- problems in internalizing the new knowledge needed.

Breakdowns add burdens to the usage of technology. Possible consequences are users deciding to substitute new technologies with well-known practice. To overcome this, we need approaches to create an environment for a constructive dialectical confrontation between all kinds of stakeholders such as process participants and developers. Designing and mutual learning need to be acknowledged as inseparable parts of development processes, requiring methods for an active creation of opportunities for drift.

The active creation of dialectical confrontation needs actors to engage in mutual learning processes, as a foundation for co-evolution of understanding and technology (Ciborra, 1996). We were able to observe such processes in a project for ambient assisted living, in which users, staff and technology developers were participating in designing service processes and supporting technology. "Drift" in the sense of an open co-evolution process is the focus of the analysis of this project presented in this paper. In the project, we combined the method of Sociotechnical Walkthrough (STWT) and several other methods for sociotechnical design for the development of services and technology to support elderly people in their homes. We will first give a brief introduction to the applied methods and then present several examples from the case.

## METHODOLOGY OF SOCIOTECHNICAL Walkthrough

The Sociotechnical Walkthrough (STWT) is a result of continuous method development, which lasted about thirteen years and was driven by experiences from various projects (for a list of projects see Herrmann, 2009). As an initial reason for this ongoing action research effort, we found available modeling methods to be inappropriate in describing the characteristics of sociotechnical systems and their technical, organizational, and personal viewpoints. At the same time, the documentation of requirements and concepts which accompany the design process do not usually sufficiently support an integrated view on technical and organizational structures. The theoretical background of these developments is a perspective on sociotechnical systems, which were first described in the 1950s (Emery & Trist, 1960), and it also includes newer systems theory. Following Luhmann (1995), social systems are interconnected acts of communication, which reproduce themselves. The foundations include the concept of contingency as part of human behavior and which makes them different to technical parts of sociotechnical systems. The consequence of contingency is that human beings communicating can never be sure of the results of an act. Freedom of decision and interpretation are part of the nature of communication. Resulting phenomena are well known e.g. the differentiation of plans vs. situated action (Suchman, 1987) or non-anticipated changes (Orlikowski, 1996). Based on this theory our perspective on organizational adoption of technology is mainly intended to create appropriate structures for communication about technology usage, communication mediated by technology and technical structures in systems supporting these communications.

The STWT is a method for sociotechnical design. Earlier experience within practical projects revealed that available approaches like prototyping, use case diagrams, storyboards, mock-ups, and different visualization techniques do not sufficiently support the communication about the interrelationships between the various aspects of sociotechnical systems. For example, using prototypes usually only allows for the collection of feedback on issues of screen design, leaving out highly relevant aspects of work processes and coordination. The SeeMe notation is the result of this development. It is based on more technical modeling notations and extends them with means to express vagueness such as incompleteness and uncertainty. Vagueness in SeeMe is especially related to a qualitative lack of information (Herrmann & Loser, 1999). The method furthermore supports

- the visualization of complex interdependencies between the activities of people, between human work and supportive (IT) technology, and—if needed—between technical components.
- the integration of technical and social aspects, including their respective characteristics of formality or contingency, in the same view.
- switching between various levels of detail from abstract overviews (maps for orientation) to very concrete descriptions with rich details.
- the depiction of various levels of knowledge explication, ranging from vague and questionable to stable and agreed upon.
- acknowledging the various perspectives already existing, including established conventions and different interests.

Our approach of facilitating communication is based on the SeeMe notation and called the Sociotechnical Walkthrough (STWT). For a detailed description of the method, we point the reader to Herrmann, Kunau, Loser, and Menold (2004) and Herrmann (2009). The term walkthrough is meant as a reference to a step-by-step approach in a collaborative modeling workshop, in which models of processes are methodically questioned and presented along a prepared thread of questions in comprehensible chunks. Part of the concept is also the fact that for complex real-world situations a modeling process is not limited to a single workshop but needs repeating work. For example, we had a project in which eleven two-hour meetings were needed to negotiate and describe a work process reorganization task. In follow-up meetings, the presentation of the current state is necessary to catch up, especially for people who are new to the process. This also gives an opportunity to rethink results in a dialectical process.

The workshops are the main forum for collaborative reflection and negotiation. The work with the models as visible explications has various consequences:

- Models are oftentimes used as boundary objects (Star, 1990) representing different perspectives for different stakeholders. The models are the used as a common reference, in which participants can see their own perspective and others, enabling them to understand and discuss the influence of others on their behavior.

- Models usually include the integration of the technical and organizational aspects. Decisions and changes on either of these aspects lead to changes to the respective area, and discussing on these changes helps to make better decisions by reflecting on consequences to other areas.
- The modeling process provides a chance for agreement, if it is wanted or needed. At the same time, the flexibility of the method also provides opportunities to explicate areas where (intentionally or unintentionally) free behavior is kept to the decisions of the actors. Similar to approaches for the introduction of software such as Adaptive Structuration Theory (AST), groups perform an appropriation process in which structures are examined and agreed upon (Poole & DeSanctis, 1990).

Models and the STWT are used during all stages of development cycles, including idea explication of new technology development, refinement of technical and organizational concepts, explication of technically needed decisions and usage for training purposes. However, the method should not replace appropriation processes initiated by participants. In addition, there are blind spots of the method, which need further support. Usability of applications is one aspect for which discussions with prospective users are of limited use. Another aspect is concerned with manual activities in a process, which cannot be clarified in a walkthrough, but need to be processed in place. Moreover, utterances in group meetings may not always be representative. Thus, meetings should be accompanied with other empirical methods.

## PENBASED SERVICES FOR ELDERLY PEOPLE

In this chapter, we will refer to experiences from the project service4home[1]. During this project, we applied an approach combining various methods such as usability testing and interviews with the STWT. The resulting models and continuous refinement served as central artifacts documenting both intermediate results of the project and processes to be implemented in the project.

The project offers domestic services to elderly people and supports their ordering by Pen and Paper technology (see Figure 1). Besides ordering, this technology is also used for the coordination of different service orders and giving feedback after service conduction. The pens used for this resemble usual pens and allow for instant digital processing of handwriting, including paper notes and filled forms. To enable the capturing of handwriting, the paper used for the forms is imprinted with a special pattern, which is almost unnoticeable to the human eye, but can be recognized by the pen. This way, the content of forms can be digitally sent to a central server and processed by handwriting recognition, bitmap processing, and the detection of pre-defined user actions such as ticking a "send form" button. In the project, these digitally transferred orders are sent to a service agency established during the project, which then receives them and dispatches corresponding services such as a shopping escort or repair services. Additionally, it bundles the requested services of multiple customers and coordinates the individual service providers' work.

The application of these pens is not a novel approach, but has been done in other areas such as health care and quality management. However, we experienced that due to the characteristics of service processes and the special client target group, several issues such as the balance between formality and flexibility in form design had to be solved which had not been found for the application of the technology before. Accord-

*Figure 1. Usage of pen and paper technology for service orderings in the project service4home*

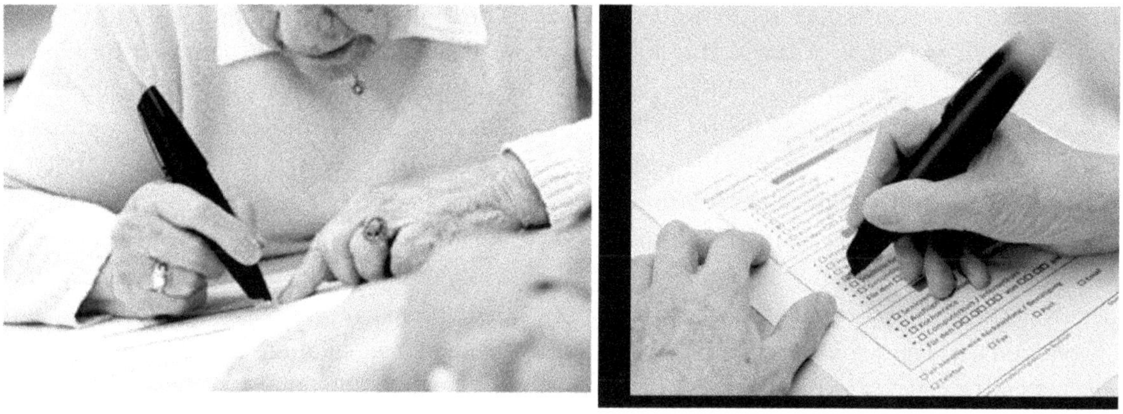

ing to the complex setting, we involved several stakeholders to design and implement services for the elderly. Central actors of the planning and design phase were scientific partners from sociological, economical, and engineering disciplines as well as a housing association and a charitable voluntary association for health care services, each contributing their specific viewpoint into modeling the sociotechnical system envisioned.

## OCCURRENCE OF DRIFT

In the following description, we confine ourselves on aspects, which illustrate the practice of designing sociotechnical systems during the course of a project of developing pen-based services. The following four examples occurred during the action research and show different aspects of drift. We show how our methodology is provoking, coping and resolving the existence of drifting domains. In addition, we discuss the limitations of up-front planning in sociotechnical design.

### Sociotechnical Views on an Infrastructure

The broad solution we developed for the coordination in the project of services has to be seen as an infrastructure rather than a tool, forming a shared resource used by different actors (cf. Hanseth & Lundberg, 2001) and built for extension and change. It consists of various loosely coupled technology fragments such as the pen and paper device combination used by clients, servers to enable data capturing, transmission and exchange from these devices to a database and a client spreadsheet based system to use and manipulate data for service coordination. Additionally, its components and functionality cannot be separated from the social systems using it (cf. Hanseth & Lundberg, 2001; Herrmann, 2009). As a typical feature for an infrastructure, its components – social and technical – are used by different groups of people and sometimes are also hidden to them. To give an example for that, the pen devices are only used by clients of the agency, but these clients are aware of neither the server infrastructure nor the client software used in the agency. All they know is that their forms are (reliably) transferred to the agency after they filled them out and ticked the "send form" button on the form.

Like other infrastructures, our solution poses specific needs for its introduction and sustainable operation. These needs foremost include the establishment of trust in its functionality by making transparent how the transmission works and providing knowledge on infrastructure details in order to cope with problems such as problems with sending the data to the agency. To establish trust in the transmission of data, we found it vital to show clients how their form data is represented on the agency spreadsheets. This way, clients can compare their paper form to its digital representation and thus directly see that the system works. As an example in which the user model of the technology broke down, we had to explain to clients that once they filled in an unwanted order by accident, they should better abandon the form and use a new one. In order for clients to understand, we had to explain to them basics of the technology. The other way around, agency staff had to learn to trust in this transmission, as they had to rely on updates of their spreadsheets instead of talking to clients directly on the phone.

As can be seen from the description above, our case includes characteristic aspects of infrastructures such as perspectives in perception and use of the infrastructure as well as trust in its functionality and behavior. Additionally, it shows the aspect of abstraction and limited view of users on an infrastructure (cf. Bowker & Star, 2000): While we fully explained the respective tools to staff and clients, we only explained little about the staff's tools or the server part of the infrastructure to clients. It should be noted that this abstraction is not only convenient and efficient, as users do not need to know technical details to use the technology, but also necessary in order to avoid information overload caused by these details.

The STWT and the modeling method SeeMe supports this kind of abstraction. There is a hide and show mechanism available that allows users to focus on relevant aspects for the current discussion yet having specifics available if needed. Details can easily be made visible if discussions lead to the need for further information.

## Learning in Collaborative Process Modeling

The main challenge of the described project was to find and design services to support elderly people in living in their own home for as long as possible. In order to cope with this challenge we identified potential experts to contribute their experience in various fields. Some of them were researchers in process design, others were domain experts such as service providers or service professionals working in nursing homes. Thus, they all had an understanding of the needs of elderly people. Furthermore, we invited IT-experts who are particularly experienced in implementing Pen and Paper solutions and their infrastructure. Future process participants such as the staff of the service agency and elderly people were involved later in the process as well. All this was backed up with an iterative development approach where prototypes where quickly tested with customers and process stakeholders.

We started the design process by involving the experts in several modeling and design workshops, as the direct integration of all parties proves to be beneficial for the modeling outcome (cf. Ackermann, Franco, Gallupe, & Parent, 2005). These workshops were aimed at finding innovative yet suitable services, designing the corresponding service processes with special respect to the underlying technology, and identifying potential service providers. The approach included creativity workshops (Carell & Herrmann, 2010), process brainstorming (Herrmann & Nolte, 2010) and modeling workshops facilitated using the method of the STWT (Herrmann, et al., 2004).

After finishing the first series of workshops, we did a pretest. During this phase, we conducted multiple instances of the service tested, involving all process stakeholders (e.g. the future staff of the service agency, different service providers and elderly people) to test its coordination. However, we abstained from using electronic transmission of forms in order to reduce complexity in the test. Instead, we simulated the scenario similar to a 'Wizard of Oz' experiment. We requested the customers to fill out the new paper forms with the electronic pens and went to the homes of the customers to collect the forms. During this phase, all process stakeholders and designers were able to collect firsthand experience with the coordination between the different parties involved. Afterwards we went into another series of workshops leading to various adjustments with respect to the tuning of the process of which we will elaborate in the following section.

During the whole process of conducting STWT workshops, creating models of the coordination process and testing it in practice a drift of knowledge could be observed. The different parties gained deeper insight into the domain of the respective other participants, resulting in a continuous refinement of the process as well as the underlying model. In consequence, the process designers gained a deeper understanding of the needs of elderly people. In addition, the elderly people learned to incorporate the activity of electronically ordering services into their everyday life. This process of gaining insights into the domains of each other however cannot be considered final at a certain point, but has to be continuously adapted to the changing conditions and requirements inevitably resulting from continuous drift of mutual knowledge.

## Limiting the Control Aspects of Modeling

One of the processes we developed during the project was part of a newly-created service offering elderly people company during their shopping. In this and other processes, the service agency serves as a central point of contact for the customers (elderly people) as well as different service providers. Customers order services by filling out a request form, which is transferred electronically to the service agency. The service agency then bundles the requests and prepares the service by passing the required information to relevant service providers. After that, the service is conducted by service providers such as taxi companies and public transport, supermarkets, and different delivery services. Furthermore, there are volunteers involved who accompany the elderly people during their shopping trip. During the execution of the services as well as afterwards further administrative tasks are triggered like billing and quality management.

The model, which was jointly developed during multiple modeling workshops, describes the process of filling out the form, transferring it to the service agency, coordinating the interplay between the different parties, preparing the service and conducting it in several different variants. These variants range from shopping with and without other customers or company, traveling to the supermarket by foot, taxi or public transportation and taking the groceries directly home or receiving them later via delivery (see Figure 2 for an excerpt of the service conduction part of the model).

After some weeks, it turned out that process conduction was done differently than it was specified in the process model and that people were using workarounds of the specified process, as some aspects of it simply did not work the way they were originally planned. For example, ordering a taxi after the shopping was more complex than it had been foreseen: it took the driver a considerable amount of time before arriving at the supermarket and the elderly people and the volunteer had to wait. This problem was handled quite differently by different volunteers resulting in different processes. Some simply ordered the taxi early during the shopping itself while others set a specified time when the taxi should pick them up right after their arrival at the supermarket. Furthermore ordering the services did not work out as originally planned: Due to an accidental server breakdown, some customers had concerns about the technology, which caused some people to always verify offers via phone.

In addition to problems that emerged during the execution of the process itself, we had considerable trouble when trying to alter it by inserting additional options, combining different services, or exchanging e.g. the shuttle service provider.

*Figure 2. Excerpt of the service conduction part of the model*

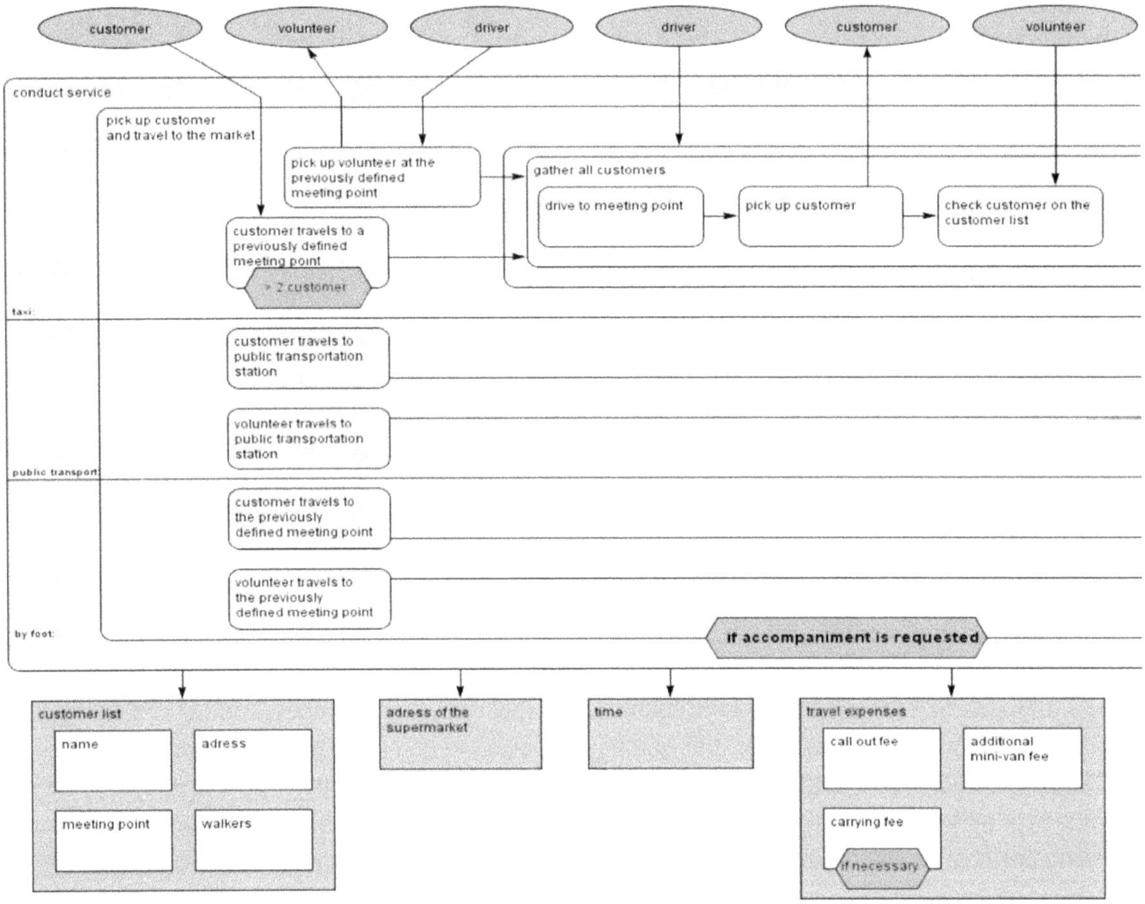


This added even more complexity to the model and the process and made the resulting process harder to handle and control. At that point and instead of inserting more and more details into the model, we used the feature of vagueness available in SeeMe to reduce complexity: As displayed in Figure 3 we reduced the model to five high-level activities showing the involvement of the various process participants and left the details of each process to the parties involved. This is displayed e.g. by marking the activity *deliver groceries, etc.* with a condition (green hexagon) that is not further specified as the stakeholders involved (*service provider* and *customer*) decide whether or not this activity is executed. Furthermore, we did not specify all the activities required to conduct a service but rather inserted only those activities that involve coordination between the several process stakeholders. In order to show that there are other activities remaining, we used another option of vagueness that is provided by the modeling method. The semi-circle at the bottom of the activity *conduct service* in Figure 3 is basically used as a wildcard for content not represented, making leaving out of possible content visible.

Control over the whole process is maintained via an additional sheet (see bottom-left part of Figure 3) that is handed to the volunteers at the beginning of each service. It contains all relevant information to organize the service such as the points of pickup, the address of the supermarket, the method of transport, etc. So, rather than con-

*Figure 3. Revised service conduction part of the model combined with a volunteer sheet*

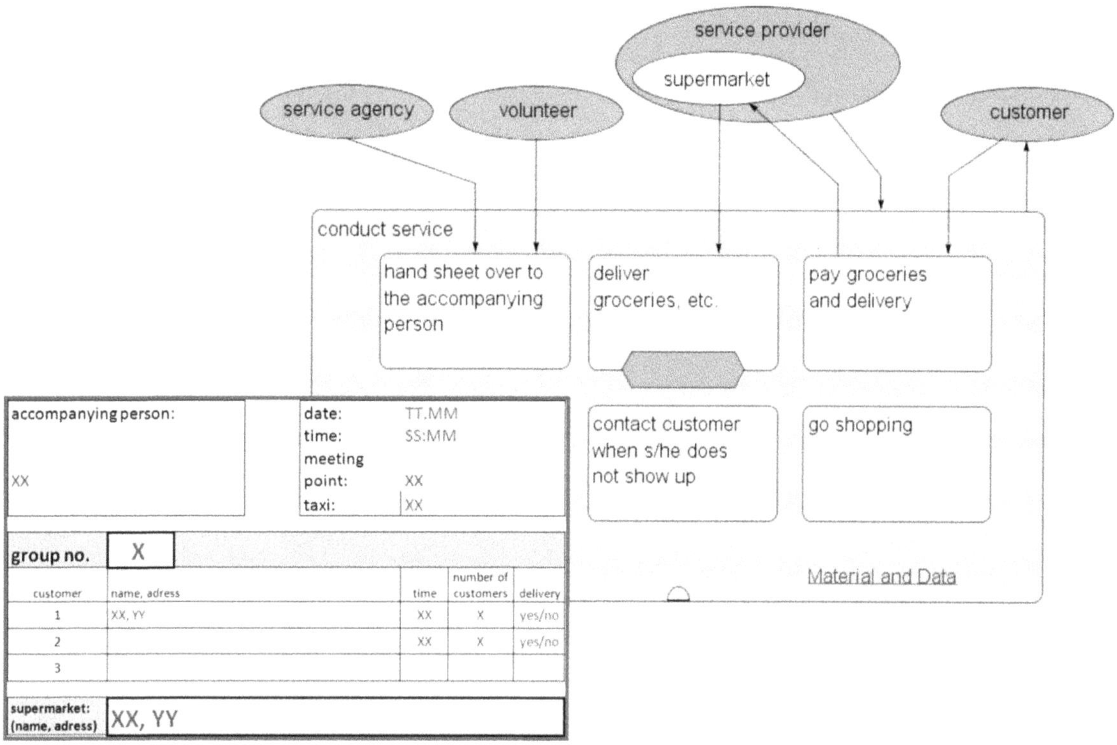


trolling the process through specifying every detail in a model, we left the control to the process participants.

Overall, these changes in the model led to a more generic process enabling further adaptations to an evolving environment while still serving as guidance for process participants during their work. In this context, models have to be seen as a point of orientation rather than a strict guidance—one cannot plan and control every aspect of a complex process with different stakeholders involved in changing environments anyway. As a consequence, one should abstain from over-specifying the process and leave parts of the process model vague, indicating that process details are left to people performing it.

## Drift Phenomena in Forms Design

Another example of the need to embrace drift and flexibility rather than pursuing control can be found in our experiences related to the design of forms for services. In this project, we needed paper forms that could be processed automatically. This poses two challenges for design and implementation: First, forms cannot be changed quickly once they have been defined, mapped to a data structure representing clients' orders and printed. Therefore, form design needs formalized knowledge on the service to be represented and careful upfront planning concerning information needed for service conduction. However, we found that there are parts of services that cannot be planned upfront and fixed eternally such as shuttle times and shopping providers—either because it can be assumed that they will change over time or because there is not enough experience and expertise concerning the impact of fixing them. Second, the representation of services on forms should be complete, meaning that all the information categories needed to conduct a service should be part of the form and the form includes all options the real service includes. This, however, excludes some services such as the delivery of goods from the representation by forms because of their complexity or the amount of information needed to be known upfront.

The two challenges described above impose the need of a special design approach. Rather than implying that design has to be complete and intended to find a durable solution, it should embrace change and include it into the process of design. An example for such an approach can be found e.g. in (Fischer, Giaccardi, Ye, Sutcliffe, & Mehandjiev, 2004), who pursue an approach of designing for design (which they call 'meta-design'). Following their spirit, we developed solutions for the challenges described above. In what follows, we will use an example of a shopping service to describe these solutions and the need to include drift explicitly in their design.

Starting with the second challenge of complete service representation, we found that some of the services we intended to support were not suitable for forms. For one service, we had planned to provide clients with a form resembling a shopping list, enabling them to order goods from home. Their orderings would then be delivered to their homes. Unfortunately, we found that the corresponding number of items would have been very high, causing either the forms or additional material such as catalogues to be very long and, in fact, unusable. Additionally, choosing from items would have meant to leave some others out, thus lessening the value included in the service for many people. Therefore, we decided not to implement this service, but to transform it into a service of accompanied shopping. This way, we could implement the intended benefit of helping people to do their shopping and make it available in a form. This means abstaining from a technology driven solution of including all items in a form and leaving the process of shopping to social interaction. While this might seem less concrete and useful at first sight, it enables remote ordering and has the side effect of engaging people into social interactions. In fact, we now know that this service is the most popular one among the clients of the agency.

Besides the sheer amount of items, the problem of including shopping items of a supermarket on a form also shows the first challenge mentioned above. In particular, we had problems of gathering and especially fixing all information for a service in advance. Obviously, items in supermarkets and also their prices change, causing the need to change forms continuously. Even worse, also in the transformed service of booking an accompanied shopping service, this problem had to be overcome: the choice of supermarkets has to be adapted to the requests of clients and opening or closure of supermarkets. Therefore, we developed a concept of generic forms. In this concept, instead of specifying all fields in a form, we included generic fields to be imprinted on demand with e.g. different names of supermarkets. This is a good example for meta-design and embracing drift, as it anticipates change and leaves space for spontaneously reacting to it. In the beginning, it was hard to clarify the idea to the developers, who had no prior experience with dynamic parts of forms. Thus, the developers considered the idea not to be viable. However, after we managed to get a shared understanding of our intention with them, they agreed to it.

Taken together, these observations show the phenomenon of drift and the need to include it into the design of technology to support people in their daily life and work. Additionally, it shows the need for incomplete and vague specification in models in order to leave room for adaptations caused by changing requirements.

## CONCLUSION

In this contribution, we described a project showing typical characteristics of complex sociotechnical environments such as drift phenomena. Additionally, we described our approach in designing a special **sociotechnical system** of Pen and Paper technology enabled services for the elderly. Based on our observations of the application of this approach, we showed how it supported the design and where we found limitations to it.

One of the major insights we derived from our work is that for sociotechnical design, the power of up-front planning and process design is limited due to the rapidly changing requirements of reality and the lack of a priori knowledge on process conduction. This, however, should not lead to the conclusion to omit process design completely. In contrast, it shows the need to both include all relevant stakeholders in such design efforts early on and to intertwine sociotechnical process design closely with real world applications of the process. This should result in a circle of mutual influence of these phases. In addition, people participating in such design processes should always be aware of the drift their design artifact inevitably will be in.

Another insight is that process design under the circumstances of drift can neither be prescriptive nor complete. In contrast, it has to include incomplete parts of the process in order to allow spontaneous additions and adaptations by process participants. However, there is a need for specifications of critical process parts such as interfaces between process participants and for a process scaffold framing its conduction.

As another insight, our example has shown that for sociotechnical solutions to become accepted and supported by everyone, they have to overcome the modeling stage and show their utility in practice. In the case of the project described, the relevance of prototypes in relation to the more abstract models showing the coordinated activities of the various stakeholders was higher than in other projects. The pen interaction was new to many participants, resulting in the need for hands on experience. The experiences then served as the foundation for a deeper understanding of the sociotechnical system under design.

Analyzing these findings, our approach of collaborative process design with the STWT allowed for bringing together relevant stakeholders and focus their viewpoints on a commonly agreed upon process model. During modeling workshops, divergent perceptions of activities in the process became obvious and enabled people to create a common basis. However, we also recognized that this approach has to be complemented with other methods supporting e.g. the feedback of findings from real world process application to such process design workshops. These methods cannot be determined a priori, but have to complement an approach of sociotechnical design on demand. This way, getting aware of changes and affordances stemming from practice, process models can be utilized as an element to cope with drift phenomena, pushing forward the intertwining of IT usage and human action.

## ACKNOWLEDGMENT

We especially thank Alexandra Frerichs, Dennis Liedtke, and Gergana Nalbantova for their indispensable contribution to this effort.

## REFERENCES

Ackermann, F., Franco, L., Gallupe, B., & Parent, M. (2005). GSS for multi-organizational collaboration: Reflections on process and content. *Group Decision and Negotiation*, *14*(4), 307–331. doi:10.1007/s10726-005-0317-4

Beyer, H., & Holtzblatt, K. (1998). *Contextual design: Defining customer-centered systems*. San Francisco, CA: Morgan Kaufmann Publishing.

Bowker, G. C., & Star, S. L. (2000). *Sorting things out: Classification and its consequences*. Cambridge, MA: MIT Press.

Carell, A., & Herrmann, T. (2010). Interaction and collaboration modes for integrating inspiring information into technology-enhanced creativity workshops. In *Proceedings of the 2010 43rd Hawaii International Conference on System Sciences,* (pp. 1-10). Washington, DC: IEEE Computer Society.

Ciborra, C. U. (Ed.). (1996). *Groupware and teamwork: Invisible aid or technical hindrance*. New York, NY: John Wiley and Sons.

Emery, F., & Trist, E. (1960). Socio-technical systems. In C. W. Churchman & M. Verhulst (Eds), *Management Sciences Models and Techniques: Proceedings of the 6th International Meeting of the Institute of Management Science,* (pp. 83-97). New York, NY: Pergamon Press.

Fischer, G., Giaccardi, E., Ye, Y., Sutcliffe, A. G., & Mehandjiev, N. (2004). Meta-design: A manifesto for end-user development. *Communications of the ACM*, *47*, 33–37. doi:10.1145/1015864.1015884

Hanseth, O., & Lundberg, N. (2001). Designing work oriented infrastructures. *Computer Supported Cooperative Work*, *10*(3), 347–372. doi:10.1023/A:1012727708439

Herrmann, T. (2009). Systems design with the socio-technical walkthrough. In Whitworth, B., & De Moor, A. (Eds.), *Handbook of Research on Socio-Technical Design and Social Networking Systems* (pp. 336–351). Hershey, PA: IGI Global. doi:10.4018/978-1-60566-264-0.ch023

Herrmann, T., Kunau, G., Loser, K.-U., & Menold, N. (2004). Sociotechnical walkthrough: Designing technology along work processes. In A. Clement, F. de Cindio, A.-M. Oostveen, D. Schuler, & P. van den Besselaar (Eds.), *Artful Integration: Interweaving Media, Materials and Practices: Proceedings of the 8th Participatory Design Conference PDC 2004,* (pp. 132-141). New York, NY: ACM.

Herrmann, T., & Loser, K.-U. (1999). Vagueness in models of socio-technical systems. *Behaviour & Information Technology*, *18*(5), 313–323. doi:10.1080/014492999118904

Herrmann, T., & Nolte, A. (2010). The integration of collaborative process modeling and electronic brainstorming in co-located meetings. In G. Kolfschoten, T. Herrmann, & S. Lukosch (Eds.), *CRIWG '10 Proceedings of the 16th International Conference on Collaboration and Technology,* (pp. 145-160). Berlin, Germany: Springer.

Luhmann, N. (1995). *Social Systems*. Palo Alto, CA: Stanford University Press.

Orlikowski, W. J. (1996). Improvising organisational transformation over time: A situated change perspective. *Information Systems Research*, *7*(1), 63–92. doi:10.1287/isre.7.1.63

Poole, M., & DeSanctis, G. (1990). Understanding the use of group decision support systems: The theory of adaptive structuration. *Organizations and Communication Technology*, *173*, 191.

Star, S. (1990). *The structure of ill-structured solutions: Boundary objects and heterogeneous distributed problem solving: Distributed artificial intelligence* (*Vol. 2*). San Francisco, CA: Morgan Kaufmann Publishing.

Strauss, A. (1988). The articulation of project work: An organizational process. *The Sociological Quarterly*, *29*(2), 163–178. doi:10.1111/j.1533-8525.1988.tb01249.x

Suchman, L. (1987). *Plans and situated actions: The problem of human machine communication*. Cambridge, MA: Cambridge University Press.

## KEY TERMS AND DEFINITIONS

**Ambient Assisted Living (AAL):** Ambient assisted living refers to technology support in the everyday life of people, preferably for the elderly.

Ambience is realized by embedding such technology smoothly into daily tasks and facilities. Its aim is to support and prolong the well-being of people.

**Collaborative Modeling:** We refer to collaborative modeling as the joint creation of a shared graphical representation of a system.

**Participatory Design:** Participatory design is an approach where all stakeholders are actively involved in the design of a system to ensure that it fits their needs and is usable.

**Semistructured Modeling:** Systematic description of a sociotechnical system which combines context-free specifications with vague descriptions.

**Sociotechnical System:** A sociotechnical system consists of one or more social and technical subsystems, which are inseparably connected with each other – without the technical subsystem the social system would be existing and the other way round. Both systems mutually influence and shape each other.

**Sociotechnical Walkthrough:** The sociotechnical walkthrough consists of a series of collaborative modeling workshops anticipating future processes and visualizing the discussion of these processes (c.f. Herrmann et al., 2004)

**Vagueness:** Vagueness refers to the incompleteness of specifications, which might be caused by uncertainty or the lack of knowledge on certain elements of a specification. Vagueness should not be considered to be a problem and thus to be diminished, but rather be embraced and explicated, as it is a ubiquitous phenomenon on socio-technical systems.

**Walkthrough:** Deliberate inspection of the documentation of a system which applies step-by-step a selected set of prepared questions.

## ENDNOTE

1 The project service4home has been funded since 2008 by the Federal Research Ministry of Germany and will report its final findings in 2011.

# Chapter 18
# Organizational Learning and Action Research: The Organization of Individuals

**Roberto Albano**[1]
*University of Turin, Italy*

**Tommaso M. Fabbri**[2]
*University of Modena and Reggio Emilia, Italy*

**Ylenia Curzi**[3]
*Marco Biagi Foundation, Italy & University of Modena and Reggio Emilia, Italy*

## ABSTRACT

*This chapter focuses on organizational change, its control and regulation, from the standpoint of a non-dualistic conception of the relationship between individuals and organizations, rooted in the social theory of Norbert Elias, in a theory of organizational learning (Fabbri, 2003), and in a conception of action research (Albano, 2010) in support of organizational learning (Albano & Fabbri, 2010), which are consistent with each other. The combination of those theoretical and methodological references allows an original interpretation of organizational change and the identification of some prior rules for organized collective actions helpful at guiding change accordingly to organizational members' needs.*

*The chapter shows that conflicts may arise in the regulation of empirical organizational processes, due to the incongruence between the existing structure of interdependencies and human needs. Under such circumstances and to the extent that empirical organizational processes are allowed to develop according to those conflicts, organizational learning becomes an actual cause of structural change. However, the chances to overcome conflicts and to identify more satisfying structural choices will be greater if the individuals involved in the analysis of organizational structure, can rely not only on the knowledge drawn from their everyday practice of work, but also on theoretical and methodological knowledge provided by organization theory. A peculiar conception of action research may support the re-composition of the two forms of organizational knowledge, which, to its turn, nurtures organizational learning, thereby positively affecting the relative position of the process/organization with respect to other natural or social processes/organizations.*

DOI: 10.4018/978-1-4666-0303-5.ch018

## INTRODUCTION

*The person acting within the flow may have a better chance to see how much can depend on individual people in individual situations, despite the fixed general direction. (Elias, 1991, p. 48)*

*Mainstream* theories of organization have always faced interpretative difficulties, which in great part are traceable back to one common postulate, i.e. the hypostasis of organization.

The impasse is particularly evident when researches focus on organizational change, its control, and regulation by individuals. The morphostatic aspects of the organization, and—with all the more reason—the morphogenic ones are considered separately from the dynamics and structures of individuals' personality, and from individuals' choices and actions. This is the case when organizations are seen as systems that, by analogy with natural ones, are regulated according to general laws, as well as the case when they are conceived as entities that unintentionally emerge from individual interactions. In both cases, individuals are not considered protagonists of their history. On the one side, a structural determinism holds fast them; on the other, their individual rationalities disperse in collective artifacts that are totally indeterminable beforehand, and escape any design aspiration.

Being convinced that human beings are competent protagonists in the process of reproduction[4] of many of their systems, regulations, and institutions, even if with many limits on which we will dwell upon further, for years we have been investigating organizational phenomena rejecting their reification and conceiving them as processes of action and decision guided by intentional and bounded rationality towards *satisfycing* results. This view of the organizational action is deeply rooted in social and human sciences. The Italian scholar Bruno Maggi (1990) tracks it down through an original and systematic review of some classic authors of the organizational thinking, that he called "Theory of Organizational Action"[5]. Without going too far back in time, the notion can be found in H. A. Simon's works. Starting from his famous work *Administrative Behavior* (1976[3]), he clarified that a fertile organizational theory treats organizational phenomena such as products of intentional and bounded rationality of human beings; consequently, according to Simon, organizational artifacts are characterized by the same type of rationality. In this perspective, J. D. Thompson (1967) pinpointed that each organizational process always faces uncertainty, albeit in varying degrees and forms. This uncertainty is generated by lack of knowledge and/or consensus about desired outcomes, as well as about the tools and techniques to achieve them. At the same time, however, the process seeks to reduce uncertainty through a rational structuration of actions. Each empirical organizational process, guided by intentional and bounded rationality, expresses in such a way its relatively autonomous capacity of regulation (structuration), the validity of which is then evaluated on the basis of the achievement of the desired outcomes.

The cited contributions, as well as some others not mentioned here, share with the general social theory some important issues: the relationship between individual and organization is, in fact, nothing other than a special case of the relationship between individual and society.

A non-dualistic and processual conception of the society has been advanced in the past by classical sociologists such as Weber and Simmel, and—in the recent decades—by many other famous authors (namely, Giddens, Archer, Bourdieu, and Touraine, etc.). Particularly interesting for us is the Social Theory of Norbert Elias, because of its systematic nature—that can be encountered in a number of his theoretical and empirical research works—its heuristic capacity and, last but not least, its striking similarities with the point of view of the Theory of Organizational Action[6]. This chapter shall address the issue of organizational change drawing from Elias's Social Theory and the Theory of Organizational Action—more precisely from the notions of organizational learning and action research in support of organizational learning that we have developed from it. The analysis of a case study will then allow us to highlight the complementary nature of the two approaches with respect to A) the interpretation of social/ organizational change, and B) the possibility of orienting organizational change, by means of prior rules of coordination, towards the members' desired outcomes (including well-being, i.e. the so called occupational health).

## SOCIAL CHANGE IN A NON-DUALISTIC AND PROCESSUAL CONCEPTION

The starting point is the conception of the society as an ever-changing process. This conception rejects both the holistic and individualistic views, which, although in different ways, inevitably lead to static views of the society. This is made clear by Norbert Elias throughout his prominent studies, since his explicit intention, back in the 1930s, to study the 'society of individuals.' According to the author, the terms 'society' and 'individual' do not denote static and opposite polarities. The society should be understood as a wide and stratified intertwining of processes in which agents are linked to each other by dense patterns of interdependence. To the hegemony of the Eleatic principle *Operari sequitur esse*, Elias opposes the contrary principle of Heraclitus *Esse sequitur operari*.

According to Elias, humans are part of a social order in which "processes and transformations take place that are not pre-programmed in human nature" (Elias, 1991, p. 41). Social order and change are therefore inseparable from individuals: "human beings [...] are normally involved in structural transformation" (Elias, 2000, p. 455).

Contemporary humanity, both from a social and a psychic point of view, is the result of a long-term process, so if the social scientist sets out to discover universal, non-spatial, a-temporal laws, if he renounces to historicize institutions, habits, social norms, etc., he will encounter unclear and unrealistic images of the society. The same happens with the reification of the opposite sign, which leads to conceive institutions, habits, etc. as stipulations of élites that somehow fell on the society, rather than as changing conditions in an entire community or society, local or national.

We can grasp this anti-idealistic stance in various passages of Elias's works. Particularly striking to us, both for the clarity and the organizational resonance of the subject, is the one in *The Genesis of the Naval Profession* (Elias, 2007), where the author points out that the study of the genesis of a profession is not simply the study of individuals that were first to perform some functions for others and then entered certain relations with others. Professions continue to exist also when their single representatives die; as languages, they presuppose the existence of a group. If they change, if new occupations emerge in the community, again these changes are not simply due to acts and thinking of a certain individual, as it is the changing situation of the entire community that creates the conditions for the emergence of new occupations, and determines the course of their development.

Elias (1978) works at a reformulation of the sociological lexicon in order to purify it from the concepts and argumentative structures of other sciences, on the one hand, and from the concepts widespread in everyday language, on the other hand, as both inadequate for the study of the social reality.

In this reformulation, the notion of *figuration* is pivotal. It "serves as a simple conceptual tool to loosen this social constraint to speak and think as if "the individual" and "society" were antagonistic as well as different" (Elias, 1978, p. 130). It "draws attention to people's interdependencies. [... S]ocieties [are] figurations formed by interdependent people" (Elias, 1978, p. 132).

Ultimately, in Elias's view, the meaningful human action cannot be other than a social relation: accordingly, there is neither an opposition between the system and the actor, nor a prevalence of one over the other. The subjective meaning of the action changes (psychogenesis) concurrently with the objective networks of meanings (sociogenesis).

Social scientists have no need to reconstruct the link between two separate entities, because the individual and the society are two sides of same coin, always mixed in empirical processes. However, the analytical decomposition is always possible. For instance, in the study of social change, where the impossibility of a clear-cut distinction between sociogenesis and psychogenesis in the flow of empirical processes still allows a useful analytical distinction between the two, or, similarly, in the study of social structures, where the distinction between micro and macro is still analytical, and does not denote separate empirical units to be investigated independently[7].

Elias's reflection provides us not only with a general frame of reference, but also with some theoretical 'building blocks' to analyze the change of institutions and habits. In particular, we refer to his reflections on power, which is linked to the ability of control and regulation of an ongoing process.

Power is "a structural characteristic of *all* human relationships" (Elias, 1978, p. 74) and a "relational concept" (Elias, 1978, p. 93). It should be understood as the possibility to influence the decisional space and self-constraint of other individuals[8], based on a certain form of justification valid within a figuration.

Power is not something 'possessed' by some individuals, but a changeable feature of the interdependences of social relations. In other words, power is itself an ever-changing process, or a becoming, not a 'state.' Indeed, intentionally or not, new room for decision and new options always open up, which impact more or less extensively on ourselves and on the others. At the same time, however, a single individual, or the dominant coalition of a collective process, are always bounded by their interdependencies with others, no matter how wide their control over power sources is: "Belief in the unlimited power of individual people over the course of history is wishful thinking" (Elias, 1991, p. 54).

To sharpen this combination of relationality and processuality, Elias often uses expressions like 'power differentials,' 'balance of power,' 'mobility (fluidity) of the balance of power,' 'distribution of power.' In practice, power differential is never completely and steadily in favor of only one part of a social relationship. To some extent, even in a situation of consolidated dominance, the subordinate does not cease to exercise power over his superior.

The interdependencies thus bind individuals or coalitions, but at the same time provide space for action that would be unthinkable without a society of interdependent people[9]. In the flow of empirical processes, power expresses itself as the (always limited) ability to control and regulate the process according to the needs and goals of the agents. Along with the process of civilization, power is increasingly related to the self-constraint of individuals. The latter refers to an "individual who controls himself and can be controlled or regulated by no one else if he does not do so himself" (Elias, 1991, p. 26); the self-constraint of individual people shapes, changes, and re-shapes continuously in their relation to each other, and through their relationship to each other.

In his studies on the civilizing process, Elias (2000) shows that the self-constraint of the individual—conscious and unconscious—develops in close connection with the change of the social structure, in particular with the process of monopolization and centralization of the chances of social strength (the exercise of physical violence and taxation firstly). The latter is recursively associated with a differentiation and specialization of social functions that is the tasks that each person performs for others and that others perform for him/her. The more social functions are differentiated, the greater is their number and therefore the number of those on whom the individual depends in all his/her activities. The behaviors of a growing number of individuals must then be coordinated, and the intertwining of their actions must be organized in a more precise and rigorous way to ensure that each single action can perform its social function.

In this increasing social interdependence, the satisfaction of individual needs no longer relies on the brutal use of physical force and on sudden explosions of aggressive impulses, because the individual who indulges in them, would see threatened his/her social existence, and therefore the possibility of satisfying his/her needs. As a consequence, the satisfaction of individual needs is more and more linked to self-constraint, to the restriction of the mostly unsocial part of the emotional sphere, and to the ability to reflect retrospectively, on the prior causal chains, and prospectively, on the chains of consequences of his/her and others' actions and intentions. The monopolization of physical violence, then, makes it possible *and* requires more cooperation among people. Structural change requires that the individual, in interpersonal relationships, make greater use of self-constraint. This opens up a wider set of possible behaviors than that offered by "the power of emotions." These behavioral abilities are therefore a prerequisite of the individual's possibility to maintain, and/or strengthen his/her social position in the pattern of interdependencies in which s/he is involved, and to influence the others.

Within this conceptual framework, conflict becomes crucial in the understanding of structural change.

In the studies on the genesis of the naval profession, Elias (2007) clarifies that conflicts among rival groups, who compete for the title over goods, social values or tasks to perform, always have an institutional character, because they are generated by the incongruence of the existing structure (of interdependent functions) with respect to the human needs served by it. Therefore, conflicts can only be overcome through some structural transformation. Disagreements and conflicts signal and raise the problem (structural incongruence), and at the same time foster the search for a solution (structural change). Elias (1978, p. 130) pinpoints that social figurations are steadily under strain because the inevitable interdependence of individuals may take the form of alliances, competition or antagonism. When incongruence between the structure of interdependencies and agents' needs appears, structural changes can evolve towards higher levels of civilization, or involve and deteriorate. The first direction has more chances to occur if conflicts, disputes, competitions, and antagonisms among groups are recognized and regulated, rather than denied and repressed[10].

On these premises, and bearing in mind that the "individuality" and "social conditioning" of the individual "are in fact [...] expressions for his function as both *die* and *coin*" in the social figuration (Elias, 1991, p. 60), we argue that conflict—not less than cooperation—is a significant social relationship which should be aimed at incorporating heterogeneity, and plurality of possible points of view in the structuration of both the psychic habits and the collective action. The non repression of conflicts by an external source (political, economic, or cultural) allows the figuration to resort to new knowledge developed in practice, by trial and error, to self-regulate its change, solve the conflicts, and overcome the difficulties.

We now use these latter reflections on knowledge development in social figurations as a link between social theory and organizational theory.

## ORGANIZATIONAL LEARNING, ACTION RESEARCH, AND ORGANIZATIONAL CHANGE IN A NON-DUALISTIC AND PROCESSUAL CONCEPTION

The development of knowledge in organizations, as the systematic attempt of a process to improve the power differential with respect to other natural and social processes, is the subject of various streams of researches and studies, which address the issue of organizational learning.

Here we investigate the conditions that foster organizational learning and the ways to support it, from an anthropocentric and processual standpoint, akin to the one underlying Elias's social theory, but before that we need to state our theoretical reference points.

The theorization on organizational learning was inaugurated on the pages of Administrative Science Quarterly in 1965 (Cangelosi & Dill, 1965), and has significantly grown in recent decades. On the basis of an epistemological review of the theoretical proposals following since then (Fabbri, 2003), we have highlighted that they express the two prevailing conceptions of organization, and we have developed two ideal-typical conceptions of organizational learning, logically associated to them (Fabbri, 2003). Given their roots in reifying views of organization, the conceptions of organizational learning that prevail today are based on the dichotomy between individual and organization; consequently they have problems in thematizing organizational learning properly and effectively. Here we just briefly recall the two ideal-typical conceptions in order to mark the difference from the theory put forward for the first time in Fabbri (2003). This theory expresses a non-dualistic and processual conception of organizational learning. Therefore, in our opinion, it is syntonic with Elias's conception of the society.

Theories which conceive organization as a predetermined entity (with respect to individuals), define organizational learning as an increase in the stock of organizational knowledge (organizational knowledge base, organizational memory), driven by the needs of the organization to adapt to a given environment. According to this definition, the organizational system is a cognitive system, and knowledge is symbolic representation, that is a mentalistic phenomenon that materializes in cognitive structures (i.e. schemata, scripts, exemplars, etc. [Sims & Gioia, 1986]). Knowledge defined in such a way becomes *organizational* knowledge through the sharing and the institutionalization (organizational codes, routines) of individual knowledge bases. In other words, organizational knowledge is the subset of all individual knowledge bases that have been shared by the members of the organization. In light of that, organizational learning can be further defined as the modification of the cognitive structures of the organization (organizational memory), which is realized through the sharing and institutionalization of individuals' cognitive structures.

On this basis, the intervention in support of organizational learning consists in the design of knowledge sharing, and is usually carried out through the so-called "knowledge management systems." These are information and communication infrastructures, increasingly based on information and communication technologies, aimed at increasing the frequency and intensity of the communication among organizational members, assuming that the greater is the intensity of the communication, the greater is knowledge sharing among individuals, and consequently the greater is organizational learning. Setting aside a discussion of the factors that make this hypothesized sequence unlikely (see Albano & Fabbri, 2010), here we wish to emphasize two aspects. First, even if knowledge were entirely communicable and explicit, this would not mean that the holders are willing to share it, especially if its possession gives them power on other individuals or groups. The first ideal-typical conception of organizational learning thus faces the same interpretative difficulties that Elias points out in relation to the approach to the study of historical and social formations which considers them as the product of anonymous or supra-individual forces and denies that individual people have any role: "it is no clearer, in the latter camp, how the forces producing the formations are to be linked to the goals and actions of individual people" (Elias,

1991, p. 5). The second point to underline is the possible outcome of a super-efficient system of knowledge management: the instantaneous convergence of all individual knowledge bases, in particular that of new members, towards the organizational knowledge base (organizational memory, codes, routines, etc.) clearly contrasts the notion of organizational learning as the increase of the repertoire of actions and responses available to the organization (Weick, 1991). Here the notion of individual knowledge sharing, presented as the solution for the problem of connecting the individual level and the organizational one, comes to a paradox: the greater is the organizational learning, the lower is the organization's ability to change its repertoire of actions (March, 1991).

The conception of organization as an entity that unintentionally emerges from individual interactions is associated with a socio-constructivist conception of learning. Knowledge, in this case, is not a symbolic representation but a phenomenon embodied and embedded in a relational and physical context. Moreover, knowledge does not precede action but is situated in the action itself, which therefore, assumes an epistemic status. Consequently, learning is not the acquisition of new symbolic representations, but of the ability of taking part in a new practice (becoming a practitioner). In organized settings, this generally occurs within collective formations which are spontaneous, informal, unofficial, and often transversal to hierarchies, and to which Lave and Wenger (1991) gave the name of communities-of-practice. We believe that it is difficult to maintain that this is a conception of *organizational* learning. It seems rather an idea of *individual* learning *within organizations*, associated with a different conception of knowledge (knowing rather than knowledge), according to which the organization is the lay-out where individual learning occurs, moreover, in spite of the formal organizational structure.

Given the intrinsic spontaneity of the collective formations that foster individual learning, no intervention in support of learning is conceivable. At the most, one can refrain from hindering it, by legitimizing[11] certain informal dynamics, if they show the recurrent features of a community-of-practice, but it will always be individual learning, and its possible "macro" outcome, which might be observed retrospectively as a change in the practices/theories-in-use in the organization (Argyris & Schon, 1978), could only be conceived as an unintentional result.

Having noted the inability of the available theories to thematise organizational learning properly and effectively, and having attributed this incapacity to the dichotomy between individual and organization that underlies reifying conceptions of organization, in Fabbri (2003) we have thus proposed an original reflection on organizational learning starting form a conception of organization as a process of actions and decisions (Maggi, 1990), guided by intentional and bounded rationality towards satisfycing results (Simon, 1976[3]); a form of meaningful social action in which the individual and the organization are empirically inseparable.

According to the Theory of Organizational Action (Maggi, 1990), three levels can be distinguished in the analysis of empirical organizational processes (cf. Maggi, 1990, p. 75): the institutional level (choices, that is actions and decisions concerning the desired outcomes of the process), the technical level (choices concerning the means- technical knowledge- to be used to achieve the desired outcomes), and the structural level (choices concerning the rules of control and coordination). Due to the bounded rationality that characterizes organizational action[12], the choices concerning each level rely on (and generate) imperfect knowledge. Consequently, there is not a deterministic and unidirectional relation among the three levels of analysis but a constant interdependency that should be controlled over time and evaluated according to the criterion of mutual congruency.

Since structural choices represent the elective research subject of organizational theory (Maggi, 1990, 2003), we have defined organizational knowledge *stricto sensu* as the knowledge—both representational and practical—which underlies structuring choices. Structural choices concern the order of the process, namely the identification of the rules of coordination and control that are able to reduce, as far as possible, the uncertainty introduced by institutional and technical choices (Thompson, 1967). As a result, organizational knowledge *stricto sensu* is adequate to the extent to which it permits to identify rules of coordination and control that can satisfactorily cope with the uncertainties self-generated by the process. Organizational knowledge thus is at the basis of the ability of the process to self-regulate, to maintain and change its order, to self-coordinate and self-control over time congruently in relation to the desired outcomes and available technical knowledge. Since the variability of these last two is both of form and in time (cf. Maggi, 1990, p. 55), also structural choices describe a heuristic path, a process of production of rules guided by intentional and bounded rationality (accordingly, the rules are always sub-optimal). Then, the rules and the order of the organizational process change over time (organizational change), and can be made the subject of assessment and improvement through self-reflection (Giddens, 1984). Organizational learning can be defined as the development of organizational knowledge *stricto sensu*, that is the development of the ability of a process to regulate its own (continuous) change (Fabbri, 2003). Therefore, organizational learning is:

- a "secondary" process, meaning a reflection on the regulation of primary processes;
- a condition that makes organizational change possible;
- a process which can be consciously ruled in advance, according to bounded rationality and with possible unexpected consequences.

On these premises, an intervention in support of organizational learning is indeed conceivable, as an empirical support to the process of structuration/production of rules for organizational action.

Since 2003, we have been commissioned on some organizational re-design projects that have given us the opportunity both to put the present theory[13] into use and to identify the type of action-research consistent with the theory.

In fact, action research can support organizational learning (as defined above) if it moves from the epistemological assumption that the knowledge generated by those individuals involved in the organizational processes under investigation is complementary to the knowledge, theoretical and methodological, developed by organization experts and/or scholars. This assumption grounds a peculiar type of action-research named "analytical re-composition" (Albano, 2010): individuals involved in the organizational process learn "expert" categories for organizational analysis and apply them to the assessment and (re)design[14] of their same primary organizational process. It is a re-composition of different forms of organizational knowledge so that the dichotomy between ETIC/observer and EMIC/observed object[15] is overcome, in favour of their fruitful cross-fertilization. The "analytical" character (as opposed to "synthetic") of this re-composition consists in the acknowledgement of the legitimacy and relative autonomy of both forms of organizational knowledge, albeit on different levels of organizational action: EMIC knowledge developed by the agents within the primary processes, and ETIC knowledge, developed by organizational scholars and learned by the agents in the context of secondary organizational learning processes.

The description and discussion of a case study will further illustrate the consistency of the "analytical re-composition" with the present theory of organizational learning and with Elias's sociological perspective. Before that, it might be useful to depict organizational learning and action research as regulation matters. Maggi (2003) distinguished rules on the basis of the *mode* of production: rules prior to action and rules intrinsic to action. Each and every process of action and decision has or uses rules decided a priori or in advance of the accomplishment of the action, as well as rules generated simultaneously or during the course of the action. Because of the variability in the composition of prior and intrinsic rules (i.e. integration, specification, substitution, etc.) it is impossible that the order of a process coincides with its *a priori* regulation. In the same manner, it is impossible that a process is totally deprived of prior rules, whether not formal or tacit. This means that (the change of) the regulation of a process can never be decided ex ante, perhaps by an agent who is not directly involved in the primary process to be changed, leaving aside the contribution to the regulation which intrinsically belongs to the agents involved in the process[16]. Here is the room, and in our opinion, the opportunity for a "re-composition." There are no recipients of organizational change who are not at the same time protagonists of organizational change, but, if requested, an external intervention, as long as it is methodological and not substantial, can be useful. The external expert has no qualifications to prescribe "organizational recipes"; instead, he/she can provide a method that agents can use to analyse and change the structuration of their work processes, thereby developing competences of organizational (re)design to be used at present and in the future. In so doing, the capacity of the process to consciously control its own continuous change, i.e. organizational learning, is possibly and rationally supported.

## A CASE OF ACTION RESEARCH

In March 2006, the Board of Directors of an Italian medium sized University (hereinafter, the Board) set up a Commission for the simplification of the administrative procedures used at the Central Administration, the Departments and Faculties and other decentralized offices (e.g. the University Library System). The Board gave the Commission (hereinafter, CeSPA) the task to change the regulation of administrative work processes in order to improve their effectiveness and efficiency.

This section will dwell upon the aspects of this still ongoing action research project, which are the most relevant for the purposes of the present chapter.

The first aspect concerns the research design. It should be noted that a few years before that CeSPA was set up, its president[17] took part in another ad hoc University Committee appointed by the Board to carry out a project of reorganization, computerization, and communication of the University. The project led to the identification of new organizational solutions on the overall management structure and its functional units. The academic bodies and union representatives approved and implemented them. The president, however, highlighted that those solutions were, at least in part, unsatisfactory, especially with respect to the requirements of the social system studied (Fabbri & Neri, 2004). He pointed out that this was due to the fact that they were identified through a "consulting" approach, i.e. by two actors (the president himself and another professor of "Organization" of the University) not directly working in the administrative structures and processes. For this reason, when CeSPA entered upon office, the president proposed that administrative and technical staff analysed administrative procedures and identified new organizational choices. He motivated his proposal explaining that only those who directly work in the processes under analysis and intervention have the specific knowledge, experience, and capacity to identify effective and feasible changes. According to the president, the analysis and intervention ought to be run by working groups consisting of both employees and heads of the central and decentralized administrative offices, involved in the administrative process under consideration. He also suggested that the technical and administrative staff involved in the groups should use a method of analysis of the work processes in order to effectively re-configure administrative procedures. The members of CeSPA, including the Administrative Director of the university, agreed on this proposal. On this basis, some administrative procedures, that the Administrative Director, union representatives, and the core assessment group of the university unanimously considered as priorities, were selected, and as many working groups as administrative procedures were formed. Each group was composed on voluntary basis, ensuring that any office involved in the procedure was represented. Finally, the commission decided that a training process on organizational analysis was to be carried out under the supervision of the president and a researcher in the field of organization studies, who were supposed to act as "methodologists."

The two methodologists started by illustrating the Method of Organizational Congruencies (MOC) (Maggi, 1990). This is derived from the Theory of Organizational Action and can be used to interpret and evaluate an empirical organizational process through the decomposition of its analytical components. They, then, presented some examples of how to use the MOC in order to help working group members in familiarizing with its interpretative categories. Finally, they invited group members to use it in order to describe, evaluate and propose changes in their work processes. Empirical processes (e.g. the procedure for the "Allocation of scholarships, and research and advanced training grants") are thus decomposed in the following analytical categories: "desired outcomes"; "required technical knowledge" concerning tools, objects of transformation, and processes of transformation; "control and coordination of the tasks" (or technical actions); "control and coordination of tasks' assignments and accomplishments" (i.e. social structure)[18]. In doing that, groups members also use their knowledge, skills, and abilities, which an outsider (i.e. an actor not directly involved in the process) does not possess. By describing the analytical components of the process, groups members can assess more precisely the congruence or incongruence among the organizational choices actually made at the various analytical levels of action and decision. By that, it is relatively easier for them to identify different choices of regulation aimed at improving the process internal congruence, and the degree of achievement of its desired outcomes.

The methodologists have often taken part in several meetings of analysis and intervention. However, they have never replaced group members in the identification of new administrative work process organizational choices, nor they have ever judged those proposed by the group. Their "task" was rather to ascertain whether (or not) group members had correctly understood the method, for example the analytical distinction between the tasks and their assignment to individuals. Moreover, the methodologists tried to dialogically verify that agents had become fully aware of two crucial issues.

The first issue is the possibility to orient the process regulation towards desired outcomes, which are different from the current ones, and are able to include agents' demands. With regard to this issue, it is worth pointing out two aspects. First: assessing the congruence of the process implies to consider the mutual relations among all its analytical components, including the relationship between process desired outcomes and individual interests, personal motivations, subjective meanings, and agent's identity. Second: the agents of the process have needs, interests, and preferences, the (dis)satisfaction of which can be generally defined as (ill)well-being in the organization. From this point of view, the assessment of the congruence may suggest that the constraints that are generated by the interdependencies among tasks, and among individuals performing the tasks might be re-distributed in the space/time knots of the process or among processes. In this regard, an example drawn from the analysis of the procedure for the "Allocation of scholarships, and research and advanced training grants" can be of help. According to the members of the working group, the procedure desired outcomes are "to run, manage, and renew forms of temporary, paid, and graduate-reserved training in order to contribute to the scientific goals of the University, and according to the Department research activities schedules." A task, which serves these purposes, is the "payment of the grant instalment to the grant recipient." The Accounting and Salaries office of the central administration performs the task. This is related to other tasks of payment performed by the same office in the ambit of other administrative procedures, for a number of other professional roles (e.g. researchers, research fellows, first and second level professors, etc.) and in different periods. The accomplishments of all these tasks should be properly coordinated in order both to meet the deadlines for payment, and to avoid—or to limit—emerging organizational constraints for the Accounting and Salaries office employees. The task under analysis further depends on another one, i.e. "to submit the grant recipient's attendance certificate to the Accounting and Salaries office." The tutor of the grant recipient performs this task. In light of that, group members decided that the tutor must send the attendance certificate on or before a deadline which ensures that the office may adequately coordinate the accomplishment of the "payment of the grant instalment to the grant recipient" with that of other tasks of payment that it performs in favour of different professional roles. The deadline to submit the attendance certificate was thus defined by considering those for performing the other tasks of payment and the office need of receiving the attendance certificate in advance with respect to the deadline for paying the grant instalment. The group made the decision in order to pay the instalment to the grant recipient on time and, simultaneously, to avoid temporal overlapping among tasks' accomplishments which might deteriorate the working conditions in the office.

With regard to the second crucial issue, the methodologists tried to verify that group members had realized that the possibility of changing current regulation choices in order to meet their goals and needs is neither unlimited, nor immediate. Not unlimited, because, as Elias (1991) points out, the space for action of individual people is always precisely bounded by their interdependencies with others. Not immediate, because the extent to which constraints can be actually redistributed in space and time depends on the ability to assert one's own point of view, to defend one's own interests, and to sustain one's own choices in the regulation of the interdependencies within the process and between processes. This capacity varies from case to case. In what follows, three incidents occurred in the action research will be compared to clarify this aspect.

The first concerns the interdependencies between the reconfiguration of the "Master" procedure by working group members and the work of a Commission of teachers, which was purposely appointed by the Board to amend the Rules to set up Master Courses. At first, this commission worked separately from the group. It thus formulated new Rules that could have adversely affected the internal congruency of the administrative procedure. Some of the incongruencies, however, were avoided, as the group managed to get that the new Rules were jointly assessed by the "Master" Commission and the group itself, and this led to the revision of some of the initial amendments proposed by the former[19].

The second incident deals with the new Rules for "taking stock of chattels." In accordance with the new guideline (see endnote number 16), a group was formed to analyse their procedural "feasibility." It was composed by the two drafters, from the Economics and Finance Direction of the University, and the representatives of other administrative structures affected by the new Rules. By using the MOC, group members jointly identified some incongruencies of the new rules, and the consequent proposals for their revision.

The third incident concerns the introduction of changes in the Rules for the procurement of goods and services by the Direction for Legal, General and Institutional Affairs[20]. Also, in this case, a group was formed to analyse their procedural "feasibility." At least theoretically, it was composed by representatives from the above Direction and the administrative structures that were affected by, but didn't write the new Rules. However, the analysis was actually carried out by the latter only. By using the method of organizational analysis that they had recently learned and the skills developed over time by buying goods and services, they found that the new Rules provided incomplete, contradictory, ambiguous, and incorrect instructions. Therefore they asked that the Direction attended the meetings of the group to clarify, supplement, and jointly (i.e. dialogically) amend the new Rules. This request, however, was not satisfied. The Direction and particularly its Head were thus claiming to have exclusive competence on the administrative work process regulation, legitimising this claim by the possession of the legal knowledge necessary to regulate the process. Unable to continue because of the lack of knowledge on legal matters, the group stopped the analysis of procedural "feasibility." In the meantime, due to the behaviour of the chief executive, tensions were growing even more intense between the latter and all administrative and technical staff involved in the purchase of goods and services. Someone threatened to stop buying until the chief executive had provided the required clarifications and amendments. In a number of meetings of the Department Boards, the secretaries of the Departments openly and lively contested the new Rules for the procurement of goods and services.

## DISCUSSION

In a non-dualistic view, organizational learning is the development of the capacity of a process to self-regulate competently. Learning is an analytical dimension of organizational change that is a process itself, oriented towards desired outcomes according to intentional and bounded rationality. It is not necessarily planned and ruled consciously. It can be a tacit aspect of the primary process. When it is made explicit, it takes the form of a secondary process that is relatively autonomous from the primary process enacting it. Such is the case when action research projects support primary processes. In the case study described above, action research is consciously aimed at going beyond the dualisms between individual and organization, and between researcher and research "object." Overcoming these dualisms implies that spaces for an autonomous research are actually available for the research "object." It further involves considering that knowledge and skills developed in the everyday practice of work, and theoretical and methodological knowledge (i.e. disciplinary knowledge) needed to analyse and change work processes are complementary, and both can be learned and used (i.e. in this sense, recomposed) by the agents of the processes.

In the case study in point, there is neither a consultant who defines a shared set of values, solves the problems, and makes the decisions for change, nor a researcher who understands, and then comes back to the agents of the process with the subjective meanings underlying their actions. Rather, there are some methodologists who assist agents in learning a method consistent with the non-dualistic view[21]. In so doing, a secondary process supports the competent self-regulation of the primary process. The latter is always possible in any process, even when an explicit secondary process is not there. What is interesting though is that where the secondary process is present, it is required by the primary one, rather than being imposed from the outside.

As we have seen, describing the primary process analytical components and identifying hypotheses for change demand that working group members be able to relate two "discourses" about work to each other. A discourse which relies on theoretical concepts/abstract ideal-types (ETIC knowledge) and one that depends on knowledge, experience, and evaluations that people develop in the course of the work process (EMIC knowledge). The theoretical apparatus is a means of thought and "imagination" which should strengthen workers' ability to re-frame knowledge, experience, and evaluations concerning the daily work process. One could metaphorically say that if workers learn to merge these two forms of knowledge, their work will "talk" to them, even though they, as well as their interdependencies, remain the protagonists of work.

In so doing, a new awareness can develop. The process regulation can be modified to redistribute constraints in the space and time knots of the process. It can be changed to satisfy the needs, including those of well-being, of people at work. As Elias (1991) would say, by that it is possible to reconcile social requirements and individual needs and inclinations.

Notwithstanding, in the present case of action research, agents also realized that constraints redistribution is a possibility, because the space for action of individual people is always bounded in a well-defined manner by their interdependence with others (Elias, 1991), for example, other commissions, teachers, chief executives. Event though in individual situations a lot can depend on individual people, the identification of different regulation choices implies to reflect on how to combine and coordinate a number of individual identities, purposes, and actions which may be antagonistic and not easily reconcilable by a top down imposition[22].

This last issue is related to the final remarkable aspect of the present case study. Faced with the conflicts arisen in the groups for the analysis and intervention on processes regulation (i.e. the tensions between the chief executive and the technical and administrative staff of other central and decentralized offices for the regulation of the purchase of goods and services), the methodologists never spent their reputation as negotiators among parties, nor did they try to resolve tensions and take sides in conflicts among offices, between administrative staff and teachers, subordinates and superiors, and so on. Rather, they provided group members with tools and dialogical spaces to be used to bring process incongruencies and possible more satisfycing organizational choices to light. Their behaviour may be rationalized by combining hypotheses drawn from the social theory of Elias and our previous works on organizational learning and action research in its support.

According to Elias, conflict occurs when there is some incongruence between the existing structure of interdependent functions and human needs. Therefore, it can be overcome through some structural transformation. This change, in turn, will be possible if social figuration is allowed to develop according to the conflict. In so doing, in fact, the structuration of both the psychic habits and the collective action can comprise heterogeneity and plurality of possible viewpoints. In addition, social figuration can resort to new knowledge, which is developed in practice by trial and error, to self-regulate its own change and resolve tensions.

The above interpretation can be of help to clarify why the present theory of organizational learning assumes that the development of the capacity of a process to regulate its own continuous change is not an efficient cause, but an enabling condition that makes it possible organizational change directed towards certain desired outcomes. In the circumstances where tensions occur because of the inconsistency between the structure of interdependent functions and human needs, the fact that organizational learning becomes an actual cause of structural change depends on the extent that social figuration is allowed to develop according to the conflict generated by that inconsistency.

Accordingly, even the present theory of organizational learning and the connected conception of action research in support of organizational learning regard conflict as an eventuality far from being pathological for the regulation of empirical organizational processes. Rather, for them the central question is whether individuals may orient conflict towards a positive organizational innovation for the figuration involving them that is towards a power ratio surplus for all individuals involved in the interdependency relations where conflict occurs.

Our previous works allows us to put forward the following answer. Conflicts and incongruencies are expression of uncertainty and thus of bounded and intentional rationality. They can be overcome through a more rational (adequate) structuration of the interdependencies. The chances of reaching this result might be greater if the current regulation and its consequences are made the subject of a reflection aimed at bringing the inconsistencies underlying conflicts to light, and at identifying more satisfycing organizational choices. This reflection can benefit not only from the knowledge that individuals develop in the everyday practice of work, but also from theoretical and methodological knowledge provided by organization theory to analyse and change processes structuration. Disciplinary knowledge is developed by organizational scholars. However, the agents of empirical organizational processes may learn it. A peculiar conception of action research may support the re-composition of the two forms of organizational knowledge, thereby strengthening organizational learning, i.e. the development of organizational knowledge, and of the capacity of the process to self-regulate competently. In so doing, it is possible to positively affect the relative position of the process with respect to other natural or social processes, and thus producing a power ratio surplus for all individuals involved in the interdependency relations where conflict occurred.

## FUTURE (ACTION) RESEARCH DIRECTIONS

The previous sections drew attention on organizational change without addressing this issue in connection with the design, adoption and use of Information and Communication Technologies (hereafter, ICTs) in organizations. From our standpoint, however, the latter is conceivable as a special case of the former.

As Masino (1997) highlights, mainstream researches and studies, which focus on individuals' control over ICTs-related organizational change, face the same interpretative difficulties as those that concern the wider issue of organizational change do. Many scholars usually consider organizational change as technologically determined, and thus neglect the role of individuals' actions and decisions. Other scholars, on the opposite, see organizations and ICTs as collective artefacts whose rationality is recognizable only ex post facto, and thus escape any design aspiration; such collective artefacts form in a way that often avoids individuals' rationality, as they become reified. In so doing, they prevail over individuals, thereby binding rather than enabling their actions and decisions.

The complementary approaches outlined in this chapter could be of help to overcome the theoretical and practical problem connected with the control, regulation and orientation of ICTs-related organizational change; furthermore, they are consistent—and can be used in synergy—with specific theorizations on ICTs based on the same process conception outlined in this chapter (e.g. Masino & Zamarian, 2003)[23].

The combined use of these organizational frames could provide new room for interdisciplinary action research projects where individuals may control ICTs-related organizational change. In such projects, ICTs designers, adopters, and users should be all involved in the assessment of the organizational (re)design, which might be associated with the introduction of the new working tools. Consequently, the assessment necessarily concerns the extent that change is desirable from the standpoint of organizational members' needs, interests, preferences (i.e. their well-being). Having Elias's social theory as a guide ensures that possible emerging conflicts would not be treated as pathological events, while the reference to the present theory of organizational learning, and to the conception of action research consistent with it, would provide the agents of empirical organizational processes with prior rules to orient conflict towards a positive organizational innovation. In so doing, one might come close to action research projects that provide greater chances that ICTs-related organizational change might be consistent with organizational members' needs.

## CONCLUSION

The Social Theory of Norbert Elias, the theory of organizational learning (Fabbri, 2003) and the conception of action research (Albano, 2010) in support of organizational learning (Albano & Fabbri, 2010) appear to be of help in the effort to address the issue of individuals' control over social and organizational change.

After having highlighted the consistency of the two theoretical discourses on social systems, and thus on organizational systems, the chapter discussed their complementarity with respect to the interpretation of social and organizational change and the identification of prior rules for organized collective actions helpful at guiding changes accordingly to organizational members' needs (including well-being, i.e. the so called occupational health).

Our analysis suggested that conflicts occur because of the incongruence between the existing structure of interdependent functions and human needs. In such cases, the fact of turning organizational learning (here seen as an enabling condition) into an actual cause of structural change depends on the extent that empirical organizational processes are allowed to develop on the basis of the conflict generated by that incongruence. However, the chances to overcome conflicts and to identify more satisfycing structural choices will be greater if social figurations can jointly rely on knowledge developed in the everyday practice of work, and on theoretical and methodological knowledge provided by organization theory to analyse the structure of interdependencies and the inconsistencies underlying the conflicts.

In so doing, one might come close to outline an "organization of individuals" that moves forward according to intentional and bounded rationality, in this way running its course: "From plans arising, yet not predetermined; by ends moved, yet endless" (Elias, 1991, p. 64).

## REFERENCES

Albano, R. (2010). La ricerca intervento. In Fabbri, T. M. (Ed.), *L'organizzazione: Concetti e Metodi* (pp. 285–310). Roma, Italy: Carocci.

Albano, R., & Fabbri, T. M. (2010). Competenze per l'apprendimento organizzativo: Un approccio simoniano. *Rassegna Italiana di Sociologia*, *51*(2), 255–282.

Argyris, C., & Schon, D. A. (1978). *Organizational Learning*. Reading, MA: Addison- Wesley.

Cangelosi, V. E., & Dill, W. R. (1965). Organizational learning: Observations toward a theory. *Administrative Science Quarterly*, *10*(2), 175–203. doi:10.2307/2391412

Elias, N. (1978). *What is sociology?* London, UK: Hutchinson & Co.

Elias, N. (1991). *The society of individuals*. Oxford, UK: Basil Blackwell.

Elias, N. (2000). *The civilizing process*. Oxford, UK: Blackwell.

Elias, N. (2007). *The genesis of the naval profession*. Dublin, Ireland: University College Dublin Press.

Fabbri, T. M. (2003). *L'apprendimento organizzativo – Teoria e progettazione*. Roma, Italy: Carocci.

Fabbri, T. M. (2004). L'apprendimento organizzativo – Un esperimento di progettazione. In D'atri, A. (Ed.), *Innovazione Organizzativa e Tecnologie Innovative* (pp. 113–121). Milano, Italy: Etas.

Fabbri, T. M., & Neri, M. (2004). l'organizzazione dei sistemi informatici nell'università. In *Quaderni del Dipartimento di Scienze Sociali, Cognitive e Quantitative*. Reggio Emilia, Italy: University of Modena and Reggio Emilia, Italy.

Feldman, M. S., & Orlikowski, W. J. (2011). Theorizing practice and practicing theory. *Organization Science*, *1100*(0612).

Giddens, A. (1984). *The constitution of society: Outline of the theory of structuration*. Cambridge, MA: Polity Press.

Lave, J., & Wenger, E. (1991). *Situated learning: Legitimate peripheral participation*. Cambridge, MA: Cambridge University Press.

Maggi, B. (1990). *Razionalità e benessere: Studio interdisciplinare dell'organizzazione*. Milano, Italy: Etaslibri.

Maggi, B. (2003). *De l'agir organisationnel: Un point de vue sur le travail, le bien-être, l'apprentissage*. Toulouse, France: Octarès Éditions.

March, J. G. (1991). Exploration and exploitation in organizational learning. *Organization Science*, *2*(1), 71–87. doi:10.1287/orsc.2.1.71

Masino, G. (1997). *Nuove regole di progettazione: Opportunità tecnologiche e scelte organizzative*. Roma, Italy: Carocci.

Masino, G. (2004). Cambiamento della progettazione organizzativa e progettazione del cambiamento. In Maggi, B., & Masino, G. (Eds.), *Imprese in Cambiamento: Officina di Organizzazione* (pp. 17–28). Bologna, Italy: Bononia University Press.

Masino, G., & Zamarian, M. (2003). Information technology artefacts as structuring devices in organizations: Design, appropriation and use issues. *Interacting with Computers*, *15*(5), 693–707. doi:10.1016/S0953-5438(03)00059-6

Reynaud, J.-D. (1989). *Les règles du jeu: L'action collective et la régulation sociale*. Paris, France: A. Colin.

Simon, H. A. (1976). *Administrative behavior*. New York, NY: Free Press.

Sims, H. P., & Gioia, D. A. (Eds.). (1986). *The thinking organizations*. San Francisco, CA: Jossey-Bass.

Thompson, J. D. (1967). *Organizations in action*. New York, NY: McGraw-Hill.

Weick, K. E. (1991). The non traditional quality of organizational learning. *Organization Science*, *2*(1), 116–124. doi:10.1287/orsc.2.1.116

Zamarian, M. (2010). Le routine organizzative tra dimensione operativa e dimensione generativa: Modi di riproduzione dell'agire organizzativo. *Rassegna Italiana di Sociologia*, *51*(2), 233–253.

## KEY TERMS AND DEFINITIONS

**Action Research:** a form of applied research in which EMIC and ETIC knowledge are strongly intertwined (Albano, 2010).

**Bounded Rationality:** any attempt to relate means to an end *despite* that decision-making processes take place under uncertain conditions which are due to the fact that individuals make decisions on the basis of preferences that are partly inconsistent, not totally comparable with each other, and change over time; and of knowledge, about desired outcomes, adequate technical actions to reach them, actions expected consequences, and preferences, which is not complete (Simon, 1976[3]).

**EMIC/ETIC:** a couple of terms drawn from anthropology; in this chapter, the term 'EMIC' specifically refers to knowledge, experience, beliefs, skills, abilities that people daily develop by taking part in work processes, while the term 'ETIC' specifically refers to the scientific knowledge provided by the organizational disciplines.

**Non Dualistic View:** a view which does not separate/reify individuals and society and individuals and organization.

**Organizational Change:** the temporal variability of any process of actions and decisions, namely the variability in time of institutional, technical and structural actions and decisions.

**Organizational Learning:** the development of organizational knowledge *stricto sensu*; the development of the capacity of a process of actions and decisions to regulate its own continuous change (Fabbri, 2003).

**Power:** according to Elias's point of view, it is a structural characteristic of all human relationships, and refers to the possibility of influencing the decisional space and self-constraint of other individuals, on the basis of a certain form of justification valid within a figuration.

**Processual Conception:** any attempt to explain why and, contextually, to understand how a given course of actions and decisions has (or might) become thus and not otherwise, by focusing on the conditioning role played by past actions and decisions which enable and, at the same time, bind future actions and decisions.

**Self-Constraint:** In the civilizing process, Norbert Elias emphasises, among other things, a long-term shifting in the social control of individual behaviours: in more civilized societies, external constraints tend to diminish in favour of self-constraint.

**Self-Regulation:** the capacity of a process to self-produce the rules for its unfolding; opposite to the situation in which the process draws the rules for its unfolding from the outside (Maggi, 2003).

## ENDNOTES

1. Roberto Albano is author of the sections "Introduction" and "Social Change in a Non-Dualistic and Processual conception." He is also equal collaborator in writing the sections "Future (Action) Research Directions" and "Conclusion."
2. Tommaso M. Fabbri is author of the section "Organizational learning, Action research, and Organizational change in a Non-Dualistic and Processual conception." He is also equal collaborator in writing the sections "Discussion," "Future (Action) Research Directions" and "Conclusion."
3. Ylenia Curzi is author of the section "A Case of Action research." She is also equal collaborator in writing the sections "Discussion," "Future (Action) Research Directions" and "Conclusion."
4. Change concerns also organizational routines, that is the sequences of activities which, according to the *mainstreams* of organizational theories, are constant reproductions, determined ex ante or recognisable ex post, but always the same. For this point see Zamarian (2010) with whom we share Simon's perspective on organizational rationality.
5. Maggi (2003) redefined later his personal theory as "Théorie de l'*Agir* Organisationnel" (emphasis added) to stress its Weberian epistemological ground. English translation however remains "Theory of Organizational Action." Indeed, "agir" is usually translated in English by "to act" or "acting." While the former hinders the possibility of distinguishing the verbal from the noun, the latter would be probably misleading. In organization studies, in fact, "acting," and more precisely "organizing," "knowing," "strategizing," etc. are pivotal concepts-words of theoretical schemes rooted in the epistemological viewpoint which—differently from the Weberian one—assumes that organizational phenomena

are completely undeterminable ex ante. They can be recognized ex post facto only, as the result of people's recurrent (inter-)actions that gradually become objectified and institutionalized. Thus, the above-mentioned theoretical frameworks are grounded in the study of everyday interactions, which, to their turn, are researched by employing synthetic and sensitizing concepts (as opposite of analytical ones), such as that of "everyday activity" (Feldman & Orlikowski, 2011). The latter is the accomplishment, an instance of action by specific people in specific times and places, and thus a practice (as opposite of the "process of actions and decisions" [Maggi, 1990, 2003]).

6 Apart from few exceptions that will be pointed out hereinafter, Elias's reflection is absolutely unrelated to organizational theory. The latter, in turn, has only recently discovered Elias: see the special issue of Organization, August 2001, 8(3), entitled *Elias and Organization.*

7 One can consider for example the empirical research by Elias (2007) on the genesis of the naval profession in the British navy. On the one side, the author accounts for the latter by considering the broader economic, social, and political macrostructures where the naval profession develops. On the other side, he sheds light on the macro level (national policies and international relations, social stratification and socio-economic conditions etc. in England between XVI and XVIII centuries) by examining the tensions and innovations characterising the micro level (in particular the professional relations between *gentlemen* and *tarpaulins)*.

8 The theme of power is already mentioned in *The Society of Individuals,* within a theory of decision and action ("...What we call 'power' is really nothing other than a somewhat rigid and undifferentiated expression for the special extent of the individual scope for action associated with certain social positions" [Elias, 1991, p. 52]). Then, it is systematically analysed in *What Is Sociology?* (Elias, 1978).

9 Just as other classic authors like Smith, Marx, and Durkheim, Elias places great emphasis on the division of social labour as a driving force in the progress of human society.

10 Reynaud (1989) makes a similar point in his *theory of joint regulation,* although he seems more optimistic than Elias, as he assumes that conflict is always followed by a negotiation between the norms prescribed at the top and the local rules produced at the bottom level, and that this negotiation always generates a more effective regulation of the work processes.

11 Certainly not by "designing" communities of practice as some consulting proposals suggest.

12 Meaningful action may be rational in relation to both the purpose and the value.

13 See for example Fabbri (2004) and here the next paragraph.

14 By adopting a processual point of view, Fabbri (2003) defines "organizational design" as the process of actions and decisions aimed at ordering the primary process in accordance with bounded and intentional rationality. It is embedded in the process and distributed among all those who are directly involved in it; furthermore, it is intertwined with power. As Masino (2004) highlighted later, a processual conception views design as a search under condition of uncertainty, and particularly as a process that co-defines means and ends in accordance with a logic of mutual consistency.

15 As Albano (2010) points out, this dichotomy has characterised the great part of the reflection on action research in organizations and can assume various forms: prevalence of the disciplinary knowledge (ETIC knowledge) of the "observer" (the consultant) on the "knowledge-in-action" (EMIC knowledge) of the "natives" (the organizational members); the contrary prevalence; the coexistence of

different and mutually "impermeable" forms of knowledge.

16 In this regard, one can refer to the broad theme of *change management*, understood as a useful tool to assure that organizations accept the new organizational solutions proposed by the consultancy.

17 One of the authors of this chapter and a professor of "Organization" at the University.

18 According to the MOC, social structure may be analytically decomposed into: agents who perform tasks, and their mutual interdependencies; places, time and ways of tasks' accomplishment; and ways of agents' involvement, such as forms of recruitment and selection, education and training, compensation and career development, knowledge held by the agents, individual interests, subjective meanings, personal motivations, and identities.

19 In light of this incident, CeSPA proposed that the Board adopted the following decision-making guideline: all new rules, which are submitted to the Board for deliberation by any person or Commission, must be accompanied by a verification of their "feasibility." This consists in the analysis of the changes they imply in the existing administrative procedures or in the conception of the new procedures they require to be implemented. The Direction for Planning, Evaluation, and Training of the University, that had just been formed at that time, was given these tasks. The Board contextually gave the Direction also the task of controlling and coordinating the action research on the administrative procedures, which was formerly assigned to CeSPA. As a consequence, an administrative structure of the University became responsible for it.

20 In accordance with the legislative decree of September 11 2008, n. 152: "Further integrations and changes to the legislative decree of April 12, 2006, n. 163, concerning the code of public contracts regarding work, services and supplies."

21 In the present case study, the Method of Organizational Congruencies, but it could be another one.

22 Some Elias's comments on the development of a special training to the profession of naval officer, which took the double function of seaman and military gentleman into consideration (Elias, 2007), can be of help to clarify this point. This suitable solution concerned the so called "voluntaries of first class," derived from a long fight and was not due to the ideas of few members of the Admiralty or of the naval administration. Top down regulations merely ratified the practices and uses directly developed on shipboard by trial and error. It would be a serious error considering government orders or the King and naval administrators' ideas as the main propulsive forces for the development of the English naval profession. The latter developed through the interaction of several forces. Government laws determined, in some ways, the practices and relations of the Navy. These last two, however, had their weight, because their development often brought about legislative changes.

23 According to these authors, organizational change is the consequence of the way in which ICTs are used in organizations. ICTs use, in turn, is the result of complex, dynamic, and often conflicting interactions among different decisions guided by bounded and intentional rationality, i.e. design, adoption, and use decisions.

# Compilation of References


Ackerman, M. S. (1997). Augmenting the organizational memory: A field study of answer garden. In T. Malone (Ed.), *Proceedings of the Conference on Computer-Supported Cooperative Work,* (pp. 243-252). Chapel Hill, NC: ACM Press.

Ackerman, M. S., & Malone, T. W. (1990). Plans and situated actions: The problem of human-machine communication. In *Proceedings of the Conference on Office Information Systems*, (pp. 31-49). Cambridge, MA: ACM Press.

Ackerman, M. S. (2000). The intellectual challenge of CSCW: The gap between social requirements and technical feasibility. *Human-Computer Interaction, 15*(2-3), 179–203. doi:10.1207/S15327051HCI1523_5

Ackermann, F., Franco, L., Gallupe, B., & Parent, M. (2005). GSS for multi-organizational collaboration: Reflections on process and content. *Group Decision and Negotiation, 14*(4), 307–331. doi:10.1007/s10726-005-0317-4

Adler, P. S., & Winograd, T. A. (Eds.). (1992). *Usability: Turning technologies into tools*. New York, NY: Oxford University Press.

Agamben, G. (1990a). *La comunità che viene*. Torino, Italy: Einaudi.

Agamben, G. (1990b). *The Coming Community*. Minneapolis, MN: The University of Minnesota Press.

Agre, P. (1995). The soul gained and lost: Artificial intelligence as a philosophical project. *Stanford Humanities Review, 4*(2), 1–19.

Akdere, M. (2003). Action research paradigm in the field of training and development. *Journal of European Industrial Training, 27*(8), 413–422. doi:10.1108/03090590310498559

Akrich, M. (1992). The de-scription of technical objects. In Bijker, W. E., & Law, J. (Eds.), *Shaping Technology/ Building Society: Studies in Sociotechnical Change* (pp. 205–224). Cambridge, MA: MIT Press.

Akrich, M. (1995). User representations: Practices, methods and sociology. In Rip, A., Misa, T. J., & Schot, J. (Eds.), *Managing technology in society: The approach of constructive technology assessment* (pp. 167–184). London, UK: Cassel Publishers Ltd.

Albano, R. (2010). La ricerca intervento. In Fabbri, T. M. (Ed.), *L'organizzazione: Concetti e Metodi* (pp. 285–310). Roma, Italy: Carocci.

Albano, R., & Fabbri, T. M. (2010). Competenze per l'apprendimento organizzativo: Un approccio simoniano. *Rassegna Italiana di Sociologia, 51*(2), 255–282.

Alby, F., & Zucchermaglio, C. (2005). *Gruppi e tecnologie al lavoro*. Bari, Italy: Laterza.

Al-Mashari, M. (2003). Enterprise resource planning (ERP) systems: A research agenda. *Industrial Management & Data Systems, 102*(3), 165–170. doi:10.1108/02635570210421354

Almklov, P. G. (2008). Standardized data and singular situations. *Social Studies of Science, 38*(6), 873–897. doi:10.1177/0306312708098606

Alvesson, M., & Sköldberg, K. (2009). *Reflexive methodology: New vistas for qualitative research*. London, UK: Sage.

American Hospital Association. (2001). *Patients or paperwork? The regulatory burden facing America's hospitals. PricewaterhouseCoopers Report*. New York, NY: PricewaterhouseCoopers.

Anderson, C. (2008, July). The end of theory. *Wired.*

Anderson, B. (2000). Where the rubber hits the road: notes on the deployment problem in workplace studies. In Luff, P., Hindmarsh, J., & Heath, C. (Eds.), *Workplace Studies: Recovering Work Practice and Informing System Design*. Cambridge, UK: Cambridge University Press. doi:10.1017/CBO9780511628122.011

Anderson, J. R. (1983). *The architecture of cognition*. Cambridge, MA: Harvard University Press.

Anderson, R. J. (1994). Representations and requirements: The value of ethnography in system design. *Human-Computer Interaction*, *9*(1), 151–182. doi:10.1207/s15327051hci0902_1

Anderson, R. J. (1997). Work, ethnography and system design. In Kent, A., & Williams, J. G. (Eds.), *Encyclopedia of Microcomputers* (*Vol. 20*, pp. 159–183). New York, NY: Marcel Dekker.

Anderson, S., Hardstone, G., Procter, R., & Williams, R. (2006). Down in the (data)base(ment): Supporting configuration in organisational information systems. In Ackerman, M. S., Halverson, C. A., Erickson, T., & Kellogg, W. A. (Eds.), *Resources, Co-Evolution and Artifacts: Theory in CSCW* (pp. 221–254). Dordrecht, The Netherlands: Kluwer. doi:10.1007/978-1-84628-901-9_9

Anessi-Pessina, E., Nasi, G., & Steccolini, I. (2008). Accounting reforms: Determinants of local government's choices. *Financial Accountability & Management*, *24*(3), 321–342. doi:10.1111/j.1468-0408.2008.00455.x

Argyris, C., & Schon, D. A. (1978). *Organizational Learning*. Reading, MA: Addison- Wesley.

Aristotle,. (1976). *The nicomachean ethics* (2nd ed.). London, UK: Penguin Classics.

Arrow, K. J. (1962). The economic implications of learning by doing. *The Review of Economic Studies*, *29*, 155–173. doi:10.2307/2295952

Artman, H., & Wærn, Y. (1999). Distributed cognition in an emergency co-ordination centre. *Cognition Technology and Work*, *1*, 237–246. doi:10.1007/s101110050020

Arts, D. G., De Keizer, N. F., & Scheffer, G. J. (2002). Defining and improving data quality in medical registries. *Journal of the American Medical Informatics Association*, *9*(6), 600–611. doi:10.1197/jamia.M1087

Ashby, W. R. (1956). *An introduction to cybernetics*. London, UK: Methuen.

Atkinson, C. J., & Checkland, P. (1988). Extending the metaphor system. *Human Relations*, *41*, 709–725. doi:10.1177/001872678804101001

Atkinson, P., Coffey, A., & Delamont, S. (2003). *Key themes in qualitative research: Continuities and change*. Walnut Creek, CA: Alta Mira Press/Rowman and Littlefield.

Avgerou, C., Ciborra, C., & Land, F. F. (Eds.). (2004). *The social study of information and communications technology: Innovation, actors and context*. Oxford, UK: Oxford University Press.

Avison, D. E., & Fitzgerald, G. (1995). *Information systems development: Methodologies, techniques and tools*. London, UK: McGraw-Hill International UK limited.

Avison, D., Elliot, S., Krogstie, J., & Pries-Heje, J. (Eds.). (2006). The past and future of information systems: 1976-2006 and beyond. In *Proceedings of the IFIP 19th World Computer Congress*, (vol. 214). Berlin, Germany: Springer.

Avison, D., & Fitzgerald, G. (2003). *Information systems development: Methodologies, techniques and tools* (3rd ed.). Maidenhead, UK: McGraw Hill.

Ayres, I. (2007). *Super crunchers: How everything can be predicted*. London, UK: Murray.

Baba, M. (2006). Anthropology and business. In Birx, H. J. (Ed.), *Encyclopedia of Anthropology* (pp. 83–117). Thousand Oaks, CA: Sage.

Ballou, D. P., Wang, R. Y., Pazer, H., & Tayi, G. K. (1998). Modelling information manufacturing systems to determine information product quality. *Management Science*, *44*(4), 462–484. doi:10.1287/mnsc.44.4.462

Bannon, L. (1992). Perspectives on CSCW: From HCI and CMC to CSCW. In *Proceedings of the International Conference on Human-Computer Interaction (EW-HCI 1992)*, (pp. 148-158). St. Petersburg, Russia: EW-HCI.

Bannon, L. (1997). Constructing common information spaces. In *Proceedings of the Fifth European Conference on Computer Supported Cooperative Work*, (pp. 81-96). IEEE.

Bannon, L. J., & Bødker, S. (1997). Constructing common information spaces. In *Proceedings of ECSCW 2007, The 5th European CSCW Conference*. Lancaster, UK: CSCW.

Bannon, L. (1990). A pilgrim's progress: From cognitive science to cooperative design. *AI & Society*, *4*(4), 259–275. doi:10.1007/BF01894031

Bannon, L. (1993). CSCW: An initial exploration. *Scandinavian Journal of Information Systems*, *5*, 3–24.

Bannon, L. J., & Bødker, S. (1991). Designing interaction: Psychology at the human-computer nterface. In *Beyond the Interface: Encountering Artifacts in Use*. Cambridge, UK: Cambridge University Press.

Bannon, L., & Schmidt, K. (1991). CSCW: Four characters in search of a context. In Bowers, J., & Benford, S. (Eds.), *Studies in Computer Supported Cooperative Work: Theory, Practice and Design* (pp. 3–16). Amsterdam, The Netherlands: North-Holland.

Bardram, J. E. (1997). I love the system - I just don't use it. In *Proceedings of the ACM Conference on Group Work (GROUP 1997)*, (pp. 251-260). ACM Press.

Bardram, J. E. (1997). Plans as situated action: An activity theory approach to workflow systems. In *Proceedings of the Fifth European Conference on Computer Supported Cooperative Work*, (pp. 17–32). IEEE.

Bardram, J. E., & Bossen, C. (2005). A web of coordinative artifacts: Collaborative work at a hospital ward. In *Proceedings of GROUP*, *2005*, 168–176.

Barley, S. R., Dutton, W. H., Kiesler, S., Resnick, P., Kraut, R. E., & Yates, J. (2004). Does CSCW need organization theory? In *Proceedings of the 2004 ACM Conference on Computer Supported Cooperative Work (CSCW 2004)*, (pp. 122-124). New York, NY: ACM.

Baskerville, R. (2011). Design theorizing individual information systems. In *Proceedings of the Fifteenth Pacific-Asia Conference on Information Systems (PACIS 2011)* on *Quality Research in Pacific Asia*. PACIS.

Baskerville, R. (1999). Investigating information systems with action research. *Communications of AIS*, *2*, 19.

Baskerville, R. L., & Myers, M. D. (2002). Information systems as a reference discipline. *Management Information Systems Quarterly*, *26*(1), 1–14. doi:10.2307/4132338

Baskerville, R., & Myers, M. D. (2004). Making IS research relevant to practice – Foreword. *Management Information Systems Quarterly*, *28*(3), 329–335.

Baskerville, R., & Wood-Harper, A. T. (1998). Diversity in information systems action research methods. *European Journal of Information Systems*, *7*(2), 90–107. doi:10.1057/palgrave.ejis.3000298

Batini, D. P., Lenzerini, & Santucci. (2002). *La progettazione concettuale dei dati*. Rome, Italy: Franco Angeli.

Batini, C., Cabitza, F., Cappiello, C., & Francalanci, C. (2008). A comprehensive data quality methodology for web and structured data. *International Journal of Innovative Computing and Applications*, *1*(3), 205–218. doi:10.1504/IJICA.2008.019688

Batini, C., & Scannapieco, M. (2006). *Data quality: Concepts, methodologies and techniques*. Berlin, Germany: Springer.

Bauchspies, W. K. (2007). Science, ethnographic studies of. In G. Ritzer (Ed.), *Blackwell Encyclopedia of Sociology*. Malden, MA: Blackwell. Retrieved from http://www.blackwellreference.com/public/tocnode?id=g9781405124331_chunk_g978140512433125_ss1-40.

Baud, R. H., Rassinoux, A. M., & Scherrer, J. R. (1992). Natural language processing and semantical representation of medical texts. *Methods of Information in Medicine*, *31*, 117–125.

Becker, H. (1967). Whose side are we on? *Social Problems*, *14*, 239–247. doi:10.1525/sp.1967.14.3.03a00010

Benders, J., Batenburg, R., & van der Blonk, H. (2006). Sticking to standards: Technical and other isomorphic pressures in deploying ERP-systems. *Information & Management*, *43*(2), 194–203. doi:10.1016/j.im.2005.06.002

Beniger, J. R. (1986). *The control revolution: Technological and economic origins of the information society*. Cambridge, MA: Harvard University Press.

Bennett, M., Dennett, D. C., Hacker, P., Searle, J. R., & Robinson, D. (2007). *Neuroscience & philosophy: Brain, mind and language*. New York, NY: Columbia University Press.

Bentley, R., Hughes, J. A., Randall, D., Rodden, T., Sawyer, P., Shapiro, D., & Sommerville, I. (1992). Ethnographically-informed systems design for air traffic control. In *Proceedings of the 1992 ACM Conference on Computer-Supported Cooperative Work*, (pp. 123-129). Toronto, Canada: ACM.

Berg, A. M., & Eikeland, O. (Eds.). (2008). *Action research and organization theory*. Frankfurt, Germany: Peter Lang.

Berger, P. (1967). *The social construction of reality: A treatise in the sociology of knowledge*. London, UK: Penguin.

Berger, P., & Luckmann, T. (1966). *The social construction of reality: A treatise in the sociology of knowledge*. London, UK: Penguin.

Berg, M. (1998). Medical work and the computer-based patient record: A sociological perspective. *Methods of Information in Medicine*, *37*, 294–301.

Berg, M. (1998). The politics of technology: On bringing social theory into technological design. *Science, Technology & Human Values*, *23*(4), 456–490. doi:10.1177/016224399802300406

Berg, M. (1999). Accumulating and coordinating: Occasions for information technologies in medical work. *Computer Supported Cooperative Work*, *8*(4), 373–401. doi:10.1023/A:1008757115404

Berg, M., & Bowker, G. (1997). The multiple bodies of the medical record: Toward a sociology of an artefact. *The Sociological Quarterly*, *38*, 511–535. doi:10.1111/j.1533-8525.1997.tb00490.x

Berg, M., & Goorman, E. (1999). The contextual nature of medical information. *International Journal of Medical Informatics*, *56*, 51–60. doi:10.1016/S1386-5056(99)00041-6

Berg, M., & Timmermans, S. (2000). Orders and their others: On the constitution of universalities in medical work. *Configurations*, *8*(1), 31–61.

Berg, M., & Winthereik, B. (2003). Waiting for Godot: Episodes from the history of patient records. In Berg, M. (Ed.), *Health Information Management: Integrating Information and Communication Technology in Healthcare Work*. London, UK: Routledge.

Berlage, T., & Sohlenkamp, M. (1999). Visualizing common artefacts to support awareness in computer-mediated cooperation. *Computer Supported Cooperative Work*, *8*, 207–238. doi:10.1023/A:1008608425504

Bernstein, R. J. (1983). *Beyond objectivism and relativism: Science, hermeneutics and praxis*. Philadelphia, PA: University of Pennsylvania Press.

Bessant, J. (1983). Management and manufacturing innovation: The case of information technology. In Winch, G. (Ed.), *Information Technology in Manufacturing Processes* (pp. 14–30). London, UK: Rossendale.

Bessant, J., & Rush, H. (1995). Building bridges for innovation: The role of consultants in technology transfer. *Research Policy*, *24*, 97–114. doi:10.1016/0048-7333(93)00751-E

Beyer, H., & Holtzblatt, K. (1998). *Contextual design: Defining customer-centered systems*. San Francisco, CA: Morgan Kaufmann Publishing.

Bijker, B. (2001). Understanding technological culture through a constructivist view of science, technology and culture. In Cutcliffe, S., & Mitcham, C. (Eds.), *Visions of STS: Counterpoints in Science, Technology and Society Studies*. New York, NY: State University of New York.

Bijker, W. E., Hughes, T. P., & Pinch, T. (Eds.). (1987). *The social construction of technological systems*. Cambridge, MA: The MIT Press.

Binder, T., Loewgren, J., & Marlborg, L. (Eds.). (2008). *(Re-)searching the digital bauhaus*. Berlin, Germany: Springer.

Blomberg, J. (1995). Ethnography: Aligning field studies of work and system design. In Monk, A. F., & Gilbert, N. (Eds.), *Perspectives on HCL: Diverse Approaches* (pp. 175–197). London, UK: Academic Press Ltd.

Blomberg, J., Giacomi, J., Mosher, A., & Swenton-Wall, P. (1993). Ethnographic field methods and their relation to design. In Schuler, D., & Namioka, A. (Eds.), *Participatory Design: Principles and Practices* (pp. 123–157). Hillsdale, NJ: Erlbaum.

Blumer, H. (1969). *Symbolic interactionism: Perspective and method*. Englewood Cliffs, NJ: Prentice Hall.

Blythin, S., Rouncefield, M., & Hughes, J. A. (1997). Never mind the ethno stuff - What does all this mean and what do we do now? Ethnography in the commercial world. *Interaction*, *4*, 38–47. doi:10.1145/255392.255400

Boden, D. (1994). *The business of talk: Organizations in actions*. Cambridge, UK: Polity Press.

Bødker, K., Kensing, F., & Simonsen, G. (2004). *Participatory IT design*. Cambridge, MA: MIT Press.

Boland, R. (1999). Accounting as a representational craft lessons for research on information systems. In Currie, W., & Galliers, B. (Eds.), *Rethinking Management Information Systems* (pp. 229–243). Oxford, UK: Oxford University Press.

Boland, R. J. (1983). The in-formation of information systems. In Boland, R. J., & Hirschheim, R. A. (Eds.), *Critical Issues in Information Systems Research* (pp. 363–394). New York, NY: John Wiley & Sons.

Boland, R. J. (1985). Phenomenology: A preferred approach to research on information systems. In Mumford, E., Hirschheim, R. A., Fitgerald, G., & Wood-Harper, A. T. (Eds.), *Research Methods on Information Systems* (pp. 193–201). Amsterdam, The Netherlands: North Holland.

Boland, R. J. (1991). Information system use as a hermeneutic process. In Nissen, H. E., Klein, H. K., & Hirschheim, R. A. (Eds.), *Information Systems Research: Contemporary Approaches and Emergent Conditions* (pp. 439–464). Amsterdam, The Netherlands: North Holland.

Borgmann, A. (1992). *Crossing the postmodern divide*. Chicago, IL: The University of Chicago Press.

Borgmann, A. (1999). *Holding on to reality: The nature of information at the end of the millennium*. Chicago, IL: The University of Chicago Press.

Bortolotti, L. (2008). *An introduction to the philosophy of science*. Cambridge, UK: Polity.

Botta-Genoulaz, V., Millet, P. A., & Grabot, B. (2005). A survey on the recent literature on ERP systems. *Computers in Industry*, *56*, 510–522. doi:10.1016/j.compind.2005.02.004

Bourdieu, P. (1984). *Distinctions: A social critique of the judgment of taste* (Nice, R., Trans.). Boston, MA: Harvard University Press.

Bowers, J., & Benford, S. (Eds.). (1991). *Studies in CSCW: Theory, practice and design*. Amsterdam, The Netherlands: North Holland.

Bowker, G. C., & Star, S. L. (1999). *Sorting things out: Classification and its consequences*. Cambridge, MA: MIT Press.

Brady, T., Tierney, M., & Williams, R. (1992). The commodification of industry-application software. *Industrial and Corporate Change*, *1*(3), 489–514. doi:10.1093/icc/1.3.489

Bragato, L., & Jacobs, K. (2003). Care pathways: The road to better health services? *Journal of Health Organization and Management*, *17*(3), 164–180. doi:10.1108/14777260310480721

Branscomb, L. M., & Kahin, B. (1996). Standards processes and objectives for the national information infrastructure. In Branscomb, L. M., & Keller, J. (Eds.), *Converging Infrastructures: Intelligent Transportation Systems and the National Information Infrastructure* (pp. 3–31). Cambridge, MA: MIT Press.

Braudel, F. (1949). *The Mediterranean and the Mediterranean world in the Age of Philip II* (*Vol. 1*). London, UK: Collins.

Brehm, L., Heinzl, A., & Markus, M. L. (2001). Tailoring ERP systems: A spectrum of choices and their implications. In *Proceedings of the 34th Annual Hawaii International Conference of System Sciences (HICSS-34)*, (vol 8), (p. 8017). IEEE.

Briggs, C. L. (1986). *Learning how to ask*. Cambridge, UK: Cambridge University Press.

Brighenti, A., Giglioli, P. P., Ravaioli, P., Turina, I., & Venturini, T. (2008). Per un'etnografia dei moderni: Intervista a Bruno Latour. *Etnografia e Ricerca Qualitativa*, *3*, 347–367.

Bringay, S., Barry, C., & Charlet, J. (2006). Annotations: A functionality to support cooperation, coordination and awareness in the electronic medical record. In *Proceedings of the 7th International Conference on the Design of Cooperative Systems*. COOP.

Brook, T. V. (1995, May 7th). UW considers guaranteeing 4-year finish. *The Milwaukee Journal*.

Brown, J. S., & Duguid, P. (1991). Organizational learning and communities–of-practice: Toward a unified view of working, learning, and innovation. *Organization Science*, *2*(1), 40–57. doi:10.1287/orsc.2.1.40

Brown, J. S., & Duguid, P. (1994). Borderline issues: Social and material aspects of design. *Human-Computer Interaction*, *9*(1), 3–36. doi:10.1207/s15327051hci0901_2

Brun-Cottan, F., & Wall, P. (1995). Using video to represent the user. *Communications of the ACM*, *38*(5), 61–71. doi:10.1145/203356.203368

Bruni, A. (2003). *Lo studio etnografico delle organizzazioni*. Roma, Italy: Carocci.

Bruni, A., & Gherardi, S. (2007). *Studiare le pratiche lavorative*. Bologna, Italy: Il Mulino.

Bucciarelli, L. L. (1988). An ethnographic perspective on engineering design. *Design Studies*, *9*(3), 159–168. doi:10.1016/0142-694X(88)90045-2

Buckingham, R. A., Hirschheim, R. A., Land, F. F., & Tully, C. J. (1987). Information systems curriculum: A basis for course design. In Buckingham, R. A., Hirschheim, R. A., Land, F. F., & Tully, C. J. (Eds.), *Information Systems Education: Reccomendations and Implementation*. Cambridge, UK: Cambridge University Press.

Bunduchi, R., Graham, I., Smart, A., & Williams, R. (2005). The tensions shaping the emergence of standard bodies: The case of a national health informatics standard standards body. *Prometheus*, *23*(2), 149–166. doi:10.1080/08109020500085544

Bunduchi, R., Williams, R., Graham, I., & Smart, A. (2006). XML based clinical data standardisation in the national health service Scotland. *Informatics in Primary Care*, *14*(4), 227–234.

Burrell, G., & Morgan, G. (1979). *Sociological paradigms and organizational analysis*. London, UK: Heinemann Educational.

Büscher, M., Christensen, M., Hansen, K., Mogensen, P., & Shapiro, D. (2009). Bottom-up, top-down? Connecting software architecture design with use. In Voss, A., Hartswood, M., Procter, R., Rouncefield, M., Slack, R. S., & Büscher, M. (Eds.), *Configuring User-Designer Relations: Interdisciplinary Perspectives*. London, UK: Springer. doi:10.1007/978-1-84628-925-5_8

Büscher, M., Gill, S., Mogensen, P., & Shapiro, D. (2001). Landscapes of practice - Bricolage as a method for situated design. *Computer Supported Cooperative Work*, *10*, 1–28. doi:10.1023/A:1011293210539

Button, G., & Dourish, P. (1996). Technomethodology: Paradoxes and possibilities. In *Proceedings of CHI 96*, (pp. 19-26). Vancouver, Canada: ACM.

Button, G. (1993). *Technology in working order: Studies of work, interaction, and technology*. New York, NY: Routledge.

Button, G. (1993). The curious case of the vanishing technology. In Button, G. (Ed.), *Technology in Working Order: Studies in Work, Interaction, and Technology* (pp. 10–28). London, UK: Routledge.

Button, G. (2000). The ethnographic tradition and design. *Design Studies*, *21*, 319–332. doi:10.1016/S0142-694X(00)00005-3

Button, G. (Ed.). (1992). *Technology in working order: Studies of work, interaction and technology*. London, UK: Routledge.

Button, G., Coulter, J., Lee, J. R. E., & Sharrock, W. (1995). *Computers, minds and conduct*. Cambridge, UK: Polity.

Button, G., & Harper, R. (1995). The relevance of 'work-practice' for design. *Computer Supported Cooperative Work*, *4*, 263–280. doi:10.1007/BF01846695

Button, G., & Sharrock, W. (1998). The organizational accountability of technological work. *Social Studies of Science*, *28*(1), 75–102. doi:10.1177/030631298028001003

Byrd, T. A. (1992). A synthesis of research on requirements analysis and knowledge acquisition techniques. *Management Information Systems Quarterly*, *16*(1), 117–138. doi:10.2307/249704

Cabitza, F., & Simone, C. (2008). Supporting practices of positive redundancy for seamless care. In *Proceedings of the 21th IEEE International Symposium on Computer-Based Medical Systems*. Jyväskylä, Finland: IEEE Press.

Cabitza, F., & Simone, C. (2009). Active artifacts as bridges between context and community knowledge sources. In *Proceedings of the 4th International Conference on Communities and Technologies*. University Park, PA: ACM Press.

Cabitza, F., Sarini, M., Simone, C., & Telaro, M. (2005). When once is not enough: The role of redundancy in a hospital ward setting. In *Proceedings of GROUP 2005*. ACM.

Cabitza, F., Simone, C., & Zorzato, G. (2009). PRODOC: An electronic patient record to foster process-oriented practices. In *Proceedings of the 11th European Conference on Computer Supported Cooperative Work*. Vienna, Austria: Springer.

Cabitza, F., & Gesso, I. (2011). Web of active documents: An architecture for flexible electronic patient records. In Fred, A., Filipe, J., & Gamboa, H. (Eds.), *Communications in Computer and Information Science* (*Vol. 147*). Berlin, Germany: Springer. doi:10.1007/978-3-642-18472-7_4

Cabitza, F., Sarini, M., Simone, C., & Telaro, M. (2006). Torres: A conceptual framework for articulation work across boundaries. In *Frontiers in Artificial Intelligence and Applications* (pp. 102–117). Dordrecht, The Netherlands: IOS Press.

Cabitza, F., Simone, C., & Sarini, M. (2009). Leveraging coordinative conventions to promote collaboration awareness. *Computer Supported Cooperative Work*, *18*(4), 301–330. doi:10.1007/s10606-009-9093-z

Cabitza, F., & Simone, F. (2009). WOAD: A framework to enable the end-user development of coordination oriented functionalities. *Journal of Organizational and End User Computing*, *22*(1), 1546–5012.

Callon, M., & Latour, B. (1981). Unscrewing the big leviathan: How actors macrostructure reality and how sociologists help them to do so. In Knorr-Cetina, K. D., & Cicourel, A. V. (Eds.), *Advances in Social Theory and Methodology: Toward an Integration of Micro- and Macro-Sociologies*. Boston, MA: Routledge and Kegan Paul.

Cangelosi, V. E., & Dill, W. R. (1965). Organizational learning: Observations toward a theory. *Administrative Science Quarterly*, *10*(2), 175–203. doi:10.2307/2391412

Carell, A., & Herrmann, T. (2010). Interaction and collaboration modes for integrating inspiring information into technology-enhanced creativity workshops. In *Proceedings of the 2010 43rd Hawaii International Conference on System Sciences,* (pp. 1-10). Washington, DC: IEEE Computer Society.

Carr, N. (2008). *The big switch: Rewiring the world, from Edison to Google*. New York, NY: Norton.

Carstensen, P. (1996). *Computer supported coordination*. Roskilde, Denmark: Risø National Laboratory.

Cassirer, E. (1955). *The philosophy of symbolic forms*. New Haven, CT: Yale University Press.

Chae, B., & Poole, M. S. (2005). The surface of emergence in systems development: Agency, institutions, and large-scale information systems. *European Journal of Information Systems*, *14*(1), 19–36. doi:10.1057/palgrave.ejis.3000519

Chapoulie, J.-M. (1987). Everett C. Hughes and the development of fieldwork in sociology. *Urban Life*, *15*(3-4), 259–298.

Chapoulie, J.-M. (1996). Everett Hughes and the Chicago tradition. *Sociological Theory*, *14*(1), 3–29. doi:10.2307/202150

Checkland, P. (1981). *Systems thinking, systems practice*. Chichester, UK: John Wiley & Sons.

Checkland, P. (2000). Soft systems methodology: A thirty year retrospective. *Systems Research and Behavioral Science*, *17*(1), 11. doi:10.1002/1099-1743(200011)17:1+<::AID-SRES374>3.0.CO;2-O

Checkland, P., & Scholes, J. (1990). *Soft systems methodology in action*. Chichester, UK: John Wiley & Sons.

Chengalur-Smith, I., Ballou, D. P., & Pazer, H. L. (1999). The impact of data quality information on decision making: An exploratory analysis. *IEEE Transactions on Knowledge and Data Engineering*, *11*(6), 853–864. doi:10.1109/69.824597

Chen, W., & Hirschheim, R. (2004). A paradigmatic and methodological examination of information systems research from 1991 to 2001. *Information Systems Journal*, *14*, 197–235. doi:10.1111/j.1365-2575.2004.00173.x

Ciborra, C. (2004). *Getting to the heart of the situation: The phenomenological roots of situatedness*. Retrieved from http://projectsfinal.interactionivrea.org/2004-2005/SYMPOSIUM%202005/communication%20material/Getting%20to%20the%20Heart%20of%20the%20Situation_C_Ciborra.pdf.

Ciborra, C. (2000). *From control to drift: The dynamics of corporate information infrastructures*. Oxford, UK: Oxford University Press.

Ciborra, C. (2002). *The labyrinths of information - Challenging the wisdom of systems*. Oxford, UK: Oxford University Press.

Ciborra, C. (2006). The mind or the heart? It depends on the (definition of) situation. *Journal of Information Technology*, *21*(3), 129–139. doi:10.1057/palgrave.jit.2000062

Ciborra, C. U. (2000). *From control to drift: The dynamics of corporate information infrastructures*. Oxford, UK: Oxford University Press.

Ciborra, C. U. (Ed.). (1996). *Groupware and teamwork: Invisible aid or technical hindrance*. New York, NY: John Wiley and Sons.

Ciborra, C. U., Braa, K., Cordella, A., Dahlbom, B., Failla, A., Hanseth, O., & Simon, K. A. (2001). *From control to drift: The dynamics of corporate information infrastructures*. Oxford, UK: Oxford University Press.

Ciborra, C., & Hanseth, O. (1998). From tool to gestell - Agendas for managing the information infrastructure. *Information Technology & People*, *11*(4), 305–327. doi:10.1108/09593849810246129

Clarke, A. E. (1991). Social worlds /arenas theory as organizational theory. In Maines, D. (Ed.), *Social Organization and Social Process: Essays in Honor of Anselm Strauss* (pp. 119–158). New York, NY: Aldine.

Clarke, A. E. (2005). *Situational analysis: Grounded theory after the postmodern turn*. Thousand Oaks, CA: SAGE.

Clarke, A. E. (2006). Social worlds. In Ritzer, G. (Ed.), *The Blackwell Encyclopedia of Sociology* (pp. 4547–4549). Malden, MA: Blackwell.

Clarke, A. E., & Star, S. L. (2007). The social worlds/arenas/discourse framework as a theory-methods package. In Hackett, E. (Eds.), *Handbook of Science and Technology Studies* (pp. 113–137). Cambridge, MA: MIT Press.

Clarke, K., Hartswood, M., Procter, R., Rouncefield, M., & Slack, R. (2003). Trusting the record. *Methods of Information in Medicine*, *42*(4), 345–352.

Clark, P., & Staunton, N. (1989). *Innovation in technology and organisation*. London, UK: Routledge.

Clausen, C., & Williams, R. (Eds.). (1997). The social shaping of computer-aided production management and computer integrated manufacture. In *Proceedings of International Conference, COST A4*. Luxembourg: European Commission DGXIII.

Clausen, C., & Koch, C. (1999). The role of spaces and occasions in the transformation of information technologies: Lessons from the social shaping of IT systems for manufacturing in a Danish context. *Technology Analysis and Strategic Management*, *11*(3), 463–482. doi:10.1080/095373299107456

Coghlan, D., & Rashford, N. S. (2006). *Organizational change and strategy: An interlevel dynamics approach*. New York, NY: Routledge.

Coiera, E. (1997). *Guide to medical informatics: The internet and telemedicine*. London, UK: Chapman and Hall Medical.

Cole, I. (1982). Human aspects of office filing: Implications for the electronic office. In *Proceedings of the Human Factors Society 26th Annual Meeting*, (pp. 59-63). Santa Monica, CA: ACM.

Collins, P., Shukla, S., & Redmiles, D. (2002). Activity theory and system design: A view from the trenches. *Computer Supported Cooperative Work: The Journal of Collaborative Computing*, *11*(1-2), 55–80. doi:10.1023/A:1015219918601

Conklin, J. (1989). Design rationale and maintainability. In B. Shriver (Ed.), *Proceedings of 22nd Annual Hawaii International Conference on System Sciences,* (Vol. 2), (pp. 533-539). IEEE Computer Society.

Conklin, J., & Begeman, M. L. (1988). gIBIS: A hypertext tool for exploratory policy discussion. In *Proceedings of the Conference on Computer-Supported Cooperative Work*, (pp. 140-152). Portland, OR: ACM Press.

Conklin, J. (1993). Capturing organizational knowledge. In *Readings in Groupware and Computer-Supported Cooperative Work: Assisting Human-Human Collaboration* (pp. 561–565). San Mateo, CA: Morgan Kaufmann Publishers.

Cooke, D. P., Gelman, L., & Peterson, W. J. (2001). *ERP trends*. Report No. R-1292-01-RR. New York, NY: The Conference Board.

Corbin, J., & Strauss, A. L. (1993). The articulation of work through interaction. *Qualitative Sociology*, *34*, 71–83.

Crabtree, A. (1998). Ethnography in participatory design. In *Proceedings of the 1998 Participatory Design Conference*, (pp. 93-105). Seattle, WA. Retrieved from http://www.mrl.nott.ac.uk/~axc/documents/papers/PDC98.pdf.

Crabtree, A. (2001). *Wild sociology: Ethnography and design*. Ph.D. Thesis. Lancaster, UK: Lancaster University.

Crabtree, A. (2003). *Designing collaborative systems: A practical guide to ethnography*. London, UK: Springer.

Crabtree, A. (2004). Taking technomethodology seriously: Hybrid change in the ethnomethodology-design relationship. *European Journal of Information Systems*, *13*(3), 195–209. doi:10.1057/palgrave.ejis.3000500

Crabtree, A., Nichols, D. M., O'Brien, J., Rouncefield, M., & Twidale, M. B. (2000). Ethnomethodologically informed ethnography and information system design. *Journal of the American Society for Information Science American Society for Information Science*, *51*(7), 666–682. doi:10.1002/(SICI)1097-4571(2000)51:7<666::AID-ASI8>3.0.CO;2-5

Cross, N. (2000). *Engineering design methods - Strategies for product design*. Chichester, UK: John Wiley & Sons.

Crotty, M. (1998). *The foundations of social research - Meaning and perspective in the research process*. London, UK: Sage.

Currie, W. L., & Guah, M. W. (2007). Conflicting institutional logics: A national programme for IT in the organisational field of healthcare. *Journal of Information Technology*, *22*(3), 235–247. doi:10.1057/palgrave.jit.2000102

Czarniawska, B. (1998). *A narrative approach to organization studies*. London, UK: Sage.

Dahlbom, B., & Mathiassen, L. (1993). *Computers in context: The philosophy and practice of systems design*. Cambridge, MA: Blackwell Publishers.

Damasio, A. R. (1995). *Descartes' error: Emotion, reason, and the human brain*. London, UK: Harper.

Davenport, E. (2008). Social informatics and sociotechnical research: A view from the UK. *Journal of Information Science*, *34*(4), 519–530. doi:10.1177/0165551508091011

Davenport, T. H. (1998). Putting the enterprise into the enterprise system. *Harvard Business Review*, *76*(4), 121–131.

Davidson, D. (2001). *Inquiries into truth and interpretation* (2nd ed.). Oxford, UK: Clarendon Press. doi:10.1093/0199246297.001.0001

Davis, M. W. (1971). That's interesting. *Philosophy of the Social Sciences*, *1*(4), 309–344. doi:10.1177/004839317100100211

De Marco, M. (2005). A portrait of a scientist. *European Journal of Information Systems*, *15*(5), 442–445. doi:10.1057/palgrave.ejis.3000553

De Michelis, G., Dubois, E., Jarke, M., Matthes, F., Mylopoulos, J., & Papazoglou, M. … Yu, E. (1998). Cooperative information systems: A manifesto. In M. P. Papazoglou & G. Schlageter (Eds.), *Cooperative Information Systems: Trends & Directions*, (pp. 315-363). New York, NY: Academic-Press.

De Michelis, G. (1998). *Aperto, molteplice, continuo*. Milano, Italy: Dunod Italia.

De Michelis, G. (2008). The phenomenological stance of the designer. In Binder, (Eds.), *Re-)Searching the Digital Bauhaus* (pp. 145–162). Berlin, Germany: Springer.

De Michelis, G., Dubois, E., Jarke, M., Matthes, F., Mylopoulos, J., & Schmidt, J. (1998). A three-faceted view of information systems: The challenge of change. *Communications of the ACM*, *41*(12), 64–70. doi:10.1145/290133.290150

De Michelis, G., Loregian, M., & Moderini, C. (2009). Itsme: Interaction design innovating workstations. *Knowledge, Technology & Policy*, *22*, 71–78. doi:10.1007/s12130-009-9069-9

Deetz, S. (1996). Describing differences in approaches to organization science: Rethinking Burrell and Morgan and their legacy. *Organization Science*, *7*(2), 191–207. doi:10.1287/orsc.7.2.191

Deleuze, G. (1969a). *Difference et repetition*. Paris, France: Presses Universitaires de France.

Deleuze, G. (1969b). *Logique du sens*. Paris, France: Editions de Minuit.

DeLone, W. H., & McLean, E. R. (1992). Information systems success: The quest for the dependent variable. *Information Systems Research*, *3*(1), 60–95. doi:10.1287/isre.3.1.60

Detienne, M., & Vernant, J.-P. (1974). *Les ruses de l'intelligence: La métis des Grecs*. Paris, France: Flammarion.

Dick, B. (2004). Action research literature: Themes and trends. *Action Research*, *2*(4), 425–444. doi:10.1177/1476750304047985

Dick, B. (2006). Action research literature 2004-2006: Themes and trends. *Action Research, 4*(4), 439–458. doi:10.1177/1476750306070105

Dick, B. (2009). Action research literature 2006-2008: Themes and trends. *Action Research, 7*(4), 423–441. doi:10.1177/1476750309350701

Dick, R. S., & Steen, E. B. (Eds.). (1991). *The computer-based patient record: An essential technology for health care*. Washington, DC: National Academy Press.

Dieng, R., Giboin, A., & Karsenty, L. (Eds.). (2000). Designing cooperative systems: The use of theories and models. In *Proceedings of the 5th International Conference on the Design of Cooperative Systems (COOP 2000)*. Amsterdam, The Netherlands: IOS Press.

Diggins, T., & Tolmie, P. (2003). The 'adequate' design of ethnographic outputs for practice: Some explorations of the characteristics of design resources. *Personal and Ubiquitous Computing, 7*, 147–158. doi:10.1007/s00779-003-0226-y

Dillon, A. (1998). Cultural analysis and what designers need to know: A case of sometimes too much, sometimes too little, and always too late. *ACM Journal of Computer Documentation, 22*(1), 13–17.

DiMaggio, P. J., & Powell, W. W. (1983). The iron cage revisited: Institutional isomorphism and collective rationality in organisational fields. *American Sociological Review, 48*(2), 147–160. doi:10.2307/2095101

Divitini, M., Simone, C., & Schmidt, K. (1996). ABACO: Coordination mechanisms in a multi-agent perspective. In *Proceedings of the Second International Conference on the Design of Cooperative Systems*, (pp. 103-122). Antibes-Juan- les-Pins, France: INRIA Sophia Antipolis.

Doolin, B. (2002). To reveal is to critique: Actor–network theory and critical information systems research. *Journal of Information Technology, 17*, 69–78. doi:10.1080/02683960210145986

Dourish, P. (2001). *Where the action is: The foundation of embodied interaction*. Cambridge, MA: MIT Press.

Dourish, P., & Button, G. (1998). On "technomethodology": Foundational relationships between ethnomethodology and system design. *Human-Computer Interaction, 13*, 395–432. doi:10.1207/s15327051hci1304_2

Downey, G. L. (1998). *The machine in me: An anthropologist sits among computer engineers*. New York, NY: Routledge.

Dreyfus, H. (1992). *What computers still can't do: A critique of artifical reason*. Cambridge, MA: MIT Press.

Dreyfus, H. (2001). *On the internet*. London, UK: Routledge.

Dreyfus, H. L. (1979). *What computers can't do: The limits of artificial intelligence*. San Francisco, CA: Harper & Row.

Dreyfus, H. L. (1991). *Being-in-the-world*. Cambridge, MA: The MIT Press.

Dreyfus, H. L. (2003). Being and power revisited. In Milchman, A., & Rosenberg, A. (Eds.), *Foucault and Heidegger: Critical Enconters*. Minneapolis, MN: University of Minnesota Press.

Dreyfus, H., & Dreyfus, S. (1986). *Mind over machine*. New York, NY: Free Press.

Eason, K. E. (2007). Local sociotechnical system development in the NHS national programme for information technology. *Journal of Information Technology, 22*(3), 257–264. doi:10.1057/palgrave.jit.2000101

Edwards, P. N., Bowker, G. C., Jackson, S. J., & Williams, R. (2009). Introduction: An agenda for infrastructure studies. *Journal of the Association for Information Systems, 10*(5), 155–173. Retrieved from http://aisel.aisnet.org/jais/vol10/iss5/

Edwards, P. N., Jackson, S. J., Bowker, G. C., & Knobel, C. P. (2007). *Understanding infrastructure: Dynamics, tensions, and design*. Ann Arbor, MI: Deep Blue.

Ehn, P. (2006). *Invited talk*. Retrieved from http://www.mcis2006.org.

Ehn, P. (1990). *Work-oriented design of computer artifacts*. Hillsdale, NJ: Lawrence Erlbaum Associates.

Ehn, P. (1993). Scandinavian deign: On participation and skill. In Schuler, D., & Namioka, A. (Eds.), *Participatory Design: Principles and Practices* (pp. 41–78). Hillsdale, NJ: Lawrence Erlbaum Associates.

Einstein, A. (1954). *Relativity*. London, UK: Routledge.

Elias, N. (1978). *What is sociology?*London, UK: Hutchinson & Co.

Elias, N. (1991). *The society of individuals*. Oxford, UK: Basil Blackwell.

Elias, N. (2000). *The civilizing process*. Oxford, UK: Blackwell.

Elias, N. (2007). *The genesis of the naval profession*. Dublin, Ireland: University College Dublin Press.

Eliot, T. S. (1969). Four quartets: Little gidding. In *The Complete Poems & Plays*. London, UK: Faber & Faber.

Ellingsen, G., & Monteiro, E. (2003). A patchwork planet integration and cooperation in hospitals. *Computer Supported Cooperative Work*, *12*(1), 71–95. doi:10.1023/A:1022469522932

Emery, F., & Trist, E. (1960). Socio-technical systems. In C. W. Churchman & M. Verhulst (Eds), *Management Sciences Models and Techniques: Proceedings of the 6th International Meeting of the Institute of Management Science,* (pp. 83-97). New York, NY: Pergamon Press.

Engeström, Y. (1990). When is a tool? Multiple meanings of artifacts in human activity. In *Learning, Working and Imagining: Twelve Studies in Activity Theory* (pp. 171–195). Helsinki, Finland: Orienta-Konsultit Oy.

Engeström, Y. (1999). Innovative learning in work teams: Analyzing cycles of knowledge creation in practice. In *Perspectives on Activity Theory* (pp. 377–404). Cambridge, UK: Cambridge University Press.

Erickson, T., Huang, W., Danis, C., & Kellogg, W. A. (2004). A social proxy for distributed tasks: Design and evaluation of a prototype. In *Proceedings of the CHI 2004*, (pp. 559-566). ACM Press.

Erickson, F. (1988). Ethnographic description. In Ammon, U. (Ed.), *Sociolinguistics: An International Handbook of the Science of Language and Society* (pp. 1081–1095). Berlin, Germany: De Gruyter.

Ericsson, K. A., & Simon, H. A. (1980). Verbal reports as data. *Psychological Review*, *87*(3), 215–251. doi:10.1037/0033-295X.87.3.215

Eriksen, S. (2002). Designing for accountability. In O. W. Bertelsen, S. Bødker, & K. Kutti (Eds.), *Nordichi 2002 – Proceedings of the Second Nordic Conference on Human-Computer Interaction,* (pp. 177-186). Aarhus, Denmark: Nordichi.

ERP Research Group. (2006). ERP systems and related issues bibliography: 600 articles. *Victoria University*. Retrieved from http://www.business.vu.edu.au/sap/Research.html#ERP_Systems_and_Related_Issues_Bibliogr.

Esposito, E. (2004). The arts of contingency. *Critical Inquiry*, *31*, n.d. Retrieved from http://criticalinquiry.uchicago.edu/features/artsstatements/arts.esposito.htm-doi:10.1086/427300

Esposito, R. (1998). *Communitas: Origine e destino della comunità*. Torino, Italy: Einaudi.

Esteves, J., & Pastor, J. (2001). Enterprise resource planning systems research: An annotated bibliography. *Communications of AIS*, *7*(8), 1–52.

Evans, R. J. (1997). *In defence of history*. London, UK: Granta.

Evan, W. F. (1964). On the margin: The engineering technician. In Berger, P. L. (Ed.), *The Human Shape of Work: Studies in the Sociology of Occupations* (pp. 83–112). New York, NY: MacMillan.

Fabbri, T. M., & Neri, M. (2004). l'organizzazione dei sistemi informatici nell'università. In *Quaderni del Dipartimento di Scienze Sociali, Cognitive e Quantitative.* Reggio Emilia, Italy: University of Modena and Reggio Emilia, Italy.

Fabbri, T. M. (2003). *L'apprendimento organizzativo – Teoria e progettazione*. Roma, Italy: Carocci.

Fabbri, T. M. (2004). L'apprendimento organizzativo – Un esperimento di progettazione. In D'atri, A. (Ed.), *Innovazione Organizzative e Tecnologie Innovative* (pp. 113–121). Milano, Italy: Etas.

Faraj, S. (2004). Contested artifact: Technology sensemaking, actor networks, and the shaping of the web browser. *Information Technology & People*, *17*(2), 186–209. doi:10.1108/09593840410542501

Feldman, M. S., & Orlikowski, W. J. (2011). Theorizing practice and practicing theory. *Organization Science*, *1100*(0612).

Fele, G. (2005). *Interaction and collaborative work in a medical emergency dispatch centre*. Paper presented at the 100th Meeting of the American Sociological Association. Philadelphia, PA.

Fele, G. (2008). In praise of thin descriptions: Ethnography, ethnomethodology and the study of practice. In *Proceedings of ALPIS 2008 – Alpine Ski Seminar on Information Systems,* (pp. 51-58). Carisolo, Italy: University of Trento Press.

Fenwick, T. (2005). Conception of critical HROD: Dilemmas for theory and practice. *Human Resource Development International, 8*(2), 225–238. doi:10.1080/13678860500100541

Ferraris, M. (2005). *Dove sei? Ontologia del telefonino.* Milano, Italy: Bompiani.

Ferraris, M. (2007). *La fidanzata automatica.* Milano, Italy: Bompiani.

Finken, S. (2000). Bringing ethnography home: Reflections upon a style of enquiry. In *Proceedings of IRIS,* (vol 23), (pp. 267-277). IRIS.

Fischer, G., Giaccardi, E., Ye, Y., Sutcliffe, A. G., & Mehandjiev, N. (2004). Meta-design: A manifesto for end-user development. *Communications of the ACM, 47,* 33–37. doi:10.1145/1015864.1015884

Fitzpatrick, G. (2000). Understanding the paper health record in practice: Implications for EHRs. In *Proceedings of Health Informatics Conference (HIC 2000)*. Adelaide.

Fitzpatrick, G. (2004). Integrated care and the working record. *Health Informatics Journal, 10*(4), 291–302. doi:10.1177/1460458204048507

Fitzpatrick, G. (Ed.). (2003). *The locales framework: Understanding and designing for wicked problems*. Dordrecht, The Netherlands: Kluwer Academic Publishers.

Fjuk, A., Nurminen, M. I., & Smørdal, O. (1997). *Taking articulation work seriously: An activity theoretical approach. TUCS Technical Report, N. 120, August*. Turku, Turkey: Turku Centre for Computer Science.

Flachbart, G., & Weibel, P. (Eds.). (2005). *Disappearing architecture*. Basel, Switzerland: Birkhauser. doi:10.1007/3-7643-7674-0

Fleck, J. (1988). Innofusion or diffusation? *The nature of technological development in robotics*. Working Paper. Edinburgh, UK: The University of Edinburgh.

Fleck, J., Webster, J., & Williams, R. (1990). The dynamics of IT implementation: A reassessment of paradigms and trajectories of development. *Futures, 22,* 618–640. doi:10.1016/0016-3287(90)90131-Z

Floridi, L. (2011). Semantic conceptions of information. In Zaltz, E. N. (Ed.), *The Stanford Encyclopedia of Philosophy*. Palo Alto, CA: Stanford University Press.

Floyd, C. (1987). Outline of a paradigm change in software engineering. In G. Bjerknes, P. Ehn, & M. Kyng (eds.), *Computers and Democracy: A Scandinavian Challenge*, (pp. 191-208). Avebury, UK: Aldershot.

Flusser, V. (2000). *Towards a philosophy of photography*. London, UK: Reaktion Books.

Forsythe, D. E. (1999). It's just a matter of common sense: Ethnography as invisible work. *Computer Supported Cooperative Work, 8*(1), 127–145. doi:10.1023/A:1008692231284

Foucault, M. (1966). *Les mots et les choses*. Paris, France: Gallimard.

Foucault, M. (1970). *The order of things*. London, UK: Tavistock.

Foucault, M. (1971). Nietzsche, la généalogie, l'histoire. In Bachelard, S. (Eds.), *Hommage à J. Hyppolite*. Paris, France: Presses Universitaires de France.

Foucault, M. (1972). *The archaeology of knowledge*. London, UK: Tavistock.

Foucault, M. (1988). Technologies of the self. In Martin, L. H., Gutman, G., & Huttton, P. H. (Eds.), *Technologies of the Self.* London, UK: Tavistock.

Frantz, Pollyanne S., Arthur R. Southerland, and James T. Johnson. 2002. "ERP Software: Implementation Best-Practices." *EDUCAUSE Quarterly November 4*:38-45.

French, W., & Bell, C. (1999). *Organization development: Behavioural science interventions for organization improvement*. Upper Saddle River, NJ: Prentice Hall.

Fujimura, J. (1992). Crafting science: Standardized packages, boundary objects, and "translation". In Pickering, A. (Ed.), *Science and Culture and Practice* (pp. 168–211). Chicago, IL: University of Chicago Press.

Gadamer, H.-G. (1988). *Truth and method*. New York, NY: Crossroad.

Gamst, F. C. (1980). *The hoghead: An industrial ethnology of the locomotive engineer*. New York, NY: Holt, Rinehart & Winston.

Gantz, J. F., et al. (2008). *The diverse and exploding digital universe: A forecast of worldwide information growth through 2011*. IDC White Paper. Retrieved from www.emc.com/digital_universe.

Gap Analysis. (2006). *Oxford reference online*. Retrieved from http://www.oxfordreference.com.

Gardner, S. (1999). *Routledge philosophy guidebook to Kant and the critique of pure reason*. London, UK: Routledge.

Garfinkel, H. (1963). A conception of, and experiments with, 'trust' as a condition of stable concerted actions. In Harvey, O. J. (Ed.), *Motivation and Social Interaction* (pp. 187–238). New York, NY: The Ronald Press.

Garfinkel, H. (1967). *Studies in ethnomethodology*. Engelwood Cliffs, NJ: Prentice Hall.

Garfinkel, H. (1986). Introduction. In *Ethnomethodological Studies of Work* (pp. vii–viii). London, UK: Routledge.

Garfinkel, H. (2002). *Ethnomethodology's Program: Working out Durkheim's aphorism*. Lanham, MD: Roman & Littlefield.

Garfinkel, H. (2006). *Seeing sociologically: The routine grounds of social action*. Boulder, CO: Paradigm.

Garfinkel, H., & Rawls, A. W. (2008). *Toward a sociological theory of information*. Boulder, CO: Paradigm Publishing.

Gartner, J., & Wagner, I. (1996). Mapping actors and agendas: Political frameworks of system design and participation. *Human-Computer Interaction, 11*(3), 187–214. doi:10.1207/s15327051hci1103_1

Gasser, L. (1986). The integration of computing and routine work. *ACM Transactions on Office Information Systems, 4*, 257–270. doi:10.1145/214427.214429

Geertz, C. (1983). From the native's point of view. In *Local knowledge* (pp. 55–72). New York, NY: Basic Books.

Gendlin, E. T. (1978). Benfindlichkeit: Heidegger and the philosophy of psychology. *Review of Existential Psychology & Psychiatry, 14*(1-3), 43–71.

Gendron, M., & D'Onofrio, M. (2001). Data quality in the healthcare industry. *Data Quality, 7*(1).

Gerson, E. M., & Star, S. L. (1986). Analyzing due process in the work place. *ACM Transactions on Office Information Systems, 4*(3), 257–270. doi:10.1145/214427.214431

Gerson, E., & Star, S. L. (1986). Analyzing due process in the workplace. *ACM Transactions on Office Information Systems, 4*, 70–78. doi:10.1145/214427.214431

Gerst, M. (2006). *The socio-technical dynamics of ICT innovation: A social shaping analysis of portals*. Unpublished PhD Thesis. Edinburgh, UK: University of Edinburgh.

Gherardi, S. (2006). *Organizational knowledge: The texture of workplace learning*. Oxford, UK: Blackwell.

Gherardi, S. (2008). Dalla comunità di pratica alle pratiche della comunità: Breve storia di un concetto in viaggio. *Studi Organizzativi, 10*(1), 49–72.

Gibson, J. J. (1979). *The ecological approach to visual perception*. Boston, MA: Houghton-Mifflin.

Giddens, A. (1984). *The constitution of society: Outline of the theory of structuration*. Cambridge, MA: Polity Press.

Gilat, D. (2005). Autonomic computing-building self-managing computing systems. In Flachbart, G., & Weibel, P. (Eds.), *Disappearing Architecture*. Basel, Switzerland: Birkhauser.

Goguen, J. (1996). Formality and informality in requirements engineering. In *Proceedings of the 2nd International Conference on Requirements Engineering*. Washington, DC: IEEE Computer Society.

Goguen, J. A. (1994). Requirements engineering as the reconciliation of social and technical issues. In Jirotka, M., & Goguen, J. A. (Eds.), *Requirements Engineering: Social and Technical Issues* (pp. 165–200). San Diego, CA: Academic Press.

Goguen, J. A. (1997). Towards a social, ethical theory of information. In Bowker, G., Gasser, L., Star, L., & Turner, W. (Eds.), *Social Science Research, Technical Systems and Cooperative Work: Beyond the Great Divide* (pp. 27–56). Upper Saddle River, NJ: Erlbaum.

Goodwin, C. (1996). Transparent vision. In Thompson, S. (Ed.), *Interaction and Grammar* (pp. 370–404). Cambridge, UK: Cambridge University Press. doi:10.1017/CBO9780511620874.008

Goodwin, C. (2000). Action and embodiment within situated human interaction. *Journal of Pragmatics, 32*, 1489–1522. doi:10.1016/S0378-2166(99)00096-X

Goodwin, C., & Goodwin, M. H. (1996). Seeing as situated activity: Formulating planes. In Engeström, Y., & Middleton, D. (Eds.), *Cognition and Communication at Work* (pp. 61–95). Cambridge, UK: Cambridge University Press.

Goodwin, M. H. (1995). Assembling a response: Setting and collaboratively constructed work talk. In Ten Have, P., & Psathas, G. (Eds.), *Situated Order: Studies in the Social Organization of Talk and Embodied Activities* (pp. 173–186). Washington, DC: University Press of America.

Goodwin, M. H. (1996). Informings and announcements in their environment: Prosody within a multi-activity work setting. In Couper-Kuhlen, E., & Selting, M. (Eds.), *Prosody in Conversation: Interactional Studies* (pp. 436–461). Cambridge, UK: Cambridge University Press. doi:10.1017/CBO9780511597862.013

Graham, I., Lloyd, A., Slack, R., & Williams, R. (2003). *Uptake of BPR*. PRECEPT Working Paper. Lyngby, Denmark: Danish Technical University.

Graham, I., Spinardi, G., & Williams, R. (1996). Diversity in the emergence of electronic commerce. *Journal of Information Technology, 11*(1), 161–172. doi:10.1080/026839696345360

Graham, I., Spinardi, G., Williams, R., & Webster, J. (1995). The dynamics of EDI standards development. *Technology Analysis and Strategic Management, 7*(1), 3–20. doi:10.1080/09537329508524192

Gregor, S. (2002). A theory of theories in information systems. In Gregor, S., & Hart, D. (Eds.), *Information Systems Foundations: Building the Theoretical Base* (pp. 1–20). Canberra, Australia: Australian National University.

Gregor, S., & Iivari, J. (2007). Designing for mutability in information systems artifacts. In Hart, D. N., & Gregor, S. D. (Eds.), *Information Systems Foundations: Theory, Representation and Reality*. Sydney, Australia: Australian National University Press.

Gribbin, J. (2003). *Science: A history 1543-2001*. London, UK: Penguin.

Grinter, R. E. (1996). Supporting articulation work using software configuration management systems. *Computer Supported Cooperative Work: The Journal of Collaborative Computing, 5*(4), 447–465. doi:10.1007/BF00136714

Grudin, J., & Grinter, R. (1994). Ethnography and design. *Computer Supported Cooperative Work, 3*(1), 55–59. doi:10.1007/BF01305846

Gubrium, J. F. (1988). *Analyzing field reality*. Thousand Oaks, CA: Sage.

Gubrium, J. F., & Holstein, J. A. (2003). *Postmodern interviewing*. Thousand Oaks, CA: Sage.

Guy, E. S. (2003). Patterns as artifacts in user-developer collaborative design. In *Proceedings of ECSCW 2003*. ECSCW.

Haas, C. (1996). *Writing technology: Studies on the materiality of literacy*. Mahwah, NJ: Lawrence Erlbaum.

Hacking, I. (1985). Making up people. In Heller, T. L., Sosna, M., & Wellbery, T. E. (Eds.), *Reconstructing Individualism*. Palo Alto, CA: Stanford University Press.

Hacking, I. (1999). *The social construction of what?* Cambridge, MA: The Harvard University Press.

Hakken, D. (1999). *Cyborgs@Cyberspace?: An ethnographer looks to the future*. New York, NY: Routeledge.

Halverson, C. A., Ellis, J. B., Danis, C., & Kellogg, W. A. (2006). Designing task visualizations to support the coordination of work in software development. In *Proceedings of the 20th Anniversary Conference on Computer Supported Cooperative Work,* (pp. 39-48). New York, NY: ACM Press.

Halverson, C. A. (2002). Activity theory and distributed cognition: Or what does CSCW need to do with theories? *Computer Supported Cooperative Work, 11*(1-2), 243–267. doi:10.1023/A:1015298005381

Hanseth, O., & Braa, K. (1998). Technology as traitor: Emergent SAP infrastructure in a global organization. In *Proceedings of the International Conference on Information Systems,* (pp. 188-196). Helsinki, Finland: Association for Information Systems.

Hanseth, O., & Monteiro, E. (1997). *Understanding information infrastructures*. Unpublished Manuscript. Retrieved from http://www.ifi.uio.no/~oleha/Publications/bok.pdf.

Hanseth, O. (2000). The economics of standards. In Ciborra, C. (Ed.), *From Control to Drift: The Dynamics of Corporate Information Infrastructures* (pp. 56–70). Oxford, UK: Oxford University Press.

Hanseth, O. (2001). Gateways - Just as important as standards: How the internet won the "religious war" about standards in Scandinavia. *Knowledge, Technology & Policy, 14*(3), 71–89. doi:10.1007/s12130-001-1017-2

Hanseth, O., & Aanestad, M. (2003). Design as bootstrapping: On the evolution of ICT networks in health care. *Methods of Information in Medicine, 42*(4), 385–391.

Hanseth, O., Jacucci, E., Grisot, M., & Aanestad, M. (2006). Reflexive standardization: Side effects and complexity in standard-making. *Management Information Systems Quarterly, 30*, 563–581.

Hanseth, O., & Lundberg, N. (2001). Designing work oriented infrastructures. *Computer Supported Cooperative Work, 10*(3), 347–372. doi:10.1023/A:1012727708439

Hanseth, O., & Lyyttinen, K. (2010). Theorizing about the design of information infrastructures: Design kernel theories and principles. *Journal of Information Technology, 25*(1), 1–19. doi:10.1057/jit.2009.19

Hanseth, O., Monteiro, E., & Hatling, M. (1996). Developing information infrastructure standards: The tension between standardization and flexibility. *Science, Technology & Human Values, 21*(4), 407–426. doi:10.1177/016224399602100402

Haraway, D. (1988). Situated knowledges: The science question in feminism and the privilege of partial perspective. In *Her Simians, Cyborgs, and Women: The Reinvention of Nature* (pp. 183–202). New York, NY: Routledge. doi:10.2307/3178066

Haraway, D. (1991). *Simians, cyborgs, and women: The reinvention of nature*. New York, NY: Routledge.

Hardstone, G., D'Adderio, L., & Williams, R. (2006). Standardization, trust and dependability. In Clarke, K., Hardstone, G., Rouncefield, M., & Sommerville, I. (Eds.), *Trust in Technology: A Socio-Technical Perspective* (pp. 69–103). Dordrecht, The Netherlands: Springer. doi:10.1007/1-4020-4258-2_4

Hardstone, G., Hartswood, M., Procter, R., Slack, R., Voss, A., & Rees, G. (2004). Supporting informality: Team working and integrated care records. In *Proceedings of the CSCW, 2004*, 142–151. doi:10.1145/1031607.1031632

Harper, R. H. R. (1996). *The ethnographic turn: Why it has come about and how to do it*. Technical Report EPC-1996-109. Cambridge, UK: Rank Xerox.

Harper, R. H. R. (2000). The organization in ethnography: A discussion of ethnographic fieldwork programs in CSCW. *Computer Supported Cooperative Work, 9*(2), 239–264. doi:10.1023/A:1008793124669

Harper, R. H. R., & Hughes, J. A. (1992). What a f ing system: Send 'em all to the same place and then expect us to stop 'em hitting: Making technology work in air traffic control. In Button, G. (Ed.), *Technology in Working Order: Studies of Work, Interaction and Technology* (pp. 127–144). London, UK: Routeledge.

Harper, R. H. R., O'Hara, K. P. A., Sellen, A. J., & Duthie, D. J. R. (1997). Toward the paperless hospital? A case study of document use by anaesthetists. *British Journal of Anaesthesia, 78*, 762–767.

Harper, R. H., O'Hara, K., Sellen, A., & Duthie, D. (1997). Towards the paperless hospital? *British Journal of Anaesthesia, 78*, 762–767.

Harper, R., & Carter, K. (1994). Keeping people apart: A research note. *Computer Supported Cooperative Work, 2*(3), 199–207. doi:10.1007/BF00749017

Harper, R., & Hughes, J. A. (1993). What a f-ing system! Send 'em all to the same place and then expect us to stop 'em hitting: Managing technology work in air traffic control. In Button, G. (Ed.), *Technology in Working Order: Studies of Work, Interaction, and Technology* (pp. 127–144). London, UK: Routledge.

Hartswood, M., Procter, R., Rouncefield, M., & Slack, R. (2003). Making a case in medical work: Implications for the electronic medical record. *Computer Supported Cooperative Work, 12*, 241–266. doi:10.1023/A:1025055829026

Hartswood, M., Procter, R., Slack, R., Voss, A., Büscher, M., Rouncefield, M., & Rouchy, R. (2002). Co-realisation: Towards a principled synthesis of ethnomethodology and participatory design. *Scandinavian Journal of Information Systems, 14*(2), 9–30.

Harvey, D. (1990). *The condition of postmodernity*. Oxford, UK: Blackwell.

Harvey, L. (1997). A discourse on ethnography. In Lee, A. S., Liebenau, J., & DeGross, J. I. (Eds.), *Information Systems and Qualitative Research* (pp. 207–224). London, UK: Chapman and Hall.

Harvey, L., & Myers, M. D. (1995). Scholarship and practice: The contribution of ethnographic research methods to bridging the gap. *Information Technology & People, 8*(3), 13–27. doi:10.1108/09593849510098244

Hayles, C. (2005). Computing the human. *Theory, Culture & Society, 22*(1), 131–151. doi:10.1177/0263276405048438

Hayles, C. (2006). Unfinished work: From cyborg to cognisphere. *Theory, Culture & Society, 23*(7-8), 159–166. doi:10.1177/0263276406069229

Heath, C. (1984). Everett Cherrington Hughes (1897-1983): A note on his approach and influence. *Sociology of Health & Illness, 6*(2), 218–237. doi:10.1111/1467-9566.ep10778393

Heath, C., Hindmarsh, J., & Luff, P. (2010). *Video in qualitative analysis*. London, UK: Sage.

Heath, C., Knoblauch, H., & Luff, P. (2000). Technology and social interaction: The emergence of workplace studies. *The British Journal of Sociology, 51*(2), 299–320. doi:10.1080/00071310050030190

Heath, C., & Luff, P. (1996). Convergent activities: Line control and passenger information on the London underground. In Engeström, Y., & Middleton, D. (Eds.), *Cognition and Communication at Work* (pp. 96–129). Cambridge, UK: Cambridge University Press.

Heath, C., & Luff, P. (1996). Documents and professional practice: 'Bad' organisational reasons for 'good' clinical records. In *Proceedings of CSCW, 1996*, 354–362.

Heath, C., & Luff, P. (2000). *Technology in action*. Cambridge, UK: Cambridge University Press. doi:10.1017/CBO9780511489839

Heath, R. P. (1997). Seeing is believing: Ethnography gets consumers where they live and work, and play, and shop. *Marketing Tools, 3*, 4–9.

Heeks, R., Mundy, D., & Salazar, A. (1999). *Why health care information systems succeed or fail: Information systems for public sector management*. Working Paper. Manchester, UK: University of Manchester.

Heidegger, M. (1962). *Being and time* (Macquarrie, J., & Robinson, E., Trans.). Oxford, UK: Basil Blackwell.

Heidegger, M. (1966). Conversation on a country path about thinking. In *Discourse on Thinking* (Anderson, J. M., & Freund, E. H., Trans.). New York, NY: Harper and Row.

Heidegger, M. (1966). *Discourse on thinking*. New York, NY: Harper and Row Publishers.

Heidegger, M. (1977). *The question concerning technology and other essays* (Lovitt, W., Trans.). New York, NY: Harper & Row.

Heidegger, M. (1982). *The basic problems of phenomenology*. Bloomington, IN: Indiana University Press.

Heidegger, M. (1992). *Gelassenheit*. Pfullingen, Germany: Neske.

Heidegger, M. (1993a). The question concerning technology. In Krell, D. F. (Ed.), *Basic Writings*. London, UK: Routledge.

Heidegger, M. (1993b). The way to language. In Krell, D. F. (Ed.), *Basic Writings*. London, UK: Routledge.

Heidegger, M. (1993c). The end of philosophy and the task of thinking. In Krell, D. F. (Ed.), *Basic Writings*. London, UK: Routledge.

Heidegger, M. (1994). *Bremer und freiburger fortraege*. Frankfurt, Germany: Klostermann.

Heidegger, M. (1995). *The fundamental concepts of metaphysics*. Bloomington, IN: Indiana University Press.

Heidegger, M. (1996). *Being and time* (Stambaugh, J., Trans.). Albany, NY: SUNY Press.

Heidegger, M. (1997). *Vom wesen der wahrheit: Zu platons höhlengleichnis und theätet*. Berlin, Germany: Klostermann.

Heidegger, M. (1998). Traditional language and technological language. *Journal of Philosophical Research*, *23*, 129–145.

Heidegger, M. (2006). *Essere e tempo and sein und zeit*. Milan, Italy: Mondadori.

Heidegger, M. (2006). *Sein und Zeit*. Tübingen, Germany: Niemeyer.

Heims, S. J. (1980). *John Von Neumann and Norbert Wiener: From mathematics to the technologies of life and death*. Cambridge, MA: MIT Press.

Heims, S. J. (1991). *The cybernetics group*. Cambridge, MA: MIT Press.

Heims, S. J. (1993). *Constructing a social science for postwar America: The cybernetics group, 1946-1953*. Cambridge, MA: MIT Press.

Heller, A. (1999). *A theory of modernity*. Oxford, UK: Balckwell.

Helmes-Hayes, R. C. (1998). Everett Hughes: Theorist of the second Chicago school. *International Journal of Politics Culture and Society*, *11*(4), 621–673. doi:10.1023/A:1025157424093

Henderson, K. (1999). *On line and on paper: Visual representations, visual culture, and computer graphics in design engineering*. Cambridge, MA: MIT Press.

Hepsø, V., Monteiro, E., & Rolland, K. H. (2009). Ecologies of e-infrastructures. *Journal of the Association for Information Systems*, *10*(5), 430–446. Retrieved from http://aisel.aisnet.org/jais/vol10/iss5/

Heritage, J. (1984). *Garfinkel and ethnomethodology*. Cambridge, UK: Polity Press.

Herrmann, T., & Nolte, A. (2010). The integration of collaborative process modeling and electronic brainstorming in co-located meetings. In G. Kolfschoten, T. Herrmann, & S. Lukosch (Eds.), *CRIWG '10 Proceedings of the 16th International Conference on Collaboration and Technology,* (pp. 145-160). Berlin, Germany: Springer.

Herrmann, T., Kunau, G., Loser, K.-U., & Menold, N. (2004). Sociotechnical walkthrough: Designing technology along work processes. In A. Clement, F. de Cindio, A.-M. Oostveen, D. Schuler, & P. van den Besselaar (Eds.), *Artful Integration: Interweaving Media, Materials and Practices: Proceedings of the 8th Participatory Design Conference PDC 2004,* (pp. 132-141). New York, NY: ACM.

Herrmann, T. (2009). Systems design with the sociotechnical walkthrough. In Whitworth, B., & De Moor, A. (Eds.), *Handbook of Research on Socio-Technical Design and Social Networking Systems* (pp. 336–351). Hershey, PA: IGI Global. doi:10.4018/978-1-60566-264-0.ch023

Herrmann, T., & Loser, K.-U. (1999). Vagueness in models of socio-technical systems. *Behaviour & Information Technology*, *18*(5), 313–323. doi:10.1080/014492999118904

Hertzum, M. (1993). Information retrieval in a work setting: A case study of the documentation part of chemists' work. In J. P. Bansler et al. (Eds.), *Proceedings of the 16th IRIS Information Systems Research Seminar in Scandinavia,* (pp. 786-798). Copenhagen, Denmark: IRIS.

Hertzum, M. (1999). Six roles of documents in professional's work. In S. Bœdker, M. Kyng, & K. Schmidt (Eds.), *Proceedings of the Sixth European Conference on Computer-Supported Cooperative Work,* (pp. 41-60). Dordrecht, The Netherlands: Kluwer Academic Publishers.

Hevner, A. R., March, S. T., & Park, J. (2004). Design science in information systems research. *Management Information Systems Quarterly*, *28*(1), 75–105.

Hevner, A. R., March, S. T., Park, J., & Ram, S. (2004). Design science in information systems research. *Management Information Systems Quarterly*, *28*(1), 75–106.

Hildebrandt, M., & Gutwirth, S. (Eds.). (2008). *Profiling the European citizen*. Berlin, Germany: Springer. doi:10.1007/978-1-4020-6914-7

Hildebrandt, M., & Rouvroy, A. (Eds.). (2011). *Law, human agency, and autonomic computing: The philosophy of law meets the philosophy of technology*. London, UK: Routledge.

Hine, C. (2007). Multi-sited ethnography as a middle range methodology for contemporary STS. *Science, Technology & Human Values*, *32*(6), 652–671. doi:10.1177/0162243907303598

Hirschauer, S. (2007). Putting things into words - Ethnographic description and the silence of the social. *Human Studies*, *29*, 413–441. doi:10.1007/s10746-007-9041-1

Hislop, D. (2002). The client role in consultancy relations during the appropriation of technological innovations. *Research Policy*, *31*(5), 657–671. doi:10.1016/S0048-7333(01)00135-4

Hodson, R. (2004). A meta-analysis of workplace ethnographies: Race, gender, and employee attitudes and behaviours. *Journal of Contemporary Ethnography*, *33*(1), 4–38. doi:10.1177/0891241603259808

Honderich, T. (Ed.). (1999). *The philosophers: Introducing great western thinkers*. Oxford, UK: Oxford University Press.

Hong, K.-K., & Kim, Y.-G. (2002). The critical success factors for ERP implementation: An organizational fit perspective. *Information & Management*, *40*, 24–40. doi:10.1016/S0378-7206(01)00134-3

Hood, C. (1995). The new public management in the 1980's: Variations on a theme. *Accounting, Organizations and Society*, *20*(2/3), 93–101. doi:10.1016/0361-3682(93)E0001-W

Howells, J. (2006). Intermediation and the role of intermediaries in innovation. *Research Policy*, *35*(5), 715–728. doi:10.1016/j.respol.2006.03.005

Hughes, J. A., King, V., Rodden, T., & Andersen, H. (1994). Moving out from the control room: Ethnography in system design. In *Proceedings of the 1994 ACM Conference on Computer Supported Cooperative Work*, (pp. 429-439). New York, NY: ACM Press.

Hughes, J. A., Randall, D., & Shapiro, D. (1992). Faltering from ethnography to design. In *Proceedings of the 1992 ACM Conference on Computer Supported Cooperative Work (CSCW 1992)*, (pp. 115-122). Toronto, Canada: ACM.

Hughes, J., O'Brien, J., Rodden, J., & Rouncefield, M. (1997b). *Ethnography, communication and support for design*. Retrieved from http://www.comp.lancs.ac.uk/computing/research/cseg/97rep.html.

Hughes, J., O'Brien, J., Rodden, J., Rouncefield, M., & Blythin, S. (1997a). Designing with ethnography: A presentation framework for design. In *Proceedings of DIS 1997: Designing Interactive Systems: Processes, Practices, Methods, & Techniques*, (pp. 147-158). DIS.

Hughes, E. C. (1952). The sociological study of work. *American Journal of Sociology*, *57*(5), 423–426. doi:10.1086/221010

Hughes, E. C. (1970). *The sociological eye*. Chicago, IL: Aldine.

Hughes, H. S. (1958). *Consciousness and society*. New York, NY: Alfred Knopf.

Hughes, J. A., King, V., Rodden, T., & Andersen, H. (1995). The role of ethnography in interactive system design. *Interaction*, *2*(2), 56–65. doi:10.1145/205350.205358

Hughes, J. A., Randall, D., & Shapiro, D. (1993). From ethnographic record to system design: Some experiences from the field. *Computer Supported Cooperative Work*, *1*(3), 123–147. doi:10.1007/BF00752435

Husserl, E. (1950). *Cartesian meditations: An introduction to phenomenology* (Cairns, D., Trans.). The Hague, The Netherlands: Martinus Nijhoff.

Husserl, E. (1970). *The crisis of European sciences and transcendental phenomenology: An introduction to phenomenological philosophy* (Carr, D., Trans.). Evanston, IL: Northwestern University Press.

Husserl, E. (1973). *Experience and judgement* (1st ed.). (Churchill, S., Trans.). Evanston, IL: Northwestern.

Husserl, E. (1983). *Ideas pertaining to a pure phenomenology and to a phenomenological philoshophy, first book* (Kersten, F., Trans.). Dordrecht, The Netherlands: Kluwer.

Hutchins, E., & Palen, L. (1998). *Constructing meaning from space, gesture and speech*. Retrieved from http://goldin-meadow-lab.uchicago.edu/embodiedcogrg/Hutchins.pdf.

Hutchins, E. (1996). *Cognition in the wild*. Cambridge, MA: The MIT Press.

Hyppolite, J. (1974). *Genesis and structure of Hegel's "phenomenology of spirit"* (Cherniak, S., & Heckman, J., Trans.). Evanston, IL: Northwestern University Press.

Hyysalo, S. (2004). *Uses of innovation: Wristcare in the practices of engineers and elderly*. Academic Dissertation. Helsinki, Finland: University of Helsinki.

Hyysalo, S. (2006). The role of learning-by-using in the design of healthcare technologies: A case study. *The Information Society*, *22*(2), 89–100. doi:10.1080/01972240600567196

Hyysalo, S. (2009). Review: How packaged software conquers the organization: Neil Pollock and Robin Williams, software and organizations: The biography of the packaged enterprise system or how SAP conquered the world. *Social Studies of Science*, *39*, 635–642. doi:10.1177/0306312709335847

Hyysalo, S. (2010). *Health technology development and use*. New York, NY: Routledge.

Institute of Medicine. (1997). *The computer-based patient record: An essential technology for health care*. Washington, DC: National Academy Press.

Introna, L. D. (2005). Claudio Ciborra's way of being: Authenticity and the world of information systems. *European Journal of Information Systems*, *15*(5), 513–517. doi:10.1057/palgrave.ejis.3000574

Introna, L. D., & Ilharco, F. M. (2004). Phenomenology, screens and the world: A journey with Husserl and Heidegger into phenomenology. In Mingers, J., & Wilcocks, L. (Eds.), *Social Theory and Philosophy for Information Systems*. Chichester, UK: Wiley.

Introna, L. D., & Whitley, E. A. (1997). Against method-ism. *Information Technology & People*, *10*(1), 31–45. doi:10.1108/09593849710166147

Introna, L. D., & Whittaker, L. (2002). The phenomenology of information systems evaluation: Overcoming the subject object dualism. In Wynn, E. H., Myers, M. D., & DeGross, J. I. (Eds.), *Global Organizational Discourse about Information Technology* (pp. 155–175). Dordrecht, The Netherlands: Kluwer Academic Publishers.

Iqbal, R., Jamesb, A., & Gatward, R. (2005). Designing with ethnography: An integrative approach to CSCW design. *Advanced Engineering Informatics*, *19*, 81–92. doi:10.1016/j.aei.2005.05.003

Jacucci, G. (2007). Social practice design (SPD), pathos, improvisation, mood, and bricolage: The Mediterranean way to make place for IT? In I. Benbasat, M. Limayem, D. Te'eni, G. Jacucci, & E. Monod (Eds.), *Proceedings of the Eighth Mediterranean Conference on Information Systems,* (pp. 89-103). Venice, Italy: University of Trento Press.

Jacucci, G., Cattani, C., Campagnolo, G. M., Wagner, I., & Tellioglu, H. (2008). *Facilitating co-construction by social practice design: A case of geodistributed employees facing the design of model-based enterprise systems*. Working Paper, AIS eLibrary MCIS2007. Retrieved from http://www.aisel.aisnet.org.

Jacucci, G., Tellioglu, H., & Wagner, I. (2006). *Practices and cultures of knowledge management in global virtual (software) and real component manufacturing*. Working Paper, AIS eLibrary MCIS2006. Retrieved from http://www.aisel.aisnet.org.

Jacucci, G., Tellioglu, H., & Wagner, I. (2007). Design games as a part of social practice design: A case of employees elaborating on organizational problems. In I. Benbasat, M. Limayem, D. Te'eni, G. Jacucci, & E. Monod (Eds.), *Proceedings of the Eighth Mediterranean Conference on Information Systems,* (vol 1), (pp. 141-155). Venice, Italy: University of Trento Press.

Jaeger, B., Slack, R., & Williams, R. (2000). Europe experiments with multimedia: An overview of social experiments and trials. *The Information Society*, *16*(4), 277–302. doi:10.1080/019722400457252

James, W. (1912). *Essays in radical empiricism*. New York, NY: Longmans, Green, & Co.

Jay, M. (1994). *Downcast eyes: The denigration of vision in twentieth-century French thought*. Berkeley, CA: University of California Press.

Jefferson, G. (1984). Transcription notation. In Atkinson, J., & Heritage, J. (Eds.), *Structures of Social Interaction* (pp. ix–xvi). New York, NY: Cambridge University Press.

Jenkins, H. (2006). *Convergence culture: How new and old media collide*. New York, NY: New York University Press.

Johannessen, L. K., & Ellingsen, G. (2009). Integration and generification--Agile software development in the healthcare market. *Computer Supported Cooperative Work*, *18*(5-6), 607–634. doi:10.1007/s10606-009-9097-8

Johnson, C. W. (2009). Politics and patient safety don't mix: Understanding the failure of large-scale software procurement for healthcare systems. In P. Casely & C. W. Johnson (Eds.), Proceedings, *Fourth IET Systems Safety Conference*. London, UK: IET Conference Publications.

Johnson, C. W. (2011). *Case studies in the failure of healthcare information systems*. Retrieved from http://www.dcs.gla.ac.uk/~johnson/papers/AHRQ/case_study.pdf.

Jones, M. R. (2008). Gidden's structuration theory and information systems research. *Management Information Systems Quarterly*, *32*(1), 127–157.

Jordan, B., & Lambert, M. (2009). Working in corporate jungles: Reflections on ethnographic praxis in industry. In M. Cefkin (Ed.), *Ethnography and the Corporate Encounter: Reflections on Research in and of Corporations*. New York, NY: Berghahn Books. Retrieved from http://www.lifescapes.org/Papers/7_Ch_4%20%20Jordan.doc.

Jordan, B. (1996). Ethnographic workplace studies and computer supported cooperative work. In Shapiro, D., Tauber, M., & Traunmüller, R. (Eds.), *The Design of Computer-Supported Cooperative Work and Groupware Systems* (pp. 17–42). Amsterdam, The Netherlands: Elsevier Science. doi:10.1016/S0923-8433(96)80005-0

Jordan, B. (1997). Transforming ethnography - Reinventing research. *Cultural Anthropology Methods Journal*, *9*(3), 12–17.

Jordan, B., & Dalal, B. (2006). Persuasive encounters: Ethnography in the corporation. *Field Methods*, *18*(4), 359–381. doi:10.1177/1525822X06292976

Jordan, B., & Henderson, A. (1995). Interaction analysis: Foundations and practice. *Journal of the Learning Sciences*, *4*(1), 39–103. doi:10.1207/s15327809jls0401_2

Jordan, B., & Yamauchi, Y. (2008). Beyond the university: Teaching ethnographic methods in the corporation. *Anthropology News*, *49*(6), 35. doi:10.1111/an.2008.49.6.35

Jørgensen, U., & Sørensen, O. (1999). Arenas of development: A space populated by actor-worlds, artefacts, and surprises. *Technology Analysis and Strategic Management*, *11*(3), 409–429. doi:10.1080/095373299107438

Kahn, B. K., Strong, D. M., & Wang, R. Y. (2002). Information quality benchmarks: Product and service performance. *Communications of the ACM*, ▪▪▪, 184–192. doi:10.1145/505248.506007

Kallinikos, J. (1995). The architecture of the invisible: Technology is representation. *Organization*, *2*(1), 117–140. doi:10.1177/135050849521006

Kallinikos, J. (2004a). Farewell to constructivism: Technology and context-embedded action. In Avgerou, C., Ciborra, C., & Land, F. (Eds.), *The Social Study of Information and Communication Technology: Innovation, Actors, and Contexts*. Oxford, UK: Oxford University Press.

Kallinikos, J. (2004b). Deconstructing information packages: Organizational and behavioural implications of ERP systems. *Information Technology & People*, *17*(1), 8–30. doi:10.1108/09593840410522152

Kallinikos, J. (2005). The spirit never dies. *European Journal of Information Systems*, *15*(5), 467–469. doi:10.1057/palgrave.ejis.3000582

Kallinikos, J. (2006). *The consequences of information: Institutional implications of technological change*. Cheltenham, UK: Elgar.

Kallinikos, J. (2009). On the computational rendition of reality: Artefacts and human agency. *Organization*, *16*(2), 183–202. doi:10.1177/1350508408100474

Kallinikos, J. (2011). *Governing through technology: Information artefacts and social practice*. New York, NY: Palgrave/MacMillan.

Kallinikos, J., & Hasselbladh, H. (2009). Work, control and computation: Rethinking the legacy of neo-institutionalism. *Research in the Sociology of Organizations*, *27*, 257–282. doi:10.1108/S0733-558X(2009)0000027010

Kanellis, P., Lycett, M., & Paul, R. J. (1999). Evaluating business information systems fit: From concept to practical application. *European Journal of Information Systems*, *8*, 65–76. doi:10.1057/palgrave.ejis.3000316

Kanigel, R. (2005). *The one best way: Frederick Winslow Taylor and the enigma of efficiency*. Cambridge, MA: MIT Press.

Kant, I. (1900). *Critique of pure reason* (1st ed.). (Meiklejohn, J. M. D., Trans.). London, UK: Colonial Press.

Kant, I. (1988). *Logic* (1st ed.). (Hartzman, R. S., & Schwarz, W., Trans.). New York, NY: Dover.

Kaplan, D. (2008). *Reading Ricoeur*. New York, NY: SUNY.

Karasti, H. (2001). Bridging work practice and system design: Integrating systemic analysis, appreciative intervention and practitioner participation. *Computer Supported Cooperative Work*, *10*, 211–246. doi:10.1023/A:1011239126617

Kasvi, J. J. J., Vartiainen, M., Pulkkis, A., & Nieminen, M. (2000). The role of information support systems in the joint optimization of work systems. *Human Factors and Ergonomics in Manufacturing*, *10*, 193–221. doi:10.1002/(SICI)1520-6564(200021)10:2<193::AID-HFM5>3.0.CO;2-H

Katz, D., & Kahn, R. L. (1969). Common characteristics of open systems. In Emery, F. E. (Ed.), *Systems Thinking* (pp. 86–104). Harmondsworth, UK: Penguin.

Kayas, O. G., McLean, R., Hines, T., & Wright, G. H. (2008). The panoptic gaze: Analysing the interaction between enterprise resource planning technology and organisational culture. *International Journal of Information Management*, *28*, 446–452. doi:10.1016/j.ijinfomgt.2008.08.005

Keen, P. G. W. (1980). MIS research: Reference traditions and a cumulative tradition. In *Proceedings of 1st International Conference on Information Systems*, (pp. 9–18). ACM.

Kemmis, S. (2001). Exploring the relevance of critical theory for action research: Emancipatory action research in the footsteps of Jurgen Habermas. In Reason, P., & Bradbury, H. (Eds.), *Handbook of Action Research: Participative Inquiry and Practice* (pp. 91–102). London, UK: Sage.

Kensing, F. (1998). Participatory design: Issues and concerns. *Computer Supported Cooperative Work*, *7*(3-4), 167–185. doi:10.1023/A:1008689307411

Kensing, F., & Blomberg, J. (1998). Participatory design: Issues and concerns. *Computer Supported Cooperative Work*, *7*, 167–185. doi:10.1023/A:1008689307411

Khandani, S. (2005). *Engineering design process: Education transfer plan*. Retrieved from http://iisme.org/etp/HS%20Engineering-%20Engineering.pdf.

Kittler, F. (1997). *Literature, media, information systems*. Amsterdam, The Netherlands: OPA.

Kitto, S., & Higgins, V. (2010). Working around ERPs in technological universities. *Science, Technology & Human Values*, *35*, 29–54. doi:10.1177/0162243908329535

Klein, H. K., & Myers, M. D. (1999). A set of principles for conducting and evaluating interpretive field studies in information systems. *Management Information Systems Quarterly*, *23*(1), 67–94. doi:10.2307/249410

Kling, R., & Scacchi, W. (1982). The web of computing: Computer technology as social organization. *Advances in Computing*, *21*, 1–90. doi:10.1016/S0065-2458(08)60567-7

Knoblauch, H. (2005). Focused ethnography. *Forum Qualitative Sozial Forschung*, *6*(3). Retrieved from http://www.qualitative-research.net/fqs-texte/3-05/05-3-44-e.htm

Knoblauch, H. (2006). Videography: Focused ethnography and video-analysis. In Knoblauch, H., Schnettler, B., & Raab, J. (Eds.), *Video-Analysis: Methodological Aspects of Interpretive Audiovisual Analysis in Social Research* (pp. 69–84). Frankfurt, Germany: Lang.

Knorr Cetina, K. D. (1983). The ethnographic study of scientific work: Toward a constructivist interpretation of science. In Knorr Cetina, K., & Mulkay, M. (Eds.), *Analytical Perspectives in Science Studies*. London, UK: Sage.

Knorr-Cetina, K. (1999). *Epistemic cultures: How the sciences make knowledge*. Cambridge, MA: Harvard University Press.

Koch, C. (2003). 'ERP-software packages: Between mass production communities and intraorganisational political processes. In Preece, D., & Laurila, J. (Eds.), *Technological Change and Organizational Action* (pp. 56–76). London, UK: Routledge.

Koch, C. (2007). ERP: A moving target. *International Journal of Business Information Systems*, *2*(4), 426–443. doi:10.1504/IJBIS.2007.012544

Kock, N. (Ed.). (2007). *Information systems action research: An applied view of emerging concepts and methods*. New York, NY: Springer.

Kock, N., & Lau, F. (2001). Information systems action research: Serving two demanding masters. *Information Technology & People*, *4*(1), 6–11.

Korpman, R. A. (1990). Patient care automation: The future is now. *Nursing Economics*, *8*(5), 345–349.

Koschmann, T., Stahl, G., & Zemel, A. (2006). The video analyst's manifesto (or the implications of Garfinkel's policies for studying practice within design-based research). In R. Goldman, B. Barron, S. Derry, & R. Pea (Eds.), *Video Research in the Learning Sciences*. Mahway, NJ: Lawrence Erlbaum Associates. Retrieved from http://edaff.siumed.edu/tk/manifesto.pdf.

Kotro, T. (2005). *Hobbyist knowing in product development - Desirable objects and passion for sports in Suunto corporation.* Academic Dissertation. Helsinki, Finland: National Consumer Research Center and University of Art and Design Helsinki.

Krogstie, J., Opdahl, A. L., & Brinkkemper, O. S. (Eds.). (2007). *Conceptual modelling in information systems engineering*. Berlin, Germany: Springer Verlag. doi:10.1007/978-3-540-72677-7

Kumar, K., Maheshwari, U., & Kumar, U. (2003). An investigation of critical management issues in ERP implementation: Empirical evidence from Canadian organizations. *Technovation, 23*, 793–807. doi:10.1016/S0166-4972(02)00015-9

Kumar, K., & van Hillegersberg, J. (2000). ERP experiences and evolution. *Communications of the ACM, 43*, 23–26.

Kunda, G. (1992). *Engineering culture: Control and commitment in a high-tech corporation.* Philadelphia, PA: Temple University Press.

Kuutti, K. (1999). Activity theory, transformation of work and information systems design. In Engestrog, Y., Miettinen, M. R., & Punamag Ki, R. (Eds.), *Perspectives on Activity Theory* (pp. 360–376). Cambridge, UK: Cambridge University Press.

Kvavik, R. B., Katz, R. N., Beecher, K., Caruso, J., King, P., Voloudakis, J., & Williams, L.-A. (2002). *The promise and performance of enterprise systems for higher education: Educause center for applied research.* Retrieved from http://www.educause.edu.

Kyng, M., & Mathiassen, L. (Eds.). (1997). *Computers and Design in Context*. Cambridge, MA: MIT press.

Lamb, R., & Kling, R. (2003). Reconceptualizing users as social actors in information systems research. *Management Information Systems Quarterly, 27*(2), 197–235.

Langenwalter, G. A. (1999). *Enterprise resources planning and beyond integrating your entire organization.* Boca Raton, FL: St. Lucie Press.

Latour, B. (1986). *Visualization and cognition: Thinking with eyes and hands. Knowledge and Society: Studies in the Sociology of Culture Past and Present* (pp. 1–40). Greenwich, CT: Jai Press.

Latour, B. (1987). *Science in action.* New York, NY: Open University.

Latour, B. (1991). Technology is society made durable. In Law, J. (Ed.), *A Sociology of Monsters: Essays on Power, Technology and Domination* (pp. 103–131). London, UK: Routledge.

Latour, B. (1992). Where are the missing masses? The sociology of a few mundane artifacts. In Bijker, W. E., & Law, J. (Eds.), *Shaping Technology/Building Society: Studies in Sociotechnical Change* (pp. 225–258). Cambridge, MA: MIT Press.

Latour, B. (1992a). *Aramis ou l'amour des techniques.* Paris, France: La Découverte.

Latour, B. (1994). On technical mediation – Philosophy, sociology, genealogy. *Common Knowledge, 3*(2), 29–64.

Latour, B. (2004). On using ANT for studying information systems: A (somewhat) Socratic dialogue. In Avgerou, C., Ciborra, C., & Land, F. (Eds.), *The Social Study of Information and Communication Technologies* (pp. 62–75). Oxford, UK: Oxford University Press.

Latour, B., & Woolgar, S. (1979). *Laboratory life: The social construction of scientific facts.* Los Angeles, CA: Sage.

Lau, F. (1997). A review on the use of action research in information systems studies. In Lee, J., Liebenau, J., & DeGross, J. (Eds.), *Information Systems and Qualitative Research* (pp. 31–68). London, UK: Chapman & Hall.

Lave, J., & Wenger, E. (1991). *Situated learning: Legitimate peripheral participation.* Cambridge, MA: Cambridge University Press.

Law, J. (2004). *After method: Mess in social science research.* London, UK: Routledge.

Law, J., & Callon, M. (1992). The life and death of an aircraft: A network analysis of technical change. In Bijker, W. E., & Law, J. (Eds.), *Shaping Technology/Building Society: Studies in Sociotechnical Change* (pp. 21–52). Cambridge, MA: MIT Press.

Law, J., & Hassard, J. (Eds.). (1999). *Actor network theory and after*. Oxford, UK: Blackwell.

Law, J., & Singleton, V. (2000). Performing technology's stories: On social constructivism, performance, and performativity. *Technology and Culture, 41*(4), 765–775.

Lee, A. S. (1991). Integrating positivist and interpretive approaches to organizational research. *Organization Science*, *2*(4), 342–365. doi:10.1287/orsc.2.4.342

Lee, A. S., & Baskerville, R. L. (2003). Generalizing generalizability in information systems research. *Information Systems Research*, *14*(3), 1–14. doi:10.1287/isre.14.3.221.16560

Leonard-Barton, D. (1988). Implementation as mutual adaptation of technology and organization. *Research Policy*, *17*, 251–267. doi:10.1016/0048-7333(88)90006-6

Leont'ev, A. N. (1981). *Problems of the development of the mind.* Moscow, Russia: Progress Publishers.

Lessig, L. (2006). *Code 2.0.* New York, NY: Basic Books. Retrieved from http://codev2.cc/download+remix/.

Lessig, L. (2008). *Remix: Making art and commerce thrive in the hybrid economy*. London, UK: Bloomsbury. Retrieved from http://codev2.cc/download+remix/.

Letrilliart, L., Viboud, C., Boelle, P., & Flahault, A. (2000). Automatic coding of reasons for hospital referral from general medicine free-text reports. In *Proceedings of the AMIA Symposium*, (pp. 487-491). AMIA.

Lévi-Strauss, C. (1962). *La pensée sauvage*. Paris, France: Plon.

Lévy-Bruhl, L. (1922). *La mentalité primitive*. Paris, France: Alcan.

Lewin, K. (1946). Action research and minority problems. *The Journal of Social Issues*, *2*, 34–46. doi:10.1111/j.1540-4560.1946.tb02295.x

Liang, H., Xue, Y., Boulton, W., & Byrd, T. (2004). Why western vendors don't dominate china's ERP market: Examining cases of failed ERP system implementation in China and explaining the unique circumstances. *Communications of the ACM*, *47*(7), 69–72. doi:10.1145/1005817.1005824

Light, B. (2003). An alternative theory of legacy information systems. In *Proceedings of the ECIS*. ECIS.

Light, B. (2005). Going beyond "misfit" as a reason for ERP package customisation. *Computers in Industry*, *56*, 606. doi:10.1016/j.compind.2005.02.008

Light, B., Holland, C., & Wills, K. (2001). ERP and best of breed: A comparative analysis. *Business Process Management Journal*, *7*(3), 216–224. doi:10.1108/14637150110392683

Lilienfeld, R. (1988). *The rise of systems theory: An ideological analysis*. Malabar, FL: Robert E. Krieger Publishing.

Loveridge, R., Willman, P., & Deery, S. (2007). 60 years of human relations. *Human Relations*, *60*(12), 1873–1888. doi:10.1177/0018726707084917

Luff, P., Hindmarsh, J., & Heath, C. (Eds.). (2000). *Workplace studies: Recovering work practices and informing system design*. Cambridge, UK: Cambridge University Press. doi:10.1017/CBO9780511628122

Luff, P., Jirotka, M., Heath, C. C., & Greatbatch, D. (1993). Tasks and social interaction: The relevance of naturalistic analyses of conduct for requirements engineering. In Fickas, S., & Finkelstein, A. (Eds.), *Requirements Engineering* (pp. 187–190). IEEE Press. doi:10.1109/ISRE.1993.324818

Luhmann, N. (1995). *Social Systems*. Palo Alto, CA: Stanford University Press.

Lundberg, N., & Sandahl, T. I. (1998). Artifacts in work practice. In *Proceedings of IRIS 1998*. IRIS.

Lutters, W. G., & Ackerman, M. S. (2002). Achieving safety: A field study of boundary objects in aircraft technical support. In *Proceedings of the ACM Conference on Computer Supported Cooperative Work*, (pp. 266-275). New Orleans, LA: ACM Press.

Lynch, M. (1985). *Art and artifact in laboratory science: A study of shop work and shop talk in a research laboratory*. London, UK: Routledge & Kegan Paul.

Lynch, M. (1993). *Scientific practice and ordinary action: Ethnomethodology and the social studies of science*. Cambridge, UK: Cambridge University Press.

Lynch, M., & Bogen, D. (1994). Harvey Sacks' primitive natural science. *Theory, Culture & Society*, *11*, 65–104. doi:10.1177/026327694011004003

Lyytinen, K. (1987). A taxonomic perspective of informaiton systems development: Theoretical constructus and recommendations. In Boland, R. J., & Hirschheim, R. A. (Eds.), *Critical Issues in Information Systems Research* (pp. 3–41). New York, NY: John Wiley.

Mabert, V., Soni, A., & Venkataramanan, M. (2001, May). Enterprise resource planning: Common myths versus evolving reality. *Business Horizons, n.d.*, 69–76. doi:10.1016/S0007-6813(01)80037-9

Mabert, V., Soni, A., & Venkataramanan, M. A. (2003). Enterprise resource planning: Managing the implementation process. *European Journal of Operational Research, 146*, 302–314. doi:10.1016/S0377-2217(02)00551-9

Macaulay, C., Benyon, D., & Crerar, A. (2000). Ethnography, theory and systems design: From intuition to insight. *International Journal of Human-Computer Studies, 53*, 35–60. doi:10.1006/ijhc.2000.0376

MacKay, D. M. (1969). *Information, mechanism and meaning*. Cambridge, MA: MIT Press.

Mackay, H., Carne, H., Beynon-Davies, P., & Tudhope, D. (2000). Reconfiguring the user: Using rapid application development. *Social Studies of Science, 30*(5), 737–757. doi:10.1177/030631200030005004

Mackay, H., & Gillespie, G. (1992). Extending the social shaping of technology approach: Ideology and appropriation. *Social Studies of Science, 22*, 685–716. doi:10.1177/030631292022004006

MacKenzie, D., & Wajcman, J. (Eds.). (1985). *The social shaping of technology: How the refrigerator got its hum*. Milton Keynes, UK: Open University Press.

Maggi, B. (1990). *Razionalità e benessere: Studio interdisciplinare dell'organizzazione*. Milano, Italy: Etaslibri.

Maggi, B. (2003). *De l'agir organisationnel: Un point de vue sur le travail, le bien-être, l'apprentissage*. Toulouse, France: Octarès Éditions.

Makoto Su, N., Hiroko, D., Wilensky, N., & Redmiles, D. F. (2011). Doing business with theory: Communities of practice in knowledge management. *Computer Supported Cooperative Work – The Journal of Collaborative Computing*. Retrieved from http://www.normsu.com/papers/Su-COP-JCSCW11.pdf.

Malinowski, B. (1922). *Argonauts of the western pacific*. London, UK: Waveland Press.

Malone, T. W., & Crowston, K. (1994). The interdisciplinary study of coordination. *ACM Computing Surveys, 26*(1). doi:10.1145/174666.174668

Mann, R., & Williams, J. (2003). Standards in medical record keeping. *Clinical Medicine, 3*(4), 329–332.

Manovich, L. (2001). *The language of new media*. Cambridge, MA: The MIT Press.

March, J. G. (1991). Exploration and exploitation in organizational learning. *Organization Science, 2*(1), 71–87. doi:10.1287/orsc.2.1.71

Marcus, G. E. (1995). Ethnography in/of the world system: The emergence of multi-sited ethnography. *Annual Review of Anthropology, 24*, 95–117. doi:10.1146/annurev.an.24.100195.000523

Marcus, G. E. (1995). Ethnography in/of the world system: The emergence of multi-sited ethnography. *Annual Review of Anthropology, 24*, 95–117. doi:10.1146/annurev.an.24.100195.000523

Markus, M. L. (1983). Power, politics, and MIS implementation. *Communications of the ACM, 26*, 430–444. doi:10.1145/358141.358148

Markus, M. L. (1984). *Systems in organizations: Bugs and features*. Boston, MA: Pitman.

Markus, M. L. (2000). Toward an integrative theory of risk control. In Baskerville, R., Stage, J., & DeGross, J. I. (Eds.), *Organizational and Social Perspectives on Information Technology* (pp. 167–178). Boston, MA: Kluwer Academic Publishers.

Markus, M. L., & Robey, D. (1988). Information technology and organizational change: Causal structure in theory and research. *Management Science, 34*(5), 583–598. doi:10.1287/mnsc.34.5.583

Markus, M. L., Steinfield, C. W., & Wigand, R. T. (2006). The evolution of vertical IS standards: Electronic interchange standards in the US home mortgage industry. *Management Information Systems Quarterly, 30*, 439–465.

Markus, M. L., Tanis, C., & van Fenema, P. C. (2000). Enterprise resource planning: Multisite ERP implementations. *Communications of the ACM, 43*(4), 42–46. doi:10.1145/332051.332068

Marshak, R. J. (1993). Lewin meets Confucius: A review of the OD model of change. *The Journal of Applied Behavioral Science, 29*, 393–415. doi:10.1177/0021886393294002

Mårtensson, P., & Lee, A. S. (2004). Dialogical action research at omega corporation. *Management Information Systems Quarterly*, *28*(3), 507–536.

Martin, D., Procter, R., Mariani, J., & Rouncefield, M. (2007a). *Working the contract*. Paper presented at OZCHI 2007: Australian Computer-Human Interaction Conference. Adelaide, Australia.

Martin, D., Hartswood, M., Slack, M., & Voss, A. (2006). Achieving dependability in the configuration, integration and testing of healthcare technologies. *Computer Supported Cooperative Work*, *15*(5-6), 467–499. doi:10.1007/s10606-006-9032-1

Martin, D., Mariani, J., & Rouncefield, M. (2007). Managing integration work in an NHS electronic patient record (EPR) project. *Health Informatics Journal*, *13*(1), 47–56. doi:10.1177/1460458207073648

Martin, D., & Sommerville, I. (2006). Ethnography and the social structure of work. In Besnard, D., Gacek, C., & Jones, C. B. (Eds.), *Structure for Dependability: Computer-Based Systems from an Interdisciplinary Perspective* (pp. 169–188). London, UK: Springer. doi:10.1007/1-84628-111-3_9

Masino, G. (1997). *Nuove regole di progettazione: Opportunità tecnologiche e scelte organizzative*. Roma, Italy: Carocci.

Masino, G. (2004). Cambiamento della progettazione organizzativa e progettazione del cambiamento. In Maggi, B., & Masino, G. (Eds.), *Imprese in Cambiamento: Officina di Organizzazione* (pp. 17–28). Bologna, Italy: Bononia University Press.

Masino, G., & Zamarian, M. (2003). Information technology artefacts as structuring devices in organizations: Design, appropriation and use issues. *Interacting with Computers*, *15*(5), 693–707. doi:10.1016/S0953-5438(03)00059-6

Maturana, H., & Varela, F. J. (1992). *The tree of knowledge: The biological roots of human understanding*. Boston, MA: Shambhala Publications.

Maurer, M., & Githens, R. P. (2010). Toward a reframing of action research for human resource and organization development: Moving beyond problem solving and toward dialogue. *Action Research*, *8*(3), 267–292. doi:10.1177/1476750309351361

McCorduck, P. (2003). *Machines who think* (2nd ed). Natick, UK: A K Peters.

McCorduck, P. (1987). *Machines who think*. New York, NY: WH Freeman & Co.

McDermott, D. (1981). Artificial intelligence meets natural stupidity. In Haugeland, J. (Ed.), *Mind Design* (pp. 143–160). Cambridge, MA: MIT Press.

McGarty, T. P. (1992). Alternative networking architectures: Pricing, policy, and competition. In Kahin, B. (Ed.), *Building Information Infrastructure* (pp. 218–270). New York, NY: McGraw-Hill Primis.

McKinney, E., & Yoos, C. (2010). Information about information: A taxonomy of views. *Management Information Systems Quarterly*, *34*(2), 329–344.

McLaughlin, J., Rosen, P., Skinner, D., & Webster, A. (1999). *Valuing technology: Organisations, culture and change*. London, UK: Routledge. doi:10.4324/9780203019870

Merleau-Ponty, M. (1942). *La structure du comportement*. Paris, France: Presses Universitaires de France.

Merleau-Ponty, M. (1945). *Phénoménologie de la perception*. Paris, France: Gallimard.

Merleau-Ponty, M. (1958). *Phenomenology of perception* (Smith, C., Trans.). London, UK: Routledge.

Merleau-Ponty, M. (1962). *Phenomenology of perception*. London, UK: Routledge.

Merleau-Ponty, M. (1969). *Visible and the invisible*. Evanston, IL: Northwestern University Press.

Millen, D. R. (2000). Rapid ethnography: Time deepening strategies for HCI field research. In *Proceedings of DIS*, *2000*, 280–286.

Miller, & O'Leary, T. (1987). Accounting and the construction of the governable person. *Accounting, Organizations and Society*, *12*(3), 235-265.

Mingers, J. (2003). A classification of the philosophical assumptions of management science methods. *The Journal of the Operational Research Society*, *54*, 559–570. doi:10.1057/palgrave.jors.2601436

Mingers, J., & Willcocks, L. (Eds.). (2004). *Social theory and philosophy for information systems*. Chichester, UK: Wiley.

Mintzberg, H. (1990). The design school: Reconsideration of the basic premises of strategic management. *Strategic Management Journal*, *11*, 171–195. doi:10.1002/smj.4250110302

Molina, A. (1989a). *The social basis of the microelectronics revolution*. Edimburgh, UK: Edinburgh University Press.

Molina, A. (1995). Sociotechnical constituencies as processes of alignment: The rise of a large-scale European information technology initiative. *Technology in Society*, *17*(4), 385–412. doi:10.1016/0160-791X(95)00016-K

Molotch, H. (2003). *Where stuff comes from: How toaster, toilets, cars, computers and many other things come to be as they are*. New York, NY: Routledge.

Monteiro, E. (2000). Actor-network theory and information infrastructure. In Ciborra, C. (Ed.), *From Control to Drift: the Dynamics of Corporate Information Infrastructures* (pp. 71–83). Oxford, UK: Oxford University Press.

Monteiro, E. (2003). Integrating health information systems: A critical appraisal. *Methods of Information in Medicine*, *42*, 428–432.

Monteiro, E., & Hanseth, O. (1995). Social shaping of information infrastructure: On being specific about the technology. In Orlikowski, W., Walsham, G., Jones, M. R., & DeGross, J. I. (Eds.), *Information Technology and Changes in Organisational Work* (pp. 325–343). London, UK: Chapman & Hall.

Moorman, C., & Miner, A. S. (1997). The impact of organizational memory on new product performance and creativity. *JMR, Journal of Marketing Research*, *34*, 91–106. doi:10.2307/3152067

Moran, D. (2000). *Introduction to phenomenology*. London, UK: Routledge.

Morrill, C., & Fine, G. A. (1997). Ethnographic contributions to organizational sociology. *Sociological Methods & Research*, *25*, 424–451. doi:10.1177/0049124197025004003

Moustakas, C. (1994). *Phenomenological research methods*. Thousand Oaks, CA: Sage.

Muller, M., & Carey, K. (2002). Design as a minority discipline in a software company: Toward requirements for a community of practice. In *Proceedings of CHI 2002*. Minneapolis, MN: CHI.

Mumford, E. (1983). Participative systems design: Practice and theory. *Journal of Occupational Behaviour*, *4*(1), 47–57.

Mumford, L. (1934). *Technics and civilization*. San Diego, CA: HJB.

Myers, M. D. (2008). *Qualitative research in information systems*. Retrieved from http://www.qual.auckland.ac.nz/a>.

Myers, M. D. (1999). Investigating information systems with ethnographic research. *Communications of AIS*, *2*(23), 1–20.

Nancy, J. L. (1986). *La communauté désoeuvrée*. Paris, France: Christian Bourgois.

Nancy, J. L. (1988). *Die undarstellbare gemeinschaft. Stuttgart, Germany: Legueil. Nancy, J. L. (1991). The inoperative community*. Minneapolis, MN: University of Minnesota Press.

Nancy, J. L. (1992). *La comunità inoperosa*. Napoli, Italy: Cronopio.

Nancy, J. L. (1996). *Étre singulier pluriel*. Paris, France: Galilée.

Nandhakumar, J. (2010). Contrarian information systems studies. *European Journal of Information Systems*, *19*, 687–688. doi:10.1057/ejis.2010.49

Nardi, B. A. (1993). *A small matter of programming: Perspectives on end user computing*. Cambridge, MA: MIT Press.

Nardi, B. A. (1996). Activity theory and human-computer interaction. In Nardi, B. A. (Ed.), *Context and Consciousness: Activity Theory and Human-Computer Interaction* (pp. 7–16). Cambridge, MA: The MIT Press.

Nardi, B., & Kow, Y. M. (2010). Digital imaginaries: What we (think we) know about Chinese gold farming. *First Monday*, *15*, 6–7.

Natanson, M. (1963). A study in philosophy and the social sciences. In Natanson, M. (Ed.), *Philosophy of the Social Sciences* (pp. 271–285). New York, NY: Random House.

Natanson, M. (1973). *Edmund Husserl: Philosopher of infinite tasks*. Evantson, IL: Northwestern University.

National Audit Office. (2008). *The national programme for IT in the NHS: Progress since 2006 National Audit Office Value for Money Report, HC 484-I Session 2007-2008: 16 May 2008*. London, UK: National Audit Office.

Neff, W. S. (2006). *Work & human behavior*. Piscataway, NJ: Transaction Publishers.

Neisser, U. (1976). *Cognition and reality: Principles and implications of cognitive psychology*. San Francisco, CA: Freeman.

Nelson, D. (1980). *Frederick W. Taylor and the rise of scientific management*. Madison, WI: The University of Wisconsin.

Nelson, D. (1992). *A mental revolution: Scientific management since Taylor*. Columbus, OH: Ohio State University Press.

Newell, A., & Card, S. (1985). The prospects for psychological science in human computer interaction. *Human-Computer Interaction*, *1*(3), 209–242. doi:10.1207/s15327051hci0103_1

Newell, A., & Card, S. (1986). Straightening out softening up: Response to Carroll and Campbell. *Human-Computer Interaction*, *2*(3), 251–267. doi:10.1207/s15327051hci0203_4

Newman, M. W., & Landay, J. A. (2000). Sitemaps, story boards, and specifications: A sketch of web site design practice. In *Proceedings of DIS 2000*. Brooklyn, NY: ACM Press.

Newman, M., & Westrup, C. (2005). Making ERPs work: Accountants and the introduction of ERP systems. *European Journal of Information Systems*, *14*(3), 258–272. doi:10.1057/palgrave.ejis.3000539

Nielsen, K. A., & Svensson, L. G. (Eds.). (2006). *Action research and interactive research: Beyond practice and theory*. Maastricht, Germany: Shaker.

Nielsen, P., & Aanestad, M. (2006). Infrastructuralisation as design strategy: A case study of a content service platform for mobile phones in Norway. *Journal of Information Technology*, *21*(3), 185–194. doi:10.1057/palgrave.jit.2000064

Noble, D. F. (1977). *America by design: Science, technology, and the rise of corporate capitalism*. New York, NY: A.A. Knopf.

Nonaka, I., & Takeuchi, H. (1995). *The knowledge-creating company*. Oxford, UK: Oxford University Press.

Nooteboom, B. (1992). Information technology, transaction costs and the decision to 'make or buy'. *Technology Analysis and Strategic Management*, *4*, 339. doi:10.1080/09537329208524105

Norman, D. (1988). *Psychologygy of everyday things*. New York, NY: Basic Books.

Norman, D. (1988). *The psychology of everyday things*. New York, NY: Basic Books.

Norman, D. (1992). *Turn signals are the facial expressions of automobiles*. Cambridge, MA: Perseus Books.

Norman, D. (1993). *Things that make us smart: Defending human attributes in the age of the machine*. New York, NY: Addison-Wesley Publishing.

Norman, D. (1999). *The invisible computer: Why good products can fail, the personal computer is so complex and information appliances are the solution*. Cambridge, MA: MIT Press.

Norman, D. (2004). *Emotional design: Why we love (or hate) everyday things*. New York, NY: Basic Books.

Norman, D. (2007). *The design of future things*. New York, NY: Basic Books.

Norman, D. A. (1988). *The design of everyday things*. New York, NY: Doubleday.

Normark, M. (2002). *Using technology for real-time coordination of work: A study of work and artefacts in the everyday activities at SOS alarm*. Stockholm, Sweden: Kungl Tekniska Högskolan.

Norris, A. C. (2002). Current trends and challenges in health informatics. *Health Informatics Journal*, *8*, 205–213. doi:10.1177/146045820200800407

Norris, G., Hurley, J. R., Dunleavy, J. R., Balls, J. D., & Hartley, K. M. (2000). *E-business and ERP: Transforming the enterprise*. Chichester, UK: John Wiley.

Noyes, J. W. (2003). The ERP dilemma: "Plain vanilla" versus customer satisfaction. *EDUCAUSE Quarterly*, *2*, 54–55.

Nunberg, G. (1996). Farewell to the information age. In Nunberg (Ed.), *The Future of the Book*, (pp. 103-138). Berkeley, CA: University of California Press.

Nussbaum, M. C. (1956). *The fragility of goodness: Luck and ethics in Greek tragedy and philosophy*. Cambridge, UK: Cambridge University Press.

Odello, G. (2007). *Espresso Italiano tasting*. Brescia, Italy: Centro Studi Assaggiatori.

O'Leary, D. E. (2000). *Enterprise resource planning systems: Systems, life cycle, electronic commerce, and risk*. Cambridge, MA: Cambridge University Press.

Olivé, A. (2007). *Conceptual modeling of information systems*. Berlin, Germany: Springer.

Orlikowski, W. J. (1996). Improvising organisational transformation over time: A situated change perspective. *Information Systems Research, 7*(1), 63–92. doi:10.1287/isre.7.1.63

Orlikowski, W. J. (2000). Using technology and constituting structures: A practical lens for studying technology in organizations. *Organization Science, 11*, 404–428. doi:10.1287/orsc.11.4.404.14600

Orlikowski, W. J., & Barley, S. R. (2001). Technology and institutions: What can research on information technology and research on organizations learn from each other? *Management Information Systems Quarterly, 25*(2), 145–165. doi:10.2307/3250927

Orlikowski, W. J., & Baroudi, J. (1991). Studying information technology in organizations: Research approaches and assumptions. *Information Systems Research, 2*(1), 1–28. doi:10.1287/isre.2.1.1

Orlikowski, W. J., & Iacono, C. S. (2001). Research commentary: Desperately seeking the "IT" in IT research - A call to theorizing the IT artifact. *Information Systems Research, 12*(2), 121–134. doi:10.1287/isre.12.2.121.9700

Orr, J. E. (1996). *Talking about machines: An ethnography of a modern job*. Ithaca, NY: ILR Press.

Oudshoorn, N., Rommes, E., & Stienstra, M. (2004). Configuring the user as everybody: Gender and design in information and communication technologies. *Science, Technology & Human Values, 29*(1), 30–63. doi:10.1177/0162243903259190

Pahl, G., Beitz, W., Feldhusen, J., & Grote, K.-H. (2007). *Engineering design: A systematic approach* (3rd ed.). Berlin, Germany: Springer.

Papanek, V. (1972). *Design for the real world (suom. turhaa vai tarpeellista?)*. London, UK: Thames and Hudson.

Parker, M., Stofberg, C., De la Harpe, R., Venter, I., & Wills, G. (2006). *Data quality: How the flow of data influences data quality in a small to medium medical practice*. Paper presented at the Community Informatics for Developing Countries: Understanding and Organising for a Participatory Future Information Society Conference. Cape Town, South Africa.

Paynter, B. (2008, July). Feeding the masses. *Wired.*

Peltu, M., Eason, K., & Clegg, C. (2008). *How a sociotechnical approach can help NPfIT deliver better NHS patient care.* London, UK: Bayswater Institute. Retrieved from http://www.bayswaterinst.org/downloads/Sociotech%20Group%20NPfIT%20report%20May08.pdf.

Pettersson, M., Randall, D., & Helgeson, B. (2004). Ambiguities, awareness and economy: A study of emergency service work. *Computer Supported Cooperative Work, 13*, 125–154. doi:10.1023/B:COSU.0000045707.37815.d1

Pfaffenberger, B. (1992). Technological dramas. *Science, Technology & Human Values, 17*, 282–312. doi:10.1177/016224399201700302

Pickering, A. (2004). The science of the unknowable: Stafford Beer's cybernetic informatics. *Kybernetes, 33*(4), 499–521. doi:10.1108/03684920410523535

Pipek, V., & Wulf, V. (2009). Infrastructuring: Toward an integrated perspective on the design and use of information technology. *Journal of the Association for Information Systems, 10*(5). Retrieved from http://aisel.aisnet.org/jais/vol10/iss5/

Pitt, L. F., Watson, R. T., & Kavan, C. B. (1995). Service quality: A measure of information system effectiveness. *Management Information Systems Quarterly, 19*(2), 173–188. doi:10.2307/249687

Plato,. (1941). *The republic of Plato* (Cornford, F. M., Trans.). Oxford, UK: Oxford University Press.

Plowman, L., Rogers, Y., & Ramage, M. (1995). What are workplace studies for? In H. Marmolin, Y. Sundblad, & K. Schmidt (Eds.), *Proceedings of the Fourth European Conference on Computer-Supported Cooperative Work*, (pp. 309-324). Stockholm, Sweden: Kluwer Academic Publishers.

Poirier, C. (2003). *The supply chain manager's problem-solver*. Boca Raton, FL: CRC Press.

Polanyi, M. (1958). *Personal knowledge*. New York, NY: Routledge & Kegan Paul.

Polanyi, M. (1966). *The tacit dimension*. London, UK: Routledge & Kegan Paul.

Pollock, N., & Williams, R. (2009). Constructing technology markets: How industry analysts organise technological fields. *Sprouts: Working Papers on Information Systems, 9*(10).

Pollock, N. (2005). When is a work-around? Conflict and struggle in computer systems development. *Science, Technology & Human Values, 30*(4), 496–514. doi:10.1177/0162243905276501

Pollock, N., & Cornford, J. (2004). ERP systems and the university as an 'unique' organisation. *Information Technology & People, 17*(1), 31–52. doi:10.1108/09593840410522161

Pollock, N., & Williams, R. (2007). Technology choice and its performance: Towards a sociology of software package procurement. *Information and Organization, 17*(3), 131–160. doi:10.1016/j.infoandorg.2007.05.001

Pollock, N., & Williams, R. (2008). *Software and organisations: The biography of the enterprise-wide system or how SAP conquered the world*. London, UK: Routledge.

Pollock, N., Williams, R., & D'Adderio, L. (2007). Global software and its provenance: Generification work in the production of organizational software packages. *Social Studies of Science, 37*(2), 254–280. doi:10.1177/0306312706066022

Pollock, N., Williams, R., D'Adderio, L., & Grimm, C. (2009). Post local forms of repair: The (extended) situation of virtualised technical support. *Information and Organization, 19*, 253–276. doi:10.1016/j.infoandorg.2009.08.001

Poole, M., & DeSanctis, G. (1990). Understanding the use of group decision support systems: The theory of adaptive structuration. *Organizations and Communication Technology, 173*, 191.

Popper, K. R. (1945). The open society and it's enemies: *Vol. 2. Hegel and Marx*. London, UK: Routledge.

Portelli, A. (2004). Oral history as genre. In Chamberlain, M., & Thomposn, P. (Eds.), *Narrative and Genre: Contexts and Types of Communication*. Transaction Publishers.

Prasad, P. (1997). Systems of meaning: Ethnography as a methodology for the study of information technologies. In Lee, A. S., Liebenau, J., & DeGross, J. I. (Eds.), *Information Systems and Qualitative Research* (pp. 101–118). London, UK: Chapman and Hall.

Prochaska, J. O., Di Clemente, C. C., & Norcross, J. C. (1992). In search of how people change: Applications to addictive behaviors. *The American Psychologist, 47*, 1102–1114. doi:10.1037/0003-066X.47.9.1102

Prus, R. (1995). The ethnographic research tradition. In *Symbolic Interaction and Ethnographic Research: Intersubjectivity and the Study of Human Lived Experience* (pp. 103–140). New York, NY: SUNY Press.

Psathas, G. (2008). Reflections on the history of ethnomethodology: The Boston and Manchester 'schools'. *The American Sociologist, 39*(1), 38–67. doi:10.1007/s12108-008-9032-3

Ramachandran, V. S. (2003). *The emerging mind: BBC Reith lectures 2003*. London, UK: Profile Books.

Randall, D., Harper, R., & Rouncefield, M. (2007). *Fieldwork for design: Theory and practice*. London, UK: Springer.

Randell, B. (2007). A computer scientist's reactions to NPfIT. *Journal of Information Technology, 22*, 222–234. doi:10.1057/palgrave.jit.2000106

Rawls, A. W. (2002). Editor's introduction. In Garfinkel, H. (Ed.), *Ethnomethodology's Program: Working out Durkheim's Aphorism* (pp. 1–64). Lantham, UK: Rowman & Littlefield.

Rector, A. L., Nolan, W. A., & Kay, S. (1991). Foundations for an electronic medical record. *Methods of Information in Medicine, 30*, 179–186.

Redman, T. C. (1998). The impact of poor data quality on the typical enterprise. *Communications of the ACM, 41*(2), 79–82. doi:10.1145/269012.269025

Redmiles, D. (2002). Introduction. *Computer Supported Cooperative Work, 11*(1-2), 1–11. doi:10.1023/A:1015215726353

Reese, W. (2002). Behavioral scientists enter design: Seven critical histories. In Squires, S., & Byrne, B. (Eds.), *Creating Breakthrough Ideas: The Collaboration of Anthropologists and Designers in the Product Development Industry*. Westport, CT: Greenwood.

Resnick, L., Säljö, R., Pontecorvo, C., & Burge, B. (Eds.). (1997). *Discourse, tools and reasoning: Essays on situated cognition*. Berlin, Germany: Springer Verlag.

Ribes, D., & Finholt, T. A. (2009). The long now of technology infrastructure: Articulating tensions in development. *Journal of the Association for Information Systems, 10*(5), 375–398. Retrieved from http://aisel.aisnet.org/jais/vol10/iss5/

Ricoeur, P. (1991). *From text to action*. Evanston, IL: Northwestern University Press.

Riesman, D. (1983). The legacy of Everett C. Hughes. *Contemporary Sociology, 12*(5), 477–481.

Riessman, C. K. (1993). *Narrative analysis*. London, UK: Sage.

Robey, D., & Markus, M. L. (1984). Rituals in information system design. *Management Information Systems Quarterly, 8*, 5–15. doi:10.2307/249240

Robinson, M., & Bannon, L. (1991). *Questioning representations*. Paper presented at ECSCW 1991. Amsterdam, The Netherlands.

Rogers, C. R. (1951). *Client-centred therapy*. New York, NY: Constable.

Rogers, E. M. (1983). *Diffusion of innovations*. New York, NY: The Free Press.

Rogers, Y., & Bellotti, V. M. E. (1997). Grounding blue-sky research: How can ethnography help? *Interaction, 4*(3), 58–63. doi:10.1145/255392.255404

Rosenberg, D. (2000). Three steps to ethnography: A discussion of inter-disciplinary contributions. *AI & Society, 15*(1), 295–315.

Rosen, M. (1991). Coming to terms with the field: Understanding and doing organizational ethnography. *Journal of Management Studies, 28*(1), 1–24. doi:10.1111/j.1467-6486.1991.tb00268.x

Ross, J. W. (1998). *The ERP revolution: Surviving versus thriving*. Cambridge, MA: MIT.

Rouvroy, A. (2011). Technology, virtuality and utopia. In Hildebrandt, M., & Rouvroy, A. (Eds.), *The Philosophy of Law meets the Philosophy of Technology: Autonomic Computing and Transformations of Human Agency*. London, UK: Routledge.

Rowland, N. (2010). *Turning buyers into sellers: The spread and spreading of ERP in American higher education*. Paper presented at the Conference of the European Association for the Study of Science and Tecnhology. Trento, Italy.

Rowland, N. J. (2008). *Fit-gap work: Implementation, innovation, and organizational forgetting in higher education with ERPs*. Paper presented at the Annual Meeting for the American Sociological Association. Boston, MA.

Rowland, N. J., & Gieryn, T. F. (2008). Transfer troubles: Outsourcing information technology in higher education. In Swedberg, R., & Pinch, T. J. (Eds.), *Living in a Material World: Economic Sociology Meets Science and Technology Studies* (pp. 375–391). Cambridge, MA: MIT Press.

Ruhleder, K., & Jordan, B. (1997). capturing complex, distributed activities: video-based interaction analysis as a component of workplace ethnography. In A. S. Lee, J. Liebenau, & J. I. DeGross (Eds.), *Information Systems and Qualitative Research*, (pp. 246-275). London, UK: Chapman and Hall.

Rusaw, A. C. (2007). Changing public organizations: Four approaches. *International Journal of Public Administration, 30*(3), 347–361. doi:10.1080/01900690601117853

Ryle, G. (1949). *The concept of mind*. London, UK: Hutchinson.

Ryle, G. (1966). *Dilemmas*. Cambridge, UK: Cambridge University Press.

Sahay, S., Monteiro, E., & Aanestad, M. (2009). Configurable politics and asymmetric integration: Health e-infrastructures in India. *Journal of the Association for Information Systems, 10*(5), 399–414. Retrieved from http://aisel.aisnet.org/jais/vol10/iss5/

Salber, D., Dey, A. K., & Abowd, G. D. (1999). The context toolkit: Aiding the development of context-enabled applications. In *Proceedings of the SIGCHI Conference on Human Factors in Computing Systems: The CHI Is the Limit*, (pp. 434-441). Pittsburgh, PA: ACM Press.

Saleem, J. J., Russ, A. L., Justice, C. F., Hagg, H., Ebright, P. R., Woodbridge, P. A., & Doebbeling, B. N. (2009). Exploring the persistence of paper with the electronic health record. *International Journal of Medical Informatics, 78*(9), 618–628. doi:10.1016/j.ijmedinf.2009.04.001

Saltzer, J. H. (1984). End-to-end arguments in systems design. *ACM Transactions on Computer Systems, 2*(4), 277–288. doi:10.1145/357401.357402

Salvador, T., Bell, G., & Anderson, K. (1999). Design ethnography. *Design Management Journal, 19*(4), 35–41.

Sanday, P. R. (1979). The ethnographic paradigms. *Administrative Science Quarterly, 24*(4), 527–538. doi:10.2307/2392359

Sawyer, R. K. (2000). Improvisational cultures: Collaborative emergence and creativity in improvisation: Commentary on lead article by Howard Becker. *The Etiquette of Improvisation. Mind, Culture, and Activity, 7*(3), 180–185. doi:10.1207/S15327884MCA0703_05

Sawyer, S., & Southwick, R. (2002). Temporal issues in information and communication technology-enabled organizational change. *The Information Society, 18*, 263–280. doi:10.1080/01972240290075110

Schegloff, E. A. (2007). *Sequence organization in interaction*. Cambridge, UK: Cambridge University Press. doi:10.1017/CBO9780511791208

Schmidt, K. (1991). Riding a tiger, or computer supported cooperative work. In *Proceedings of the Second Conference on European Conference on Computer-Supported Cooperative Work (ECSCW 1991)*, (pp. 1-16). Norwell, MA: Kluwer Academic Publishers.

Schmidt, K. (1997). Of maps and scripts - The status of formal constructs in cooperative work. In *Proceedings of the International ACM SIGGROUP Conference on Supporting Group Work: The Integration Challenge (GROUP 1997)*, (pp. 138-147). New York, NY: ACM.

Schmidt, K. (2000). *The critical role of workplace studies in CSCW*. Retrieved from http://www.itu.dk/people/schmidt/papers/workplace_studies.pdf.

Schmidt, K. (2001). Classification schemes as material practice. In *Proceedings of the Second CISCPH Workshops on Cooperative Organization of Common Information Spaces*. Copenhagen, Denmark: CISCPH.

Schmidt, K. (2010). Keep up the good work! - The concept of work in CSCW. In *Proceedings of the 9th International Conference on Designing Cooperative Systems*, (pp. 265-286). Aix-en-Provence, France: Springer.

Schmidt, K., & Wagner, I. (2002). Coordinative artefacts in architectural practice. In M. Blay-Fornarino et al. (Eds.), *Cooperative Systems Design: A Challenge of the Mobility Age: Proceedings of the Fifth International Conference on the Design of Cooperative Systems (COOP 2002)*, (pp. 257-274). Amsterdam, The Netherlands: IOS Press.

Schmidt, K. (1994). *Modes and mechanisms of interaction in cooperative work: Outline of a conceptual framework*. Roskilde, Denmark: Risø National Laboratory.

Schmidt, K. (2000). The critical role of workplace studies in CSCW. In Luff, P., Hindmarsh, J., Heath, C., & Hinds, P. J. (Eds.), *Workplace Studies: Recovering Work Practice and Informing System Design* (pp. 141–149). Cambridge, UK: Cambridge University Press. doi:10.1017/CBO9780511628122.007

Schmidt, K. (2002). The problem with "awareness": Introductory remarks on awareness in CSCW. *Computer Supported Cooperative Work, 11*(3), 285–298. doi:10.1023/A:1021272909573

Schmidt, K., & Bannon, L. (1992). Taking CSCW seriously: Supporting articulation work. *Computer Supported Cooperative Work (CSCW). International Journal (Toronto, Ont.), 1*, 7–41.

Schmidt, K., & Bannon, L. (1992). Taking CSCW seriously: Supporting articulation work. *Computer Supported Cooperative Work, 1*, 7–40. doi:10.1007/BF00752449

Schmidt, K., & Simone, C. (1996). Coordination mechanisms: Towards a conceptual foundation of CSCW systems design. *Journal of Collaborative Computing, 5*, 155–200. doi:10.1007/BF00133655

Schmidt, K., & Wagner, I. (2004). Ordering systems: Coordinative practices and artifacts in architectural design and planning. *Computer Supported Cooperative Work*, *13*(5-6), 349–408. doi:10.1007/s10606-004-5059-3

Schon, D. (1983). *The reflective practitioner: How professionals think in action*. New York, NY: Basic Books.

Schon, D. A. (1983). *The reflective practitioner*. New York, NY: Harper Collins.

Schuetz, A. (1943). The problem of rationality in the social world. *Economica*, *10*(38), 130–149. doi:10.2307/2549460

Schuetz, A. (1951). Choosing among projects of action. *Philosophy and Phenomenological Research*, *12*(2), 161–184. doi:10.2307/2103478

Schuetz, A. (1953). Common-sense and scientific interpretation of human action. *Philosophy and Phenomenological Research*, *14*(1), 1–38. doi:10.2307/2104013

Schuler, D., & Namioka, A. (Eds.). (1993). *Participatory design: Principles and practices*. Hillsdale, NJ: Lawrence Erlbaum Associates.

Schultze, U., & Leidner, D. E. (2002). Studying knowledge management in information systems research: Discourses and theoretical assumptions. *Management Information Systems Quarterly*, *26*(3), 213–242. doi:10.2307/4132331

Schütz, A. (1962). *Collected papers: The probelm of social reality* (*Vol. 1*). The Hague, The Netherlands: Martinus Nijhoff.

Schütz, A. (1967). *The phenomenolgy of the social world* (Walsh, G., & Lehnert, F., Trans.). Evanston, IL: Northwestern University Press.

Schwartzman, H. B. (1993). *Ethnography in organizations*. Newbury Park, CA: Sage.

Scott, J. C. (1999). *Seeing like a state: How certain schemes to improve the human condition have failed*. New Haven, CT: Yale University Press.

Scott, S. V., & Wagner, E. L. (2003). Networks, negotiations and new times: The implementation of enterprise resource planning into an academic administration. *Information and Organization*, *13*, 285–313. doi:10.1016/S1471-7727(03)00012-5

Scruton, R. (2002). *A short history of modern philosophy*. Oxon, UK: Routledge.

Scruton, R. (2004). *Modern philosophy*. London, UK: Pimlico.

Sellen, A., & Harper, R. (2001). *The myth of the paperless office*. Cambridge, MA: MIT Press.

Sesé, F., Wareham, J., & Bonet, E. (2006). Words and objects in information systems development: Six paradigms of information as representation. In *Proceedings of the 14th European Conference on Information Systems*. Göteborg, Sweden: IEEE.

Shanks, G. (1999). Semiotic approach to understanding representation in information systems. In *Proceedings of the IS Foundations Workshop*. Sydney, Australia: IS Founation.

Shanks, G., & Corbitt, B. (1999). Understanding data quality: Social and cultural aspects. In *Proceedings of the 10th Australasian Conference on Information Systems*. Wellington, New Zealand: IEEE.

Shapiro, D. (1994). The limits of ethnography: Combining social sciences for CSCW. In *Proceedings of CSCW, 1994*, 417–428.

Sharrock, W., & Hughes, J. A. (2001). Ethnography in the workplace: Remarks on its theoretical bases. *Team Ethno, 1*. Retrieved from http://www.teamethno-online.org/Issue1/Wes.html.

Sharrock, W., & Anderson, R. (1994). The user as a scenic feature of the design space. *Design Studies*, *5*, 5–28. doi:10.1016/0142-694X(94)90036-1

Sharrock, W., & Randall, D. (2004). Ethnography, ethnomethodology and the problem of generalisation in design. *European Journal of Information Systems*, *13*(3), 186–194. doi:10.1057/palgrave.ejis.3000502

Sherry, J. F. (2002). Foreword: Ethnography, design, and customer experience: An anthropologist's sense of it all. In Squires, S., & Byrne, B. (Eds.), *Creating Breakthrough Ideas: The Collaboration of Anthropologists and Designers in the Product Development Industry*. Westport, CT: Greenwood.

Shirky, C. (2008). *Here comes everybody*. London, UK: Penguin.

Simone, C., Divitini, M., & Schmidt, K. (1995). A notation for malleable and interoperable coordination mechanisms for CSCW systems. In *Proceedings of Conference on Organizational Computing Systems*, (pp. 44-54). New York, NY: ACM.

Simone, C., Mark, G., & Giubbilei, D. (1999). Interoperability as a means of articulation work. In *Proceedings of WACC 1999*, (pp. 39-48). San Francisco, CA: ACM Press.

Simone, C., & Divitini, M. (1997). Ariadne: Supporting coordination through a flexible use of the knowledge on work process. *Journal of Universal Computer Science*, *3*(8), 865–898.

Simon, H. A. (1969). Rational choice and the structure of the environment. In Emery, F. E. (Ed.), *Systems Thinking* (pp. 214–229). Harmondsworth, UK: Penguin. doi:10.1037/h0042769

Simonsen, J., & Kensing, F. (1994). Take users seriously, but take a deeper look: Organizational and technical effects from designing with an ethnographically inspired approach. In *Proceedings of PDC 1994*, (pp. 47-58). Palo Alto, CA: PDC.

Simonsen, J., & Kensing, F. (1997). Using ethnography in contextual design. *Communications of the ACM*, *40*(7), 82–88. doi:10.1145/256175.256190

Simonsen, J., & Kensing, F. (1998). Make room for ethnography in design! Overlooked collaborative and educational prospects. *The Journal of Computer Documentation*, *22*(1), 20–30.

Sims, H. P., & Gioia, D. A. (Eds.). (1986). *The thinking organizations*. San Francisco, CA: Jossey-Bass.

Slack, J. D., & Fejes, F. (1987). *The ideology of the information age*. New York, NY: Ablex Publishing.

Soh, C., Kien, S. S., & Tay-Yap, J. (2000). Cultural fits and misfits: Is ERP a universal solution? *Communications of the ACM*, *43*(4), 47–51. doi:10.1145/332051.332070

Somers, T., & Nelson, K. (2004). A taxonomy of players and activities across the ERP project life cycle. *Information & Management*, *41*, 257–278. doi:10.1016/S0378-7206(03)00023-5

Sommerville, I. (2004). Making ethnography accessible. In *Proceeding of the Workshop on Software Engineering/HCI*, (pp. 36-40). Edinburgh, UK: Institute of Electrical Engineers.

Sommerville, I., Rodden, T., Bentley, R., & Sawyer, P. (1992). Sociologists can be surprisingly useful in interactive systems design. In A. Monk, D. Diaper, & M. Harrison (Eds.), *Proceedings of the HCI 1992, Conference on People and Computers*, (pp. 341-353). Cambridge, UK: Cambridge University Press.

Sommerville, I., Rodden, T., Sawyer, P., Bentley, R., & Twidale, M. (1993). Integrating ethnography into the requirements engineering process. In *Proceeding of the IEEE International Symposium on Requirements Engineering (RE 1993)*, (pp. 165-173). San Diego, CA: IEEE Press.

Sommerville, I., & Sawyer, P. (1997). *Requirements engineering: A good practice guide*. Chichester, UK: Wiley.

Sørensen, K., & Levold, N. (1992). Tacit networks, heterogeneous engineers, and embodied technology. *Science, Technology & Human Values*, *17*(1), 13–35. doi:10.1177/016224399201700102

Sprott, D. (2000). Enterprise resource planning: Componentizing the enterprise application packages tomorrow's customers will demand the ability to buy, reuse and build their competitive-edge solutions: Can the package vendors adapt in time? *Communications of the ACM*, *43*(4), 63–69. doi:10.1145/332051.332074

Srinivasan, R. (2007). Ethnomethodological architectures: The convergence between an information system and the cultural landscape. *Journal of the American Society for Information Science and Technology*, *58*(5), 723–733. doi:10.1002/asi.20544

Stamper, R. (1992). Signs, organisations, norms and information systems. In *Proceedings of the 3rd Australian Conference on Information Systems*. Wollongong, Australia: IEEE.

Star, S. L., Bowker, G. C., & Neumann, L. J. (1997). *Transparency at different levels of scale: Convergence between information artifacts and social worlds*. Retrieved 27 May 2009 from http://epl.scu.edu:16080/~gbowker/pubs.htm.

Star, S. L. (1989). The structure of ill-structured solutions: Boundary objects and heterogeneous distributed problem solving. In Gasser, L., & Huhns, M. (Eds.), *Distributed Artificial Intelligence* (*Vol. 2*, pp. 37–54). San Francisco, CA: Morgan Kaufman.

Star, S. L. (1991). The sociology of the invisible: The primacy of work in the writings of Anselm Strauss. In Maines, D. (Ed.), *Social Organization and Social Process: Essays in Honor of Anselm Strauss* (pp. 265–283). Hawthorne, NY: Aldine de Gruyter.

Star, S. L. (1997). Anselm Strauss: An appreciation. *Sociological Research Online*, *2*(1). Retrieved from http://www.socresonline.org.uk/socresonline/2/1/1.htmldoi:10.5153/sro.92

Star, S. L. (1998). Working together: Symbolic interactionism, activity theory and distributed artificial intelligence. In Engeström, Y., & Middleton, D. (Eds.), *Cognition and Communication at Work* (pp. 296–318). Cambridge, UK: Cambridge University Press.

Star, S. L. (1999). The ethnography of infrastructure. *The American Behavioral Scientist*, *43*(3), 377–391. doi:10.1177/00027649921955326

Star, S. L., & Griesemer, J. R. (1989). Institutional ecology, 'translations' and boundary objects: Amateurs and professionals in Berkley's museum of vertebrate zoology, 1907–39. *Social Studies of Science*, *19*(3), 387–420. doi:10.1177/030631289019003001

Star, S. L., & Ruhleder, K. (1996). Steps towards an ecology of infrastructure: Design and access for large information spaces. *Information Systems Research*, *7*(1), 111–134. doi:10.1287/isre.7.1.111

Star, S. L., & Strauss, A. (1999). Layers of silence, arenas of voice: The ecology of visible and invisible work. *Computer Supported Cooperative Work*, *8*, 9–30. doi:10.1023/A:1008651105359

Stewart, J., & Hyysalo, S. (2008). Intermediaries, users and social learning in technological innovation. *International Journal of Innovation Management*, *12*(3), 295–325. doi:10.1142/S1363919608002035

Stewart, J., & Williams, R. (2005). The wrong trousers? Beyond the design fallacy: Social learning and the user. In Rohracher, H. (Ed.), *User Involvement in Innovation Processes: Strategies and Limitations from a Socio-Technical Perspective*. Munich, Germany: Profil-Verlag.

Strauss, A. L. (1991). *Social worlds and spatial processes: An analytic perspective*. Working Paper. Berkeley, CA: University of California.

Strauss, A. (1978). A social world perspective. In Denzin, N. (Ed.), *Studies in Symbolic Interaction* (pp. 119–128). Greenwich, CT: JAI Press.

Strauss, A. (1985). Work and the division of labor. *The Sociological Quarterly*, *26*, 1–19. doi:10.1111/j.1533-8525.1985.tb00212.x

Strauss, A. (1988). The articulation of project work: An organizational process. *The Sociological Quarterly*, *29*(2), 163–178. doi:10.1111/j.1533-8525.1988.tb01249.x

Strauss, A. (1996). Everett Hughes: Sociology's mission. *Symbolic Interaction*, *19*(4), 271–283. doi:10.1525/si.1996.19.4.271

Strauss, A. L. (1978a). A social worlds perspective. *Studies in Symbolic Interaction*, *1*, 119–128.

Strauss, A. L. (1978b). *Negotiations: Varieties, contexts, processes and social order*. San Francisco, CA: Jossey Bass.

Strauss, A. L. (1987). *Qualitative analysis for social scientists*. New York, NY: Cambridge University Press. doi:10.1017/CBO9780511557842

Strauss, A. L. (1993). *Continual permutation of action*. New York, NY: Aldine de Gruyter.

Strauss, A. L., & Corbin, J. (1998). *Basics of qualitative research: Techniques and procedures for developing grounded theory*. Thousand Oaks, CA: Sage.

Strauss, A., Fagerhaugh, S., Suczek, B., & Wiener, C. (1985). *Social organization of medical work*. Chicago, IL: University of Chicago Press.

Strong, M. D., Lee, W. Y., & Wang, Y. R. (1997). Data quality in context. *Communications of the ACM*, *40*(5), 103–110. doi:10.1145/253769.253804

Strübing, J. (1992). Negotiation: A central aspect of collaborative work in software design. In *Proceedings of the 5ème Workshop sur la Psychologie de la Programmation*, (pp. 31-39). Paris, France: Institut National de Recherche en Informatique et en Automatique.

Strübing, J. (1998). Bridging the gap: On the collaboration between symbolic interactionism and distributed artificial intelligence in the field of multi-agent systems research. *Symbolic Interaction*, *21*(4), 441–463. doi:10.1525/si.1998.21.4.441

Sturdy, A. (1997). The consultancy process: An insecure business? *Journal of Management Studies*, *34*(3), 389–413. doi:10.1111/1467-6486.00056

Suchman, L. (1987). *Plans and situated actions: The problem of human-machine communication*. New York, NY: Cambridge University Press.

Suchman, L. (1993). Technologies of accountability: On lizards and airplanes. In Button, G. (Ed.), *Technology in Working Order: Studies of Work, Interaction, and Technology* (pp. 113–126). New York, NY: Routeledge.

Suchman, L. (1994). Working relations of technology production and use. *Computer Supported Cooperative Work*, *2*(1-2), 21–39. doi:10.1007/BF00749282

Suchman, L. (1996). Constituting shared workplaces. In Engstrom, Y., & Middleton, D. (Eds.), *Cognition and Communication at Work* (pp. 35–60). New York, NY: Cambridge University Press.

Suchman, L. (1996). Supporting articulation work. In Kling, R. (Ed.), *Computerization and Controversy: Value Conflicts and Social Choices* (pp. 407–423). San Diego, CA: Academic Press.

Suchman, L. (1997). Centres of coordination: A case and some themes. In Reisnick, L. B., Saljo, R., Pontecorvo, C., & Burge, B. (Eds.), *Discourse, Tools, and Reasoning: Essays on Situated Cognition* (pp. 41–62). Berlin, Germany: Springer-Verlag.

Suchman, L. (2002). Located accountabilities in technology production. *Scandinavian Journal of Information Systems*, *14*(2), 91–105.

Suchman, L. (2002). Practice-based design of information systems: Notes from the hyperdeveloped world. *The Information Society*, *8*(2), 139–144. doi:10.1080/01972240290075066

Suchman, L. A. (1987). *Plans and situation actions: The problem of human-machine communication*. Cambridge, UK: Cambridge University Press.

Suchman, L. A., & Trigg, R. H. (1991). Understanding practice: Video as a medium for reflection and design. In Greenbaum, J., & Kyng, M. (Eds.), *Design at Work: Cooperative Design of Computer Systems* (pp. 65–90). Hillsdale, NJ: Erlbaum.

Suchman, L., Blomberg, J., Orr, J. E., & Trigg, R. (1999). Reconstructing technologies as social practice. *The American Behavioral Scientist*, *43*(3), 392–408. doi:10.1177/00027649921955335

Suchman, L., Trigg, R., & Blomberg, J. (2002). Working artefacts: Ethnomethods of the prototype. *The British Journal of Sociology*, *53*, 163–179. doi:10.1080/00071310220133287

Sunderland, P. L., & Denny, R. M. T. (2007). *Doing anthropology in consumer research*. Walnut Creek, CA: Left Coast Press.

Susman, G., & Evered, R. (1978). An assessment of the scientific merits of action research. *Administrative Science Quarterly*, *23*(4), 582–603. doi:10.2307/2392581

Swartz, D., & Orgill, K. (2001). Higher education ERP: Lessons learned. *EDUCAUSE Quarterly*, *2*, 20–27.

Taylor, F. W. (1911). *The principles of scientific management*. New York, NY: Harper & Brothers.

Taylor, J. R., & Virgili, S. (2008). Why ERPs disappoint: The importance of getting the organisational text right. In Grabot, B., Mayère, A., & Baze, I. (Eds.), *ERP Systems and Organisational Change: A Socio-Technical Insight* (pp. 59–84). London, UK: Springer. doi:10.1007/978-1-84800-183-1_5

Telier, A. (2011). *Design things*. Cambridge, MA: MIT Press.

Tellioğlu, H. (2009). Model-based collaborative design in engineering. In L. Yuhua (Ed.), *Lecture Notes in Computer Science, Information Systems and Applications: Proceedings of the 6th International Conference CDVE 2009,* (vol 5738), (pp. 85-92). CDVE.

Tellioğlu, H. (2010a). Coordination 2.0: Using web-based technologies for coordination support. In *Proceedings of the International Workshop on Web Intelligence and Virtual Enterprises 2 (WIVE 2010)*. Saint-Etienne, France: WIVE.

Tellioğlu, H. (2010b). Coordination of work by using to-do lists: Practices and requirements. In *Proceedings of the International Conference on Information Technologies (InfoTech-2010)*. Elena, Bulgaria: InfoTech.

Tellioğlu, H., & Wagner, I. (2005). Work cultures in multimedia production. In *Proceedings of the ALP_IS Alpine Information Systems Seminar*, (pp. 143-161). ALP_IS.

Tellioğlu, H. (2003). Modeling coordinated work: Definition and application of the model coordinated work environment. *The Journal of Supercomputing: An International Journal of High-Performance Computer Design, Analysis, and Use*, *24*(2), 161–171.

Tellioğlu, H. (2007). *Cooperative work*. Wien, Austria: Fakultät für Informatik.

Terzides, K. (2005). *Algorithmic architecture*. London, UK: Elsevier.

Thalheim, B. (2007). Challenges to conceptual modelling. In Krogstie, J., Opdahl, A. L., & Brinkkemper, S. (Eds.), *Conceptual Modelling in Information Systems Engineering* (pp. 59–74). Berlin, Germany: Springer. doi:10.1007/978-3-540-72677-7_4

Theureau, J., & Filippi, G. (1994). Cours d'action et conception d'un système d'aide à la coordination: Le cas de la régulation du trafic du RER. *Sociologie du Travail*, *4*, 547–562.

Thomas, R. J. (1994). *What machines can't do*. Berkeley, CA: University of California Press.

Thompson, J. D. (1967). *Organizations in action*. New York, NY: McGraw-Hill.

Timmermans, S., & Berg, M. (1997). Standardization in action: Achieving universalism and localization in medical protocols. *Social Studies of Science*, *27*, 273–305. doi:10.1177/030631297027002003

Tingling, P., & Parent, M. (2002). Mimetic isomorphism and technology evaluation: Does imitation transcend judgement. *Journal of the Association for Information Systems*, *3*(5), 113–143.

Toennies, F. (1887). *Gemeinschaft und gesellschaft*. Leipzig, Germany: Fues's Verlag.

Toennies, F. (1912). *Gemeinschaft und gesellschaft* (2nd ed.). Leipzig, Germany: Fues's Verlag.

Toennies, F. (1935). *Gemeinschaft und gesellschaft* (8th ed.). Leipzig, Germany: Buske.

Toennies, F. (2005). *Community and society*. Darmstadt, Germany: Wissenschaftliche Buchgesellschaft.

Tuomi, I. (2002). *Networks of innovation: Change and meaning in the age of the internet*. Oxford, UK: Oxford University Press.

Turkle, S. (1995). *Life on the screen: Identity in the age of the internet*. New York, NY: Simon and Schuster.

Turnbull, D. (2000). *Masons, tricksters and cartographers: Comparative studies in the sociology of scientific and indigenous knowledge*. London, UK: Routledge. doi:10.4324/9780203304587

Turner, W., Bowker, G., Gasser, L., & Zacklad, M. (2006). Information infrastructures for distributed collective practices. *Computer Supported Cooperative Work*, *15*, 93–110. doi:10.1007/s10606-006-9014-3

Tzannes, N. S. (1968). The concept of 'meaning' in information theory. In *Warren McCulloch Papers*. Philadelphia, PA: American Philosophical Society Library.

Ulrich, K. T., & Eppinger, S. D. (1995). *Product design and development*. New York, NY: McGraw-Hill, Inc.

University of Northern Iowa. (1995). *UNIGradPact*. Retrieved 13 March 2010 from http://www.uni.edu/vpaa/gradpact.html.

Ure, J., Procter, R. N., Lin, Y., Hartswood, M., Anderson, S., & Lloyd, S. (2009). The development of data infrastructures for ehealth: A socio-technical perspective. *Journal of the Association for Information Systems*, *10*(5), 415–419. Retrieved from http://aisel.aisnet.org/jais/vol10/iss5/

van der Lei, J. (1991). Use and abuse of computer-stored medical records. *Methods of Information in Medicine*, *30*, 79–80.

van Fenema, P. C., Koppius, O. R., & van Baalen, P. J. (2007). Implementing packaged enterprise software in multi-site firms: Intensification of organizing and learning. *European Journal of Information Systems*, *16*, 584–598. doi:10.1057/palgrave.ejis.3000708

Van Maanen, J. (1998). *Tales of the field: On writing ethnography*. Chicago, IL: University of Chicago Press.

Van Maanen, J., & Barley, S. R. (1984). Occupational communities: Culture and control in organizations. *Research in Organizational Behavior*, *6*, 287–365.

Varela, F. J. (1999). The specious present: A neurophenomenology of time consciousness. In Petitot, J., Varela, F. J., Pachoud, B., & Roy, J.-M. (Eds.), *Naturalizing Phenomenology: Issues in Contemporary Phenomenology and Cognitive Science* (pp. 266–314). Palo Alto, CA: Stanford University Press.

Varpio, L., Schryer, C. F., Lehoux, P., & Lingard, L. (2006). Working off the record: Physicians and nurses transformations of electronic patient record-based patient information. *Academic Medicine: Journal of the Association of American Medical Colleges*, *81*(10), 35–39. doi:10.1097/01.ACM.0000237699.49711.35

Vaughan, D. (2005). *Employee resistance to enterprise resource planning technologies*. Paper presented at the Annual Meeting of the American Sociological Association. Philadelphia, PA.

Viller, S., & Sommerville, I. (2000). Ethnographically informed analysis for software engineers. *International Journal of Human-Computer Studies*, *53*, 169–196. doi:10.1006/ijhc.2000.0370

Vinck, D. (2003). *Everyday engineering: An ethnography of design and innovation*. Cambridge, MA: MIT Press.

Vinck, D. (Ed.). (2003). *Everyday engineering: An ethnography of design and innovation*. Cambridge, MA: MIT Press.

Virilio, P. (2000). *The information bomb*. London, UK: Verso.

von Bertalanffy, L. (1950). The theory of open systems in physics and biology. In Emery, F. E. (Ed.), *Systems Thinking* (pp. 70–85). Harmondsworth, UK: Penguin. doi:10.1126/science.111.2872.23

Voss, A. (2006). *Corealisation: A radical respecification of the working division of labour in systems development*. Unpublished PhD Thesis. Edinburgh, UK: University of Edinburgh.

Voss, A., Hartswood, M., Procter, R., Rouncefield, M., Slack, R. S., & Büscher, M. (Eds.). (2009). *Configuring user-designer relations--Interdisciplinary perspectives*. London, UK: Springer.

Voss, A., Procter, R., Slack, R. S., Hartswood, M., Procter, R., & Rouncefield, M. (2009). Design as and for collaboration: Making sense of and supporting practical action. In Voss, A., Hartswood, M., Procter, R., Rouncefield, M., Slack, R. S., & Büscher, M. (Eds.), *Configuring User-Designer Relations: Interdisciplinary Perspectives*. London, UK: Springer. doi:10.1007/978-1-84628-925-5_3

Wagner, E. L., & Newell, S. (2006). Repairing ERP: Producing social order to create a working information system. *The Journal of Applied Behavioral Science*, *42*, 40–57. doi:10.1177/0021886305284290

Wagner, E. L., Scott, S. V., & Galliers, R. D. (2006). The creation of 'best practice' software: Myth, reality and ethics. *Information and Organization*, *16*, 251–275. doi:10.1016/j.infoandorg.2006.04.001

Wagner, E., & Newell, S. (2004). Best for whom? The tension between 'best practice' ERP packages and diverse epistemic cultures in a university context. *The Journal of Strategic Information Systems*, *13*(4), 305–328. doi:10.1016/j.jsis.2004.11.002

Wagner, I. (2000). Persuasive artefacts in architectural design and planning. In *Proceedings of CoDesigning, 2000*, 379–390.

Walsham, G. (1997). Actor-network theory and IS research: Current status and future prospects. In *Proceedings of the IFIP Working Group 8.2 Conference: Information Systems and Qualitative Research*. Philadelphia, PA: Kluwer Academic Publishers.

Walsham, G. (2004). Knowledge management systems: Action and representation. In *Proceedings of ALOIS 2004*. ALOIS.

Walsham, G. (2001). *Making a world of difference: IT in a global context*. Chichester, UK: Wiley.

Wang, R. Y. (1998). A product perspective on total data quality management. *Communications of the ACM*, *41*(2), 58–65. doi:10.1145/269012.269022

Wasson, C. (2000). Ethnography in the field of design. *Human Organization*, *59*(4), 377–388.

Wasson, C. (2002). Collaborative work: Integrating the roles of ethnographers and designers. In Squires, S., & Byrne, B. (Eds.), *Creating Breakthrough Ideas: The Collaboration of Anthropologists and Designers in the Product Development Industry* (pp. 71–90). Westport, CT: Greenwood.

Weber, M. (1949). *The methodology of the social sciences*. New York, NY: Free Press.

Webster, J. (1990). *Office automation: The labour process and women's work in Britain*. Hemel Hempstead, UK: Harvester Wheatsheaf.

Webster, J. (1991). Advanced manufacturing technologies: Work organisation and social relations crystallised. In Law, J. (Ed.), *A Sociology of Monsters: Essays on Power, Technology, and Domination* (pp. 192–222). London, UK: Routledge.

Webster, J., & Williams, R. (1993). Mismatch and tension: Standard packages and non-standard users. In Quintas, P. (Ed.), *Social Dimensions of Systems Engineering*. Hemel Hempstead, UK: Ellis Horwood.

Weick, K. E. (1989). Theory construction as disciplined imagination. *Academy of Management Review*, *14*(4), 516–531.

Weick, K. E. (1991). The non traditional quality of organizational learning. *Organization Science*, *2*(1), 116–124. doi:10.1287/orsc.2.1.116

Weick, K. E. (2001). *Making sense of the organization*. Oxford, UK: Blackwell Publishers.

Weick, K., & Quinn, R. (1999). Organizational change and development. *Annual Review of Psychology*, *50*, 361–386. doi:10.1146/annurev.psych.50.1.361

Wei, H., Li, C., Wang, E., Li, C., & Ju, P. (2005). Understanding misalignment and cascading change of ERP implementation: A stage view of process analysis. *European Journal of Information Systems*, *14*(4), 324–334. doi:10.1057/palgrave.ejis.3000547

Weiss, R. S. (1994). *Learning from strangers*. New York, NY: The Free Press.

Welton, D. (2000). *The other Husserl: The horizons of transcendental phenomenology*. Bloomington, IN: Indiana University Press.

Wenger, E. (1998). *Communities of practice: Learning, meaning and identity*. New York, NY: Cambridge University Press.

Whalen, J., & Zimmerman, D. H. (2005). Working a call: Multiparty management and interactional infrastructure in calls for help. *Pragmatics and Beyond New Series*, *143*, 309–345.

Whitley, E. A. (2005). Visiting the red light zones with Claudio. *European Journal of Information Systems*, *15*(5), 477–479. doi:10.1057/palgrave.ejis.3000554

Williams, R., Bunduchi, R., Graham, I., Pollock, N., Procter, R., & Voss, A. (2004). Understanding the evolution of standards: Alignment and reconfiguration in standards development and implementation arenas. In *Proceedings of the 4S/EASST Conference*. Paris, France: 4S/EASST.

Williamson, O. E. (1981). The economics of organization: The transaction cost approach. *American Journal of Sociology*, *87*, 548–577. doi:10.1086/227496

Williams, R. (1997a). Universal solutions or local contingencies? Tensions and contradictions in the mutual shaping of technology and work organisation. In McLoughlin, I., & Mason, D. (Eds.), *Innovation Organizational Change and Technology* (pp. 170–185). London, UK: International Thompson Business Press.

Williams, R. (1997b). The social shaping of a failed technology: Mismatch and tension in the supply and use of computer-aided production management. In Clausen, C., & Williams, R. (Eds.), *The Social Shaping of Computer-Aided Production Management and Computer Integrated Manufacture* (*Vol. 5*, pp. 109–130). Luxembourg: European Commission DGXIII.

Williams, R. A., Stewart, J. K., & Slack, R. (2005). *Social learning in technological innovation: Experimenting with information and communication technologies*. Upper Saddle River, NJ: Edward Elgar.

Williams, R., & Edge, D. (1996). The social shaping of technology. *Research Policy*, *25*(6), 865–899. doi:10.1016/0048-7333(96)00885-2

Williams, R., Slack, R., & Stewart, J. (2005). *Social learning in technological innovation - Experimenting with information and communication technologies*. Cheltenham, UK: Edgar Elgar.

Williams, R., & Steward, F. (1985). *The role of the parties concerned in the introduction of new technology - Case studies on the implementation of collective agreements on new technology in Britain*. Dunblin, Ireland: European Foundation for the Improvement of Living and Working Conditions.

Wilson, B. (1990). *Systems: Concepts, methodologies, and applications*. Chichester, UK: John Wiley & Sons.

Winkler, W. E. (2004). Methods for evaluating and creating data quality. *Information Systems*, *29*(7), 531–550. doi:10.1016/j.is.2003.12.003

Winner, L. (1980). Do artifacts have politics? *Deadalus*, *109*, 121–136.

Winograd, T., & Flores, F. (1986). *Understanding computers and cognition: A new foundation for design*. Reading, MA: Addison Wesley.

Winthereik, B. R. (2003). We fill in our working understanding: On codes, classifications and the production of accurate data. *Methods of Information in Medicine*, *42*(4), 489–496.

Winthereik, B. R., & Vikkelsø, S. (2005). ICT and integrated care: Some dilemmas of standardising inter-organisational communication. *Computer Supported Cooperative Work*, *14*, 43–67. doi:10.1007/s10606-004-6442-9

Wittgenstein, L. (1953). *Philosophische untersuchungen*. Oxford, UK: Basil Blackwell.

Wittgenstein, L. (1998). *Philosophical investigations* (2nd ed.). (Anscombe, G. E. M., Trans.). Oxford, UK: Blackwell Publishers.

Wixon, D. (1995). Qualitative research methods in design and development. *Interaction*, *2*(4), 19–24. doi:10.1145/225362.225365

Woodruff, A., Szymanski, M. A., Grinter, R. E., & Aoki, P. M. (2002). Practical strategies for integrating a conversation analyst in an iterative design process. In *Proceedings of the 4th Conference on Designing Interactive Systems: Processes, Practices, Methods, and Techniques*, (pp. 255-264). London, UK: IEEE.

Woolgar, S. (1985). Why not a sociology of machines? The case of sociology and artificial intelligence. *Sociology*, *19*(4), 557–572. doi:10.1177/0038038585019004005

Woolgar, S. (1991). Configuring the user: The case of usability trials. In Law, J. (Ed.), *A Sociology of Monsters: Essays on Power, Technology, and Domination*. London, UK: Routledge.

Wright, M. (1996). IU trustees. *Herald Times*. Retrieved 13 March 2010 from http://www.heraldtimesonline.com/stories/1996/06/10/archive.19960610.1df54f5.sto.

Wu, J. (2008). *Critical success factors for ERP system implementation. Research and Practical Issues of Enterprise Information Systems II* (pp. 739–745). Boston, MA: Springer.

Young, D. W. (1987). Discussion paper. In Anderson, J. G., & Jay, S. J. (Eds.), *Use and Impact of Computers in Clinical Medicine* (pp. 8–14). New York, NY: Springer-Verlag.

Zamarian, M. (2010). Le routine organizzative tra dimensione operativa e dimensione generativa: Modi di riproduzione dell'agire organizzativo. *Rassegna Italiana di Sociologia*, *51*(2), 233–253.

Zuboff, S. (1988). *In the age of the smart machine: The future of work and power*. New York, NY: Basic Books.

Zuboff, S. (2005). Ciborra disclosed: *Aletheia* in the life and scholarship of Claudio Ciborra. *European Journal of Information Systems*, *15*(5), 470–473. doi:10.1057/palgrave.ejis.3000580

# About the Contributors

**Gianluigi Viscusi** obtained a Master's degree (Laurea Vecchio Ordinamento) in Philosophy (2001) from the University of Milano and a Ph.D. degree (2007) in Information and Communication Technology applied to knowledge society and to learning processes from the University of Milan Bicocca. He is currently post-doc research fellow at the Department of Informatics, Systems and Communication (DISCo) of the University of Milan Bicocca: the research title is "Strategic planning and IT alignment of eGovernment systems and services." Research interests concern methodologies for information systems planning, eGovernment, IT business value, business modeling, and IS strategy alignment, design and management of repositories, data reverse engineering. In 2010, he co-authored with Carlo Batini and Massimo Mecella the book *Information Systems for eGovernment: A Quality of Service Perspective*, published by Springer. Furthermore, he has published more than 40 referred papers in books, conference proceedings, and journals, such as *Data and Knowledge Engineering* and *Government Information Quarterly*.

**Gian Marco Campagnolo** teaches and researches in the Social Study of Information Systems at the University of Edinburgh, where he is a Lecturer in Science, Technology, and Innovation Studies within the School of Social and Political Science and a member of the Institute for the Study of Science Technology and Innovation (ISSTI). He recently participated in a national IT project on interoperability and applications cooperation among Regions (www.progettoicar.it), covering issues of local IT development in rural areas. Previously, he has also been involved in an EU project where he participated as an action researcher in the development of enterprise modeling software solutions for industrial users (mapper.eu.org). In the last few years he also did extensive fieldwork on the diffusion of Enterprise Resource Planning (ERP) systems in public sector organizations in Italy. His current research interests concern the implications of the rise of large-scale information infrastructures as an object of study for social theory (namely ethnomethodology and micro-sociology); the diffusion of generic commercial software on engineering expertise, IT consultancy, and its transformations over time.

**Ylenia Curzi** PhD is Post-Doc Research Fellow at the "Marco Biagi" Foundation and at the Department of Business Administration, University of Modena and Reggio Emilia (Italy). Visiting scholar at Cardiff Business School (Wales, UK), in 2009. Her research interests encompass: critical stances to human resource management, organizational change and Information Communication Technologies, organizational perspectives on work-life balance, the methodology of the social sciences, and particularly the social research methodology by Max Weber. Her current projects concern an original approach to action research aiming at improving organizational effectiveness and people's wellbeing in the workplace. She has contributed to international conferences, Italian and international journals, and collective books.

* * *

**Roberto Albano**, PhD in sociology, is an Associate Professor of Sociology and Social Research at the Department of Cultures, Politics, and Society, University of Torino, Italy, since 2006 (reseacher from 1999 to 2006). His research interests range widely from epistemology and methodology of social research, especially for organizational analysis, to public institutions legitimacy, to socialization processes. His recent publications include: *Social Capital, Welfare State, and Political Legitimacy*, 'American Behavioral Scientist', 5/2010 (with F. Barbera); *The Relevance of Giddens' Structuration Theory for Organizational Research*, TAO Digital Library, Bologna 2010 (with G. Masino, B. Maggi); "La Socialisation Dans la Famille," in B.Maggi (s/d), *Interpréter l'agir: Un Défi Théorique*, Presses Universitaire de France, Paris 2011

**Chiara Bassetti** is a Post-Doc Research Fellow at the Department of Disciplines of Communication of the University of Bologna, and collaborates with the Department of Sociology and Social Research of the University of Trento, where she took her Ph.D. in March 2010 with a thesis based on a (self-) ethnographic research on the world of dance (*Dance as a professional, bodily, and artistic acting. Paths and trajectories, knowledge, and everyday practice in the Italian dance field)*. She is now conducting the research project "Social inequalities, cultural consumes, and reading/literacy practices. Citizenship, school, and territory". Her larger focus being on everyday situated interaction, and her preference for qualitative methods, her research interests include: Body and Embodiment; Dance and Arts in genre; Media and Communication; Information Systems and ICTs; Organisational Studies; Cultural Studies; Gender Studies; Biographical Narratives; Ethnography; Ethnomethodology; Conversation Analysis.

**Federico Cabitza** graduated in ICT Engineering from the Politecnico of Milan (Italy) in 2001; from 2001 to 2004 he worked as business process analyst, quality manager, and software engineer coordinator in the private sector. In 2007 he received a PhD degree in Computer Science with a thesis regarding a design-oriented framework for the deployment of flexible groupware applications in organizational settings. Currently, he is research Associate Fellow at the Università degli Studi di Milano-Bicocca and a requirement engineering consultant. He is author of about 50 scientific papers in international journals and conference proceedings. His current research interests regard the design of computer-based systems for cooperative and knowledge work, especially in the healthcare domain, and the inclusion of ethnographic methods in the collection of socially-informed requirements for such systems.

**Giorgio De Michelis** teaches Informatics for Organizations and Interaction Design at the University of Milano – Bicocca. His research focuses on models of concurrent systems (Petri Nets), and computer supported cooperative work, community-ware, knowledge management, and interaction design. His group has developed several prototypes of support systems for cooperative processes and knowledge management systems (e.g.: CHAOS, UTUCS, MILANO, CAMPIELLO, KLEE&CO, MILK, ATELIER). Giorgio De Michelis has authored four books and more than 160 papers in the areas of his interest. MIT Press has announced for October 2011, the book, *Design Things*, which he co-authored with Thomas Binder, Pelle Ehn, Giulio Jacucci, Per Linde, and Ina Wagner, under the pseudonym of A. Telier. Giorgio De Michelis has created in 2008 *Itsme*, a spin-off of the University of Milano – Bicocca, devoted to the design and construction of a newly conceived Linux-based operating system for workstation.

**Paolo Depaoli** is a Lecturer on Organization at the Department of Economics, LUISS "Guido Carli" in Rome and a Researcher at the Department of Earth, Life, and Environmental Sciences, University "Carlo Bo" in Urbino, Italy. His research interests concern organizational issues relating to innovation and knowledge and technology transfer. He is involved in research areas concerning the design and implementation of both environmental (energy saving and renewable energy) and information systems policies. He has worked as a consultant in change management and organisational learning projects of several Italian companies. He has contributed to international conferences, to Italian and international journals, and to collective books.

**Tommaso M. Fabbri** is Associate Professor of Organization and Human Resources Management at the University of Modena and Reggio Emilia (Italy) and member of the scientific committee of the International Doctoral School in Labour Relations at Foundation "Marco Biagi." He has been Visiting Professor at Penn State University (USA) and has served as a consultant to the Italian Ministry of Economy and Finance, the Italian Ministry for Economic Development, other national, regional and local institutions and private companies. He is the author of several books and articles. His main research interests are organizational learning and innovation, work and well-being, human resources management, public policy, and organization.

**Giolo Fele** (born in Rome, 1958) is Professor of Culture and Communication Studies and teaches undergraduate and graduate courses in Qualitative Methods for Social Research at the University of Trento (Italy). He graduated with a degree in Sociology from the University of Rome and was awarded a research doctorate in 'Sociology and Social Research' by the Faculty of Sociology, University of Trento, in 1988. In 1995 he was appointed to a post as Researcher at the Department of Communication Studies, Faculty of Letters and Philosophy, University of Bologna. In 2000 he was appointed to a post as Associate Professor by the same Faculty. In October 2002 he was appointed to a post as interim full Professor at the Faculty of Sociology of the University of Trento. Fele has been Visiting Professor at UCLA (2004-2005) with a Marie Curie OIF grant of the European Commission. His research interests include interaction and communication in work settings using ethnographic and naturalistic research methodologies. He has analysed various contexts, such as family therapy sessions, courts of law, classrooms, as well as forms of media communication. He is interested in the phenomenological basis of social theory; the social organization of talk, discourse and activities, the place of symbolism and ritual in social life. He has worked on the institutional basis of public discourse, with special reference to the degradation ceremonies and political scandals. He is currently working on the use of communication technologies in work settings, like call and dispatch centers, in which medical emergencies are controlled and managed. He is one of the founders of the journal *Etnografia e Ricerca Qualitativa* (Il Mulino Publishers).

**Thomas Herrmann** is a Professor of Information and Technology-Management at the Institute of Applied Work Science (IAW), University of Bochum, Germany, since 2004, and a fellow of the Electrical Engineering Department. Current research interests include design methods for socio-technical systems in the areas of knowledge management, (work-)process management, computer supported collaborative learning, and concepts of social software for the support of creativity. He teaches courses in Groupware, Knowledge Management, Socio-Technical Systems Design, Information Systems and Privacy, Human-Computer Interaction, Organizational Communication, and Process Management. He was an Associate

Professor from 1992-2004 at the Computer Science Department at the University of Dortmund and was in charge of the development of infrastructure and new media for the University of Dortmund as a vice president from 2002-2004. He holds a PhD in Computer Science of the Technical University of Berlin (1986) and a Master of Art in Communication Science of the University of Bonn (1983).

**Tony Hines** is Professor of Marketing at Manchester Metropolitan University. His research makes sense of organization strategies, structures, and relationships focusing attention on management decisions. He is author of the highly acclaimed *Supply Chain Strategies,* published by Routledge, *Fashion Marketing,* published by Elsevier, and has published research papers in the *International Journal of Information Management, European Journal of Marketing, and the Journal of Marketing Management.* He was lead consultant on a number of international projects and has specific expertise in supply chain management. Prior to his academic career, he worked in management systems and consultancy in different industrial sectors.

**Sampsa Hyysalo** Professor of Co-Design in Aalto University School of Art and Design, Helsinki, Finland. His research focuses on user involvement in innovation and the co-evolution of technologies, practices, and organizations. His recent book, *Health Technology Development and Use: From Practice-Bound Imagination to Evolving Impacts* (Routledge, 2010), sums up his twelve year research engagement with these topics on various health care technologies. His work can be accessed also through twenty peer-reviewed articles. Hyysalo received his Ph.D in Behavioral Sciences in the University of Helsinki and holds a Docentship in Information Systems, Specialising in User-Centered Design.

**Gianni Jacucci** is retired Professor of Information Systems. He has been director and scientific responsible of the Laboratory for the Social Study of Information Systems and Professor of Information Systems at the Sociology doctoral school of the University of Trento. His scientific activity comprehends Participatory Design, design for accoutability and the design for end user configurability. He also conducts consultancy activity in SME and public sector organizations, with particular reference to the virtual enterprise, IT for tourism, and health care.

**Jannis Kallinikos** is Professor and PhD programme Director in the Information Systems and Innovation Group, Department of Management at the London School of Economics. His research covers a wide range of topics on the interpenetration of technology with the administrative and institutional arrangements of contemporary societies. More recently, and as the result of the technological developments that mark our time, his research has shifted to the investigation of the diffusion of technological information in all its breeds (text, voice, image) and digital formats and the implications these developments carry for organizations and institutions but also for the patterns of interaction characteristic of everyday living. Recent books include *The Consequences of Information: Institutional Implications of Technological Change*, Edward Elgar, 2006, and *Governing Through Technology: Information Artefacts and Social Practice*, Palgrave, 2011.

**Kenneth Liberman** teaches Sociology at the University of Oregon and has been lecturing on Ethnomethodology in Italy, Brazil, India, and China. A student of Peter Berger and Herbert Marcuse, but especially of Harold Garfinkel (2002) and Hubert Dreyfus (1990), Liberman invites us to explore from an ethnomethodological perspective the depths of our ordinary social world, the primitive place of our experience. As an anthropologist who spent two years with some Australian Aboriginal tribes and three years in a Tibetan monastery, he encourages us to reflect on that world taken for granted that we call reality. Professor Liberman's 2004 book, *Dialectical Practice in Tibetan Philosophical Culture*, has been translated into Chinese and is being read by Chinese academics and by Tibetans. He has currently completed a new book, *More Studies in Ethnomethodology: Studies of the In Vivo Organization of Sense*, which is based upon his last decade of teaching.

**Kai-Uwe Loser** is a Researcher at the Information and Technology Management group of the Institute of Applied Work Science at the University of Bochum. His research is focused on privacy and data protection and on participation of end users in complex software projects. Topics include the communication of organizational knowledge using process models, the participatory creation of models mainly in facilitated meetings. Recently his topics were related to meeting support and data protection in cooperative software. He holds a diploma in computer science and a PhD in informatics of the University of Dortmund. His doctoral thesis is titled "Supporting the Adoption of Commercial off the Shelf Products using Diagrams." Kai-Uwe has (co-) authored more than 40 international conferences, books, and journal contributions.

**Petter Nielsen** is currently working with business development in the Telenor group and holds an Associated Professor position at the Department of Informatics, University of Oslo, Norway. His research interest is on the design and evolution of information infrastructures within the field of mobile phones and the mobile Internet. Focusing on both the development of underlying infrastructures and the user perspective and experiences, he has published and worked extensively within this field in Norway as well as Hungary and Pakistan.

**Alexander Nolte** is a PhD student at the Information and Technology Management group headed by Thomas Herrmann at the University of Bochum. His particular research interest lies in the collaborative development of process models in different collaboration scenarios such as collocated workshops or dislocated asynchronous settings with special respect to web based modeling tools. Additionally, he is interested in enabling non-expert modelers to directly contribute to model development. Alexander holds a diploma in Computer Science with his diploma thesis being titled "Concept and Prototype of a Usability-Oriented Web-Based Editor for Semi-Structured Modeling."

**Neil Pollock** teaches and researches in the Sociology of Information and Communication Technologies (ICTs) at the University of Edinburgh where he is a Reader in Information Systems within the Business School and a member of the Institute for the Study of Science, Technology and Innovation (ISSTI). He has also conducted research on 'virtual universities' under the auspices of the Economic &and Social Research Council (ESRC) Virtual Society Programme and has written (with James Cornford) Putting the University Online: Information, Technology, and Organisational Change. Current research interests include the sociology of technology, actor network theory, the social shaping and 'generification' of information systems, and the 'performativity' of simple business tools. He has recently completed a second book (with Robin Williams) *called Software and Organisation: The Biography of the Enterprise-Wide System Or How SAP Conquered the World,* published by Routledge.

**Michael Prilla** is a Post-Doc Researcher at the Information and Technology Management work group lead by Thomas Herrmann at the University of Bochum. Michael's research is focused on the intertwining of process management and collaboration support, including the transfer of organizational knowledge via process models, the usage of process models by non-expert users and the collaborative creation of models. He holds a diploma in Computer Science and a PhD in Engineering. His doctoral thesis is titled "Knowledge Management Support for the Development and Usage of Process Models as Artefacts for Knowledge Transfer". Michael has (co-) authored more than 20 international conference, book and journal contributions.

**Nicholas J. Rowland** is an Assistant Professor of Sociology at The Pennsylvania State University in Altoona, PA. His research emphasizes the adoption/procurement and implementation/localization of innovations among non-profit organizations. From the vantage points of sociology and STS (Science and Technology Studies), he writes about the spread of large administrative information systems in American higher education as well as the rise of eHealth in America and its implications for states and state theory. He writes for a blog on infrastructure, order, and innovation at: http://www.installingorder.org/.

**Carla Simone** took her Degree in Mathematics at the University of Torino (1972). She is currently Full Professor of Computer Science at the University of Milano-Bicocca. Her main research interests show the combination of theoretical investigations and field research with the application of their results to the design of technological supports of human activities. Her main areas of research regard process modeling, design of coordination technologies, context-aware computing, design of knowledge management systems, user-centered design with a specific interest in adaptive systems, coordination models, and interaction design. She is author of about 150 scientific papers in international journals, contributed volumes and conference proceedings. She is co-founder of the Foundation for Co-Operative Work Technology and of European Society for Socially Embedded Technologies (EUSSET), member of the steering committee of the European Conferences on Computer Supported Cooperative Work, and member of the Editorial Advisory Board of *CSCW: An International Journal*, Kluwer Academic.

**Rainer Skrotzki** has been a Researcher at the Institute of Applied Work Science (IAW), University of Bochum, Germany, for over 25 years. He is a member of the Information and Technology Management group headed by Thomas Herrmann. Rainer studied at the University of Essen and Berlin engineering design and manufacturing technology. He holds a PhD in Engineering Science of the Technical University of Bochum. His doctoral thesis dealt with technology assessment and technical design. Current research interests include design methods for socio-technical systems in the areas of industrial engineering and assisted living. He teaches courses in HCI, Communication Management, and Process Management.

**Hilda Tellioglu** is an Associate Professor at the Institute of Design and Assessment of Technology of Vienna University of Technology. She is a computer scientist. She has been involved with both research and teaching on: software engineering, knowledge management, coordination support, design and development of information technology in networked (work) environments, Computer Support for Cooperative Work (CSCW) in systems design, in health care work, and in architectural planning, data security, and integrity in databases, gender issues in computer science. Her habilitation thesis is about "Cooperative Work" with the focus on coordination.

**Robin Williams** is Professor of Social Research on Technology and Director of the Institute for the Study of Science, Technology, and Innovation (formerly the Research Centre for Social Sciences) at the University of Edinburgh. His interdisciplinary research into 'the social shaping of technology' over 20 years has focused upon the interplay between business organisation and 'technical' factors in the development and implementation of a range of Informatics applications including Enterprise systems, Electronic Commerce, Finance Sector systems, and more recently upon the more dispersed 'social learning' processes involved in innovations in Multimedia, mobile, Internet, and social networking technologies. He has been PI and Co-I on over 50 externally funded research projects since 1987.

# Index

www.ingramcontent.com/pod-product-compliance
Ingram Content Group UK Ltd.
Pitfield, Milton Keynes, MK11 3LW, UK
UKHW050952290926
15470UKWH00005B/1775

9 781466 60303